Advanced Fracturing Technology for Oil and Gas Reservoir Stimulation

Advanced Fracturing Technology for Oil and Gas Reservoir Stimulation

Guest Editors

Yuxuan Liu
Mao Sheng
Jianwei Tian
Jie Zeng
Chunming He

MDPI

Basel • Beijing • Wuhan • Barcelona • Belgrade • Novi Sad • Cluj • Manchester

Guest Editors

Yuxuan Liu
Petroleum Engineering School
Southwest Petroleum University
Chengdu
China

Mao Sheng
College of Petroleum Engineering
China University of Petroleum (Beijing)
Beijing
China

Jianwei Tian
School of Civil and Transportation Engineering
Hebei University of Technology
Tianjin
China

Jie Zeng
Petroleum Engineering School
Southwest Petroleum University
Chengdu
China

Chunming He
Research Institute of Petroleum Exploration & Development
PetroChina Corporation
Beijing
China

Editorial Office
MDPI AG
Grosspeteranlage 5
4052 Basel, Switzerland

This is a reprint of the Special Issue, published open access by the journal *Processes* (ISSN 2227-9717), freely accessible at: https://www.mdpi.com/journal/processes/special_issues/K1KB8D7SY8.

For citation purposes, cite each article independently as indicated on the article page online and as indicated below:

Lastname, A.A.; Lastname, B.B. Article Title. *Journal Name* **Year**, *Volume Number*, Page Range.

ISBN 978-3-7258-5355-7 (Hbk)
ISBN 978-3-7258-5356-4 (PDF)
https://doi.org/10.3390/books978-3-7258-5356-4

Contents

About the Editors

Yuxuan Liu

Yuxuan Liu is an Associate Professor at the School of Petroleum Engineering, Southwest Petroleum University, China. He holds a Ph.D. in Petroleum Engineering and has dedicated his career to hydraulic fracturing mechanics, fracture propagation simulation, and proppant transport optimization. To date, he has published more than 30 SCI-indexed papers in leading journals such as *Energy*, *SPE Journal*, and *Journal of Natural Gas Science and Engineering*. He has been granted 10 patents and has coordinated more than 20 national and industry projects. Beyond research, Liu actively contributes to the academic community as an outstanding reviewer for journals including *SPE Journal*, *Journal of Petroleum Science and Engineering*, and *Computers and Geotechnics*, consistently ranked among the top 1% of reviewers in engineering on the Publons platform. He currently serves as the Officer and Membership Chairperson for the SPE Chengdu Section.

Mao Sheng

Mao Sheng is a Professor and Doctoral Supervisor at the China University of Petroleum (Beijing)'s Petroleum Engineering School. His research focuses on unconventional oil/gas fracturing completion theory and AI applications in petroleum engineering. He received his PhD from the CUP and was awarded the National Science Fund for Excellent Young Scholars (2021). Dr. Sheng has published 35 first/corresponding-author papers, co-authored 3 monographs, and received three provincial/ministerial awards including the Sun Yueqi Youth Science & Technology Award (2020). His work integrates AI with fracturing optimization and downhole data analysis.

Jianwei Tian

Jianwei Tian, PhD, is a Professor at the School of Civil Engineering and Transportation, Hebei University of Technology. He earned his bachelor's and master's degrees from the China University of Mining and Technology (Beijing) and his Ph.D. from the University of Western Australia. He conducted postdoctoral research at the Technical University of Denmark and McMaster University in Canada. His main research pertains to multi-physics coupling in deep energy development and storage, the efficient mineralization sequestration of CO_2, and the application of deep learning in monitoring fluid and stress evolution during underground fluid injection/extraction. As a key researcher, he has participated in one project funded by the Danish Energy Agency and another by Canada's Natural Sciences and Engineering Research Council. He has published over 20 papers in authoritative international journals such as *CMAME*, *FUEL*, and *COMPUT GEOTECH*, serving as the first author of 7. He serves as an Editorial Board Member for Green and Smart Mining Engineering and as a Youth Editorial Board Member for the Deep Underground Science and Engineering and Rock Mechanics Bulletin.

Jie Zeng

Jie Zeng, Ph.D., is an Associate Professor at the School of Petroleum Engineering, Southwest Petroleum University. He earned his bachelor's degree from Southwest Petroleum University, master's degree from the University of Regina, and Ph.D. degree from the University of Western Australia in Petroleum Engineering. His main research focuses on multi-physics-coupled fluid transport in unconventional reservoirs. He has published approximately 60 technical papers in prestigious journals and conferences. He currently serves as the Officer and Communications and Outreach Chairperson for the SPE Chengdu Section.

Chunming He

Chunming He is a Senior Engineer and Deputy Director of the Fracturing & Acidizing Technology Center at the PetroChina Research Institute of Petroleum Exploration & Development, where he also serves as a Master's Supervisor. He earned his Ph.D. from Southwest Petroleum University (2013).

His research interests include risk exploration well stimulation, unconventional reservoir development, and mature field re-fracturing technologies. Dr. He has contributed to key shale/tight oil projects in Sichuan, Mahu, Jimsar, and Dagang fields. He has undertaken 2 national major projects, obtained 8 provincial/ministerial achievements and over 10 bureau-level achievements, and published over 40 papers (25 indexed by SCI/EI). In addition, he holds 8 authorized invention patents, 3 software copyrights, and 2 monographs and has formulated 4 enterprise/industry standards.

processes MDPI

Article

A Comprehensive Evaluation of Shale Oil Reservoir Quality

Fuchun Tian [1,2], Yongqiang Fu [1], Xuewei Liu [1,*], Dongping Li [1], Yunpeng Jia [1], Lifei Shao [1], Liyong Yang [1], Yudong Zhao [1], Tao Zhao [1], Qiwu Yin [1] and Xiaoting Gou [1]

1 Petroleum Engineering Research Institute, PetroChina Dagang Oilfield Company, Tianjin 300280, China; tianfchun@petrochina.com.cn (F.T.); fuyq@petrochina.com.cn (Y.F.); lidping@petrochina.com.cn (D.L.); jiaypeng@petrochina.com.cn (Y.J.); shaolfei@petrochina.com.cn (L.S.); yanglyong@petrochina.com.cn (L.Y.); zhaoydong@petrochina.com.cn (Y.Z.); zhaotao001@petrochina.com.cn (T.Z.); yinqwu@petrochina.com.cn (Q.Y.); gouxting@petrochina.com.cn (X.G.)

2 College of Petroleum Engineering, China University of Petroleum, Beijing 102249, China

* Correspondence: 202121000998@stu.swpu.edu.cn

Abstract: To enhance the accuracy of the comprehensive evaluation of reservoir quality in shale oil fractured horizontal wells, the Pearson correlation analysis method was employed to study the correlations between geological parameters and their relationship with production. Through principal component analysis, the original factors were linearly combined into principal components with clear and specific physical meanings, aiming to eliminate correlations among factors. Furthermore, Gaussian membership functions were applied to delineate fuzzy levels, and the entropy weight method was used to determine the weights of principal components, establishing a fuzzy comprehensive evaluation model for reservoir quality. Without using principal component analysis, the correlation coefficient between production and evaluation results for the 40 wells in the Cangdong shale oil field was only 0.7609. However, after applying principal component analysis, the correlation coefficient increased to 0.9132. Field application demonstrated that the average prediction accuracy for the cumulative oil production per kilometer of fractured length over 12 months for the 10 applied wells was 91.8%. The proposed comprehensive evaluation method for reservoir quality can guide the assessment of reservoir quality in shale oil horizontal wells.

Keywords: shale oil; principal component analysis; reservoir quality; fuzzy comprehensive evaluation

1. Introduction

Following the unconventional energy revolution, shale oil development has become a crucial support for increased petroleum production [1]. To overcome the characteristics of low porosity, low permeability, and tightness in reservoirs and achieve the economic development of shale oil, extensive hydraulic fracturing modifications are necessary [2]. Reservoir quality determines the success of hydraulic fracturing measures for production enhancement [3,4]. Many factors influence reservoir quality, which can be categorized into source rock lithology and fracability [5]. Source rock lithology reflects the oil content and organic maturity of the reservoir, while fracability determines the distribution and complexity of reservoir fractures after hydraulic fracturing. Some factors are interrelated, and their contributions to reservoir quality (production) are unequal [6]. Only a few factors cannot fully reflect the true condition of the reservoir. However, when considering multiple factors, it is challenging to establish an effective comprehensive evaluation model for accurately assessing reservoir quality [7].

To overcome the aforementioned challenges, various artificial intelligence methods have been developed for comprehensive reservoir quality assessment. These methods include Artificial Neural Networks (ANNs) [8], Support Vector Machines (SVMs) [9], Genetic Algorithms (GAs) [10], and Fuzzy Logic Systems (FLSs) [11]. ANNs have the capacity to capture intricate non-linear relationships within data. Nevertheless, an extensive corpus of annotated data is requisite for training, a task that may prove challenging in certain domains [12]. Concurrently, the model lacks interpretability, rendering comprehension of the decision-making process arduous [13]. SVMs exhibit efficacy within high-dimensional spaces and are applicable to intricate decision boundaries. However, akin to ANNs, the interpretability of the model poses challenges, especially within high-dimensional spaces. GAs are well-suited for optimization quandaries and the exploration of solution spaces. However, there exists a propensity for entrapment in local optima under specific circumstances. FLSs possess the capability to model uncertainty and imprecision within data, offering interpretability through linguistic rules [14,15]. The formulation of pertinent fuzzy rules necessitates domain-specific expertise, potentially impeding the handling of complex non-linear relationships inherent in the data [16].

Zoveidavianpoor et al. [17] used Gaussian distribution membership functions to input seven geological factors such as permeability and skin coefficient, determined the weights according to expert experience, and established a fuzzy score for preselected hydraulic fracturing wells. Davarpanah et al. [18] analyzed and compared five indicators affecting the effect of hydraulic fracturing through the Fuzzy Analytic Hierarchy Process (FAHP) and Fuzzy Technique for Order Preference by Similarity to Ideal Solution (FTOPSIS) techniques. Gou et al. [15] divided geological and construction parameters into six categories, screened and retained the parameters with the greatest correlation with production as the model input through grey clustering analysis, and established a reservoir quality evaluation model of fractured wells by using the normal distribution membership function. Xie et al. [19] established a shale hydraulic fracturing brittleness evaluation model using the FAHP, taking into account factors such as shale brittle mineral content, porosity, and confining pressure. However, the above-mentioned studies directly analyze factors without considering the reduction in model accuracy caused by multicollinearity resulting from the strong correlation among production-influencing factors. Additionally, a limited number of input parameters fail to adequately represent the geological features of the reservoir, significantly impacting the accuracy of the model [20].

In summary, the geological parameters determining reservoir quality are intricate, and they possess fuzzy characteristics that are challenging to accurately describe. In current fuzzy comprehensive evaluation models, the weights of various factors are provided by empirical coefficients or experts. Moreover, factors related to reservoir fracturing capability are seldom considered, diminishing the accuracy and scientific validity of the model.

In this research, based on data samples from the Cangdong shale oil field, geological parameters influencing reservoir quality were selected using geological information and well logging data. Through Pearson correlation analysis, the correlations between geological parameters and production were determined. Additionally, by employing the principal component analysis method, factors influencing production were linearly combined into multiple comprehensive indicators, thereby enhancing the model's accuracy. Utilizing feature data after dimension reduction, a principal component membership matrix was established. Combining Gaussian membership functions for delineating fuzzy levels and incorporating the entropy weight method to determine the weights of each principal component, a comprehensive evaluation model for the reservoir quality of shale oil fractured horizontal wells was developed, thereby facilitating the high-quality development of shale oil.

2. Analysis of Main Control Factors of Shale Oil Reservoir Quality

2.1. Extraction of Geological Feature Parameters

To accurately assess the reservoir quality of shale oil horizontal wells, it is imperative to identify the key influencing factors. The evaluation indicators for shale oil reservoir quality can be categorized into two aspects: source lithology and fracability. Source lithology reflects the physical–chemical characteristics and hydrocarbon content of shale oil reservoirs, encompassing natural gamma radiation (GR), total organic carbon content (TOC), organic matter thermal maturity (Ro), the pyrolysis parameter (S_1), and the oil saturation index (OSI). Fracability, on the other hand, signifies the capacity of shale oil reservoirs to form complex fractures after hydraulic fracturing. Characterizing indicators include the mineral brittleness index (MBI), Young's modulus (E), Poisson's ratio (ν), minimum horizontal principal stress (σ_h), and the coefficient of variation of horizontal principal stress (σ_{dif}). High-quality shale oil reservoirs necessitate both a high abundance of hydrocarbons and excellent fracability. Consequently, a comprehensive evaluation considering both organic source rock properties and fracability characteristics is imperative for assessing shale oil reservoir quality.

The maturity and abundance of organic matter are focal points in the analysis of organic source rock properties. Liquid hydrocarbons at low maturity stages exhibit higher molecular weights and increased viscosity and density values, making their migration through micro-nano pores in shale challenging and resulting in lower shale oil production. As maturity increases, oil undergoes secondary cracking, yielding hydrocarbons with lower polarity, smaller molecular weights, and improved fluidity. Natural gamma radiation (GR) effectively identifies hydrocarbon source rocks, and within the context of organic source rock properties, both GR and thermal maturity (Ro) reflect hydrocarbon maturity, information obtainable from geological data [21]. Ro indicates the hydrocarbon generation potential of shale oil and is instrumental in evaluating its fluidity. Only within the oil and gas generation window can a significant quantity of oil and gas be formed. S_1 represents the residual hydrocarbon content, and its value, relative to total organic carbon (TOC), characterizes the heterogeneity of hydrocarbon accumulation in shale formations. TOC can be extracted from geological data, while S_1 and the oil saturation index (OSI) can be obtained from well logging data. Together, they allow the elucidation of the abundance of hydrocarbons in the shale formation.

Shale reservoirs, characterized by low permeability and tightness, necessitate the construction of a complex network of fractures to establish efficient fluid pathways toward the well for economically viable shale oil development. The fracability indicators of the reservoir reflect the degree and capability of fracture network formation through hydraulic fracturing [22]. Poisson's ratio and Young's modulus are key parameters for assessing shale stress and fractures [23]. However, the prediction and computation of the brittleness index often rely on rock physics parameters obtained from well logging data, without considering the influence of brittle mineral content [24]. Research indicates that as the horizontal principal stress decreases and the coefficient of variation of horizontal principal stress increases, the primary fractures formed after hydraulic fracturing become larger and more extended. These fractures are more prone to communicating with natural fractures in the deeper reservoir, leading to the creation of a complex fracture network. A higher fracture network index indicates a more intricate network. However, a larger coefficient of variation of horizontal principal stress makes the formation of a complex fracture network near the wellbore more challenging. Among the fracability indicators, the mineral brittleness index can be estimated from mineral content, while the remaining indicators can be calculated from well logging data [25].

Production is the most direct indicator of reservoir quality. However, short-term production may not adequately reflect reservoir quality due to variations in geological parameters and production systems among wells. Simultaneously, to mitigate the impact of horizontal segment length on production, the cumulative oil production over 12 months per kilometer (Q_{12mon}) was employed as the reservoir quality assessment indicator. Forty wells from the Cangdong sag shale oil field were selected as the subjects for this study (Table 1).

Table 1. Parameters of the Cangdong sag shale oil field.

Well No.	GR (API)	TOC (%)	Ro	S_1 (mg/g)	OSI	MBI	*E* (MPa)	ν	σ_h (MPa)	σ_{dif}	Q_{12mon} (t/km)
W-1	104.87	2.67	1.12	2.30	0.86	0.80	37,671.1	0.228	83.85	0.28	4449.0
W-2	98.44	2.23	1.55	3.05	1.37	0.78	37,379.8	0.228	83.04	0.28	2920.6
W-3	107.60	1.94	1.28	2.74	1.41	0.78	37,671.1	0.228	83.85	0.28	3167.9
W-4	91.07	4.48	1.03	1.97	0.44	0.76	27,198.6	0.231	56.84	0.24	3466.0
W-5	99.36	3.35	1.19	2.38	0.71	0.80	25,681.7	0.229	91.56	0.14	3977.1
W-6	96.49	5.05	1.38	9.90	1.96	0.80	25,681.7	0.229	91.56	0.14	6744.6
W-7	85.40	4.90	1.54	12.10	2.47	0.80	27,895.0	0.220	95.90	0.13	9984.7
W-8	114.31	3.62	1.87	10.50	2.90	0.80	27,895.0	0.220	95.90	0.13	7106.0
W-9	91.76	4.04	1.67	4.95	1.23	0.81	35,254.0	0.228	83.25	0.27	8604.5
W-10	95.09	3.00	1.55	3.57	1.19	0.81	33,530.4	0.229	83.55	0.26	7499.8
W-11	92.10	3.83	1.73	4.51	1.18	0.80	36,607.8	0.228	79.76	0.28	7321.0
W-12	108.99	4.20	1.72	3.24	0.77	0.76	32,778.4	0.229	81.96	0.26	2928.3
W-13	102.99	3.70	1.12	4.67	1.26	0.76	36,914.7	0.227	88.41	0.27	2435.9
W-14	85.28	3.96	1.18	4.38	1.11	0.76	37,138.5	0.228	85.19	0.28	2271.4
W-15	91.12	3.51	1.15	2.97	0.85	0.76	34,369.7	0.228	84.48	0.26	2047.5
W-16	98.14	5.00	1.52	6.13	1.23	0.80	33,530.4	0.229	83.55	0.26	8415.2
W-17	97.30	2.94	1.33	6.76	0.97	0.77	33,332.3	0.225	76.08	0.21	4078.9
W-18	94.87	3.43	1.64	5.58	1.21	0.77	34,092.1	0.226	72.42	0.24	4539.6
W-19	97.38	3.48	1.81	7.91	0.58	0.78	30,028.6	0.229	68.53	0.22	5777.9
W-20	106.72	3.35	1.73	7.86	1.01	0.78	29,855.9	0.229	70.53	0.21	5568.2
W-21	101.61	3.36	1.54	7.17	1.08	0.78	33,080.5	0.224	76.73	0.23	5448.1
W-22	99.29	3.51	1.24	6.99	1.38	0.77	34,335.2	0.225	74.19	0.20	3826.4
W-23	105.55	3.31	1.64	6.20	1.31	0.77	33,139.1	0.224	71.22	0.21	4882.8
W-24	99.64	3.72	1.71	8.44	1.54	0.78	33,364.9	0.223	75.24	0.23	6355.9
W-25	97.01	3.19	1.39	6.65	0.94	0.78	33,379.5	0.227	70.63	0.20	5214.2
W-26	100.66	3.45	1.72	6.56	1.52	0.77	34,079.4	0.225	74.28	0.22	4514.4
W-27	100.05	3.18	1.35	6.73	1.18	0.77	32,492.1	0.226	73.43	0.21	4613.0
W-28	100.15	3.48	1.77	6.69	1.48	0.77	32,724.0	0.226	73.93	0.21	4622.5
W-29	93.73	3.42	1.24	6.13	1.04	0.77	33,020.2	0.227	70.75	0.22	4366.2
W-30	98.90	3.12	1.54	6.95	1.32	0.77	32,904.5	0.225	75.65	0.21	4063.9
W-31	95.67	3.60	1.68	8.08	0.87	0.78	32,822.4	0.226	71.35	0.21	6756.1
W-32	101.48	3.34	1.39	6.12	2.22	0.77	33,620.3	0.224	79.11	0.18	3201.2
W-33	98.76	3.73	1.72	5.90	1.75	0.78	33,324.2	0.223	76.58	0.20	5103.3
W-34	99.79	3.25	1.25	7.66	1.50	0.78	31,993.0	0.227	72.64	0.20	4262.1
W-35	102.04	3.91	1.29	8.24	1.29	0.78	32,668.4	0.227	69.38	0.20	6167.0
W-36	97.48	3.70	1.62	7.25	1.49	0.78	33,849.5	0.224	72.20	0.22	6827.0
W-37	100.43	3.83	1.48	7.70	1.14	0.79	34,857.2	0.223	74.23	0.22	7082.4
W-38	103.57	3.62	1.55	8.39	0.78	0.78	31,146.0	0.227	74.39	0.23	5514.4
W-39	94.41	3.53	1.62	6.49	1.58	0.78	33,559.9	0.224	72.81	0.22	5183.4
W-40	96.95	3.46	1.84	6.64	0.91	0.78	34,783.9	0.226	70.91	0.24	6809.9

2.2. Analysis of Main Controlling Factors

From Table 1, it is evident that the Cangdong Sag shale oil field exhibits a multitude of basic parameters with varying ranges. Some parameters demonstrate substantial variations; for instance, the natural gamma radiation spans from 85.28 to 114.31 API, while S_1 ranges from 1.97 to 12.1. However, certain parameters exhibit minimal variations, such as Poisson's ratio, which ranges from 0.22 to 0.23, and the mineral brittleness index, which varies within the range of 0.76 to 0.81. To accurately identify the correlations among geological parameters and their relationship with production, a Pearson correlation coefficient method was employed for single-factor correlation analysis.

The Pearson correlation coefficient is computed as the covariance between two factors divided by the product of their respective standard deviations. It serves to quantify the strength and direction of the linear relationship between the selected variables. The formula for calculating the Pearson correlation coefficient is [26]:

$$\rho_{X,Y} = \frac{Cov(X,Y)}{\sigma_X \sigma_Y} = \frac{m\sum\limits_{i=1}^{m} x_i y_i - \sum\limits_{i=1}^{m} x_i \sum\limits_{i=1}^{m} y_i}{\sqrt{m\sum\limits_{i=1}^{m} x_i^2 - \left(\sum\limits_{i=1}^{m} x_i\right)^2}\sqrt{m\sum\limits_{i=1}^{m} y_i^2 - \left(\sum\limits_{i=1}^{m} y_i\right)^2}} \quad (1)$$

where $\rho_{X,Y}$ is the Pearson correlation coefficient between factor X and factor Y, dimensionless; $Cov(X,Y)$ is the covariance between factor X and factor Y, dimensionless; σ_X and σ_Y are the standard deviations of factor X and factor Y, dimensionless; and m is the number of wells, dimensionless.

Figure 1 depicts a heatmap of the correlation analysis among factors. The analysis revealed a strong correlation between mineral brittleness, S_1, TOC, Ro, and production, while the correlation between minimum horizontal principal stress and natural gamma radiation was relatively weak. Simultaneously, significant correlations exist among certain production-influencing factors. For instance, the correlation coefficients between the coefficient of variation of horizontal principal stress and Young's modulus, as well as between the coefficient of variation of horizontal principal stress and S_1, are 0.801 and −0.663, respectively, indicating multicollinearity. This can lead to distortion in the comprehensive evaluation model, affecting its accuracy. Therefore, this study introduced principal component analysis to address this issue.

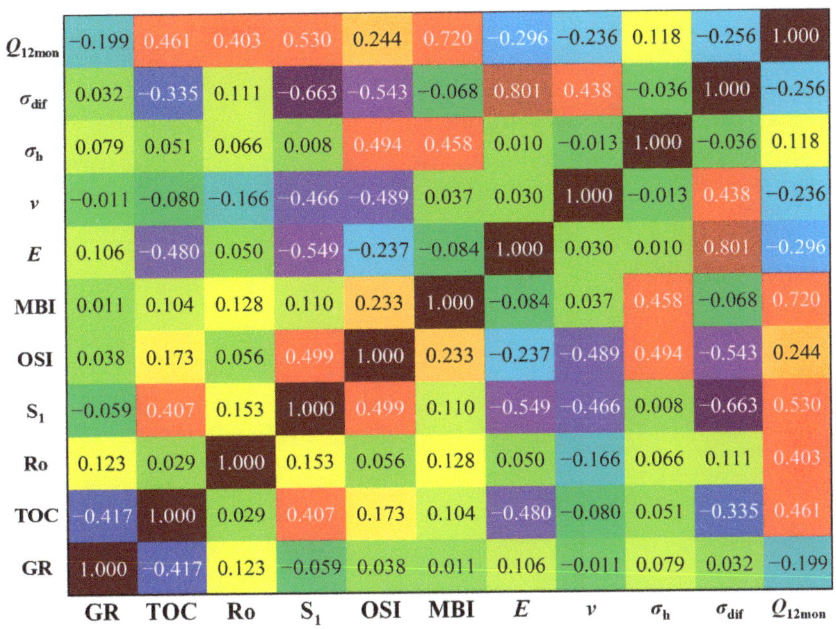

Figure 1. Pearson correlation heat map. The values at the intersection of row and column variables represent the correlation coefficient between the two factors, with a deeper shade indicating a stronger correlation between the two.

2.3. Principal Component Analysis

Principal component analysis transforms the original factors into mutually orthogonal and independent principal components through linear combinations. It selects features based on the contribution of variance from the principal components, reducing the dimensionality of the dataset and eliminating correlations among factors without losing essential information. After principal component analysis, a dataset with n factors generates n principal components. Principal Component 1 consistently represents the direction of maximum variance in the dataset, Principal Component 2 represents the second-largest variance direction, and so forth [27].

Due to the disparate scales of feature data, making direct comparisons and applying weighted processing is challenging. Prior to dimensionality reduction, it is necessary to standardize the data. Subsequently, the covariance matrix of the standardized data should be computed to delve into the interrelationships among variables. Then, one should proceed with the calculation of eigenvalues and the corresponding eigenvectors of the covariance matrix. Eigenvalues elucidate the variance explained by each principal component, while eigenvectors dictate the orientation of these components. Finally, prioritization based on eigenvalues should be undertaken, opting for a subset of principal components, capturing a pre-defined percentage of the total variance for further scrutiny.

The Z-Score normalization method is employed for this purpose:

$$x^* = \frac{x - \omega}{\delta} \tag{2}$$

where x^* is the normalized value for from 0 to 1; x is the value of the factor affecting the fracturing effect; ω is the mean of the sample data; and δ is the standard deviation.

The covariance matrix R should be calculated from the normalized dataset:

$$R = (r_{ij})_{n\times n} = \begin{bmatrix} r_{11} & r_{12} & \cdots & r_{1n} \\ r_{21} & r_{22} & \cdots & r_{2n} \\ \vdots & \vdots & \ddots & \vdots \\ r_{n1} & r_{n2} & \cdots & r_{nn} \end{bmatrix}_{n\times n} \tag{3}$$

The eigenvalue of the matrix R, $\lambda_1 \geq \lambda_2 \geq \cdots \geq \lambda_n \geq 0$, and the corresponding eigenvector $v_1, v_2, \cdots, v_n$, where $v_j = (v_{1j}, v_{2j}, \cdots, v_{nj})^{\mathrm{T}}$, v_{nj} is the n-th component of the j-th feature, should then be calculated, where the principal component after dimension reduction is as follows [28]:

$$\begin{cases} z_1 = v_{11}x_1 + v_{21}x_2 + \cdots + v_{n1}x_n \\ z_2 = v_{12}x_1 + v_{22}x_2 + \cdots + v_{n2}x_n \\ \vdots \\ z_n = v_{1n}x_1 + v_{2n}x_2 + \cdots + v_{nn}x_n \end{cases} \tag{4}$$

where z_n is the n-th principal component and x_n is the n-th factor. The variance ratio of each principal component and their cumulative variance contribution rate are calculated as follows [29]:

$$\tau_j = \frac{\lambda_j}{\sum\limits_{k=1}^{n} \lambda_k} \quad (j = 1, 2, \cdots, n) \tag{5}$$

$$\alpha_p = \frac{\sum\limits_{k=1}^{p} \lambda_k}{\sum\limits_{k=1}^{n} \lambda_k} \quad (p \leq n) \tag{6}$$

where τ_j is the variance contribution rate, dimensionless; and α_p is the cumulative contribution rate, $\alpha_p(p \leq n)$, dimensionless.

The results of principal component analysis are depicted in Figure 2. The general criteria for principal component selection are that the eigenvalue should be greater than 1, and the cumulative contribution rate of the selected principal components should exceed

85%. However, due to the relatively modest size of the sample and the dataset in this study, all 10 principal components are retained [28].

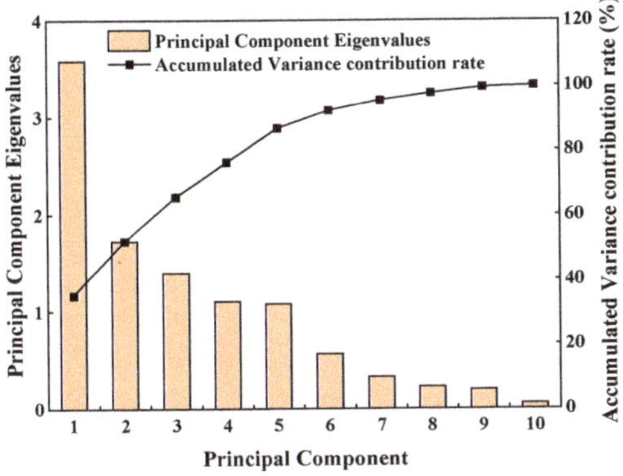

Figure 2. Initial factor feature matrix. Eigenvalues signify the extent to which the introduction of a principal component can elucidate, on average, the information contained in the original scalars. The variance contribution rate represents the proportion of variance that a principal component can explain relative to the total variance.

The principal component feature vectors are illustrated in Figure 3. Based on the coefficients of each factor corresponding to each principal component, expressions for the principal components can be constructed. For example, Principal Component 1 can be expressed as: $PC1 = -0.012GR - 0.155TOC - 0.126Ro + 0.141S_1 + 0.114OSI + 0.007MBI - 0.602E - 0.222\nu - 0.016\sigma_h + 0.708\sigma_{dif}$. Mechanical parameters such as Young's modulus and the coefficient of variation of horizontal principal stress contribute the most to Principal Component 1, indicating that production is a comprehensive reflection influenced by multiple factors. Although the correlation between reservoir mechanical parameters and production is relatively weak, it still contains a considerable amount of significant feature information.

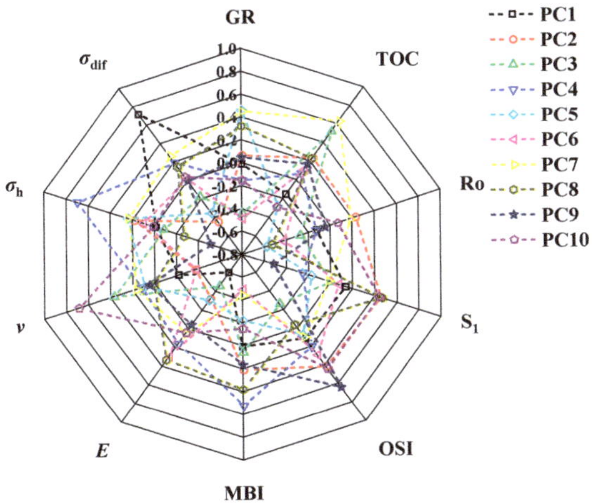

Figure 3. Principal component extraction and analysis. The numerical values corresponding to factors represent the coefficients of each factor in the process of composing principal components. A larger coefficient indicates a higher importance of the corresponding factor in the respective principal component.

3. Fuzzy Comprehensive Evaluation of Reservoir Quality

Due to the disparate contributions of individual principal components to production, precise characterization of the heterogeneity in productivity for horizontally fractured wells proves challenging with quantitative assessment methods. Fuzzy comprehensive evaluation employs membership degrees to signify the extent to which each parameter belongs to different reservoir quality grades, endowing each parameter with an inherently ambivalent nature, suitable for modeling nonlinear functions of arbitrary complexity. The fuzzy comprehensive evaluation model comprises three components: the factor set, the evaluation set, and the weight set.

3.1. Principal Component Membership

The degree of subordination is jointly determined by parameter values and membership functions. Traditional fuzzy planning problems commonly employ linear membership functions such as triangles and trapezoids. However, for the complex and nonlinear issue of reservoir quality assessment involving multiple parameters, these approaches prove inadequate. Gaussian distribution, characterized by continuity and adjustability, offers a more suitable alternative. It aptly reflects the characteristics of data variations and has demonstrated commendable application outcomes. Gaussian membership functions were assigned to each variable to capture the degree of membership of an element to a set. The functions were defined based on the mean and standard deviation of each variable. Therefore, this study employs Gaussian distribution as the membership function [30].

$$\mu(x, \lambda, \delta) = e^{-\frac{(x-\lambda)^2}{2\xi^2}} \tag{7}$$

where μ represents the degree of membership; λ is the principal component value at the peak of the Gaussian distribution; and ξ is the standard deviation.

The evaluation set is utilized to discern the excellence or inferiority of determining factors. For the sake of providing an intuitive scoring system, the following definitions were established: Grade I = "Excellent" = 100, Grade II = "Good" = 75, Grade III = "Ordinary" = 50, Grade IV = "Poor" = 25. In accordance with the evaluation set and its partitioning of reservoir quality, these four evaluation grades correspond to numerical ranges of principal component values (0 to 0.25, 0.25 to 0.50, 0.50 to 0.75, and 0.75 to 1). Based on Equation (7), Gaussian-shaped membership functions for each grade can be constructed, as illustrated in Figure 4.

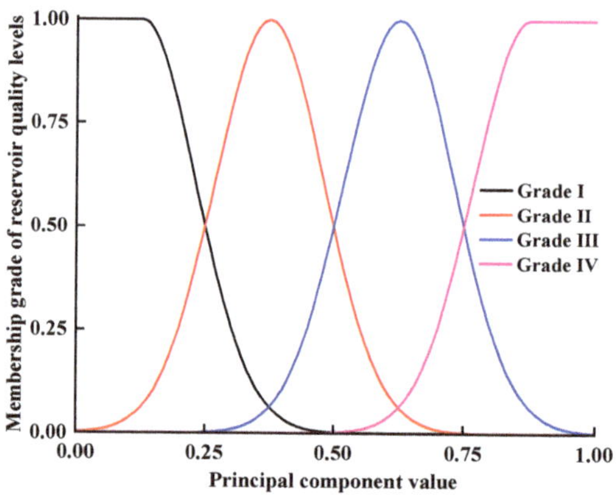

Figure 4. Gaussian membership function. The values corresponding to the intersections with the curves representing the four levels, namely "Excellent", "Good", "Ordinary", and "Poor", signify the membership degree of the principal component for the respective level.

The membership degree matrix of each principal component for each grade of each well can be constructed:

$$A = \begin{bmatrix} a_{1,1} & a_{1,2} & a_{1,3} & a_{1,4} \\ \vdots & \vdots & \vdots & \vdots \\ a_{k,1} & a_{k,2} & a_{k,3} & a_{k,4} \end{bmatrix}_{k\times 4} \tag{8}$$

where $a_{k,1}$ refers to the membership degree of the k-th principal component to its level I.

3.2. Entropy Weight Method

The entropy weight method objectively assigns weights to the overall impact of the system based on the relative change of indicators. It accurately reflects the inherent relationships of information within the system, eliminating the irrationality of subjective experiential judgment. Therefore, in this study, this method was employed to objectively and quantitatively determine the weights of the principal components based on actual data, avoiding errors introduced by the traditional fuzzy logic approach, which relies on expert experience to determine weights [31].

Based on k principal components of m data points, a standardized evaluation matrix was constructed after normalization according to Formula (3):

$$X = \{x_j(i)\}_{m\times k} = \begin{bmatrix} x_1(1) & \cdots & x_j(1) & \cdots & x_k(1) \\ \vdots & & \vdots & & \vdots \\ x_1(i) & \cdots & x_j(i) & \cdots & x_k(i) \\ \vdots & & \vdots & & \vdots \\ x_1(m) & \cdots & x_j(m) & \cdots & x_k(m) \end{bmatrix}_{m\times k} \quad i = 1,2,\ldots,m\,;\ j = 1,2,\ldots,k \tag{9}$$

The proportion of each data point in a single principal component to the sum of all data points of the principal component—that is, entropy p_{ij}—is as follows:

$$p_{ij} = \frac{x_j(i)}{\sum\limits_{i=1}^{m} x_j(i)} \tag{10}$$

The entropy of each principal component e_j is as follows [32]:

$$e_j = -\frac{1}{\ln m}\sum_{i=1}^{m} p_{ij}\cdot \ln p_{ij} \tag{11}$$

where p_{ij} is the ratio of each data point in a single principal component to the sum of all data points of the principal component; e_j is the entropy of the main component, $0 \le e_j \le 1$; and m is the number of data points.

The entropy weight method assumes that the greater the difference of the data samples of each evaluation index, the greater the weight given to the index should be. However, the higher the difference of the samples, the smaller the entropy of the index. Therefore, the entropy weight of each principal component is as follows [32]:

$$c_j = \frac{1-e_j}{\sum\limits_{j=1}^{k}(1-e_j)} j = 1,2,\cdots,k \tag{12}$$

where c_j is the weight of the j-th principal component, dimensionless; and e_j is the entropy of the j-th evaluation index, dimensionless.

Therefore, the principal component weight matrix can be established:

$$C = \begin{bmatrix} c_1 & \cdots & c_j & \cdots & c_k \end{bmatrix}_{1\times k} \tag{13}$$

Figure 5 displays the weights of the principal components. Principal Component 1 has the highest weight proportion, reaching 26.82%, followed by Principal Component 4. For Principal Component 4, the minimum horizontal principal stress and brittleness index contribute the most, further illustrating that production is influenced by a combination of geological and mechanical parameters.

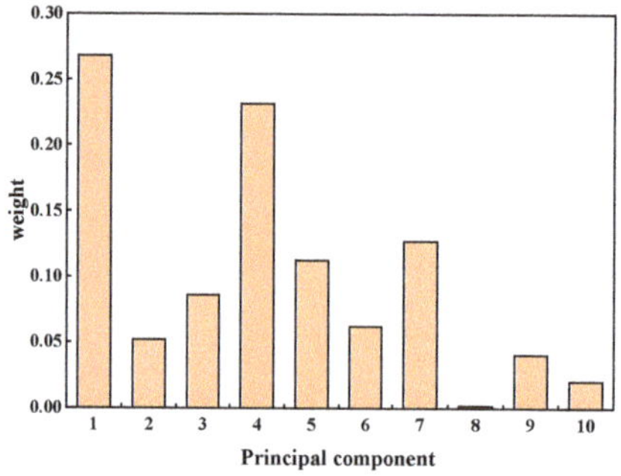

Figure 5. Principal component weight. Entropy values were calculated, reflecting the amount of uncertainty or disorder associated with each variable. Variables with higher entropy were given lower weights, emphasizing the more informative features.

3.3. *Fuzzy Comprehensive Evaluation Model*

Single-factor fuzzy evaluation only reflects the impact of one factor on the evaluation results, without capturing the comprehensive influence of all factors. Fuzzy comprehensive evaluation, based on selected parameters and production data, establishes rank divisions, membership functions, and factor weight rules. By inputting the data of a new shale oil well into these rules, the membership degrees for each grade are calculated. Considering the contribution of all factors, and employing the matrix multiplication and summation method, the weight matrix C is combined with the principal component membership matrix A, resulting in the fuzzy comprehensive evaluation matrix B for reservoir quality assessment [33]:

$$B = \begin{bmatrix} c_1 & \cdots & c_j & \cdots & c_k \end{bmatrix}_{1\times k} \circ \begin{bmatrix} a_{1,1} & a_{1,2} & a_{1,2} & a_{1,4} \\ a_{2,1} & a_{2,2} & a_{2,3} & a_{2,4} \\ \vdots & \vdots & \vdots & \vdots \\ a_{k,1} & a_{k,2} & a_{k,3} & a_{k,4} \end{bmatrix}_{k\times 4} = \begin{bmatrix} b_1 & b_2 & b_3 & b_4 \end{bmatrix}_{1\times 4} \tag{14}$$

where b_1 is the membership of the well to grade I. Further, according to the corresponding score of each grade, the fuzzy comprehensive evaluation results are transformed into fuzzy comprehensive scores [34]:

$$r = \frac{\sum_{i=1}^{4}(b_i \times u_i)}{\sum_{i=1}^{4} u_i} \tag{15}$$

The scores of b_i corresponding comment sets I, II, III and IV are 100, 75, 50 and 25 respectively.

Table 2 presents the membership degrees and fuzzy comprehensive evaluation results for the Cangdong shale oil wells. Higher membership degrees for higher grades indicate better reservoir quality for the shale oil well. Consequently, the higher the fuzzy comprehensive evaluation score, the better the reservoir quality. The correspondence between the production and evaluation score reflects the accuracy of the model.

Table 2. Evaluation results of Cangdong sag.

No. Well	Membership Degree				Score
	I	II	III	IV	
W-1	0.339	0.093	0.250	0.318	61.30
W-2	0.047	0.295	0.174	0.483	47.69
W-3	0.032	0.205	0.296	0.468	45.04
W-4	0.189	0.183	0.358	0.270	57.28
W-5	0.249	0.103	0.245	0.403	54.93
W-6	0.150	0.438	0.354	0.058	67.01
W-7	0.411	0.346	0.117	0.126	76.05
W-8	0.322	0.157	0.317	0.205	64.87
W-9	0.376	0.244	0.259	0.121	71.90
W-10	0.283	0.233	0.279	0.205	64.86
W-11	0.286	0.256	0.250	0.208	65.51
W-12	0.091	0.166	0.271	0.471	46.93
W-13	0.019	0.141	0.400	0.440	43.46
W-14	0.061	0.273	0.234	0.432	49.09
W-15	0.000	0.200	0.281	0.520	42.00
W-16	0.340	0.296	0.295	0.069	72.69
W-17	0.062	0.158	0.645	0.135	53.66
W-18	0.138	0.241	0.414	0.208	57.71
W-19	0.340	0.197	0.301	0.162	67.91
W-20	0.154	0.302	0.470	0.073	63.43
W-21	0.059	0.409	0.479	0.054	61.81
W-22	0.053	0.246	0.569	0.132	55.48
W-23	0.079	0.371	0.491	0.059	61.75
W-24	0.121	0.430	0.426	0.023	66.23
W-25	0.062	0.399	0.389	0.150	59.31
W-26	0.035	0.378	0.504	0.084	59.08
W-27	0.049	0.299	0.565	0.087	57.76
W-28	0.024	0.440	0.413	0.123	59.14
W-29	0.060	0.295	0.457	0.188	55.67
W-30	0.040	0.258	0.614	0.088	56.26
W-31	0.196	0.375	0.387	0.042	68.14
W-32	0.013	0.128	0.667	0.192	49.02
W-33	0.076	0.377	0.389	0.158	59.26
W-34	0.092	0.198	0.614	0.096	57.16
W-35	0.141	0.310	0.504	0.045	63.66
W-36	0.167	0.343	0.478	0.013	66.57
W-37	0.249	0.265	0.438	0.048	67.89
W-38	0.119	0.475	0.343	0.064	66.20
W-39	0.160	0.196	0.592	0.053	61.55
W-40	0.258	0.247	0.305	0.191	64.30

Figure 6 depicts the relationship between Q_{12mon} and the comprehensive evaluation score without using principal component analysis [34]. When compared to the analysis model using PCA (Figure 7), it is evident that the direct analysis of factors on reservoir quality has a correlation coefficient of only 0.7609. In contrast, the model developed in this study achieves a higher correlation coefficient of 0.9132. The good consistency between reservoir quality evaluation results and production indicates that the model built in this study can effectively reflect the enhanced production potential of the reservoir.

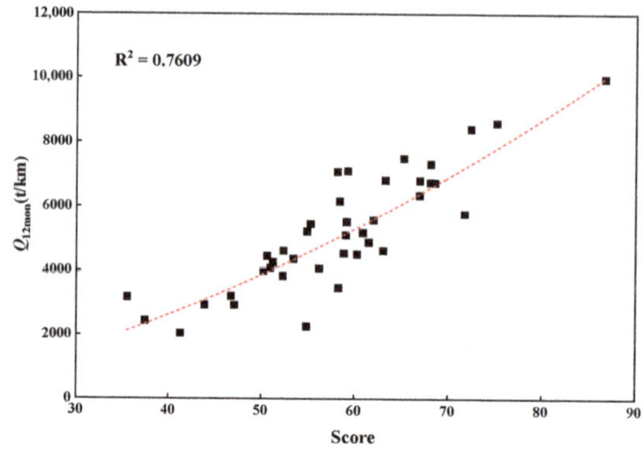

Figure 6. Q_{12mon} and Score correlation without PCA. At this juncture, without transforming the evaluation indicators into principal components, the relatively low correlation coefficient implies a significant adverse impact of inter-factor correlations on the evaluation effectiveness.

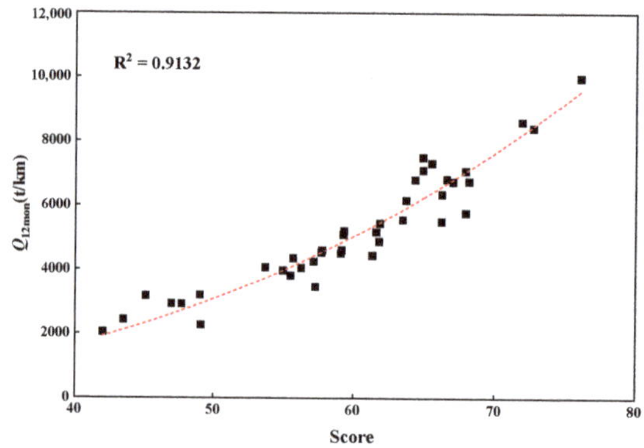

Figure 7. Q_{12mon} and Score correlation with PCA. The correlation coefficient between production and the comprehensive score, obtained by transforming multiple reservoir quality assessment indicators using this model, is 0.9132. This value indicates a robust evaluation performance of the model.

4. Field Application

The model developed in this study was applied to assess reservoir quality and predict production for 10 wells in the Cangdong shale oil field, utilizing the basic parameters of the wells as shown in Table 3. By substituting the basic parameters of the wells into the evaluation model established using the 40 sample wells in the previous analysis, comprehensive scores were obtained. Further, based on the fitted curve between production and scores shown in Figure 7, the production for the applied wells was predicted, as indicated in Table 3. Among the 10 application wells, the T-10 well exhibits the highest production, reaching 7744.5 t/km. However, the T-3 well, despite having seemingly similar parameters to other wells, demonstrates a production discrepancy close to threefold when compared to the T-10 well. This further underscores that production is the outcome of a multifactorial influence.

Figure 8 illustrates the comparative analysis between the actual cumulative oil production per kilometer of fractured length over 12 months for the 10 applied wells and the corresponding predicted results. It is evident that the relative prediction error ranges from 0.39% to 12.96%, with an average relative error of 8.2%. This indicates that the method proposed in this study can elucidate the potential correlations between source lithology, fracability, and production.

It effectively addresses the intricate issue of reservoir comprehensive evaluation, providing a quantitative assessment of shale oil reservoir quality.

Table 3. Parameters of application wells in Cangdong sag.

Well No.	GR (API)	TOC (%)	Ro	S_1 (mg/g)	OSI	MBI	E (MPa)	ν	σ_h (MPa)	σ_{dif}	Score	Q_{12mon} (t/km) Predictive	Actual
T-1	110.33	2.59	1.10	2.25	0.87	0.79	37,671.1	0.228	83.85	0.28	54.73	3901.9	3759.9
T-2	96.22	3.95	0.87	5.04	2.25	0.77	34,676.1	0.224	77.16	0.18	50.19	3196.1	3183.7
T-3	103.67	3.02	2.00	4.78	1.58	0.76	36,482.2	0.228	86.43	0.27	46.53	2720.5	2982.2
T-4	102.20	3.13	1.10	4.57	1.46	0.81	39,669.4	0.228	89.63	0.29	59.78	4873.0	4371.5
T-5	101.63	3.41	0.91	5.82	1.62	0.78	32,064.6	0.227	72.46	0.19	51.67	3410.9	3687.1
T-6	103.16	3.64	0.90	9.28	0.77	0.77	30,964.3	0.230	70.29	0.22	55.84	4098.0	4415.3
T-7	94.77	3.81	0.87	7.79	0.99	0.78	31,955.3	0.223	73.09	0.23	59.62	4839.8	5214.5
T-8	99.18	3.52	0.92	5.42	1.17	0.78	32,863.4	0.224	63.05	0.17	59.68	4852.2	5536.2
T-9	96.21	3.71	1.10	11.09	2.99	0.80	25,681.7	0.229	91.56	0.14	65.59	6292.7	7229.6
T-10	101.61	3.63	2.00	3.85	1.06	0.81	36,607.8	0.228	79.76	0.28	67.81	6937.9	7744.5

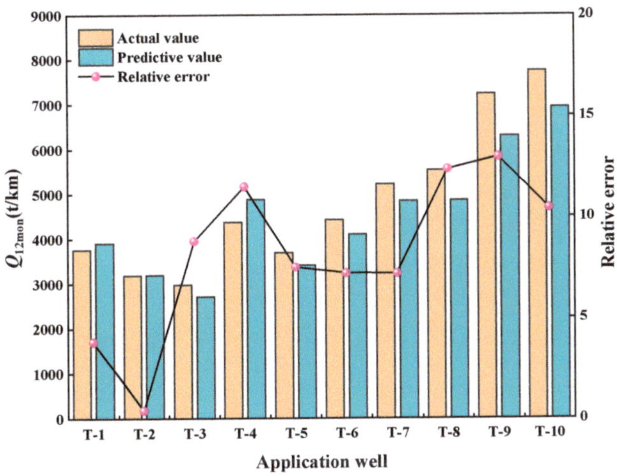

Figure 8. Comparative between actual production and predicted production. By inputting the evaluation factors of the 10 wells into the model to predict production, the relative errors range from 0.39% to 12.96% when compared with the actual production. This suggests a promising prospect for the practical application of the model in the field.

The proposed comprehensive evaluation model combines the strengths of various methods to create a versatile and interpretable framework. Its advantages lie in its ability to handle both quantitative and qualitative aspects, offering a holistic evaluation approach. However, potential limitations include the need for a comprehensive understanding of its components and potential parameter tuning. The choice between models ultimately depends on the specific characteristics and requirements of the application domain.

5. Conclusions

(1) To accurately assess the reservoir quality of shale oil fractured horizontal wells, a fuzzy comprehensive evaluation model based on principal component analysis and the entropy weight method is proposed. The contribution of parameters to production was determined through Pearson analysis. The linear combination of factors and the elimination of correlations between factors were achieved by incorporating principal component analysis. The entropy weight method was introduced to determine the weights of principal components, establishing the fuzzy comprehensive evaluation model for reservoir quality.

(2) Analysis of 40 wells in the Cangdong shale oil field revealed that the correlation coefficient between single-factor assessment of production and the evaluation result for reservoir quality is only 0.7609. However, after principal component analysis, the correlation coefficient increased to 0.9132, significantly enhancing the model's accuracy.

(3) Applying the model developed in this study to assess reservoir quality and predict production for 10 wells in the Cangdong shale oil field, the results indicate that the average relative error between the actual cumulative oil production per kilometer of fractured length over 12 months for the applied wells and the corresponding predicted results is 8.2%. This suggests that the method proposed in this study can elucidate the potential correlations between source lithology, fracability, and production.

Author Contributions: Writing—original draft, F.T., Y.F., X.L., D.L., Y.J., L.S., L.Y., Y.Z., T.Z., Q.Y. and X.G.; Writing—review and editing, F.T., Y.F., X.L. and D.L. All authors have read and agreed to the published version of the manuscript.

Funding: This research received no external funding.

Data Availability Statement: The original contributions presented in the study are included in the article, further inquiries can be directed to the corresponding authors.

Conflicts of Interest: The authors declare that the research was conducted in the absence of any commercial or financial relationships that could be construed as a potential conflict of interest.

References

1. Xu, G.J.; Ren, Z.Y.; Wang, Z.P.; Cui, L.; Su, J.Z.; Meng, X.L.; Chen, P.L.; Li, P.; Wang, N.N.; Hao, X.; et al. Petroleum-like fuels with substantially enriched branched iso-paraffins and benzenes via boehmite-assisted pyrolysis of oil shale. *Fuel* **2024**, *358*, 130324. [CrossRef]
2. Zhao, G.X.; Yao, Y.D.; Zhang, T.; Wang, L.; Adenutsi, C.D.; Nassar, N.N. An Integrated Approach for History Matching of Complex Fracture Distributions for Shale Oil Reservoirs Based on Improved Adaptive Particle Filter. *SPE J.* **2023**, *28*, 594–613. [CrossRef]
3. Lin, H.Y.; Mao, L.J.; Mai, Y. Influence of multistage fracturing in shale gas wells on the casing deformation of horizontal wells. *Pet. Sci. Technol.* **2024**, *42*, 56–80. [CrossRef]
4. Hyunjun, I.; Hyongdoo, J.; Erkan, T.; Micah, N. Long- and Short-Term Strategies for Estimation of Hydraulic Fracturing Cost Using Fuzzy Logic. *Minerals* **2022**, *12*, 715–735.
5. Iyare, U.C. Fracability evaluation of the upper Cretaceous Naparima Hill Formation, Trinidad. *J. Pet. Sci. Eng.* **2022**, *208*, 109599. [CrossRef]
6. Zheng, D.Z.; Miska, S.; Ozbayoglu, E.; Zhang, J.G. Combined Experimental and Well Log Study of Anisotropic Strength of Shale. In Proceedings of the SPE Annual Technical Conference and Exhibition, San Antonio, TX, USA, 16–18 October 2023.
7. Huang, Y.Y.; Zhang, Y.; Xi, J.H.; Huang, L.L.; Wang, S.; Zhang, Y.L.; Lai, J.; Jiang, C.Z. Logging evaluation of pore structure and reservoir quality in shale oil reservoir: The Fengcheng Formation in Mahu Sag, Junggar Basin, China. *Mar. Pet. Geol.* **2023**, *156*, 106454. [CrossRef]
8. Liu, Y.Z.; Zeng, J.H.; Qiao, J.C.; Yang, G.Q.; Liu, S.N.; Cao, W.F. An advanced prediction model of shale oil production profile based on source-reservoir assemblages and artificial neural networks. *Appl. Energy* **2023**, *333*, 120604. [CrossRef]
9. Wen, S.Q.; Wei, B.; He, Y.J.; Xin, J.; Varfolomeev, M.A. Forecasting oil production in unconventional reservoirs using long short term memory network coupled support vector regression method: A case study. *Petroleum* **2023**, *9*, 647–657. [CrossRef]
10. Qin, X.Z.; Hu, X.H.; Liu, H.; Shi, W.Y.; Cui, J.S. A Combined Gated Recurrent Unit and Multi-Layer Perception Neural Network Model for Predicting Shale Gas Production. *Processes* **2023**, *11*, 806. [CrossRef]
11. Liu, W.C.; Yang, Y.J.; Qiao, C.C.; Liu, C.; Lian, B.; Yuan, Q.W. Progress of Seepage Law and Development Technologies for Shale Condensate Gas Reservoirs. *Energies* **2023**, *16*, 2446. [CrossRef]
12. Mu, J.F.; Qiao, H.J.; Guo, Q. A Fracture Toughness-Based Evaluation Method for Deep Shale Oil Reservoir Compressibility. *Chem. Technol. Fuels Oils* **2022**, *58*, 880–886. [CrossRef]
13. Feng, C.; Feng, Z.Y.; Mao, R.; Li, G.L.; Zhong, Y.T.; Ling, K.G. Prediction of vitrinite reflectance of shale oil reservoirs using nuclear magnetic resonance and conventional log data. *Fuel* **2023**, *339*, 127422. [CrossRef]
14. You, X.T.; Liu, J.Y.; Jia, C.S.; Li, J.; Liao, X.Y.; Zheng, A.W. Production data analysis of shale gas using fractal model and fuzzy theory: Evaluating fracturing heterogeneity. *Appl. Energy* **2019**, *250*, 1246–1259. [CrossRef]
15. Gou, B.; Wang, C.; Yu, T.; Wang, K.J. Fuzzy logic and grey clustering analysis hybrid intelligence model applied to candidate-well selection for hydraulic fracturing in hydrocarbon reservoir. *Arab. J. Geosci.* **2020**, *13*, 1–13. [CrossRef]
16. Okwu, M.O.; Nwachukwu, A.N. A review of fuzzy logic applications in petroleum exploration, production and distribution operations(Review). *J. Pet. Explor. Prod. Technol.* **2019**, *9*, 1555–1568. [CrossRef]

17. Zoveidavianpoor, M.; Gharibi, A. Applications of type-2 fuzzy logic system: Handling the uncertainty associated with candidate-well selection for hydraulic fracturing. *Neural Comput. Appl.* **2016**, *27*, 1831–1851. [CrossRef]
18. Davarpanah, A.; Shirmohammadi, R.; Mirshekari, B.; Aslani, A. Analysis of hydraulic fracturing techniques: Hybrid fuzzy approaches. *Arab. J. Geosci.* **2019**, *12*, 1–8. [CrossRef]
19. Xie, J.Y.; Zhang, J.J.; Fang, Y.P.; Cao, J.X.; Deng, J.X. Quantitative Evaluation of Shale Brittleness Based on Brittle-Sensitive Index and Energy Evolution-Based Fuzzy Analytic Hierarchy Process. *Rock Mech. Rock Eng.* **2023**, *56*, 3003–3021. [CrossRef]
20. Verma, A.K.; Singh, T.N. A neuro-fuzzy approach for prediction of longitudinal wave velocity. *Neural Comput. Appl.* **2013**, *22*, 1685–1693. [CrossRef]
21. Sylwan, C.; Vernengo, L.; Sharma, R.K.; Chopra, S.; Trinchero, E. Reducing uncertainty in characterization of Vaca Muerta Formation Shale with poststack seismic data. *Lead. Edge* **2015**, *34*, 1462–1467.
22. Foster, D.; Yenugu, M.; Sayers, C.M. Introduction to this special section: Resource plays II: Geophysics. *Lead. Edge* **2015**, *34*, 1440–1441.
23. Rickman, R.; Mullen, M.; Petre, E.; Grieser, B.; Kundert, D. A Practical Use of Shale Petrophysics for Stimulation Design Optimization: All Shale Plays Are Not Clones of the Barnett Shale. In Proceedings of the SPE Annual Technical Conference & Exhibition, Denver, CO, USA, 21–24 September 2008.
24. Zhang, W.; Huang, F.; Li, Z.; He, X.; He, Y. Shale-gas reservoir-prediction study in Daanzhai, Eastern Sichuan Basin. *Lead. Edge* **2014**, *33*, 526–534.
25. Dashtian, H.; Jafari, G.R.; Sahimi, M.; Masihi, M. Scaling, multifractality, and long-range correlations in well log data of large-scale porous media. *Phys. A Stat. Mech. Its Appl.* **2011**, *390*, 2096–2111. [CrossRef]
26. Fu, Y.K.; Dehghanpour, H.; Ezulike, D.O.; Jones, R.S. Estimating Effective Fracture Pore Volume From Flowback Data and Evaluating Its Relationship to Design Parameters of Multistage-Fracture Completion. *SPE Prod. Oper.* **2017**, *32*, 423–439. [CrossRef]
27. Rios, E.H.; Azeredo, R.B.D.V.; Moss, A.K.; Pritchard, T.N.; Domingues, A.B.G. Estimating the Permeability of Rocks by Principal Component Regressions of NMR and MICP Data. *Petrophysics* **2022**, *63*, 442–453.
28. Wang, L.; Yao, Y.; Wang, K.; Adenutsi, C.D.; Zhao, G. Combined Application of Unsupervised and Deep Learning in Absolute Open Flow Potential Prediction: A Case Study of the Weiyuan Shale Gas Reservoir. In Proceedings of the SPE/AAPG/SEG Asia Pacific Unconventional Resources Technology Conference, Virtual, 16–18 November 2021.
29. Li, Y.; Li, B.Z.; Liu, L.; Xiong, L.H.; Luo, H.; Peng, H.; Wang, D.G. Case study of a super-giant field rejuvenation. In Proceedings of the 78th EAGE Conference and Exhibition 2016: Efficient Use of Technology–Unlocking Potential, Vienna, Austria, 30 May–2 June 2016.
30. Liu, Z.Q.; Wang, X. The distributivity of extended uninorms over extended overlap functions on the membership functions of type-2 fuzzy sets. *Fuzzy Sets Syst.* **2022**, *448*, 94–106. [CrossRef]
31. Gao, C.L.; Li, S.C.; Wang, J.; Li, L.P.; Lin, P. The Risk Assessment of Tunnels Based on Grey Correlation and Entropy Weight Method. *Geotech. Geol. Eng.* **2017**, *36*, 1621–1631. [CrossRef]
32. Ju, Y.; Wu, G.J.; Wang, Y.L.; Liu, P.; Yang, Y.M. 3D Numerical Model for Hydraulic Fracture Propagation in Tight Ductile Reservoirs, Considering Multiple Influencing Factors via the Entropy Weight Method. *SPE J.* **2021**, *26*, 2685–2702. [CrossRef]
33. Zeng, F.H.; Cheng, X.Z.; Guo, J.C.; Tao, L.; Chen, Z.X. Hybridising human judgment, ahp, grey theory, and fuzzy expert systems for candidate well selection in fractured reservoirs. *Energies* **2017**, *10*, 447. [CrossRef]
34. Zeng, F.H.; Guo, J.C.; Long, C. A hybrid model of fuzzy logic and grey relation analysis to evaluate tight gas formation quality comprehensively. *J. Grey Syst.* **2015**, *27*, 87–98.

processes

MDPI

Article

CFD–DEM Simulation of a Jamming Mechanism and Influencing Factors of a Fracture-Shrinking Model

Jiabin Zhang, Cong Lu *, Tao Zhang and Jianchun Guo

National Key Laboratory of Oil and Gas Reservoir Geology and Exploitation, Southwest Petroleum University, Chengdu 610500, China; zjb17748499290@163.com (J.Z.); zhangt@swpu.edu.cn (T.Z.)
* Correspondence: lucong@swpu.edu.cn

Abstract: Fractured-vuggy reservoirs are crucial for increasing unconventional oil storage and production, but the controlling mechanism of this dominant flow channel remains vague, and the jamming mechanism of modulator particles is unclear. This study explores the filling and jamming processes of particles in the fractures by conducting a computational fluid dynamics–discrete element method (CFD–DEM) coupled simulation, considering the variation of fracture width, fluid velocity, particle size, and concentration. Results suggest that four sealing modes are proposed: normal filling, local jamming, complete sealing, and sealing in the main fracture. The ratio of particle size to the main fracture width exerts the primary role, with the ratio having a range of $0.625 < D/W \leq 0.77$ revealing complete jamming. Furthermore, an optimal particle size for achieving stable sealing is observed when the particle size varies from 2 to 2.5 mm. A higher concentration of particles yields better results in the fracture-shrinking model. Conversely, a greater velocity worsens the sealing effect on fractures. This research can offer technical support for the large-scale dissemination of flow regulation technology.

Keywords: fractured-vuggy reservoir; fracture-shrinking; jamming mechanism; sealing mode; CFD–DEM

1. Introduction

Carbonate reservoirs have been the utmost source of oil and gas production in many regions of the world, including South America, the Middle East, and Central Asia [1,2]. Based on statistical data from IHS Markit, carbonate oil and gas resources have a proportion of about 70% of the total oil and gas resources globally, proven recoverable reserves of about 50%, and production of about 60%. These reservoirs are usually huge in areal extent and thickness and heterogeneous in pore structure [3], mainly distributed in the Jurassic, Cretaceous, and Neogene strata. China is also abundant in carbonate oil and gas resources, especially in the western part [4,5]. Recently, the production of carbonate oil and gas has increased rapidly; thus, it has become an important field of oil and gas exploration and development and oil and gas storage and production in China [6]. Compared with carbonate reservoirs in other countries, carbonate reservoirs in China are more difficult to exploit because of their old geological age, deep location and multi-stage tectonic transformation [7], with the primary oil reservoirs of fissure-pore type and fracture-cavity type, and the gas reservoirs of mainly porous reef dolomite.

The Ordovician fracture-vuggy carbonate reservoir belonging to Tahe oilfield is located in Tarim Basin, being buried at a depth of over 4500 m; it is a complex reservoir zone of combined fractures and vughs and varying reservoir size [8,9]. The reservoir zone is dominated by karst caves and large fractures, in which the caves are large in scale and have various connectivity forms, complex spatial distribution, and strong heterogeneity (for geological information, please refer to Figure 2 in [10]). After long-term, water-driving exploitation, the fractured-vuggy reservoirs have issues with low response and low recovery degrees [11,12]. Consequently, it is urgent to expand the water-driving-related volume of fractured-vuggy reservoirs and improve the oil recovery efficiency of fractured-vuggy reservoirs. The implementation of water injection has emerged as the foremost approach for ensuring stable production and enhancing recovery efficiency in fractured-vuggy carbonate reservoirs [13]. Nevertheless, in the fractured-vuggy reservoir of Tahe, water injection exploitation often leads to the formation of distinct preferential pathways within the injection−production well group [14,15]. These channels facilitate the rapid breakthrough of injected water, resulting in premature water breakthrough in oil wells and limited effectiveness of water flooding, thereby compromising the desired outcomes of water injection.

By implementing measures to adjust particle flow channels, significant changes have been observed in the flow characteristics of fractured reservoirs, and positive production feedback has been obtained onsite [16,17]. However, due to the intricate nature of karst caves and fractures as well as their interconnections within fractured reservoirs, the field implementation of flow channel adjustments for different well conditions requires distinct process parameters [18]. This constraint hinders the widespread adoption and application of particle flow channel adjustment technology [19]. Notably, local jamming caused by flow control agent particles is a crucial factor influencing the effectiveness of flow control [20]. The bridging between particles serves as the primary mechanism for jamming and shrinking fractures, thus significantly impacting design parameters such as particle concentration, size distribution, and injection volume.

Exploring the migration law of granular flow modulators is crucial, and computational fluid dynamics (CFD) has proven to be an effective tool for accurately simulating liquid flow at a macroscopic scale [21–24]. In addition, the discrete element method (DEM) is a widely used particle simulation technology that can accurately describe the real-time motion state of individual particles [25–28]. The combination of CFD and DEM technology provides an ideal method for simulating the migration behavior of granular modulators in fracture-cavity reservoirs [29–33]. Furthermore, numerical simulation has the advantage of obtaining the relationship between the type, size, and concentration of the plugging agent and the fracture width, and can directly extend the experimental results to the strata conditions. For instance, Zhang et al. [34] used DEM technology to describe the migration path of thin fluid as particles but ignored the influence of particles on fluid flow and particle−particle interactions. Varga et al. [35] utilized the CFD−DEM coupled technique to simulate particle flow in a feed pipe, with prediction results for local particle packing density consistent with laboratory test results. Chu and Yu [36,37] successfully combined the DEM code with CFD software to simulate the particle−fluid flow in complex three-dimensional systems, demonstrating that the CFD−DEM method is a potential tool for analyzing internal particle flow and dynamic particle−fluid flow.

This paper conducts the CFD−DEM coupled simulation to investigate the sealing process of particles in the fracture-shrinking model and examine the influences of particle concentration, size, flow velocity, and density on the sealing process. The influences of particle size, fluid velocity, and fracture structure on pressure drop and bridging morphology during particle jamming are unraveled through analysis of simulation results, subsequently proposing a predictive approach for fractures and vugh sealing. The goal is to provide a constructive guide for the fractured-vuggy carbonate reservoir to improve production performance.

2. Methodology

The coupled CFD−DEM numerical method implemented by the CFD program based on the mass conservation equation, and the Navier−Stokes (N−S) and k-epsilon equations, contributing to simulate fluid flow and the Newton's second law-based DEM module, which is utilized to derive the particle motion, is conducted to model the fluid−particle two-phase behavior. The forces involving particle–particle, particle–wall, and fluid–particle interaction are considered, and the two solid−liquid two phases are coupled by volume fraction and interaction force. In the following, the governing equations are briefly described.

2.1. CFD-Based Fluid Flow Model

The fluid flow through the fracture of an incompressible property with a secondary particulate phase can be mathematically described by the volume-averaged N−S equations [38].

$$\frac{\partial(\varepsilon_l\rho_l)}{\partial t} + \nabla\cdot(\varepsilon_l\rho_l u_l) = 0 \quad (1)$$

$$\frac{\partial}{\partial t}(\rho_l\varepsilon_l u_l) + \nabla\cdot(\varepsilon_l\rho_l u_l u_l) = -\varepsilon_l\nabla p + \nabla\cdot(\varepsilon_l\tau_l) + \rho_l\varepsilon_l g + M_{pl} \quad (2)$$

where ρ_l is the density of the liquid, kg/m^3. ε_l is the volume coefficient of the liquid, $(\varepsilon_l = 1 - \varepsilon_p)$. M_{pl} is the momentum exchange source term between the modulator and the liquid phase, $\left(M_{pl} = \frac{1}{V_{cell}}\sum_{i=1}^{m} F_{lp,i}\right)$.

2.2. DEM-Based Particle Model

DEM can calculate the trajectory of particles considering the interaction between each particle and the influence of container walls. The rotational and translational components are included in the motion of particles, and Newton's second law-derived momentum balance equation with differential form is introduced to describe the motion of each individual particle [39,40].

$$M_P\frac{\partial u_p}{\partial t} = Fc + \left(\sum_q F_{c,q}{}^p\right) + F_D + F_S + F_M + F_P \quad (3)$$

where M_P is the mass of the particle, kg. $F_{c,q}{}^p$ is the contact force between particles, N. F_D is the drag force of the fluid, N. F_S is the shearing lift force, N. F_M is the rotational lift force or Magnus force, N. F_P is the gradient force of the fluid pressure, N.

The particle rotational motion is described as:

$$\frac{d}{dt}I_P\omega_P = \sum_q\left(T^P_{t,q} + T^P_{r,q}\right) + T^P_{DT} \quad (4)$$

where $T^P_{t,q}$ is the torque vector generated by the tangential contact force of particle q acting on particle p. $T^P_{r,q}$ is the torque vector generated by the normal contact force of particle q acting on particle p. I_P is the inertia tensor. ω_P is the torque of the rotational velocity of particle p. The rotation movement is also influenced by the drag torque T^P_{DT}, which is produced by the sliding rotation.

In order to simplify the model for practical application and calculation, the following assumptions are made: (1) the liquid phase is assumed to be incompressible and inelastic fluid and is in an isothermal state, with relatively stable rheological properties that do not change with time and location; (2) the particles are assumed to be spherical rigid particles with uniform density, particle size, geometric size and friction angle, and the particles will not react with each other; (3) the particles that collide in the transport process of the modulator obey Newton's second law, and their motion can be decomposed into two states of motion: translational state and rotational state.

3. DEM−CFD Model Setup

3.1. Geometry of Fracture-Shrinking Model

Based on the actual characteristics of the strata, the rock fracture width of carbonate reservoirs in the Tahe oil field varies from 0.001 mm to tens of millimeters [41]. A fracture-shrinking structure has been established, as depicted in Figure 1 to serve as a foundation for simulating fracture plugging under varying conditions such as particle size, fluid velocity, and particle concentration. The main fracture measures 1.2 m in length with a width of 8 mm and height of 0.2 m, transitioning into a narrow fracture structure through localized necking. The narrow fracture is 0.4 m in length with a height of 0.2 m. The main fracture is undergoing a transition to a narrow fracture structure via localized necking, while the joint angle of the three distinct joint combinations remains fixed at 7.125°. Table 1 lists the geometric parameters of the models.

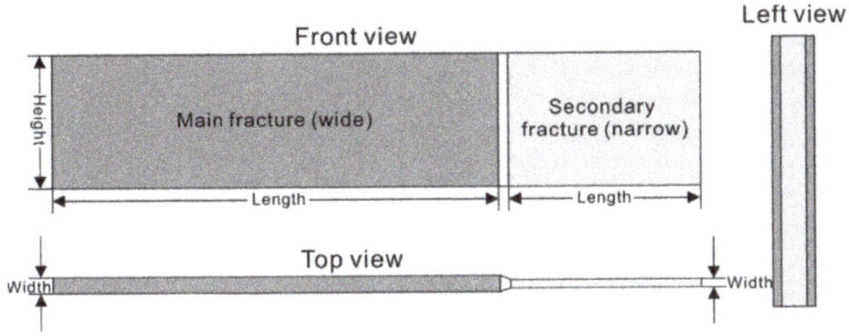

Figure 1. Geometry of the fracture-shrinking model.

Table 1. Geometry parameters of three models.

Models	Main Fracture Length (m)	Main Fracture Width (mm)	Main Fracture Height (m)	Secondary Fracture Length (m)	Secondary Fracture Width (mm)	Secondary Fracture Height (m)
Model I	0.8	8	0.2	0.4	3	0.2
Model II	0.8	8	0.2	0.4	4	0.2
Model III	0.8	8	0.2	0.4	5	0.2

3.2. Grid Model

ICEM CFD embedded in ANSYS version 22.2 (ANSYS, Inc., Canonsburg, PA, USA) is utilized to model and grid the aforementioned geometric models. Initially, the geometric modeling tool establishes the fundamental geometric model; subsequently, block grid division technology and local grid encryption technology are employed to partition the primary joint, transition segment, and secondary joint. The resulting grid system flawlessly maps onto the geometry. Then, in order to strike a balance between calculation velocity and accuracy, local nodes are established within the grid. Subsequently, the validity of the grid is verified before outputting a fluent-compatible file for solid–liquid two-phase flow calculations.

3.3. Experiment Validation

Utilizing the experimental results of the model with unchanged fracture height and suddenly contracted fracture width to 12 mm as the subject, the parameters employed in the numerical simulation are consistent with those used in the physical experiment. The simulation results are depicted in Figure 2. It is evident that the simulation results exhibit a jamming similar to that observed in the experiment, indicating that the established numerical model can accurately reflect the actual physical process.

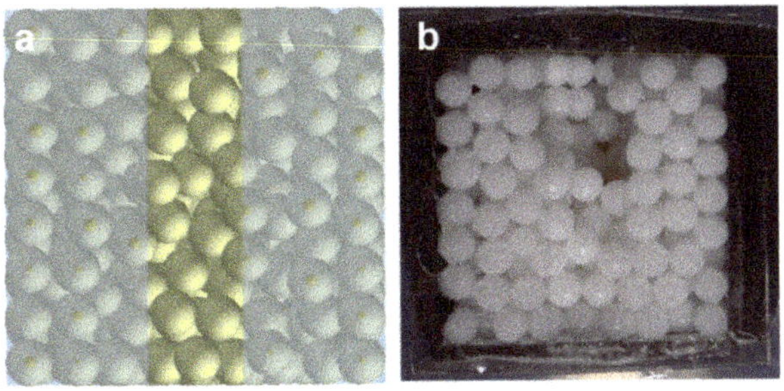

Figure 2. Comparison and validation of numerical simulation (**a**) and physical experiment (**b**).

3.4. Simulation Scenarios

Based on the actual geological characteristics and physical model experiments, four factors are selected to explore their influences on the jamming process in the simulation, including fracture size, injection velocity, particle volume concentration and particle diameter. Table 2 lists the parameters used in the simulation.

Table 2. Simulation scenarios.

Scenario	Fracture Width Combination (Wide/Narrow) (mm)	Injection Velocity (m/s)	Particle Volume Concentration (%)	Particle Diameter (mm)
1	8 mm–3 mm	0.6	4	2
2	8 mm–3 mm	0.6	4	3.5
3	8 mm–3 mm	0.6	4	3
4	8 mm–3 mm	0.6	4	2.5
5	8 mm–3 mm	0.8	4	3
7	8 mm–4 mm	0.6	4	2.66
8	8 mm–4 mm	0.6	4	1
9	8 mm–4 mm	0.6	4	1.5
10	8 mm–4 mm	0.6	4	2
11	8 mm–4 mm	0.6	4	2.5
12	8 mm–4 mm	0.6	4	3
13	8 mm–4 mm	1	4	1
14	8 mm–4 mm	1	4	1.5
15	8 mm–4 mm	1	4	2
16	8 mm–4 mm	1	4	2.5
17	8 mm–4 mm	1	4	3
18	8 mm–4 mm	0.83	8	3
19	8 mm–4 mm	0.6	6	3
20	8 mm–4 mm	0.83	4	1
21	8 mm–4 mm	0.83	4	1.5
22	8 mm–4 mm	0.83	4	2
23	8 mm–4 mm	0.83	4	2.5
24	8 mm–4 mm	0.83	4	3
25	8 mm–5 mm	0.6	4	2
26	8 mm–5 mm	0.6	4	3
27	8 mm–5 mm	0.6	4	3.33
28	8 mm–5 mm	0.6	4	4
29	8 mm–5 mm	0.6	4	5
30	8 mm–5 mm	0.6	4	5.5

4. Results

4.1. Filling and Jamming Process

After conducting the abovementioned numerical simulations, we have identified four sealing processes in the fracture-shrinking structure. The primary influencing factor is the relationship between particle size and the main and secondary fractures. By comprehending these four sealing processes and analyzing their underlying mechanisms, we thus can effectively elucidate the principle of particle sealing in fractures and determine conditions conducive to such sealings.

4.2. Sealing Mode 1: Normal Filling

Sealing mode 1 refers to the normal filling phenomenon, which occurs when the ratio of diameter to fracture width (D/W) is less than or equal to 0.4 under simulated conditions of concentration and velocity. For instance, the simulation scenario, including a diameter of 2 mm, a fracture width ranging from 8 mm to 5 mm, a concentration of 4%, and a velocity of 0.6 m/s, is introduced to unravel the characteristics of its filling process. The main filling and sealing process can be divided into two stages (Figure 3).

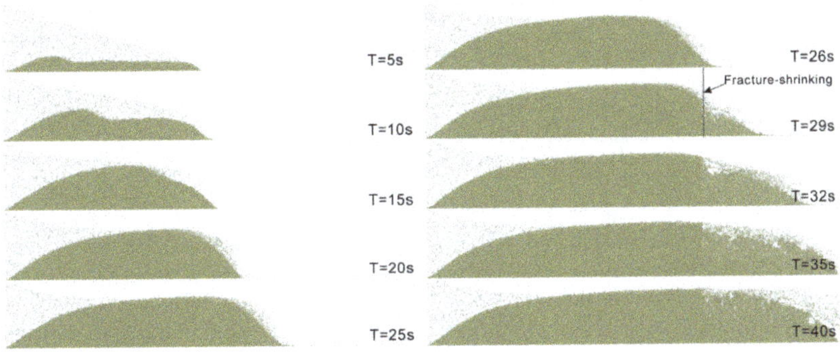

Figure 3. Filling process of sealing mode 1.

- Accumulation in the main fracture: When the particles did not reach the shrinking position of the fracture, regulator particles subjected to a combined action of fluid drag force and gravity within the fracture subsided in large quantities, forming an accumulation of sandpile, due to gravitational differences. The entire stacking process is comprised of an initial settling of the modulator, establishment of equilibrium height for modulator accumulation, forward growth of the modulator bed, and arrival at the shrinking position (Figure 3).
- Accumulation in the narrow fracture: Upon reaching the shrinking position of the fracture, due to the smaller particle size of the regulator compared to that of the secondary fracture (2.5 times smaller), its migration state through the secondary fracture is less affected by the shrinkage neck, allowing for smooth passage of regulator particles through the shrinkage position and achieving normal accumulation. Despite this, a few bridging structures may still form in the local position between particles, hindering the movement of local regulator particles. However, these unstable local bridging structures cannot withstand high-velocity fluid erosion and are easily destroyed, thus failing to jam the flow effectively. From another perspective, the sealing process is characterized by dynamic, and the extent of particle sealing at the neck is closely related to the formation and destruction of particle bridging structures. When the particle bridging structure is robust and stable, its failure and disintegration rate are lower than that of bridging structure formation, resulting in an effective particle sealing effect. However, if the particle bridging structure is unstable, its destruction and disintegration velocity will exceed the formation velocity of the bridge structure, rendering it ineffective in sealing. Considering the current simulation scenario, a relatively stable sealing structure cannot be formed due to the particle size being smaller than that of the narrow fracture. Consequently, an effective seal cannot be achieved, and a normal filling process ensues.

As indicated by Figure 4, the accumulation of particles results in a significant pressure drop within the fracture due to increased fluid resistance. Figure 4 depicts that the primary pressure drop occurs at the rear of the fracture due to the narrow fracture width. Proper particle accumulation reduces fluid resistance and increases pressure drop through the main crack. In contrast, in the case of proper particle accumulation, the fluid resistance through the main fracture is lower, resulting in a significant pressure drop. This leads to a more uniform pressure distribution throughout the entire fracture and eliminates any noticeable pressure drops caused by sealing.

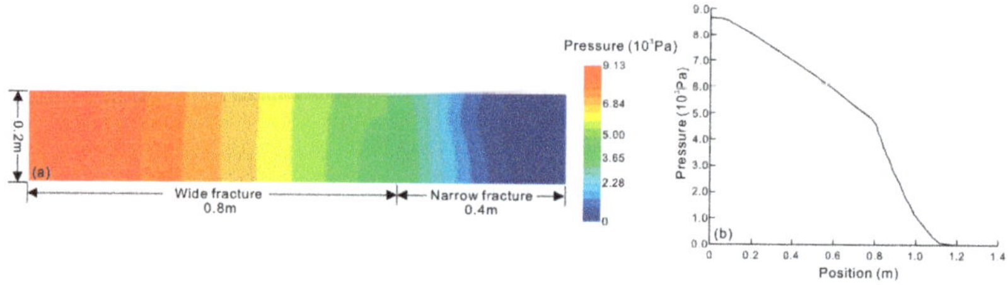

Figure 4. Significant pressure drop caused by accumulation of particles. (**a**) Fluid pressure drop in the fracture, (**b**) pressure curve along the middle section of the model (sealing mode 1).

4.3. Sealing Mode 2: Local Jamming

This mode refers to local blockage, which occurs when the ratio of particle diameter to fracture width is between 0.4 and 0.625 under simulated conditions of concentration and velocity. In the following, we describe the characteristics of its filling process with a given scenario which has a particle diameter of 2 mm, a fracture width ranging from 8 mm to 4 mm, a concentration of 4%, and a velocity of 0.6 m/s. The main filling and jamming process can be divided into two stages as follows (Figure 5).

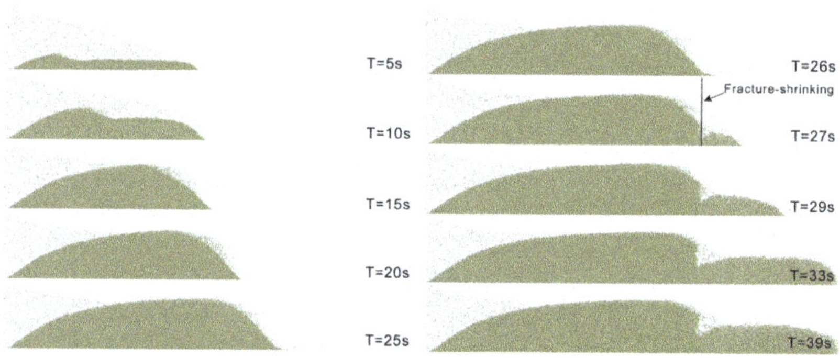

Figure 5. Accumulating and jamming process of sealing mode 2.

- Accumulation in the main fracture: The accumulation in the main fracture is similar to that of the first type. Before reaching the fracture shrinking position, the particles of the flow regulator in the fracture are subjected to both fluid drag and gravity, resulting in a large number of deposits in the fracture and forming a sand dune. The entire process is consistent with sealing mode 1.
- Local jamming: The particles with a size of 2 mm can pass through shrinkage fracture with a change width from 8 mm to 4 mm, resulting in an apparently local jamming effect, while flow-regulating particles can form local bridging structures. However, due to the relatively large width of the narrow fracture (twice the particle size), the formed bridge structure is not completely stable and is constantly generated and destroyed by particle bridging until reaching dynamic equilibrium after a certain period of time. Therefore, it cannot effectively seal the extension. Thus, it does not form the sealing extension along the vertical direction, resulting in incomplete sealing.

Figure 6 illustrates the pressure distribution, indicating that the accumulation of particles results in a significant pressure drop within the fracture due to increased resistance of fluid flowing through the particles. Since no particle accumulation occurs at the half height of the fracture in the flow regulator, pressure drop primarily occurs due to the necking effect. The flow resistance is essentially equivalent to that in the wide fracture, resulting in uniform pressure throughout and no significant sealing.

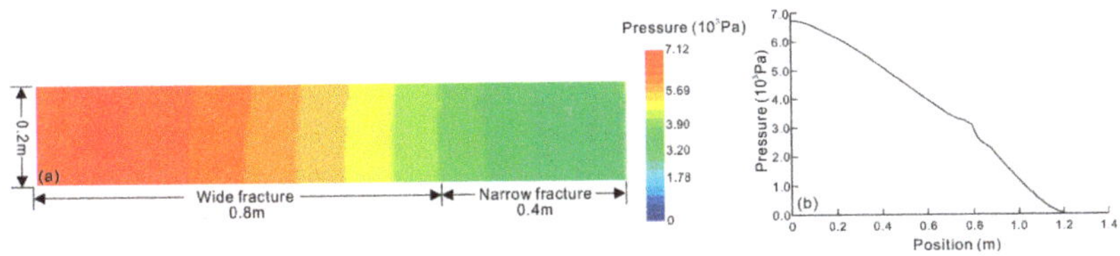

Figure 6. Pressure distribution due to the accumulation of particles. (**a**) Fluid pressure drop in the fracture, (**b**) pressure curve along the middle section of the model (sealing mode 2).

4.4. Sealing Mode 3: Complete Jamming

A complete sealing is observed at the shrinking position of this mode. Under simulated conditions of concentration and velocity, it manifests as a particle diameter to fracture width ratio with a range of $0.625 < D/W \leq 0.77$. To illustrate this phenomenon, hereafter, an instance is described with a particle diameter of 2 mm, fracture width varying from 8 mm to 3 mm, a concentration of 4%, and a velocity of 0.6 m/s. In this scenario, we delve into the characteristics exhibited during its filling process, which can be effectively divided into three distinct stages (Figure 7).

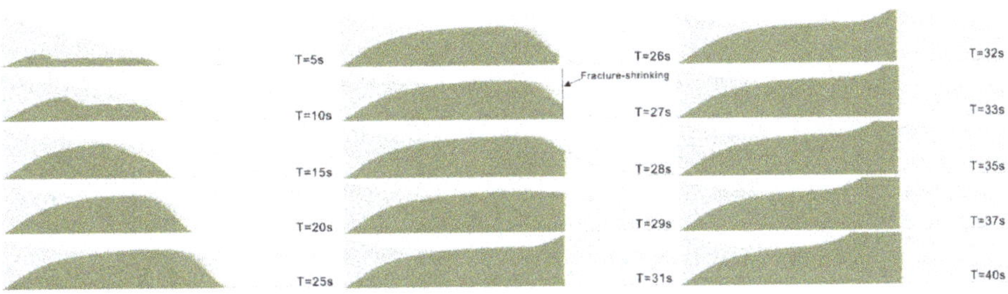

Figure 7. Accumulating and jamming process of the model.

- Accumulation in the main fracture: This stage is similar to the previous two modes.
- Jamming process at the fracture-shrinking position: When the regulator particles reach the constriction of the fracture, a significant number of them become trapped in the wide fracture and are unable to migrate to the narrow fracture due to particle bridging with the wall. Nevertheless, due to the particle size being larger than the width of the narrow fracture, a relatively stable local bridging structure is formed between particles and between particles and the wall, which can also maintain its stability under fluid erosion. Hence, the particles flowing through the shrinkage position of fracture are significantly fewer in number than those involved in bridging to establish a stable structure. Consequently, regulatory particles gradually accumulate and obstruct the fracture-shrinking position until it is completely sealed.
- Accumulating backward in the main fracture: After the complete sealing of the shrinkage position, a small amount of local bridging structure is eroded by fluid and enters the narrow fracture. However, due to the structural stability, subsequent injection particles rapidly replenish the damaged bridging structure, resulting in significantly fewer particles entering the narrow fracture compared to the number of bridging particles. Thus, the extra particles accumulate in reverse order within the wide fracture, causing the particle bed to grow backward as a whole until it completely fills up the main fracture (Figure 7).

The accumulation of particles results in a significant pressure drop within the fracture due to increased fluid flow resistance through the particle pack (Figure 8). However, minimal particle accumulation occurs at the narrow fracture, and thus, any pressure drop generated by fluid flow is primarily attributed to heightened flow resistance after the fracture shrank. The pressure drop in the main fracture is evidently composed of two distinct curves with different slopes. This phenomenon can be attributed to the accumulation of particles and high porosity near the injection port, which results in minimal impact on fluid flow through this region. During the process of backward accumulation, the main fracture at 0.6–0.8 m will undergo continuous compaction, resulting in localized particle voids and a significant pressure drop for fluid flow. This leads to the formation of two distinct sections with varying slopes on the pressure drop curve. This indicates that the primary pressure drop during the sealing process of the modulator particles is not solely caused by the fluid passing through the accumulated particles but also vastly influenced by compaction during backward accumulation.

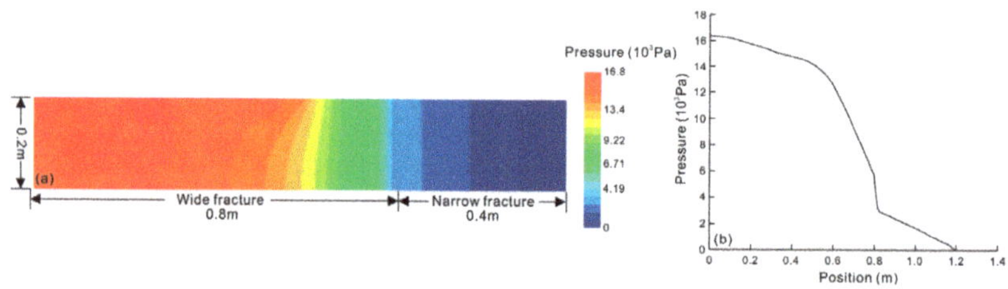

Figure 8. Pressure distribution due to the accumulation of particles. (**a**) Fluid pressure drop in the fracture, (**b**) pressure curve along the middle section of the model (sealing mode 3).

4.5. Sealing Mode 4: Sealing in the Main Fracture

This mode refers to a local sealing occurring in the main fracture when the particle size is less than one-third of the main fracture. In this case, with a particle size of 3 mm, fracture width varying from 8 mm to 5 mm, concentration of 4%, and velocity of 0.6 m/s, the following section demonstrates the characteristics of its filling process. The primary filling and sealing process can be divided into two stages (Figure 9).

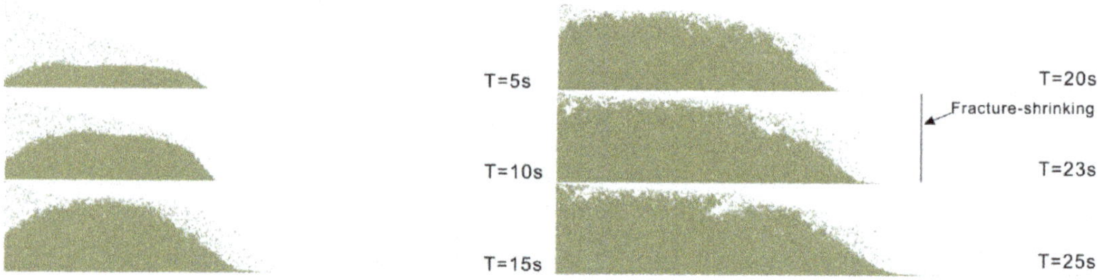

Figure 9. Accumulating and local jamming in the main fracture.

- Accumulation in the main fracture: When the particles within a fracture are rare, the modulator particles in fracture experience a joint effect of fluid drag force and gravity. As such, these modulator particles settle normally within the fracture and accumulate significantly at its entrance of narrow fracture. However, there is a minimal forward extension of the sand embankment. It forms a deposit at the entrance. Thus, accumulating processes differ from the abovementioned modes.
- Local jamming in the main fracture: With the increase in the injection of flow regulator particles, the effect of particle–particle and particle–wall interactions is enhanced. This leads to a hindered movement of particles within the fracture, preventing them

from advancing normally. Thus, the particles gradually accumulate within the main fracture and progressively block it. Once the front of the fracture is blocked, the modulator particles slowly advance, ultimately achieving complete blockage within the main fracture.

Figure 10 depicts the pressure distribution of this mode. Particles predominantly accumulate within the main fracture, resulting in a significant pressure drop across the fracture due to heightened fluid resistance through the particle-filled layer. Conversely, rare particles are observed in the narrow fracture, where the pressure drop primarily arises from increased flow resistance subsequent to fracture necking. The pressure drop across the main fracture exhibits a linear trend, indicating uniform particle distribution within the fracture and the absence of local extrusion. The sealing performance is primarily influenced by the particle accumulation state in the main fracture.

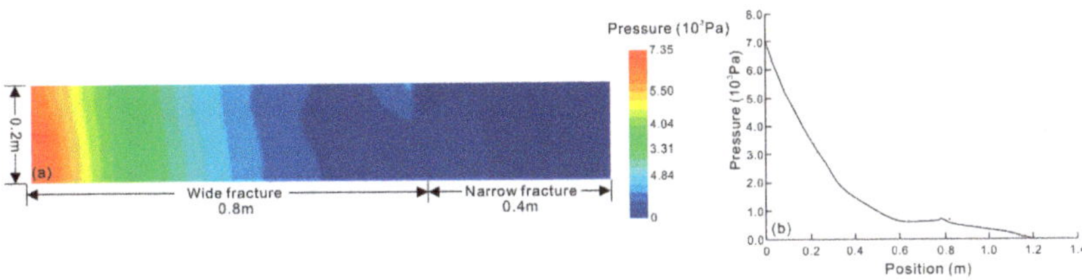

Figure 10. Pressure distribution due to the accumulation of particles. (**a**) Fluid pressure drop in the fracture, (**b**) pressure curve along the middle section of the model (sealing mode 4).

5. Microscopic Analysis of Local Accumulation

Figure 11 shows the local sealing structure of particles corresponding to different sealing modes. Figure 11a shows the typical filling behavior of mode 1, where no local bridging structure is formed in the main fracture due to the presence of small particles. The particle size to narrow fracture width ratio at the neck is 0.4, which prevents effective bridging structure formation during fracture contraction and allows for easier flow into a narrow fracture. However, this flow cannot achieve three particles flowing side by side in the same plane and will be hindered, resulting in the uneven filling of the local regulator. Figure 11b illustrates sealing mode 2, which also lacks a localized bridging structure in the main fracture. The ratio of particle size to narrow fracture width is 0.5, resulting in the formation of a bridge structure with a certain degree of stability at the crack neck, impeding the flow of particles. However, due to the relatively large width of the narrow fracture, its structural stability is insufficient, resulting in a lower sealing height and only local sealing being achieved. Figure 11c depicts the complete sealing of mode 3. It is evident that no local bridging structure forms in the main fracture due to small particles. Nevertheless, at neck shrinkage, where the ratio of particle size to narrow fracture width is 0.66, an effective bridging structure forms and remains relatively stable even under fluid erosion. Figure 11d demonstrates sealing mode 4, which forms a localized bridging structure in the main fracture due to the ratio of particle size to the width of the main fracture, leading to uneven filling of regulator particles and resulting in local jamming.

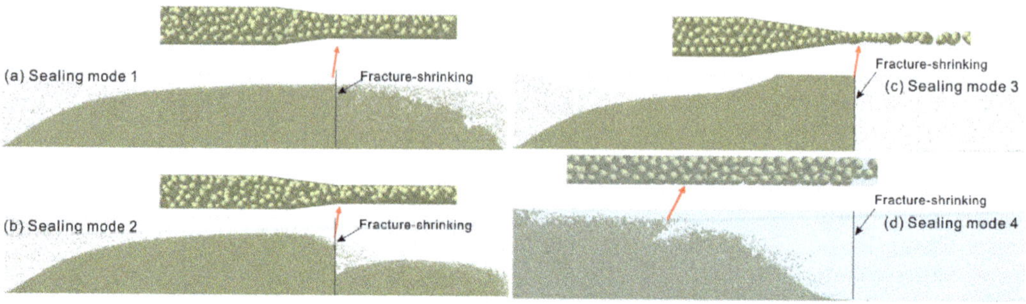

Figure 11. Local microscopic views of four modes of sealing.

Figure 12 shows the specific sealing structures corresponding to different ratios of particle to fracture width. As depicted in the figure, for mode 1, the small particle size compared to the fracture width results in only the boundary layer particles experiencing strong forces, while the inner particles are subject to weak wall constraints and a large number of particles on the cross-section. Therefore, a stable bridging structure cannot be formed to achieve sealing of the crack.

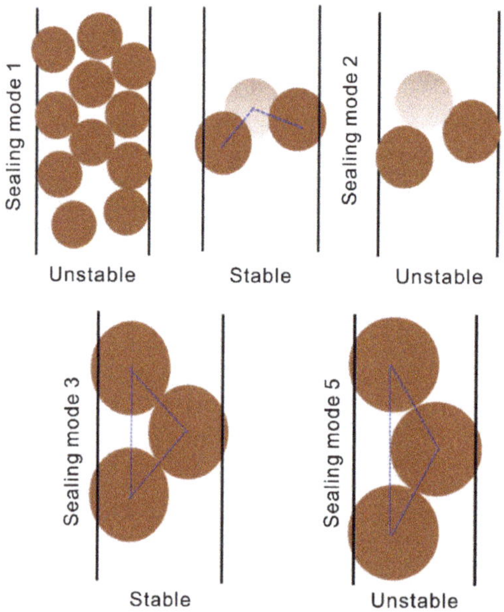

Figure 12. Jamming mechanisms corresponding to different D/W. Blue line is the connection between particles while the different colors of particle mean the locations of particles.

For mode 2, the ratio of particle size to fracture width is greater than that of type 1, resulting in an enhanced wall force on particles located in the middle of the crack section. However, since the critical value for a sufficiently large ratio has not been reached, only certain angles can form stable structures, while other areas struggle to achieve stable blocking structures. Therefore, the entire sealing remains difficult to realize.

For mode 3, the particle size distribution range is $0.625 < D/W \leq 0.77$. Due to its larger ratio of particle size to fracture width, it can form a stable triangular structure at the joint with an angle range close to 45°, thereby creating a structurally robust formation that resists destruction. This enables entire sealing at the joint during particle transportation.

Conversely, particles larger than 0.77 times the fracture width, which are close to the fracture width, result in smaller angles formed between particles and walls or between adjacent particles. Consequently, these particles fail to establish stable blockages compared to smaller particles effectively.

6. Discussion

6.1. Influence of Particle Size

The influence analysis of particle size parameters was conducted based on the parameter settings provided in Table 3. In this section, we maintained a consistent fracture size while varying the particle size within the range of 1.5–3 mm, and ensured that the particle concentration remained constant.

Table 3. Simulated scenarios of different particle sizes.

Fracture Width (Wide) (mm)	Fracture Width (Narrow) (mm)	Particle Volume Concentration (%)	Fluid Viscosity	Injection Velocity (m/s)	Particle Size (mm)
8	4	4	1	0.6 m/s	1.5, 2, 2.5, 3

Firstly, it is important to note that due to the consistent concentration of particle volume, the quantity of particles injected into the fracture simultaneously by larger particle sizes is relatively lower compared to smaller particles. Therefore, we analyze the sealing effect not based on particle accumulation form but rather on the sealing condition at the shrinkage joint. Simulation results indicate that as the modulator's particle size gradually increases, its blocking effect improves (Figure 13). Nevertheless, 3 mm particles exhibit a weaker blocking effect than 2.5 mm particles. Preliminary analysis suggests that this may be attributed to the large size of 3 mm particles, which hinders the effective formation of a "bridge" between them and the wall at the shrinkage joint, thus preventing local stable structure formation and allowing most of them to pass through.

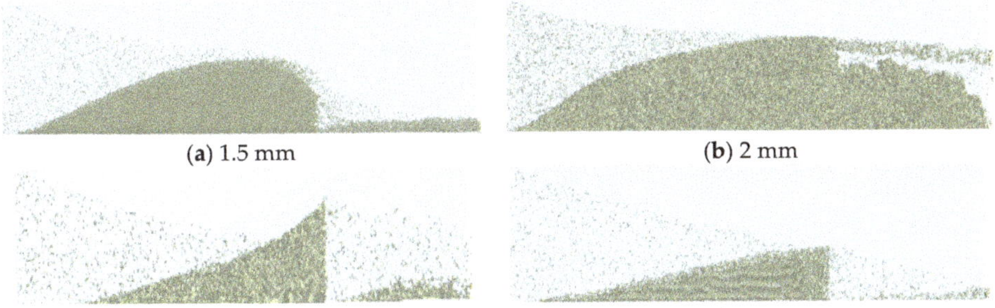

(**a**) 1.5 mm (**b**) 2 mm (**c**) 2.5 mm (**d**) 3 mm

Figure 13. Jamming morphology with different particle sizes.

The cause of this phenomenon can be elucidated by analyzing the local sealing structure (Figure 14). It is evident that when the particle size is small (2 mm), the particles only form a localized bridge and do not establish a stable structure on the fracture width, resulting in poor sealing effectiveness. As the particle size gradually increases to 2.5 mm, it forms a more stable bridging structure at the joint and avoids destruction, leading to better sealing performance of prticles. However, as particle size continues to increase (3 mm), since a single particle occupies almost the entire width of the fracture, its bridging structure with other particles and walls becomes extremely unstable and difficult to form effectively, thus resulting in a poor blocking effect.

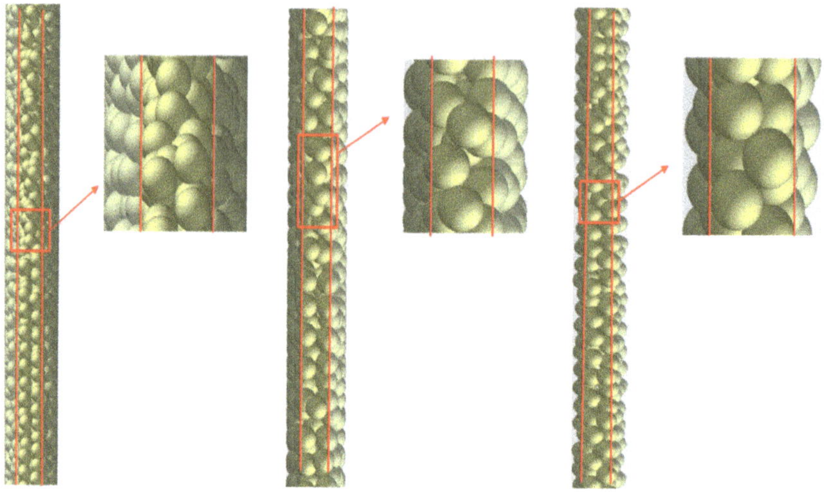

(a) 2 mm particles (b) 2.5 mm (c) 3 mm

Figure 14. Local stuck morphology under different particle sizes.

The pressure drop (Figure 15) illustrates the impact of particle size blocking. It is evident from the pressure contour plot that the local pressure drop resulting from a 2.5 mm blockage is the most significant, indicating optimal blocking efficiency.

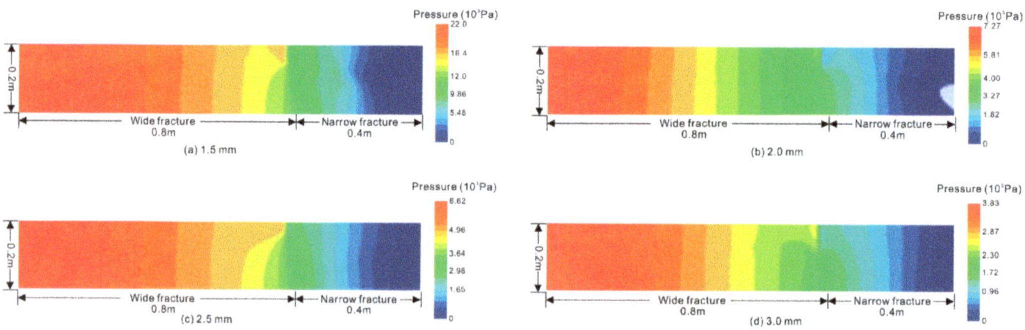

Figure 15. Fluid pressure drop in the fracture with different particle sizes.

6.2. Influence of Particle Concentration

The influence analysis of particle size parameters is conducted based on the parameter settings provided in Table 4. Herein, we maintain a consistent fracture size while varying the particle concentration within the range of 4% to 6%, ensuring that the particle size remains constant.

Table 4. Influence of different particle concentrations on simulated working conditions.

Groups	Fractures (Narrow) (mm)	Particle Volume Concentration (%)	Fluid Viscosity	Injection Velocity (m/s)	Particle Size (mm)
1	4	4, 6	1	0.6	3
2	4	4, 8	1	0.6	2

Through the simulation results of varying particle concentrations, it is noted that an increase in the concentration of the regulator within the local position leads to enhanced particle collisions, thereby resulting in greater settlement within the wide fracture (Figure 16). Similarly, at the neck of the fracture, a substantial number of particles come into contact with each other and exhibit increased interaction with the wall surface, consequently facilitating the effective formation of localized jamming structures and subsequently yielding improved sealing efficacy.

Figure 16. Blocking effect of modulator particles under different concentration conditions (Group 1).

It can be observed from the fracture pressure drop curve that the pressure drop generated in the fracture under 6% particle concentration is significantly higher than that under 4% particle concentration, indicating a superior sealing effect of 6% particles (Figure 17). Similar results were obtained through simulation in Group 2, where high-concentration groups effectively increased accumulation body within fractures and enhanced local sealing (Figure 18).

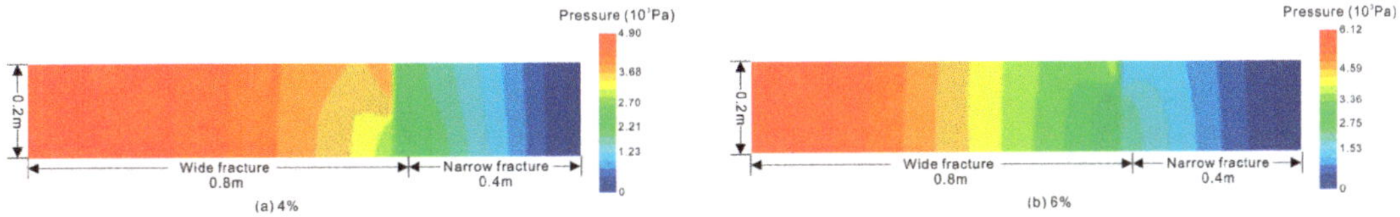

Figure 17. Pressure drop distribution under different concentration conditions.

Figure 18. Sealing effect of flow regulator particles at different volume concentrations (Group 2).

From the local particle structure (Figure 19), it is observed that low-concentration modulator particles exhibit weak local interaction and are unable to form a stable structure with the wall surface. As concentration increases, contact opportunities between particles and between particles and the wall increase, resulting in enhanced contact force and the formation of a more expansive and stable blocking structure at the fracture contraction joint.

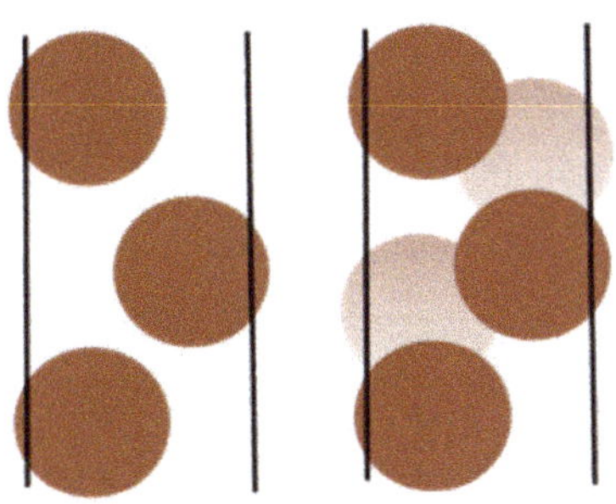

Figure 19. Schematic diagram of particle arrangement under different concentration conditions.

6.3. Influence of Injection Velocity

The influence of particle size parameters was analyzed based on the parameter settings presented in Table 5. Specifically, while maintaining a consistent fracture size and particle concentration, we varied the particle pumping velocity from 0.4 m/s to 0.8 m/s.

Table 5. Simulated scenarios with different velocities.

Groups	Fractures (Narrow) (mm)	Particle Volume Concentration (%)	Fluid Viscosity	Injection Velocity (m/s)	Particle Size (mm)
1	4	4	1	0.4~0.8	3
2	3	4	1	0.6, 1.2	2

The main manifestation for Group 1 is the second mode of local jamming. Through simulation results at various injection velocities, it can be observed that an increase in velocity leads to a deteriorating blocking effect of the flow regulator (Figure 20). This phenomenon arises due to the high shear forces generated by larger particles, which disrupts the stable structure formed between the regulator and wall at the contraction joint, consequently causing more regulator particles to advance.

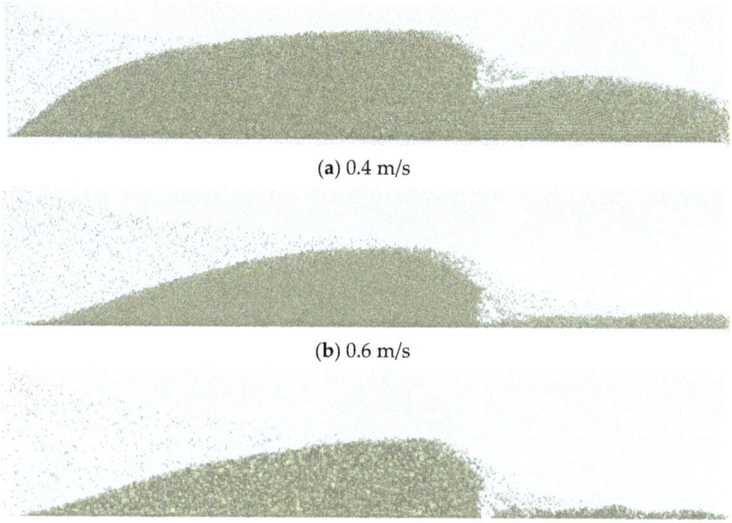

(**a**) 0.4 m/s

(**b**) 0.6 m/s

(**c**) 0.8 m/s

Figure 20. Sealing effect of flow regulator particles under different injection velocities (Group 1).

For Group 2, sealing mode 3 of complete sealing is predominant (Figure 21). The simulation results demonstrate that regulator particles can achieve full jamming in the fracture at a low velocity of 0.6 m/s, whereas an increase in velocity to 1.2 m/s fails to attain complete sealing due to insufficient fracture height.

(**a**) 0.6 m/s (**b**) 1.2 m/s

Figure 21. Sealing effect of flow regulator particles at different pumping velocities (Group 2).

The local particle sealing structure of the formed particles, as depicted in Figure 22, and the local fracture−particle structure, exhibit minimal disruption under low flow rate conditions, thereby preserving its integrity. However, as fluid velocity increases, the sealing structure experiences escalating fluid shear forces, leading to its eventual deterioration and subsequent reduction in overall sealing performance.

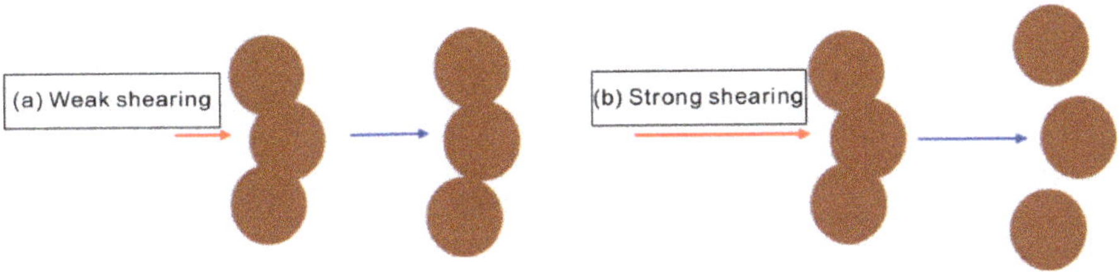

Figure 22. Variations in particle motion related to different concentration conditions. Arrows are the moving direction for flows.

7. Conclusions

This paper investigated the sealing process of particles in the fracture-shrinking model and examined the influences of particle concentration, size, flow velocity, and density on the sealing process by conducting the CFD−DEM coupled simulations. The influences of particle size, particle concentration, and injection velocity on particle jamming were analyzed, subsequently proposing a predictive approach for fractures and vugh sealing.

1. Four sealing modes are revealed: normal filling, local jamming, complete sealing, and sealing in the main fracture. Among the four sealing modes, the ratio of particle size to main fracture width exerts the primary role, while the ratio of particle diameter to fracture width with a range of $0.625 < D/W \leq 0.77$, shows the complete jamming.
2. With an increase in particle size, the sealing effect of particles is enhanced, while an optimal particle size for achieving stable sealing is observed. When the particle size varies from 2 to 2.5 mm, a more stable bridging structure at the joint and avoids destruction is observed. However, as particle size continues to increase (3 mm), the bridging structure becomes extremely unstable.
3. The influence of particles on jamming is complex and varies with the particle concentration and the injection velocity. A higher concentration of particles yields better results in the fracture-shrinking model. With the increase of the velocity, the fluid drag exerted on the clogged particles increases, leading to increased local blocking effects on damage within such structures and consequently worsening the sealing effect on fractures.

Author Contributions: Conceptualization, J.Z.; methodology, J.Z.; software, J.Z.; validation, C.L.; formal analysis, T.Z.; data curation, T.Z.; writing—original draft, J.Z.; writing—review & editing, C.L., T.Z. and J.G.; supervision, J.G.; funding acquisition, C.L. All authors have read and agreed to the published version of the manuscript.

Funding: This research was funded by the National Natural Science Foundation of China (52374044) and Sichuan Province Science and Technology Planning Projects (2023JDRC0008).

Data Availability Statement: The data presented in this study are available on request from the corresponding author.

Acknowledgments: This article was prepared under the auspices of the National Key Laboratory of Oil and Gas Reservoir Geology and Exploitation at Southwest Petroleum University.

Conflicts of Interest: The authors declare no conflict of interest.

References

1. Zhang, Y.; Zhang, L.; He, J.; Zhang, H.; Zhang, X.; Liu, X. Fracability evaluation method of a fractured-vuggy carbonate reservoir in the shunbei block. *ACS Omega* **2023**, *8*, 15810–15818. [CrossRef]
2. Martyushev, D.A.; Davoodi, S.; Kadkhodaie, A.; Riazi, M.; Kazemzadeh, Y.; Ma, T. Multiscale and diverse spatial heterogeneity analysis of void structures in reef carbonate reservoirs. *Geoenergy Sci. Eng.* **2024**, *233*, 212569. [CrossRef]
3. Chang, F.F. Acid fracturing stimulation. In *Fluid Chemistry, Drilling and Completion*; Gulf Professional Publishing: Oxford, UK, 2022; pp. 387–419.
4. Sun, Q.; Zhang, N.; Fadlelmula, M.; Wang, Y. Structural regeneration of fracture-vug network in naturally fractured vuggy reservoirs. *J. Pet. Sci. Eng.* **2018**, *165*, 28–41. [CrossRef]
5. Qu, M.; Hou, J.; Qi, P.; Zhao, F.; Ma, S.; Churchwell, L.; Wang, Q.; Li, H.; Yang, T. Experimental study of fluid behaviors from water and nitrogen floods on a 3-D visual fractured-vuggy model. *J. Pet. Sci. Eng.* **2018**, *166*, 871–879. [CrossRef]
6. He, D.; Jia, C.; Zhao, W.; Xu, F.; Luo, X.; Liu, W.; Tang, Y.; Gao, S.; Zheng, X.; Li, D.; et al. Research progress and key issues of ultra-deep oil and gas exploration in China. *Pet. Explor. Dev.* **2023**, *50*, 1333–1344. [CrossRef]
7. Yao, Y.; Zeng, L.; Mao, Z.; Han, J.; Cao, D.; Lin, B. Differential deformation of a strike-slip fault in the Paleozoic carbonate reservoirs of the Tarim Basin, China. *J. Struct. Geol.* **2023**, *173*, 104908. [CrossRef]
8. Zheng, S.; Yang, M.; Kang, Z.; Liu, Z.; Long, X.; Liu, K.; Li, X.; Zhang, S. Controlling factors of remaining oil distribution after water flooding and enhanced oil recovery methods for fracture-cavity carbonate reservoirs in Tahe Oilfield. *Pet. Explor. Dev.* **2019**, *46*, 786–795. [CrossRef]
9. Méndez, J.N.; Jin, Q.; González, M.; Zhang, X.; Lobo, C.; Boateng, C.D.; Zambrano, M. Fracture characterization and modeling of karsted carbonate reservoirs: A case study in Tahe oilfield, Tarim Basin (western China). *Mar. Pet. Geol.* **2020**, *112*, 104104. [CrossRef]
10. Zhang, Y.; Zhang, S.; Huang, B.; Lu, Z.; Ye, N.; Zhu, B.; Hou, X.; Xie, F.; Bai, X.; Zhang, X. Fluid Inclusion, Isotopic, and Elemental Geochemistry Studies of Cave-Filling Calcite in the Lower–Middle Ordovician Yingshan Formation of Tahe Oilfield, NW China: Implication for Karstification in Non-exposed Limestone. *Front. Earth Sci.* **2022**, *10*, 842386. [CrossRef]
11. Li, Y.; Sun, J.; Wei, H.; Song, S. Architectural features of fault-controlled karst reservoirs in the Tahe oilfield. *J. Pet. Sci. Eng.* **2019**, *181*, 106208. [CrossRef]
12. Xu, X.; Chen, Q.; Zhang, Y.; Wang, J.; Li, Y.; Kang, Z.; Zhang, Y. Research progress and prospect of Ordovician carbonate rocks in Tahe oilfield: Karst feature. *J. Pet. Explor. Prod. Technol.* **2021**, *11*, 3889–3902. [CrossRef]
13. Li, X.; Huang, X.; Yu, S.; Lu, D.; Du, X. Numerical study on the transient pressure response of the vug in carbonate reservoirs. *Pet. Sci. Technol.* **2023**, 1–20. [CrossRef]
14. Cheng, W.; Liu, Z.; Yang, H.; Wang, W. Non-linear seepage characteristics and influential factors of water injection in gassy seams. *Exp. Therm. Fluid Sci.* **2018**, *91*, 41–53. [CrossRef]
15. Liu, Y.; Hou, J.; Zhao, H.; Liu, X.; Xia, Z. Numerical simulation of simultaneous exploitation of geothermal energy and natural gas hydrates by water injection into a geothermal heat exchange well. *Renew. Sustain. Energy Rev.* **2019**, *109*, 467–481. [CrossRef]
16. Crandall, D.; Bromhal, G.; Karpyn, Z.T. Numerical simulations examining the relationship between wall-roughness and fluid flow in rock fractures. *Int. J. Rock Mech. Min. Sci. Geomech. Abstr.* **2010**, *47*, 784–796. [CrossRef]
17. Hawez, H.K.; Sanaee, R.; Faisal, N.H. A critical review on coupled geomechanics and fluid flow in naturally fractured reservoirs. *J. Nat. Gas Sci. Eng.* **2021**, *95*, 104150. [CrossRef]
18. Zhao, X.; Chen, Z.; Wang, B.; Liao, X.; Li, D.; Zhou, B. A Multi-medium and Multi-mechanism model for CO_2 injection and storage in fractured shale gas reservoirs. *Fuel* **2023**, *345*, 128167. [CrossRef]
19. Sun, K.; Liu, H.; Leung, J.Y.; Wang, J.; Feng, Y.; Liu, R.; Zhang, Y. Impact of effective stress on permeability for carbonate fractured-vuggy rocks. *J. Rock Mech. Geotech. Eng.* **2023**, *16*, 942–960. [CrossRef]
20. Huang, J.; Zhang, J.; Xu, D.; Zhang, S.; Tong, H.; Xu, N. From jammed solids to mechanical metamaterials: A brief review. *Curr. Opin. Solid State Mater. Sci.* **2023**, *27*, 101053. [CrossRef]
21. Behera, N.; Agarwal, V.K.; Jones, M.G.; Williams, K.C. CFD modeling and analysis of dense phase pneumatic conveying of fine particles including particle size distribution. *Powder Technol.* **2013**, *244*, 30–37. [CrossRef]
22. Raynal, L.; Augier, F.; Bazer-Bachi, F.; Haroun, Y.; da Fonte, C.P. CFD Applied to Process Development in the Oil and Gas Industry—A Review. *Oil Gas Sci. Technol. Rev. d'IFP Energies Nouv.* **2016**, *71*, 42. [CrossRef]
23. Ramakrishna, P.; Singh, J.K.; Sahoo, A.; Mohapatra, S.S. CFD simulation for coal gasification in fluidized bed gasifier. *Energy* **2023**, *281*, 128272.
24. Niu, B.; Ahmed, M.; Wen, T.; Xu, G.; Liu, X. Particle-resolved CFD modeling of the flow and heat transfer characteristics of the fluid in a fixed bed. *Powder Technol.* **2023**, *421*, 118445. [CrossRef]
25. Marquez, M.; Williams, W.; Knobles, M.M.; Bedrikovetsky, P.; You, Z. Fines Migration in Fractured Wells: Integrating Modeling with Field and Laboratory Data. *SPE Prod. Oper.* **2014**, *29*, 309–322. [CrossRef]
26. Damjanac, B.; Cundall, P. Application of distinct element methods to simulation of hydraulic fracturing in naturally fractured reservoirs. *Comput. Geotech.* **2016**, *71*, 283–294. [CrossRef]
27. Tomac, I.; Gutierrez, M. Micromechanics of hydraulic fracturing and damage in rock based on DEM modeling. *Granul. Matter* **2020**, *22*, 1–17. [CrossRef]

28. Li, Y.; Hu, W.; Wei, S.; Li, L.; Zhang, Z.; Song, S. Influence of preexisting discontinuities on hydraulic fracture complexity in a naturally fractured reservoir. *Eng. Geol.* **2022**, *311*, 106919. [CrossRef]
29. Shan, T.; Zhao, J. A coupled CFD-DEM analysis of granular flow impacting on a water reservoir. *Acta Mech.* **2014**, *225*, 2449–2470. [CrossRef]
30. Krzaczek, M.; Nitka, M.; Tejchman, J. Effect of gas content in macropores on hydraulic fracturing in rocks using a fully coupled DEM/CFD approach. *Int. J. Numer. Anal. Methods Geomech.* **2021**, *45*, 234–264. [CrossRef]
31. Wang, Z.; Hurter, S.; You, Z.; Honari, V.; Sun, Y.; Zhang, S. Influences of negative pressure on air-leakage of coalseam gas extraction: Laboratory and CFD-DEM simulations. *J. Pet. Sci. Eng.* **2021**, *196*, 107731. [CrossRef]
32. Yang, Y.; Yuan, W.; Hou, J.; You, Z. Review on physical and chemical factors affecting fines migration in porous media. *Water Res.* **2022**, *214*, 118172. [CrossRef] [PubMed]
33. Zhu, T.; Wei, X.; Zhang, Z. Numerical simulation of hydraulic-mechanical-chemical field coupled acid fracturing in complex carbonate reservoir. *Comput. Geotech.* **2023**, *156*, 105277. [CrossRef]
34. Zhang, H.; Tan, Y.; Yang, D.; Trias, F.X.; Jiang, S.; Sheng, Y.; Oliva, A. Numerical investigation of the location of maximum erosive wear damage in elbow: Effect of slurry velocity, bend orientation and angle of elbow. *Powder Technol.* **2012**, *217*, 467–476. [CrossRef]
35. Varga, M.; Goniva, C.; Adam, K.; Badisch, E. Combined experimental and numerical approach for wear prediction in feed pipes. *Tribol. Int.* **2013**, *65*, 200–206. [CrossRef]
36. Chu, K.; Yu, A. Numerical simulation of complex particle–fluid flows. *Powder Technol.* **2008**, *179*, 104–114. [CrossRef]
37. Chu, K.W.; Yu, A.B. Numerical Simulation of the Gas—Solid Flow in Three-Dimensional Pneumatic Conveying Bends. *Ind. Eng. Chem. Res.* **2008**, *47*, 7058–7071. [CrossRef]
38. Anderson, T.B.; Jackson, R.O.Y. Fluid mechanical description of fluidized beds. Equations of motion. *Ind. Eng. Chem. Fundam.* **1967**, *6*, 527–539. [CrossRef]
39. Cundall, P.A.; Strack, O.D.L. A discrete numerical model for granular assemblies. *Géotechnique* **1979**, *29*, 47–65. [CrossRef]
40. Silbert, L.E.; Ertaş, D.; Grest, G.S.; Halsey, T.C.; Levine, D.; Plimpton, S.J. Granular flow down an inclined plane: Bagnold scaling and rheology. *Phys. Rev. E* **2001**, *64*, 051302. [CrossRef]
41. Hu, W. Development technology and research direction of fractured-vuggy carbonate reservoirs in Tahe Oilfield. *Reserv. Eval. Dev.* **2020**, *10*, 1–10.

processes

MDPI

Article

Pore Fluid Movability in Fractured Shale Oil Reservoir Based on Nuclear Magnetic Resonance

Yishan Liu [1], Zhewei Chen [1], Dongqi Ji [1], Yingfeng Peng [1], Yanan Hou [2] and Zhengdong Lei [1,*]

[1] Research Institute of Petroleum Exploration and Development, PetroChina, Beijing 100083, China
[2] CNOOC Research Institute Ltd., Beijing 100028, China
* Correspondence: leizhengdong@petrochina.com.cn

Abstract: Gulong shale oil is found in a typical continental shale oil reservoir, which is different from marine shale oil reservoirs. The Gulong shale oil reservoir is a pure shale-type oil reservoir with abundantly developed nanoscale pores, making it extremely difficult to unlock fluids. Pressure drive does not easily achieve fluid unlock conditions; thus, it is necessary to utilize imbibition to unlock nanoscale pore fluids. In this study, experiments were conducted on oil displacement by high-speed centrifugal pressure and imbibition under different conditions, respectively, and simulations were used to evaluate the effects of pressure differential drive and imbibition efficiency on the utilization of crude oil following fracturing. Combined with the mixed wettability of the reservoir, the imbibition efficiency was analyzed, and the imbibition efficiency at different soaking stages was evaluated. When the fracturing pressure was higher than the matrix pore pressure, the imbibition efficiency was the most obvious, which was 27.9%. Spontaneous imbibition depending solely on capillary force had poor efficiency, at 16.8%. When the fracturing pressure was lower than the matrix pore pressure, the imbibition efficiency was the lowest, at only 1.3%. It is proposed that strengthening fracture pressure and promoting pressurized imbibition are the keys to improving shale oil development.

Keywords: shale oil reservoir; imbibition after fracturing; mixed wettability; pore fluid movability

1. Introduction

Unconventional resources, including tight oil/gas reservoirs and shale oil/gas reservoirs, have become important energy sources worldwide. Due to the abundant presence of nanoscale pores, these reservoirs have ultra-low permeabilities, which make their effective development challenging. How to effectively utilize nanoscale pore fluids is critical to shale reservoir development [1]. Compared to conventional reservoirs, the factors that lead to poor movability of shale oil include (1) shale reservoirs being relatively tight, with a complex pore–throat microstructure, low porosity and ultra-low permeability characteristics; (2) shale oil having high viscosity and large molecular size; and (3) some oil molecules being adsorbed on the surface of organic matter or minerals [2]. The adsorbed and free shale oil stored in shale is potentially movable and recoverable, while the dissolved shale oil that is miscible with organic matter is almost immobile and difficult to extract. However, under current technological conditions, free oil is the most easily recovered part [3]. Theoretically, the free oil content is the maximum recoverable amount of shale oil. Shale oil with strong movability is often dominated by the free phase, which does not flow easily when the content of the adsorbed phase is high, indicating that the occurrence state of shale oil is an important factor affecting its movability.

Due to the low porosity and especially the low permeability of shale, as well as the high density and viscosity of oil, subsurface flow does not readily occur. The key limitation restricting shale oil exploration and development is not the oil content in shale but the amount of recoverable section involved. The movability of shale oil is closely related to its occurrence characteristics (occurrence state, content, and proportion of different states) and storage space (pore and throat size, distribution, and connectivity), and is also influenced by reservoir fluid conditions (oil composition, temperature, and pressure) [4]. Theoretically speaking, shale oil movability includes at least two aspects: movable potential and movability. The saturation of movable fluid is a key parameter for evaluating the movability of shale reservoirs, and the utilization of nuclear magnetic resonance and centrifugal (displacement) testing methods is the main means of studying the saturation of movable fluid [5–7]. Combining nuclear magnetic resonance (NMR) with centrifugal experiments can reveal not only the movable fluid saturation of shale cores but also the movable fluid saturation controlled by different pore and throat properties in each core, making it possible to quantitatively analyze the distribution of the movable fluid in the reservoir and evaluate the fluid movability. After hydraulic fracturing in shale reservoirs, the fracturing fluid (water or other chemical agents) is prone to imbibition when it meets the matrix through the fracture area, and the residual oil in the fracture network is prone to imbibition with the fracturing fluid, especially during the process of sealing the well [8].

This study conducted high-speed centrifugation + NMR testing and imbibition + NMR testing, clarified the mechanism of pore fluid movability by imbibition under different pressure conditions, and clarified the influence of wettability on imbibition. This study used Gulong shale cores that had similar physical properties, high authenticity, and high comparability. Firstly, the mixed wettability of the core was evaluated and then imbibition and NMR tests were conducted under different conditions to quantitatively analyze the movability and recovery degree of different-sized pores. This work provides an experimental basis for controlling formation pressure in shale oil reservoirs and experimental data support for improving shale oil recovery.

2. Materials and Methods

2.1. High-Speed Centrifugation + NMR Test

According to the basic principle of NMR, the NMR signals of porous hydrogen-containing fluids are divided into three parts: surface relaxation, volume relaxation, and diffusion relaxation. Among them, the interaction between the surface relaxation fluid and pore surface is related, the molecular structure of the relaxation jade fluid is related to the volume, and diffusion relaxation is related to the magnetic field gradient inside the pores. The surface relaxation is dominant in light crude oil, and the lateral relaxation signal T_2 satisfies Equation (1):

$$\frac{1}{T_2} \approx \frac{F_s \rho_2}{r} \tag{1}$$

where T_2 is the transverse relaxation signal quantity, r is the scale of the fluid-containing pores, F_s is the shape factor of the pores, and ρ_2 is the lateral relaxation rate obtained from the pore surface.

The shape factor of the pore and lateral relaxation rate of the experimental rock sample are both fixed values, so the signal quantity of lateral relaxation is positively correlated with the pore radius. Based on the collected water-saturated core sample echo failure signal, a joint iterative inversion algorithm is used to obtain the sample's T_2 spectrum. The lateral relaxation time (T_2) of nuclear magnetic resonance can reflect the chemical environment in which the hydrogen protons are located inside the sample, which is related to the binding force and degree of freedom of the hydrogen protons. The degree of hydrogen proton binding is closely related to the internal structure of the sample. In porous media, the larger the pore radius, the longer the relaxation time of the water present in the pores; The smaller the pore radius, the greater the degree of binding of the water present in the pores, and the shorter the relaxation time. That is, the position of the T_2 spectrum peak is related to the size of the pore radius, and the area of the peak is related to the number of corresponding pores.

Through the T_2 spectrum, information such as core porosity and pore size distribution can be obtained to describe the complex properties of core pores more accurately, and different types and components of fluids can be identified through secondary resonance fluid identification methods. By combining mercury intrusion data and the corresponding relationship with the relaxation time of the nuclear magnetic T_2 spectrum, the pore throat radius of matrix-layered quartz core can be expressed (Equation (2)). Mercury in the non-wetting phase needs to enter smaller pores under external forces, and the pore throat radius r satisfies the Washburn equation with respect to the mercury inlet pressure p_c, the contact angle θ of mercury on the surface, and the surface tension σ.

$$r = \frac{2\sigma cos\theta}{p_c} \tag{2}$$

The frequency of pore throat distribution is set as the vertical axis and the radius of pore throat as the horizontal axis; the pore diameter distribution curve obtained by the mercury intrusion method is fitted, and the C value with the highest correlation coefficient is taken as the optimal value. This value is the conversion coefficient between the T_2 value obtained and the pore throat radius. A Gulong shale core was selected to fit the distribution frequency of pores and throats measured using the nuclear magnetic method and the mercury intrusion method. The fitting of nuclear magnetic mercury intrusion in the shale core is shown in Figure 1. The relaxation time of NMR T_2 is proportional to the pore size. The pore throat distribution map obtained from the mercury intrusion experiment was compared with the one-dimensional T_2 spectrum. The results of the two methods for determining pore structure were similar and both had a bimodal structure. Therefore, the NMR and pore throat distribution curves were overlapped and compared, and it was found that the two peaks perfectly overlapped. Through statistical calculation, the conversion coefficient between the shale's nuclear magnetic T_2 value and pore throat radius was found to be 0.008 μm/ms.

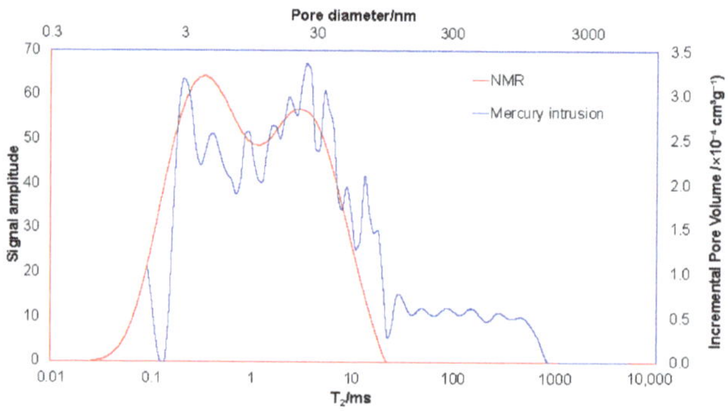

Figure 1. Fitting diagram of shale NMR and mercury intrusion.

By using centrifugal equipment and one-dimensional nuclear magnetic resonance technology (sampling frequency (KHz): 125; main frequency (KHz): 12; sampling points: 1024; 180° pulse width (μs): 13.04 (acquisition required); after center correction, we searched for the pulse width to obtain the required acquisition parameters), the threshold pressure gradient of different pore sizes was tested, and the available oil saturation under different pressures was compared. Firstly, we measured the hydrogen signal in the initial state of the rock; we saturated the rock with crude oil finally, we recorded the hydrogen signal in the saturated oil state of the core. Using 8 psi as the starting centrifugal force, the degree of crude oil recovery was recorded at 15 psi, 77 psi, 151 psi, 378 psi, and 570 psi, and the pore fluid movability was observed with NMR.

2.2. Imbibition + NMR Test

This study conducted three types of imbibition tests under different pressures, namely, lower than formation pressure, higher than formation pressure, and spontaneous imbibition. Corresponding to the imbibition effects at different development stages, self-imbibition tests were first conducted to calculate the relative wetting index to determine the mixed wettability of the shale core. Then, imbibition experiments were conducted under different pressures.

(1) Experimental study on mixed wettability

Wettability is the preferential selectivity of a solid surface covered by a certain fluid phase. The wettability of rock reservoirs can be divided into water wetting, oil wetting, and neutral wetting according to the different fluids involved [9]. Oil exists in the pores of reservoir rocks, and the wettability of the pore surface varies due to its mineral composition, thereby affecting the occurrence and flow properties of oil in it. Wettability is a relative characteristic parameter that is different from porosity and permeability, which are only related to rock properties. Wettability is related not only to the rock itself but also to the fluid properties associated with it. It affects many physical parameters of rocks, such as capillary pressure, relative permeability, adsorbed/free oil volume, residual oil saturation, bound water saturation, and fluid distribution [10].

This study used the relative index method of indirect measurement of wettability to characterize the wettability of shale in the study area. In this paper, the relative index method, which indirectly measures wettability, is used to characterize wettability at the pore scale of shale cores. The signal volume of the sample after oil absorption and drying was measured at different self-imbibition test times, and then the self-imbibition fluid volume was converted from the fluid signal volume. Finally, according to the self-imbibition method in the wettability discrimination standard [11], the relative wettability index I was used to characterize the wettability of different samples to oil and water at the same time. V is the volume of fluid self-imbibed into the core; the subscript symbol w is for water, and o is for oil. The equation and discrimination criteria are as follows:

$$I_w = \frac{V_w}{V_w + V_o} \tag{3}$$

The oil phase wetting index is shown in Equation (4):

$$I_o = \frac{V_o}{V_w + V_o} \tag{4}$$

And based on the relative wettability index calculation method (Equation (5)) and the self-imbibition method in the wettability discrimination standard (Table 1), the wettability of the rock was determined.

$$I = I_w - I_o \tag{5}$$

Table 1. Self-imbibition method of wettability discrimination standard.

Relative Wettability Index *I*	$-1.0 \leq I < -0.7$	$-0.7 \leq I < -0.3$	$-0.3 \leq I < -0.1$	$-0.1 \leq I \leq 0.1$	$0.1 < I \leq 0.3$	$0.3 < I \leq 0.7$	$0.7 < I \leq 1.0$
Wettability	Strong oil wetting	Oil wetting	Weak oil wetting	Intermediate wetting	Weak water wetting	Water wetting	Strong water wetting

(2) Spontaneous imbibition + NMR test

Soak the sample in deuterium oxide, and use capillary force to spontaneously imbibe deuterium oxide into the pores to displace crude oil. Use nuclear magnetic resonance to monitor and record the process of sample signal changes to obtain the movability efficiency of crude oil under self-imbibition. Spontaneous imbibition relies solely on capillary force, and the amount of water and oil in the core sample depends entirely on its wettability.

(3) Overpressure imbibition/underpressure imbibition + NMR test

The imbibition characteristics and pore movability characteristics of shale cores under reservoir conditions (reservoir pressure and temperature) were clarified considering the changing pressure difference in the relationship between fracture pressure and pore pressure during the soaking-backflow stage. For example, in the soaking stage, fracture pressure is higher than pore pressure. Therefore, overpressure imbibition can simulate the imbibition effect during the soaking stage, while in the later stage of backflow, fracture pressure is lower than pore pressure; underpressure imbibition can simulate the imbibition effect during the later stage of backflow. Schematic diagrams of the three imbibition methods are shown in Figure 2.

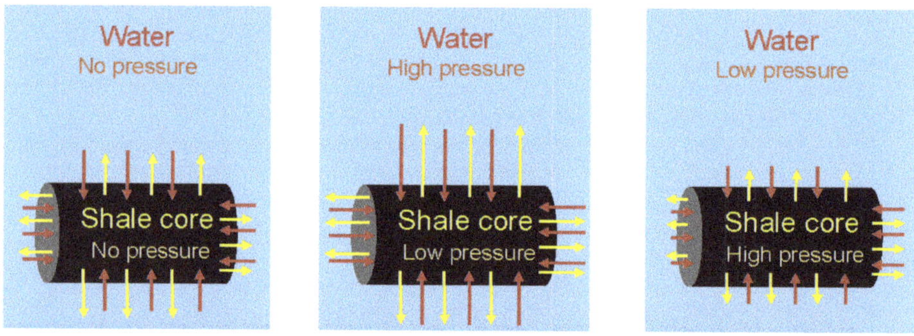

(a) (b) (c)

Figure 2. Schematic diagrams of the three imbibition methods. (**a**) Spontaneous imbibition. (**b**) Overpressure imbibition. (**c**) Underpressure imbibition.

2.3. Experimental Samples

The experimental shale core samples were obtained from the Daqing Gulong shale reservoir (Table 2). The samples were de-oiled and dried. The diameter of each experimental shale core sample was about 2.5 cm, and the length was about 5 cm. Helium gas was used to measure the porosity and permeability of the cores, and cores with similar porosity and permeability were selected for the experiment to ensure comparability. Spontaneous oil imbibition and overpressure injection were used to fully saturate the core samples. After spontaneous imbibition, the sample was saturated with crude oil; then, it was placed under normal temperature and pressure conditions for 120 h. The overpressure and underpressure imbibition samples were saturated with oil and placed under high-pressure conditions.

Table 2. Experimental core sample parameters.

Experimental Samples	Diameter (mm)	Length (mm)	Volume (cm^3)	Porosity (%)	Permeability (mD)	Photos of Shale Cores
High speed centrifugation + NMR	25.12	49.44	24.50	7.93	3.66×10^{-3}	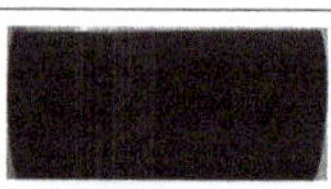
Spontaneous imbibition + NMR	25.09	54.87	27.11	6.94	3.16×10^{-3}	
Overpressure imbibition + NMR	25.05	50.33	24.80	8.23	4.05×10^{-3}	

Table 2. *Cont.*

Experimental Samples	Diameter (mm)	Length (mm)	Volume (cm^3)	Porosity (%)	Permeability (mD)	Photos of Shale Cores
Underpressure imbibition + NMR	25.28	64.83	32.54	7.20	5.82×10^{-3}	

3. Results and Discussion

3.1. Results of High-Speed Centrifugation + NMR Test

The NMR results under different centrifugal forces are shown in Figure 3. It can be observed that the T_2 spectrum shows a bimodal distribution after saturation of crude oil and a series of centrifugation steps. Research results showed that the Gulong shale reservoir develops a certain number of bedding fractures, mostly on the sub-micron to micron scale [12]. The right peak in the bimodal structure represents a micro-scale bedding fracture, and, after multiple centrifugations, this peak disappears, indicating that the fluid in the bedding fracture is more easily unlocked under centrifugation compared with the fluid in the matrix pores. The degree of crude oil recovery was calculated based on the difference in the area under the curve at different centrifugal forces, and the degree of crude oil production with different pore sizes was determined based on the relaxation time.

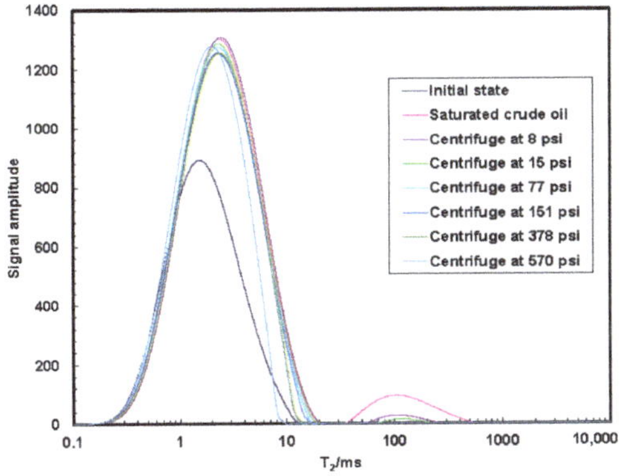

Figure 3. T_2 relaxation time measured using NMR under different centrifugal forces.

The high-speed centrifugation and one-dimensional nuclear magnetic resonance experiments showed that the available oil saturation driven by elastic energy reached 18.1%, the displacement limit was a 15 nm pore crude oil, and the corresponding pressure gradient was 1.2 MPa/cm. The bedding fracture density of Gulong shale oil was 3 pieces/cm, and within the control range of parallel fractures, the threshold pressure gradient of 15 nm pore fluid was 0.2 MPa (Figure 4).

3.2. Wettability Test Results

The imbibition characteristics of different pores are shown in the Figure 5. The core exhibits a typical bimodal pattern under water-saturated conditions, with a higher peak on the left side, indicating the development of small matrix pores in the shale. The peak value on the right side is lower, but the relaxation time is longer, indicating a lower degree of development of bedding fractures and larger pores than the matrix pores.

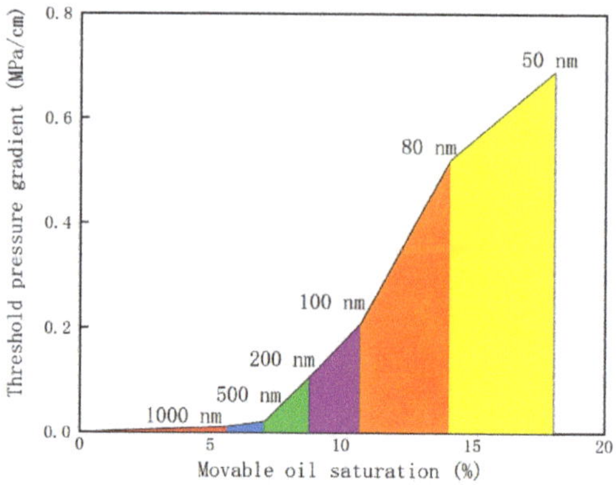

Figure 4. Fluid movability and threshold pressure gradient of shale core.

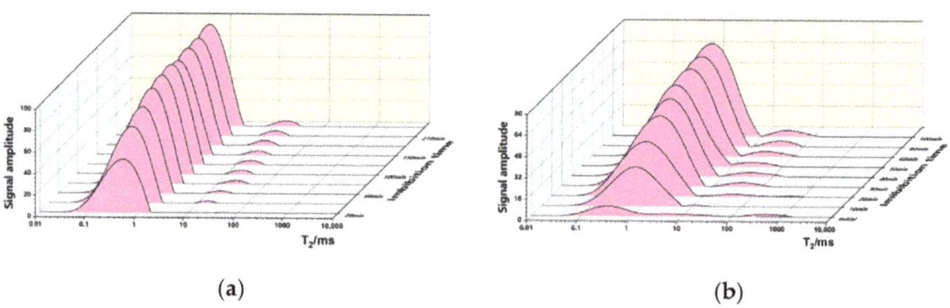

Figure 5. Shale water/oil self-imbibition characteristics. (**a**) Shale core self-imbibition of water. (**b**) Shale core self-imbibition of oil.

Comparing the self-imbibition processes of different pores (Figure 5), it was found that the shale cores had a strong self-imbibition ability for oil, and the self-imbibed amount of matrix water and oil increased rapidly in a short period of time. Moreover, compared to bedding fractures, the higher self-imbibition liquid content of the matrix pore structure reflects not only the larger pore volume of the matrix but also its stronger self-imbibition ability.

The amount of water self-imbibed by the bedding fractures slowly increased with the self-imbibition time, and the growth rate was relatively low. The amount of self-imbibed oil in the matrix pores increased as the duration of self-imbibition increased, and the difference in crude oil permeability between the shale reservoir and the matrix was significant. As mentioned earlier, the T_2 relaxation time is directly proportional to the pore size, while the large-scale structure of the Gulong shale reservoir is characterized by bedding fractures, which are formed by the interaction of a large amount of organic matter and clay minerals. The wettability of this part of the structure is lipophilic (as confirmed through the mixed wetting experiment). During the self-imbibition process, water cannot enter strongly oil-wet media, while oil can enter water-wet media [13]. This also reflects the presence of a bedding fracture structure in the shale reservoir, but the content is much smaller than that of the porous media.

Based on the calculation results of mixed wettability (Figure 6), it is shown that the overall rock exhibits intermediate wettability, with a relative wettability index between -0.1 and 0. The wettability of matrix pores and bedding fractures is different, with matrix pores being intermediate wet to weakly oil-wet, with a relative wettability index ranging from -0.1 to 0, and the bedding fracture being oil-wet to strongly oil-wet, with a relative wettability index ranging from -0.9 to -0.3. From the wettability results, it can be inferred

pressure difference-driven effect dominates; by contrast, in shale cores with extremely low permeability, capillary force dominates, while pressure-difference-driven force is relatively small, and there is an optimal range of action [14].

In reservoir formations, there is pore pressure in the matrix pores, and the efficiency of underpressure imbibition is much lower than that of spontaneous imbibition. The movability efficiency of underpressured crude oil is only about 1/10 of that of overpressured crude oil (Figure 9). The decrease in movability efficiency is mainly due to the cancellation of capillary force by the pressure in the matrix pores, making it difficult for deuterium oxide to exchange crude oil.

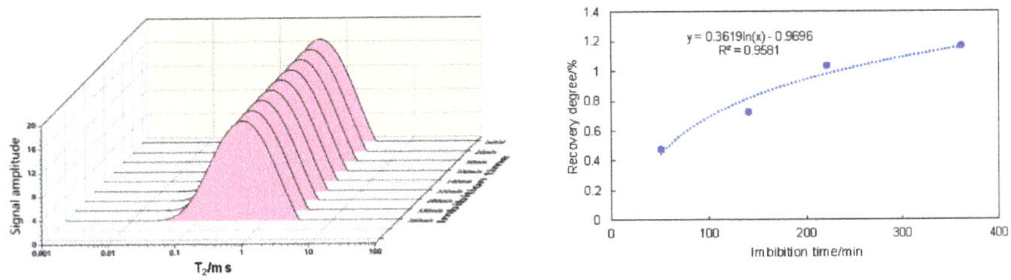

Figure 9. Underpressure imbibition fluid movability and recovery degree.

Based on the above tests, oil displacement efficiency under centrifugal pressure is significantly lower than crude oil movability efficiency under imbibition, and it can be clearly observed that the formation energy has a significant impact on the oil and water replacement efficiency during the soaking process. The main flow medium of the Gulong shale core is oil-wet to strongly oil-wet bedding fracture, which restricts the imbibition effect. Therefore, after fracturing, the pressure in the fracture media should be significantly greater than the formation pressure, promoting the effective entry of the bedding fracture fluid into the reservoir and the replacement of the crude oil.

4. Conclusions

This study used a combination of multiple experimental methods to compare and analyze the movability of matrix pore fluids during the processes of soaking and backflow in shale oil reservoirs after fracturing. The impacts of fracturing pressure at different backflow stages on fluid movability were fully considered. The movability of fluids based on nuclear magnetic resonance was quantitatively analyzed, clarifying the main contribution of the wettability of bedding fractures to fluid flow capacity.

(1) During the backflow stage, the matrix pore fluid is difficult to unlock under differential pressure drive, and under the action of high-speed centrifugal differential pressure, up to 18.1% of the fluid can be utilized; the main contribution is provided by microscale pores. The utilization of nanoscale pore fluids is almost impossible.

(2) The mixed wetting of porous and fractured dual media indicated that fluid imbibition has different effects in different media. The experimental results indicated that the fractured media are oil-wet, while the porous media show neutral wetting. This is unfavorable for flow of shale cores with the main flow channels of foliated fractures but beneficial to the flow of crude oil that mainly exists in pores.

(3) The fracturing pressure is crucial for the imbibition effect of matrix pores. When the fracture pressure is higher than matrix pore pressure, the imbibition efficiency is 27.9%; spontaneous imbibition efficiency when relying solely on capillary force is poor, at 16.8%; when the fracture pressure is lower than the matrix pore pressure, the imbibition efficiency is the lowest, at 1.3%.

(4) During the process of shut-in well backflow, it is necessary to ensure that the fracturing pressure is sufficiently higher than the pore pressure, which not only maintains the fracture morphology but also efficiently utilizes the imbibition effect to drive a continuous oil/water displacement.

Author Contributions: Methodology, Y.L. and Z.C.; validation, D.J.; formal analysis, Y.H.; investigation, Y.P.; supervision, Z.L. All authors have read and agreed to the published version of the manuscript.

Funding: This work was supported by the Chinese National Natural Science Foundations (No. U22B2075 and No. 52204053) and the China Postdoctoral Science Foundation (No. 2023M743922). This work was also conducted at the state key laboratory of enhanced oil and gas recovery.

Data Availability Statement: Due to the privacy restrictions on the experimental data and procedures, and the fact that some of the other data remain unpublished, it is not convenient to provide the data. All authors guarantee the authenticity and reliability of the data in this paper.

Conflicts of Interest: Yishan Liu, Zhewei Chen, Dongqi Ji, Yingfeng Peng and Zhengdong Lei were employed by Research Institute of Petroleum Exploration and Development, PetroChina, and Yanan Hou was employed by the company CNOOC Research Institute Ltd.

References

1. Lei, Z.; Li, J.; Chen, Z.; Dai, X.; Ji, D.; Wang, Y.; Liu, Y. Characterization of Multiphase Flow in Shale Oil Reservoirs Considering Multiscale Porous Media by High-Resolution Numerical Simulation. *SPE J.* **2023**, 1–16. [CrossRef]
2. Li, J.; Lu, S.; Zhang, J.; Zhang, P.; Xue, H. Quantitative evaluation models of adsorbed and free shale oil and its microscopic occurrence mechanism. *Oil Gas Geol.* **2019**, *40*, 583–592.
3. Jiang, Q.; Shen, B.; Yang, Y.; Li, Z.; Ma, Y.; Qian, M. Kinetic analysis of hydrocarbon generation components in shales from the Bonan sub-Sag and its significance for hydrocarbon exploration. *Pet. Geol. Exp.* **2016**, *38*, 231–239.
4. Yao, C.; Chen, H.; Onishi, T.; Datta-Gupta, A.; Mawalkar, S.; Mishra, S.; Pasumarti, A. Robust CO_2 Plume Imaging Using Joint Tomographic Inversion of Distributed Pressure and Temperature Measurements. In Proceedings of the SPE Annual Technical Conference and Exhibition, Dubai, United Arab Emirates, 21–23 September 2021.
5. Zhang, X.; Yang, S.; Zhang, Z. Optimization of injection-production parameters for CO_2 /N_2 composite gas huff and puff in fault-block reservoirs considering starting pressure gradient. *Sci. Technol. Eng.* **2021**, *21*, 14127–14132.
6. Xu, G.; Zhou, X.; Shang, D.; Zhang, G. Determination method of pressure behind blanking plug considering starting pressure gradient and its field application. *Pet. Geol. Oilfield Dev. Daqing* **2022**, *41*, 80–85.
7. Jia, R.; Shi, X.; Yu, W.; Yu, J. Simulated experimental research on starting pressure gradient of flow permeability reservoir in Nanpu Sag. *Mud Logging Eng.* **2021**, *32*, 119–123.
8. Ren, J.; Lu, B.; Ren, S. Improved pressure depletion method measures the porosity and permeability of shale. *Pet. Reserv. Eval. Dev.* **2020**, *10*, 49–55.
9. Zhao, Z. Contact angle and its application in surface chemistry research. *Chem. Res. Appl.* **2000**, *4*, 370–374.
10. Liang, C.; Xiao, L.; Zhou, C.; Guo, L.; Hu, F.; Liao, G.; Song, X.; Ge, G. Nuclear magnetic resonance characterizes rock wettability: Preliminary experimental results. *Chin. J. Geophys.* **2019**, *62*, 4472–4481.
11. *SY/T 5153-2017*; Test Method of Reservoir Rock Wettability. National Energy Administration: Beijing, China, 2017.
12. He, W.; Cui, B.; Wang, F.; Wang, Y.; Zhang, J.; Zhao, H.; Wang, R.; Bai, Y.; Lin, X.; Zhao, Y.; et al. Study on reservoir spaces and oil states of the Cretaceous Qingshankou Formation in Gulong Sag, Songliao Basin. *Geol. Rev.* **2022**, *68*, 693–741.
13. Sang, Q.; Zhang, S.; Li, Y.; Dong, M.; Bryant, S. Determination of organic and inorganic hydrocarbon saturations and effective porosities in shale using vacuum-imbibition method. *Int. J. Coal Geol.* **2018**, *200*, 123–134. [CrossRef]
14. Tian, W.; Wu, K.; Gao, Y.; Chen, Z.; Gao, Y.; Li, J. A Critical Review of Enhanced Oil Recovery by Imbibition: Theory and Practice. *Energy Fuels* **2021**, *35*, 5643–5670. [CrossRef]

processes MDPI

Article

Numerical Analyses of Perforation and Formation Damage of Sandstone Gas Reservoirs

Hao Liang [1,*], Zhihong Zhao [2] and Haozeng Jin [2]

1 China National Offshore Oil Corporation Co., Ltd., Haikou 570311, China
2 State Key Laboratory of Southwest Petroleum University, Chengdu 610500, China; zhaozhihong@swpu.edu.cn (Z.Z.); jinhaozeng@google.com (H.J.)
* Correspondence: haol_swpu@163.com

Abstract: Shaped charge perforation is an important technology for sandstone gas reservoirs. In the process of shaped charge perforating, the initial permeability and porosity of the formation are greatly reduced, directly affecting oil and gas production. This paper uses smooth particle hydrodynamics (SPH) and finite element methods (FEMs) to study the formation damage caused by the shaped charge perforation. In this perforation simulation, the impact characteristics of the metal jet formed by the liner are coupled with the damage characteristics of the sandstone. A new mathematical model is proposed to describe the damage of permeability and porosity based on the Morris and Xue models. The simulated results show that a large-scale abdominal section appears at the perforation site, and the axis of the hole tilts with the increase in the perforation depth. The porosity damage at the perforation site is the greatest, up to 60%, while the permeability recovers to 90% of its initial state after pressure relief. The degree of porosity damage decreases with the increase in negative pressure, and when the negative pressure is 30 MPa, it may cause a large amount of sand production.

Keywords: shaped charge perforation; perforation; formation damage; sandstone gas reservoirs; low permeability

1. Introduction

Shaped charge jet perforation is the key technology for a conventional oil and gas field [1–3]. Shaped charge perforation consists of four steps: detonating the explosive, the collapse of the liner, the formation and extension of the metal jet, and jet cratering and penetration. The high-pressure load produced by the high-speed metal jet hits the rock target and forms a hole. The plastic deformation of the rock is made near the hole, which creates the classical rock texture of the perforation damage area [4]. Therefore, the damage mechanism of the shaped charge jet perforation should be studied clearly, which will be helpful to improve the technology of shaped charge jet perforation and increase the production of oil and gas wells.

In recent decades, the numerical simulation of shaped charge jet perforation has been studied by many researchers. Zygmunt and Wilk [2] applied the X-ray pulse technique to research the process the jet stream formation and described the characteristics of shaped charges with metal powder liners. Karacan and Halleck [3] studied the damage of porosity and permeability for fractured zones caused by perforation in the gas-saturated and liquid-saturated Berea sandstone. Huang et al. [5] and Li et al. [6] investigated the mechanisms of and presented the results of laboratory experiments and the results of field tests on the abrasive water jet perforation for enhancing oil production. Li et al. [7] investigated the application of annular initiation to the explosively formed penetrator charge based on the LS-DYNA software (V6.5). Ayisit [8] performed numerical analyses to investigate the influence of shaped charge asymmetries on the jet characteristics.

In the above works, the researchers always used the Lagrange method, Euler method, and Arbitrary Lagrangian–Eulerian method (ALE) [9–12] to study the hole-forming process of shaped charge jet perforation. However, the Lagrange method can cause problems as the grids produce significant distortion and slip surfaces. With the Euler method, it is challenging to determine the position of the material interface, and the ALE method needs to build a more significant flow of space grids. These methods cannot sufficiently reflect the action process of metal particles on the rock. Therefore, it is better to choose the smoothed particle hydrodynamics (SPH) method [13–15] to simulate the shaped charge jet perforating metal particle flow.

In recent decades, considerable research has focused on simulating the shape of the shaped charge, the amount of ammunition, the dynamic form of the shaped charge jet, the jet velocity, and the hole-forming process through the metal target [16,17]. Gooch et al. [18] analyzed the target strength effect on penetration by shaped charge jets. Yin et al. [19] investigated the "white" etching layer on the perforation surface and adiabatic shear bands (ASBs) in the matrix of ultra-high-strength steel plates penetrated by shaped charge jets. Uhlig and Coppinger [20] used a means of the confined liquid to investigate the interplay of eroded target material and the remaining projectile to elucidate relevant penetration behaviors in solid materials. Zhu et al. [21] discussed the evolution of the response regions in front of the jet/target interface as the shaped charge jet moves into high- and ultra-high-strength concrete targets. Elshenawy and Li [22] studied the influences of concrete strength and its confinement pressure on the shaped charge jet penetration. Guo et al. [23] confirmed that the virtual origin theory could describe the interaction process between the target and the shaped jet based on the characteristic expansion hole and liquid return characteristics. However, little research has been conducted on the penetration of shaped charge jet through a sandstone target.

During the perforation process, the high-speed metal flow penetrates the formation while forming a compaction zone near the hole. The permeability and porosity of the sandstone near the hole are reduced. This extremely low permeability significantly reduces oil and gas wells' production capacities. Therefore, it is necessary to analyze the distribution of porosity and permeability in the compaction zone and evaluate the damage degree. Morris et al. [24,25] proposed an evolution model of the porosity and permeability of sandstone with a low strain rate, which used plastic strain, effective stress, and damage variables to describe the evolution of porosity and permeability. Zhang and Shifeng [26] developed a mechanical model for evaluating the perforating damage of sandstone based on the sensitivity analysis of the deformation rate effect on porosity evolution. Kemmoukhe et al. [27] used AUTODYN-2D to study the effect of the main parameters (liner material, explosive charge, stand-off distance, and wave-shaper presence) on the jet formation, jet velocity, jet length, and penetration depth. Khamitov et al. [28] conducted a coupled 3D CFD–DEM analysis to simulate the perforation damage and sand production of soft sandstone materials. Yan et al. [29] set up a numerical model based on the LS-DYNA software (V6.5). A cylindrical target penetration test was used to verify the interface damage area caused by perforation. Fu et al. [30] studied the penetration behavior of convergent, divergent, and straight linear-shaped charges (LSCs) by numerical and physical methods.

Previously, most numerical simulation studies on shaped charges mainly used the LS-DYNA and AUTODYN software (V6.5). But the calculated results are not accurate, and the phenomena are usually different from the experimental results. Therefore, developing a new efficient method is necessary to provide more accurate results. Although the previous numerical simulation research about perforation compaction obtained excellent results, the compaction damage models in 2D and 3D in previous works were established by simplifying the perforating process, and the non-symmetrical and non-uniform shape of holes and the distribution of the compaction area caused by shaped charge jet were neglected.

This paper studies the pore-forming process of the shaped charge jet perforation and the influence of the perforation damage on the formation's permeability and porosity numerically. The SPH algorithm is applied to establish the finite element model of the hole-forming process of sandstone by considerable high-speed jet particles impacting the sandstone. The ALE adaptive mesh technology is used to dynamically encrypt the mesh around the hole. The multi-node calculation is carried out by an HPC (high-performance computer) to ensure the accuracy of the results. The Hoffman damage model is used to assess the 3D hole model that is formed by the failure of sandstone. The hole-forming process and the 3D hole structure can be analyzed by the hole compaction zone, which is characterized by a custom field. Therefore, a set of mathematical modeling research methods are proposed, which has great significance in guiding the selection of perforation equipment and parameters of sandstone with low porosity and permeability, reducing the perforation damage and increasing the production of oil and gas wells. In addition, based on the model in Ref. [24], a novel mathematical model for evaluating the compaction damage in the perforation process of sandstone with low porosity and permeability is proposed to describe the permeability and porosity of the damaged formation. The distribution characteristics of compaction damage under different working conditions and the influencing factors are considered in this paper. The present work can provide a foundation for future research and has considerable referential significance.

2. Modeling Theory

The process of the shaped charge jet perforation is shown in Figure 1. The shaped charge is mainly composed of a liner and explosive, as shown in Figure 1a. When the explosion happens, the copper metal particle jet with an extremely high speed is formed, penetrating the perforating gun, well fluid, casing, and cement ring into the formation, as shown in Figure 1b. Amid the continuing explosion, the velocity of the jet becomes slower until a velocity gradient forms. A tremendous pressure is generated by the ultra-high velocity jet that creates perforation holes, as shown in Figure 1c.

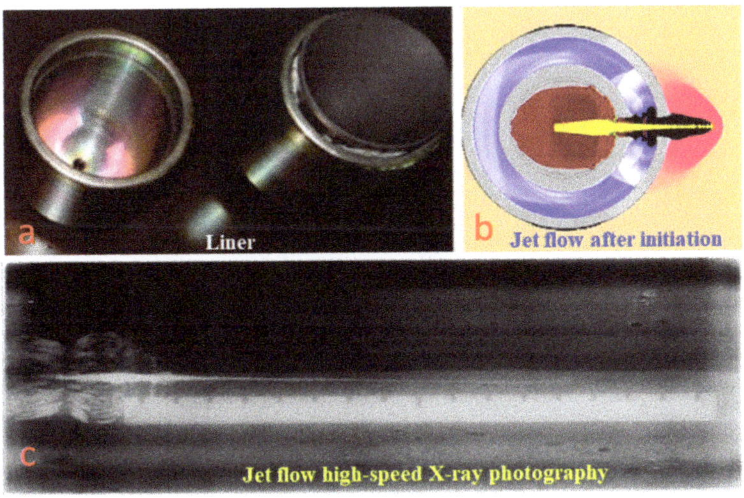

Figure 1. Shaped charge perforating jet diagram. (**a**) Detonation of the explosive. (**b**) The collapse of the liner. (**c**) Creating perforation holes.

2.1. Smoothed Particle Hydrodynamics Algorithm

The SPH method was applied to study the shaped charge jet perforation process in this paper. This method is a meshless fluid dynamic solution with pure Lagrange adaptive properties and is a clever method for discretizing continuous partial differential equations. It better reflects the process of metal particles acting on rock compared to the Lagrange method, Euler method, ALE method, and other numerical calculation methods.

The SPH method is used to solve hydrodynamics problems in two steps. The first step is the integral representation, which can be approximated by the integral over the influence region of the kernel function. The second step is the particle representation, which matches the function value of the discrete point by summing the importance of the nearest adjacent particles. SPH uses an evolutionary interpolation scheme to compare field variables at any point in the field. The kernel function is as shown:

$$\langle f(x)\rangle = \int_{\Omega} f(x')W(x-x',h)dx' \tag{1}$$

where $W(x-x',h)$ is a kernel function and is not zero; x is the spatial coordinates of points; x' is the coordinate of the point in space that contributes to x; and h is the smooth length that determines how many particles affect the interpolation of a particular point.

The kernel function W plays an important role in the SPH approximation method, which determines the accuracy and computational efficiency of function expressions. It should satisfy three conditions. Namely, the compactness condition, the regularization condition, and Dirac function property as the smooth length approaches zero. The equations are expressed as follows:

$$\int_{\Omega} W(x-x',h)dx' = 1, \tag{2}$$

$$|x-x'| > kh,\ W(x-x',h) = 0, \tag{3}$$

$$\lim_{h\to 0} W(x-x',h) = \delta(x-x'),\ \delta(x-x') = \begin{cases} 1, x = x' \\ 0, x \neq x'' \end{cases} \tag{4}$$

where k is the constant that is related to the smooth function at x and can determine the effective range (non-zero), which is called the integral in the support region of the smooth function at x. Therefore, the domain of integration Ω is generally the support region.

According to the integration by parts and the divergence theorem, Equation (1) is transformed. The derivative of $f(x)$ is estimated as:

$$\langle \nabla f(x)\rangle = -\int_{\Omega} f(x')\nabla W(x-x',h)dx. \tag{5}$$

From Equation (3), it is known that the SPH method transforms the spatial derivatives of functions into the spatial derivatives of smooth functions. Therefore, the spatial derivatives of arbitrary field functions can be obtained from the kernel functions.

After discretization, the integral expression of the function can be written into the particle approximation as follows:

$$\langle f(x_i)\rangle = \sum_{j=1}^{N} \frac{m_j}{\rho_j} f(x_j)W_{ij}, \tag{6}$$

where j and i are particle numbers; m_j and ρ_j are the mass and density of the j particles, respectively; and N is the total number of particles.

It can be seen from Equation (7) that the value of the arbitrary function at a certain position can be expressed by applying the smooth function W to the form of interpolation and summation of all particles in the compact support domain of smooth length h [31].

The SPH calculation equations of the shaped charge jet perforating metal flow simulation are [31,32]:

$$
\begin{cases}
\dfrac{d\rho_i}{dt} & = \sum\limits_{j=1}^{N} m_j \left(v_i^{\beta} - v_j^{\beta}\right) \dfrac{\partial W_{ij}}{\partial x_i^{\beta}}, \\
\dfrac{dv_i^{\alpha}}{dt} & = -\sum\limits_{j=1}^{N} m_j \left(\dfrac{P_i}{\rho_i^2} + \dfrac{P_i}{\rho_j^2} + \Pi_{ij}\right) \dfrac{\partial W_{ij}}{\partial x_i^{\alpha}} + \sum\limits_{j=1}^{N} m_j \left(\dfrac{S_i^{\alpha\beta}}{\rho_i^2} + \dfrac{S_i^{\alpha\beta}}{\rho_j^2}\right) \dfrac{\partial W_{ij}}{\partial x_i^{\beta}}, \\
\dfrac{du_i}{dt} & = \dfrac{1}{2} \sum\limits_{j=1}^{N} m_j \left(\dfrac{P_i}{\rho_i^2} + \dfrac{P_i}{\rho_j^2} + \Pi_{ij}\right) \left(v_i^{\beta} - v_j^{\beta}\right) \dfrac{\partial W_{ij}}{\partial x_i^{\beta}} + \dfrac{1}{\rho_i} S_i^{\alpha\beta} \dot{\varepsilon}_i^{\alpha\beta}, \\
\dfrac{dh}{dt} & = -\dfrac{1}{2} \dfrac{h}{p} \dfrac{d\rho}{dt}
\end{cases}
. \tag{7}
$$

where ρ is the density, m is the mass, P is the pressure, u is the internal energy, v is the velocity, S is the deviatoric stress, $\prod_{ij}$ is the artificial viscosity, ε is the strain rate, and α and β are the coordinate directions.

2.2. Damage Model

The local structure of the rock is deformed by high pressure, which is generated by the penetration of the metal particle flow of shaped charge jet perforation into the low-permeability sandstone target. The large deformation rate near the particle surface and the slow heat release rate on the sliding surface softens the material near the particle. The distance between the face and that of the local deformation surface is expressed by Δ, and the following inequalities can define the conditions for softening:

$$
\frac{\Delta^2}{\chi} > \frac{d}{v_p}, \tag{8}
$$

where Δ^2/χ is the relaxation time for the temperature caused by heat conduction, d/v_p is the characteristic time of deformation, and χ is the temperature thermal conductivity coefficient. The distance Δ between the sliding surfaces should be of about 0.1–1.0 μm order of magnitude.

The cavity will collapse in time τ as pressure p is constantly applied on the outer boundary of the grid. Therefore, the ultra-deep penetration conditions of the particles are

$$
\tau \leq \frac{d}{v_p} \leq \frac{\Delta^2}{\chi}. \tag{9}
$$

In the strongly excited state, the target material can be considered as a weakly interrelated particle whole. This weak complementarity can be regarded as being in a thermodynamic equilibrium state by comparing to the motion state of the penetration particle. Therefore, the theory of weakly interacting particles describes the interaction between the penetration particle and the target material particle. Assuming that only the force between the particle center is considered, the interaction between particles is

$$
V_{jn} = \left(\frac{2\pi}{L}\right)^3 \sum_l V_l \exp\left[il\left(x_j - x_n\right)\right], \tag{10}
$$

where l is the distance between the particles, V_l is a coefficient that only depends on the wave vector/absolute value, and x_j and x_n are the coordinates of the j_{th} particle and the n_{th} particle, respectively.

The force acting on the particle in the ultra-deep penetration state is expressed as

$$F_p = -\left[H' + 2.6\frac{\rho_s\left(\vec{v}_p - \vec{v}_1\right)^2}{\sqrt{\mathrm{Re}}}\right]\frac{\pi d^2}{4}\frac{\vec{v}_p - \vec{v}}{\left|\vec{v}_p - \vec{v}\right|}, \quad (11)$$

where $\vec{v}_1$ is the velocity of the target material, H' is the dynamic stiffness, d is the particle diameter, ρ_s is the density of the target material, and $\vec{v}_p$ is the particle velocity.

The friction generated by the contact between the moving particles and rock particles makes the energy and the velocity of the shaped charge perforating metal particle flow decrease rapidly. The friction coefficient between the particle and the target body is expressed as follows

$$\eta = \frac{16q(v)}{v}, \quad (12)$$

with two limits

$$\begin{cases} v \to 0 : q(v) \sim v,\ \eta = const, \\ v \to 0 : q(v) = \frac{1}{2v^2},\ \eta \sim \frac{1}{v^3}. \end{cases} \quad (13)$$

The Hoffman criterion describes the penetration of jet particles into sandstone more comprehensively. In the numerical analysis, the dimensionless failure coefficient F_d is introduced to represent the failure degree of the rock so that the Hoffman criterion can be written as

$$\begin{gathered} F_d = C_1(\sigma_2 - \sigma_3)^2 + C_2(\sigma_3 - \sigma_1)^2 + C_3(\sigma_2 - \sigma_1)^2 \\ + C_4\sigma_1 + C_5\sigma_2 + C_6\sigma_3 + C_7\sigma_{23}^2 + C_8\sigma_{13}^2 + C_9\sigma_{12}^2, \end{gathered} \quad (14)$$

where

$$C_1 = \frac{1}{2}\left(\frac{1}{Z_tZ_c} + \frac{1}{Y_tY_c} + \frac{1}{X_tX_c}\right),\ C_2 = \frac{1}{2}\left(\frac{1}{Z_tZ_c} - \frac{1}{Y_tY_c} + \frac{1}{X_tX_c}\right), \quad (15)$$

$$C_3 = \frac{1}{2}\left(\frac{1}{Y_tY_c} + \frac{1}{X_tX_c} - \frac{1}{Z_tZ_c}\right),\ C_4 = \frac{1}{X_t} - \frac{1}{X_c},\ C_5 = \frac{1}{Y_t} - \frac{1}{Y_c}, \quad (16)$$

$$C_6 = \frac{1}{Z_t} - \frac{1}{Z_c},\ C_7 = \frac{1}{S_{23}^2},\ C_8 = \frac{1}{S_{13}^2},\ C_9 = \frac{1}{S_{12}^2}, \quad (17)$$

where X_t, Y_t, and Z_t are the tensile strength values of the rock in the X, Y, and Z directions, respectively. X_c, Y_c, and Z_c are the compressive strength values of the rock in the X, Y, and Z directions, respectively. S_{12}^2, S_{23}^2, and S_{13}^2 are on the 12, 23, and 13 surfaces' in-plane shear strength, respectively.

Due to the complex mechanical properties of deep sandstone, establishing a damage plastic constitutive model is the key to accurately calculating the process of shaped charge jet perforation. Figure 2 plots the mechanical response curve of the damage plasticity of sandstone material. It can be seen that the unloading segment is weakened (the slope of the curve decreases) when the softened section is unloaded, which indicates that the material has been damaged (or weakened).

The damage variable d_t is a function of the plastic strain, temperature, and field variable and can be written as

$$d_t = d_t\,(\tilde{\varepsilon}_t^{pl}, \theta, f_i)\ 0 \le d_t \le 1, \quad (18)$$

where $\tilde{\varepsilon}_t^{pl}$ is the tensile equivalent plastic strain, θ is the temperature, and $f_i (i = 1, 2, \ldots)$ is another predefined field variable. $d_t = 0$ corresponds to a non-damaged material, and a completely damaged material is $d_t = 1$.

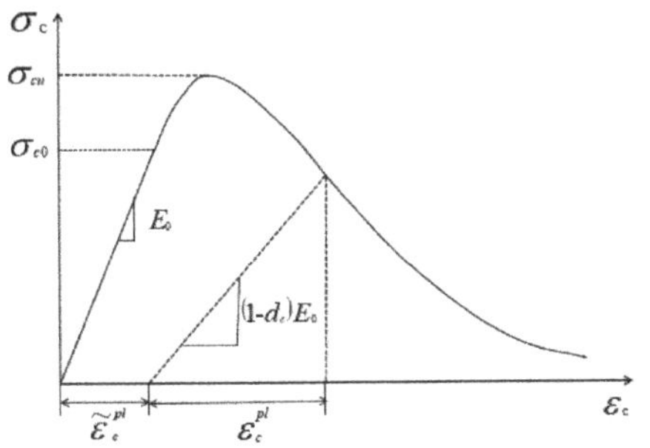

Figure 2. Mechanical response curve.

The softening section of the sandstone damage plastic constitutive model is controlled by the amount of damage and the energy released by the plastic deformation. The material damage plastic stiffness degradation model can be expressed as

$$E = (1-d)E_0,\ \sigma_t = (1-d_t)E_0\left(\varepsilon_t - \tilde{\varepsilon}_t^{pl}\right), \tag{19}$$

where E_0 is the initial (damage-free) elastic stiffness of the material and σ_t is the tensile strength.

2.3. Mathematical Model for the Perforation Compaction Damage

In this paper, based on the mechanical response characteristics of sandstone under perforation impact load, the modified evolution equation of sandstone porosity and permeability was employed, which was proposed in Refs. [30,31]. Therefore, a mathematical model of the compaction damage of perforation with low-porosity and permeability was established.

The porosity ϕ of sandstone in the compaction zone consists of two parts: the initial porosity ϕ_i and expanded porosity ϕ_{dil}. The initial porosity ϕ_i is expressed as

$$\phi_i = \frac{\phi_i^0}{1+D}, \tag{20}$$

where ϕ_i^0 and D are, respectively, the porosity of sandstone before damage and the damage variable of the rock mass element, which can be determined by Equation (16).

$$\phi_i^0 = \begin{cases} \phi_i - \sigma_{eff}(\phi_i - (\phi_* + CP_*))/P_* \\ \phi_* + C\sigma_{eff} \quad \sigma_{eff} \geq P_*, \end{cases} \tag{21}$$

$$D = \begin{cases} C_1\varepsilon_p \min\left(\sigma_{eff}, P_s^*\right) - C_2\varepsilon_v \\ 0 \ (D < 0), \end{cases} \tag{22}$$

where ϕ_* is the fluid porosity intercept; C is the porosity decay rate; P_* is the critical effective stress; σ_{eff} is the effective stress; C_1 is the plastic strain damage factor; C_2 is the hydrostatic pressure damage factor; P_s^* is the maximum plastic strain damage stress; ε_p is the plastic strain of rock mass element; and ε_v is the volume strain of rock mass element.

The expanded porosity ϕ_{dil} can be expressed as follows:

$$\phi_{dil} = \phi_{dil}^{\max} - \exp\left(\eta - \frac{m_d\sigma_{vm}\varepsilon_p}{\max(P_m, P_{\min})}\right), \tag{23}$$

where $\phi_{dil}^{\max}$ is the maximum expanded porosity; P_m is the average stress; $P_{\min}$ is the minimum expansion stress; m_d is the coefficient of expansion; and σ_{vm} is the von Mises stress.

According to the effective stress, volumetric strain, and plastic strain of the rock mass unit, the distribution of porosity and permeability in the compacted zone can be quantitatively analyzed through the damage evaluation mathematical model of the porosity and permeability in the sandstone perforated compacted zone. The expressions of the porosity ϕ and permeability k are

$$\phi = \phi_{dil}^{\max} - \exp\left(\eta - \frac{m_d \sigma_{vm} \varepsilon_p}{\max(P_m, P_{\min})}\right) + \frac{\phi_i^0}{1+D}, \tag{24}$$

$$k = k_0 \exp\left(\frac{k_1 \phi_i^0 \left(\sigma_{eff}, D\right)}{1+D} - B\min(D, D_{\max})\right), \tag{25}$$

where k_0 is the initial permeability; k_1 is the permeability index; B is the permeability damage attenuation factor; and $D_{\max}$ is the maximum permeability damage.

The evaluation index for the damage degree of the porosity and permeability can be expressed as follows:

$$\phi_{cd} = \frac{\phi_i - \overline{\phi}}{\phi_i} \times 100\%,\ K_{cd} = \frac{K_0 - \overline{K}}{K_0} \times 100\%, \tag{26}$$

where ϕ_{cd} and K_{cd} are, respectively, the damage degree of the porosity and permeability; and $\overline{\phi}$ and $\overline{K}$ are the average porosity and permeability in the compacted zone, respectively.

In addition, the stress of, the strain of, and the changes in the porosity and permeability of any node under different negative pressure differences can be calculated by using the secondary development of ABAQUS. Stress and strain parameters can be automatically extracted through the secondary development of ABAQUS, thus obtaining the maximum principal strain ε_1, intermediate principal strain ε_2, and minimum principal strain ε_3 of a node through numerical simulation. The volumetric strain ε_v is the sum of the three. Similarly, the maximum principal stress σ_1, intermediate principal stress σ_2, and minimum principal stress σ_3 of a node can be obtained. The average stress P_m is the average of the three. The effective stress σ_{eff} can be expressed as follows:

$$\sigma_{eff} = P_m - P_c = (\sigma_1 + \sigma_2 + \sigma_3)/3 - P_c, \tag{27}$$

where P_c is the confining pressure.

Finally, the plastic strain ε_p and von Mises stress σ_{vm} of the same node are acquired. Substituting these parameters and other initial values into the formula for calculating the porosity and permeability yields the porosity and permeability of any node and the degree of damage.

3. Numerical Modeling and Analyses

This section firstly establishes the finite element model (FEM) of the hole-forming process and then compares the results of the sandstone target in penetration simulation with the experiment, thus verifying the reliability of the numerical method.

3.1. Material Parameters

The relevant material models and parameters were established and are briefly introduced in this section.

3.1.1. Material Parameters of the Sandstone Target

The hardness and tensile experiments of the sandstone required the following parameters: elasticity modulus of the rock was 30,217 MPa; the triaxial compressive strength was 249.1 MPa; the tensile strength was 2 MPa; the hardness was 1754 MPa; the uniaxial compressive strength was 70 MPa; and the Poisson's ratio was 0.2216. The definitions of the model material parameters are shown in Table 1.

Table 1. Model material parameter.

Material Name	Elastic Modulus/GPa	Density/kg·m^{-3}	Yield Strength/MPa	Fracture Strain	Poisson's Ratio
Casing (N80)	210	7800	551	0.0026	0.3
Cement ring	20	2500	50	0.0020	0.2
Stratigraphic rocks	15	2300	40	0.0020	0.25

3.1.2. Initial Values of the Parameters in the Mathematical Model

Combined with the values of the parameters in the mathematical model of perforation compaction damage in Refs. [30,31], the initial values of the parameters in this work are shown in Table 2.

Table 2. Initial values of the parameters in the mathematical model of perforation compaction damage.

Parameters	Initial Values	Parameters	Initial Values
Confining pressure P_c/MPa	30	Fluid porosity intercept ϕ_*	0.122
Porosity decay rate C	−0.103	Maximum permeability damage $D_{\max}$	0.15
Coefficient of expansion m_d	1.3	Maximum expanded porosity $\phi_{dil}^{\max}$	0.15
Maximum pressure for hydrostatic damage P_{h2}/GPa	0.42	Minimum permeability K_0/m^2	1.00×10^{-17}
Constant η	−1.8971	Hydrostatic pressure damage factor C_2	27.8
Minimum expanded stress $P_{\min}$/GPa	0.03	Permeability damage attenuation factor B	12.5
Maximum plastic strain damage stress P_s^*/GPa	0.25	Critical effective stress P_*/GPa	0.01
Plastic strain damage factor C_1	28.5	Initial permeability K_k/m^2	2.00×10^{-15}
Permeability index K_1	44.434	Initial porosity ϕ_i	0.12

3.2. FEM and Simulation Results

3.2.1. FEM of Sandstone Target Penetration

Based on the penetration mechanism of shaped charge perforating particles, the SPH method and Hoffman judgment criterion were used to establish the 3D FEM of the hole-forming process and sandstone compaction damage model. The VUSDFLD user subroutines created damage variables. The boundary conditions and load parameters were set by the target experiment of the perforation and the environmental parameters of temperature and pressure in the reservoir.

Primary hypothesis: The copper and tungsten of the perforating projectile are converted into particles; the velocity of the jet particles is much higher than the average thermomechanical velocity of the target body; the particles are not deformed or damaged and disappear when the velocity is lower than the critical value; the local high-temperature target material softens after impact; and the target material of the casing, cement ring, and sandstone are damaged.

Figure 3 plots the 3D FEM of the hole-forming process. The same mechanical parameters were used in the target shooting experiment. The thickness values of the casing, cement, sandstone, and shaped charge were 12 mm, 28 mm, 960 mm, and 250 mm, respectively. At the same time, the velocity, friction coefficient, confining pressure, and effective stress of the rock were set simulate the environment. In addition, the balance between the accuracy of the calculation result and the calculation time should be considered when selecting an appropriate grid size.

Figure 4 plots the initial velocity of the particles in different jet segments. The velocity can be divided into five stages: 7300 m/s → 7800 m/s → 7300 m/s → 1900 m/s → 800 m/s, according to the boundary conditions of the model. It can be seen from Figure 3 that the number of metal particles is different in different velocity segments.

Due to a large number of particles, the calculation efficiency of SPH is much lower than that of FEM. Therefore, FEM was applied to model the initial structure when the process of three-dimensional shaped charge perforation was analyzed. The FEM mesh units in the large deformation area were automatically converted into SPH particles in the calculation process. As shown in Figure 5, at a specific moment (7–10 s), before the jet beam touches the casing wall, all the mesh units are converted into particles that collide with the casing, cement ring, and formation rock.

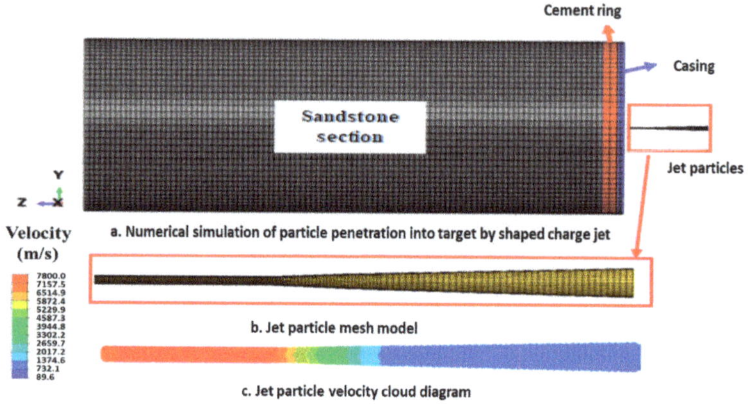

Figure 3. Three-dimensional FEM of the hole-forming process.

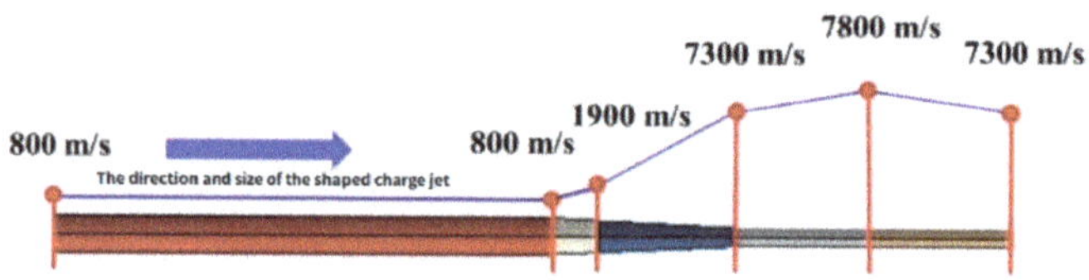

Figure 4. Initial velocity of the particles in different jet segments.

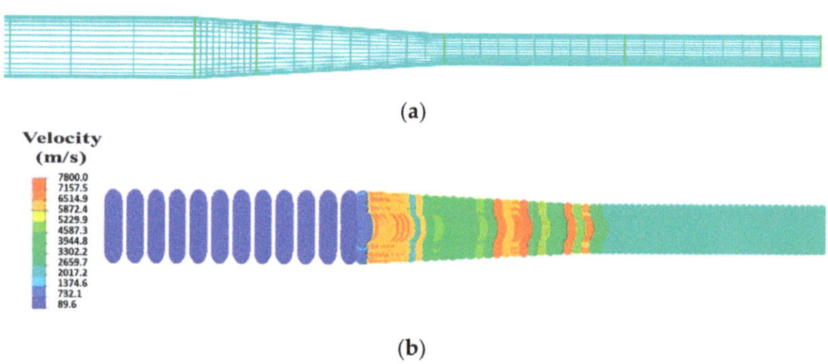

Figure 5. Mesh element model of the metal flow (**a**) and SPH model after metal flow collision (**b**).

3.2.2. Simulation of the Hole-Forming Process of the Sandstone Target

According to the mechanical parameters of sandstone and the shaped charge jet perforating metal particle flow when the confining pressure was zero, the 3D model of shaped charge jet perforating metal particle flow penetrating the casing–cement–sandstone target was established. After more than 100 h of calculation, the simulation results of the hole forming process were obtained. Figure 6 plots the sandstone stress distribution at 1.26 ms of the jet flow impact. It can be seen that the stress at the hole tip is larger than the residual stress of the rock around the perforation hole. However, the stress of the rock far away from the hole is minimal, which indicates that it is not disturbed by the perforation.

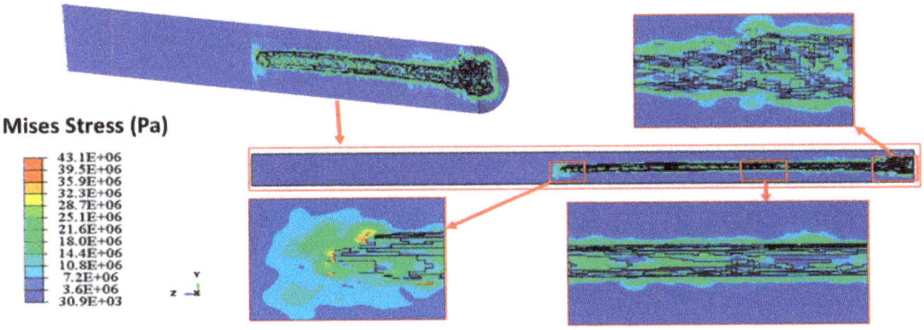

Figure 6. Sandstone stress nephogram at 1.26 ms of the jet flow impact.

The tip of the jet flow stops moving at 1.26 ms, and a 3D hole with an unsmooth wall of about 56 cm in depth is formed. The residual stress distribution of the casing, cement stone, and rock near the hole is approximately 2–5 mm thick, and the axis is deflected with the depth of the hole.

3.3. Experimental Verification

According to the characteristics of the low porosity and low permeability of the target reservoir, the white sandstone, as shown in Figure 7, was selected to make the sandstone target for the perforation experiment. The porosity of the reservoir was 8.5–9.5%, and the permeability was 0.15–0.25 mD. The diameter of the cylindrical sandstone target was 200 mm, and the length was about 0.8 mm. The reservoir environment was simulated according to the API RP 19B standard [33]. The confining pressure, pore pressure, and wellbore pressure were applied, and the shaped charge was fired to complete the shaped charge jet perforation experiment.

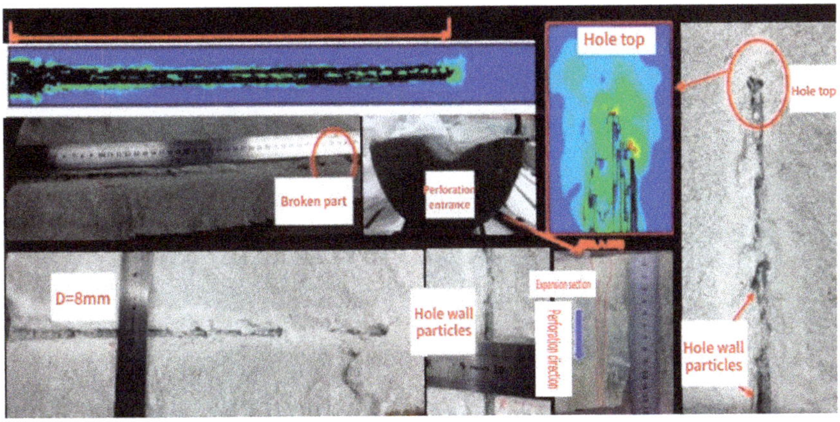

Figure 7. Comparison diagram of the simulation and experiment.

Figure 7 plots the comparison diagram of the simulation and experiment. It can be seen from the profile that the perforating depth of the split target is greater than 50 cm, and the perforating depth obtained through simulation is 56 cm. Meanwhile, there is a silty loose fracture zone around the target hole with a thickness of about 10 mm. The particle size is significantly smaller than that of the original sandstone. Moreover, the black metal particles attached to the surface of the hole wall can be seen, which is denser than that of the original sandstone. The experiment results are in good agreement with the numerical simulation results. The error is less than 8%, which indicates that the established perforation compaction damage evaluation model is correct and reproduces the damage process of the shaped charge jet perforation.

4. The Analysis of the Compaction Damage

4.1. Instantaneous and Unloading High Pressure

In the perforation process, high pressure (1–2 GPa) is generated at the moment of shaped charge jet perforation. Therefore, an internal pressure of 1000 MPa was applied to the inner wall of the hole to simulate the immediate impact of the metal jet. Figure 8 plots the distribution nephograms of stress (a), density (b), and strain rate (c) at the moment of perforation. The result shows that the rock within 30 mm of the hole wall is significantly deformed by extrusion, and the deformation at the tip of the hole is the largest. The density increased from 2.3 g/cm^3 to 2.5 g/cm^3, which is caused by compaction. Then, the residual deformation and residual stress that appear due to the sandstone cannot be recovered completely after internal pressure is unloaded.

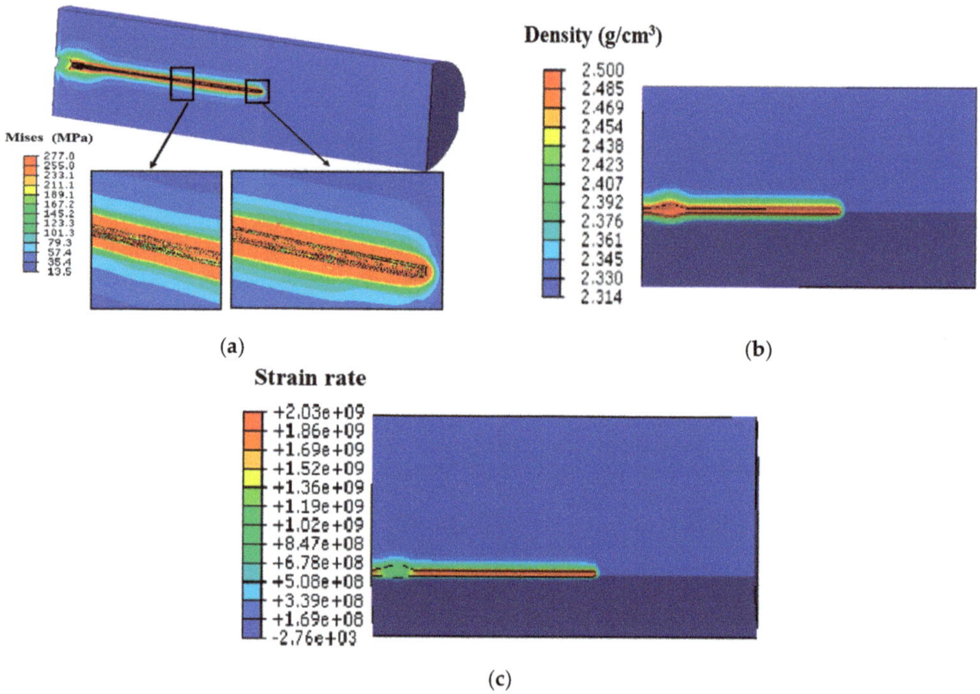

Figure 8. Distribution nephograms of stress (**a**), density (**b**), and strain rate (**c**) at the moment of perforation.

Figure 9 plots the distribution nephograms of the permeability (a) and porosity (b) at the moment of perforation. As shown in Figure 9, the permeability within 10 mm of the hole wall (blue part) is significantly reduced to 0.016 mD, and the maximum permeability is 1.9 mD. The porosity in the 5 mm range (blue region) is significantly reduced to 5%, and the tips are about 10.8%. The damage degree of the porosity and permeability can be calculated by the ratio of the variable of porosity and permeability near the hole to the initial porosity and permeability, which are shown in Figure 10.

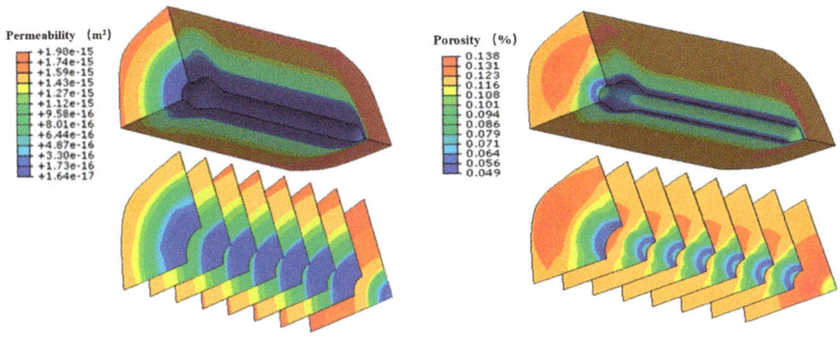

(a) (b)

Figure 9. Distribution nephograms of the permeability (**a**) and porosity (**b**) at the moment of perforation.

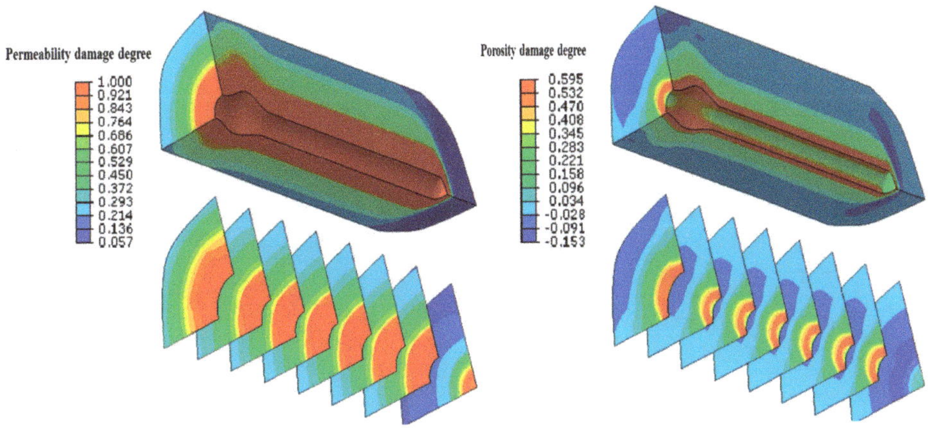

(a) (b)

Figure 10. Distribution nephograms of the damage degree of the permeability (**a**) and porosity (**b**) at the moment of perforation.

Figure 11 shows the distribution nephograms of the permeability (a) and porosity (b) when the internal pressure is unloaded. After unloading the internal pressure in the hole, the partial elastic of the wall recovers. The permeability of the hole wall decreases to less than 0.2 mD and gradually increases in the radial direction to the original permeability. The porosity of the hole wall increases from 12% to 20%, which decreases first to 6.7% and then gradually increases in the radial direction to the original porosity.

The distribution nephogram of the damage degree is show in Figure 12. It shows that the damage degree of the permeability at hole wall exceeds 90%, while the porosity is 66%. The maximum porosity increases more than 60%, and the minimum damage degree of the porosity near the hole tip exceeds 44%. Thus, the distribution area of the porosity can be divided into the area of pore expansion and fracture, the area of pore damage and compaction, and the area of the nondestructive pore, which is similar to the experimental results.

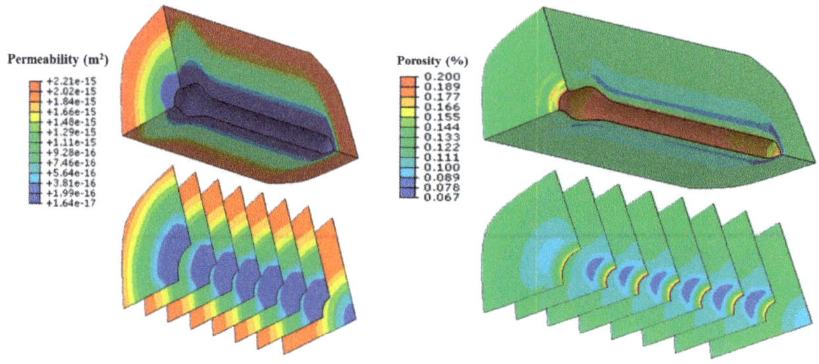

(a) (b)

Figure 11. Distribution nephograms of the permeability (**a**) and porosity (**b**) when the internal pressure is unloaded.

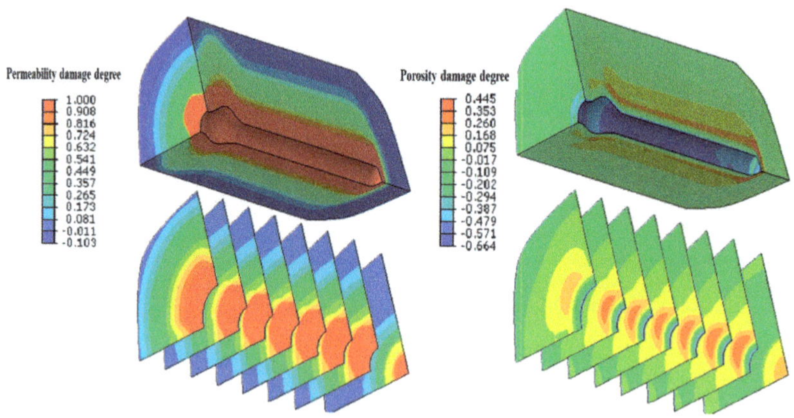

(a) (b)

Figure 12. Distribution nephograms of the damage degree of the permeability (**a**) and porosity (**b**) when the internal pressure is unloaded.

4.2. Effects of the Different Negative Pressure

After the wellbore is formed, when the fluid pressure within the wellbore is lower than the formation pressure, negative pressure perforation occurs, causing the fluid to quickly flow back from the formation, thus flushing the borehole and compacting zone, removing perforation contamination. This results in a better communication between the reservoir and the wellbore, significantly increasing the near-wellbore area permeability. To study the effects of the different negative pressure on rock damage, different negative pressure values (10 MPa, 20 MPa, 30 MPa, 40 MPa, and 50 MPa) were applied to the hole wall according to the negative pressure difference after the instantaneous high pressure is loaded.

Figure 13 plots the distribution nephograms of the permeability (a) and porosity (b) with the differential pressure of 10 MPa. The maximum porosity of the hole wall increases from 20% to 22% because of the action of the tensile load. However, the minimum porosity and permeability distribution and values are basically the same as those in Figure 11. It shows that the negative pressure of 10 MPa has little effect on the porosity and permeability of sandstone with low porosity and permeability.

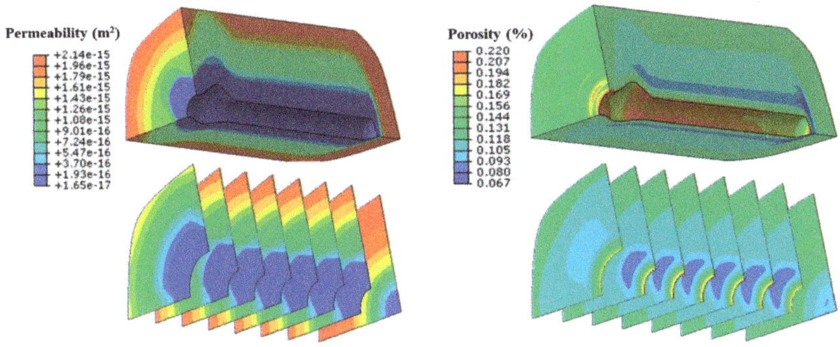

(**a**) (**b**)

Figure 13. Distribution nephograms of the permeability (**a**) and porosity (**b**) with a pressure perforation of 10 MPa.

The corresponding distribution nephogram of the damage degree is shown in Figure 14. It can be seen that the damage degree of the permeability has no noticeable change compared to that in Figure 12a. Conversely, the damage degree of the porosity in the hole tip is less than that of Figure 12b. Therefore, the pore expansion increases under the pressure difference of 10 MPa.

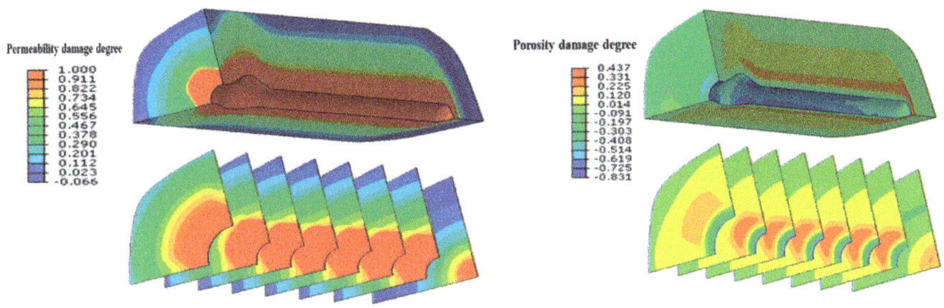

(**a**) (**b**)

Figure 14. Distribution nephograms of the damage degree of the permeability (**a**) and porosity (**b**) with a pressure perforation of 10 MPa.

As the negative pressure increases to 50 MPa, the distribution nephograms of the permeability and porosity and their damage degree are shown in Figures 15 and 16, respectively. The local porosity of the hole wall increased continuously from 22% to 23.3%. Moreover, the distribution of the porosity compaction area within the range of 10–20 mm from the hole wall in the middle of the hole underwent significant changes. In addition, the damage degree of the permeability had a few changes, which are the same as those in Figure 14a. The pore in the hole wall continued to expand with a maximum increase of 94%.

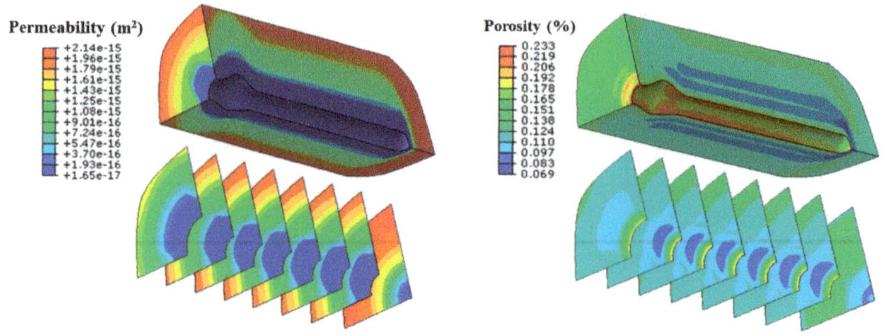

(a) (b)

Figure 15. Distribution nephograms of the permeability (**a**) and porosity (**b**) with a pressure perforation of 50 MPa.

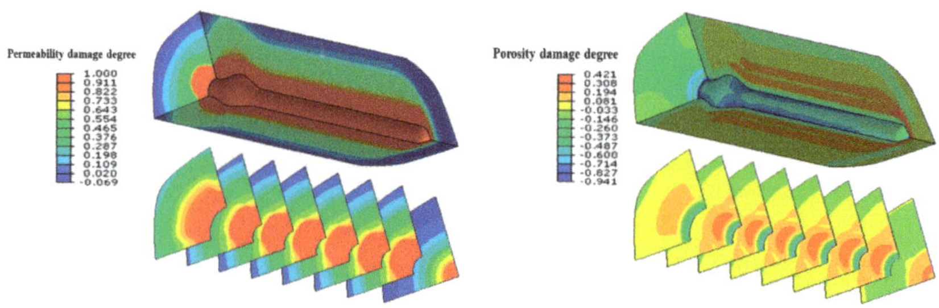

(a) (b)

Figure 16. Distribution nephograms of the damage degree of permeability (**a**) and porosity (**b**) with a pressure perforation of 50 MPa.

4.3. The Distribution Characteristics of the Compaction Damage at Different Positions around the Hole

This section considers the stress distribution around the perforation under seven conditions, including perforation torque, perforation unloading, and a negative pressure of 10–50 MPa. In the absence of an external load, the stress distribution of the different cross-sections at various positions of the perforation is similar, as shown in Figure 17. Taking the middle A-A of the perforation as an example, it is located at a distance of 9–60 mm from the perforation axis.

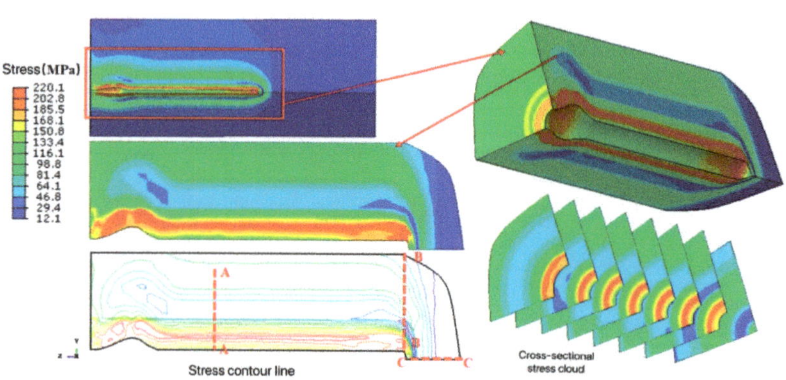

Figure 17. Nephogram of the perforation internal pressure unloading.

Figure 18 plots the distribution curves of the sandstone porosity (a) and its damage (b) in the A-A path. It can be seen that the porosity changes significantly within 20 mm from the axis of the hole, and the porosity is less than 10% at the moment of perforation. In addition, with the increase in the negative pressure in the other six working conditions, the range of the pore expansion increases continuously.

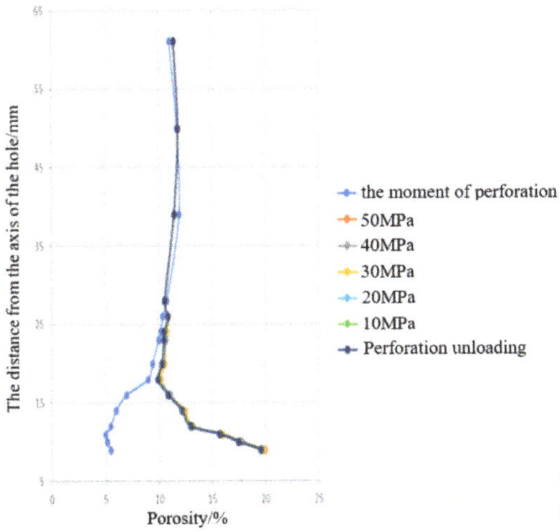

(a)

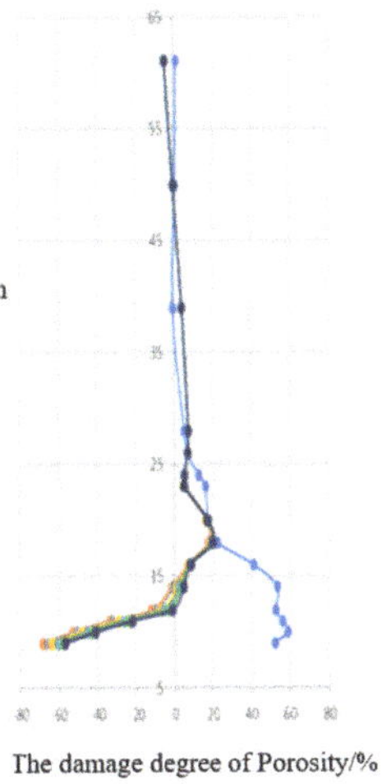

(b)

Figure 18. The distribution curves of the sandstone porosity (**a**) and damage (**b**) in the A-A path.

Figure 19 plots the distribution curves of the sandstone permeability (a) and its damage (b) in the A-A path. It can be seen that the permeability is the same within 20 mm from the distribution curve, and the permeability of the perforation moment is significantly different from that of the other six working conditions within the range of 20–30 mm. As shown in Figure 19b, the damage degree of the permeability within 20 mm from the axis of the hole is the same, and there is no damage in the permeability of the outer 60 mm.

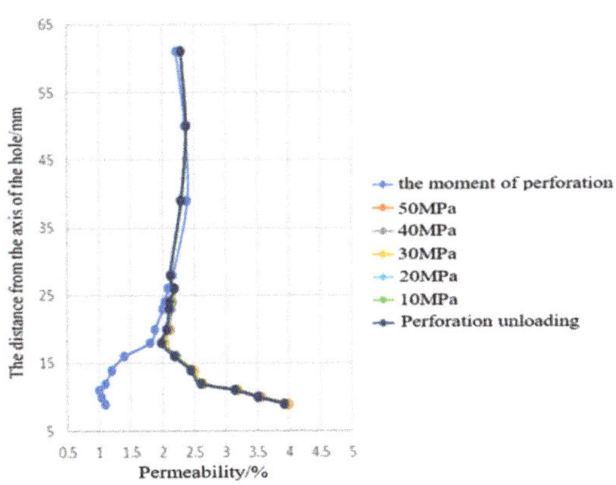

(a)

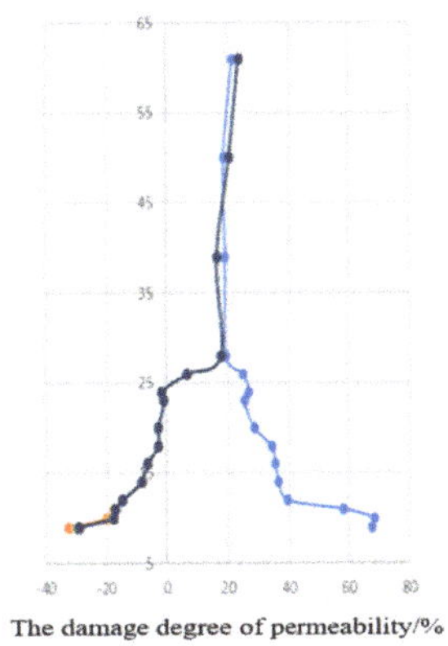

(b)

Figure 19. The distribution curves of the sandstone permeability (**a**) and its damage (**b**) in the A-A path.

5. Conclusions

In the present study, the dynamic response of a sandstone target, the damage characteristics, and the influence factors of the hole parameters during perforation are investigated, and the sand production rule is explored. Based on the numerical results, the following conclusions are drawn:

1. At the entrance of the perforation, the aperture is larger and appears at the extensive belly section. The axis of the hole is skewed with the perforation depth, and the hole wall is not smooth.
2. The porosity damage is at its maximum at the moment of perforation, reaching 60%. The permeability reached a full 90% under unloading conditions.
3. The porosity damage decreases with the increase in the negative pressure. However, a large amount of sand may be produced when the negative pressure is 30 MPa.

Although many studies have been carried out on the numerical simulation of the flow characteristics in the sandstone target perforation compacted zone at home and abroad, at present, a 3D damage mechanical model of the compaction zone considering the variation in the rock stress field and the distribution of the compaction degree under the action of non-uniform crustal stress has not been reported.

This paper aims to use the SPH method and the secondary development of ABAQUS to understand the numerical simulation of the perforation compaction process. Based on the FEM, a complete 3D impact damage-coupling dynamic perforation simulation method is established. A greater understanding of the damage, destruction, and hole-forming process after the collision of rocks and particles is obtained more realistically. This can be expanded to develop a more accurate numerical solution method for 3D flow to achieve the good simulation of the influence of the local micro-domains on multiple domains.

Author Contributions: Conceptualization, H.L.; methodology, Z.Z.; writing, H.J. All authors have read and agreed to the published version of the manuscript.

Funding: This research received no external funding.

Data Availability Statement: The data presented in this study are available on request from the corresponding author.

Conflicts of Interest: The research was conducted in the absence of any commercial or financial relationships that could be construed as a potential conflict of interest.

References

1. Dou, J.H.; Jia, X.; Huang, Z.X.; Gu, X.H.; Zheng, Y.M.; Ma, B.; Xiao, Q.Q. Theoretical and numerical simulation study on jet formation and penetration of different liner structures driven by electromagnetic pressure. *Def. Technol.* **2020**, *17*, 846–858. [CrossRef]
2. Zygmunt, B.; Wilk, Z. Formation of jets by shaped charges with metal powder liners. *Propellants Explos. Pyrotech.* **2008**, *33*, 482–487. [CrossRef]
3. Karacan, C.Ö.; Halleck, P.M. Comparison of shaped-charge perforating induced formation damage to gas-and liquid-saturated sandstone samples. *J. Pet. Sci. Eng.* **2003**, *40*, 61–75. [CrossRef]
4. Xiao, Q.Q.; Huang, Z.X.; Zu, X.D.; Jia, X.; Zhu, Q.F.; Cai, W. Shaped charge penetration into high- and ultrahigh-strength Steel-Fiber reactive powder concrete targets. *Def. Technol.* **2020**, *16*, 217–224. [CrossRef]
5. Huang, Z.; Niu, J.; Li, G.; Yuan, X.; Liu, Y. Surface experiment of abrasive water jet perforation. *Pet. Sci. Technol.* **2008**, *26*, 726–733. [CrossRef]
6. Li, G.; Niu, J.; Song, J.; Huang, Z.; Ma, L.; Tang, R.; Yu, Y. Abrasive water jet perforation—An alternative approach to enhance oil production. *Pet. Sci. Technol.* **2004**, *22*, 491–504. [CrossRef]
7. Li, W.; Wang, X.; Li, W. The effect of annular multi-point initiation on the formation and penetration of an explosively formed penetrator. *Int. J. Impact Eng.* **2010**, *37*, 414–424. [CrossRef]
8. Ayisit, O. The influence of asymmetries in shaped charge performance. *Int. J. Impact Eng.* **2008**, *35*, 1399–1404. [CrossRef]
9. Hu, H.H.; Patankar, N.A.; Zhu, M.Y. Direct Numerical Simulations of Fluid–Solid Systems Using the Arbitrary Lagrangian–Eulerian Technique. *J. Comput. Phys.* **2001**, *169*, 427–462. [CrossRef]
10. Feng, D.L.; Liu, M.B.; Li, H.Q.; Liu, G.R. Smoothed particle hydrodynamics modeling of linear shaped charge with jet formation and penetration effects. *Comput. Fluids* **2013**, *86*, 77–85. [CrossRef]

11. Nazem, M.; Carter, J.P.; Airey, D.W. Arbitrary Lagrangian–Eulerian method for dynamic analysis of geotechnical problems. *Comput. Geotech.* **2009**, *36*, 549–557. [CrossRef]
12. Tianbao, M.; Cheng, W.; Jianguo, N. Multi-material Eulerian formulations and hydrocode for the simulation of explosions. *CMES Comput. Model. Eng. Sci.* **2008**, *33*, 155–178.
13. Dong, X.W.; Liu, G.R.; Li, Z.; Zeng, W. A smoothed particle hydrodynamics (SPH) model for simulating surface erosion by impacts of foreign particles. *Tribol. Int.* **2016**, *95*, 267–278. [CrossRef]
14. Liu, M.B.; Liu, G.R. Smoothed Particle Hydrodynamics (SPH): An Overview and Recent Developments. *Arch. Comput. Methods Eng.* **2010**, *17*, 25–76. [CrossRef]
15. Yin, J.; Shi, Z.; Chen, J.; Chang, B.; Yi, J. Smooth particle hydrodynamics-based characteristics of a shaped jet from different materials. *Strength Mater.* **2019**, *51*, 85–94. [CrossRef]
16. Cheng, W.; Ma, T.B.; Ning, J.G. Design and Penetration Performance of Annular Shaped Charge Structure. *Adv. Mater. Res.* **2008**, *33*, 1175–1180.
17. Wang, F.; Ma, D.; Wang, P.; Liu, J.; Jiang, J. Experimental and numerical study on the meso-scopic characteristics of metal composites jets by a shaped charge. *J. Appl. Phys.* **2019**, *126*, 095901. [CrossRef]
18. Gooch, W.; Burkins, M.; Walters, W.; Kozhushko, A.; Sinani, A. Target strength effect on penetration by shaped charge jets. *Int. J. Impact Eng.* **2001**, *26*, 243–248. [CrossRef]
19. Yin, Z.; Ma, C.; Li, S.; Cheng, G. Perforation of an ultra-high strength steel penetrated by shaped charge jet. *Mater. Sci. Eng.* **2004**, *379*, 443–447. [CrossRef]
20. Uhlig, W.C.; Coppinger, M.J. Hydrocode simulations of liquid filled channels for understanding erosion in shaped charge jet penetration. *Procedia Eng.* **2017**, *204*, 231–238. [CrossRef]
21. Zhu, Q.F.; Huang, Z.X.; Xiao, Q.Q.; Zu, X.D.; Jia, X. Theoretical and experimental study of shaped charge jet penetration into high and ultra-high strength concrete targets. *Int. J. Impact Eng.* **2018**, *122*, 431–438. [CrossRef]
22. Elshenawy, T.; Li, Q.M. Influences of target strength and confinement on the penetration depth of an oil well perforator. *Int. J. Impact Eng.* **2013**, *54*, 130–137. [CrossRef]
23. Guo, M.; Zu, X.D.; Shen, X.J.; Huang, Z.X. Study on liquid-filled structure target with shaped charge vertical penetration. *Def. Technol.* **2019**, *15*, 861–867. [CrossRef]
24. Morris, J.P.; Lomov, I.N.; Glenn, L.A. A constitutive model for stress-induced permeability and porosity evolution of Berea sandstone. *J. Geophys. Res. Solid Earth* **2003**, *108*, 2485. [CrossRef]
25. Morris, J.P.; Lomov, I.N.; Glenn, L.A. Simulating perforation permeability damage and cleanup. *Soc. Pet. Eng.* **2001**, *23*, 120–137.
26. Zhang, L.; Shifeng, X. Evaluation of perforating damage for sandstone. *J. China Univ. Pet.* **2014**, *38*, 137–142.
27. Kemmoukhe, H.; Burzić, Z.; Savić, S.; Terzić, S.; Lisov, M. Improvement of shaped charge penetration capability and disturbation of the jet by explosive reactive armor. *Teh. Vjesn.* **2019**, *26*, 1658–1663.
28. Khamitov, F.; Minh, N.H.; Zhao, Y. Coupled CFD–DEM numerical modelling of perforation damage and sand production in weak sandstone formation. *Geomech. Energy Environ.* **2021**, *28*, 100255. [CrossRef]
29. Yan, Y.; Guan, Z.; Xu, Y.; Yan, W.; Wang, H. Numerical investigation of perforation to cementing interface damage area. *J. Pet. Sci. Eng.* **2019**, *179*, 257–265. [CrossRef]
30. Fu, Y.S.; Jiang, M.; Zhao, P.L.; Ye, X.J.; Xie, X.B.; Gao, F.Y. Physical and numerical assessments of the penetration behavior of coplanar linear shape charges. *Int. J. Mech. Mater. Des.* **2021**, *17*, 441–452. [CrossRef]
31. Wang, Y.; Li, Q.; Dong, W.; Li, Q.; Wang, F.; Bai, H.; Zhang, R.; Owusu, A. Effect of different factors on the yield of epoxy-terminated polydimethylsiloxane and evaluation of CO_2 thickening. *RSC Adv.* **2018**, *8*, 39787–39796. [CrossRef] [PubMed]
32. Li, Q.; Liu, J.; Wang, S.; Guo, Y.; Han, X.; Li, Q.; Cheng, Y.; Dong, Z.; Li, X.; Zhang, X. Numerical insights into factors affecting collapse behavior of horizontal wellbore in clayey silt hydrate-bearing sediments and the accompanying control strategy. *Ocean Eng.* **2024**, *297*, 117029. [CrossRef]
33. *API RP 19B*; Evaluation of Well perforators, Second Edition. American Petroleum Institute Standards: Washington, DC, USA, 2006.

processes

MDPI

Article

Flow and Heat Transfer of Shale Oil Reservoir during CO_2 Enhanced Pyrolysis: A Pore-Scale Modeling

Yang Shi [1], Dingwei Weng [1], Bo Cai [1], Yunpeng Zhang [2], Yaochen Zhang [2], Bin Wang [2,*] and Haizhu Wang [2]

[1] PetroChina Research Institute of Petroleum Exploration and Development, Beijing 100083, China; shy312@petrochina.com.cn (Y.S.); wendw69@petrochina.com.cn (D.W.); caibo69@petrochina.com.cn (B.C.)
[2] College of Petroleum Engineering, China University of Petroleum (Beijing), Beijing 102249, China; zyp18439703733@163.com (Y.Z.); yaochenz1214@163.com (Y.Z.); whz0001@126.com (H.W.)
* Correspondence: bin.wang@cup.edu.cn

Abstract: This study extensively investigates the influence of different pyrolysis temperatures and organic matter contents on the fluid flow and heat transfer properties in oil shale samples. Utilizing CT images to generate three-dimensional digital rock, coupled simulations of CO_2 flow and heat transfer were conducted, analyzing parameters such as velocity fields, permeability, temperature fields, average temperatures, and heat transfer coefficients. The results reveal that, for relatively homogeneous oil shale samples, the permeability exhibits a monotonous increase with rising pyrolysis temperature. While the effect of pyrolysis temperature on the distribution characteristics of velocity and temperature fields is minimal, it significantly impacts the heat transfer coefficients. Specifically, the heat transfer coefficients increase significantly in the direction perpendicular to the bedding plane, while they decrease or remain unchanged parallel to it. Additionally, the organic matter content significantly influences the fluid flow and heat transfer properties of shale samples. After undergoing heat treatment, the heterogeneity of pore structures in shale samples varies significantly, affecting the characteristics of fluid flow and heat transfer. The influence of organic matter content and pyrolysis temperature on fluid flow and heat transfer in shale primarily stems from the effect of organic matter pyrolysis on the original pore structure. The development and connectivity of pore networks are closely related to the distribution characteristics of the original organic matter and are not directly correlated with the organic matter content. These findings provide essential theoretical guidance and technical support for the development and utilization of oil shale resources, while also offering valuable references and insights for future research.

Keywords: oil shale rock; heat transfer; 3D digital rock; pyrolysis; pore-scale

1. Introduction

In the field of oil and gas exploration and development, China has achieved a leap from conventional to unconventional oil and gas, with unconventional oil and gas accounting for 41% of the country's cumulative proven oil and gas reserves [1–4]. Unconventional oil and gas are important strategic replacement resources for China. Oil shale, as a typical unconventional oil and gas resource, has low oil and gas maturity. Its development and utilization must rely on high temperatures to mature and crack the kerogen within the rock, thereby producing extractable oil and gas [5]. China is rich in oil shale resources, which are relatively concentrated in distribution. According to geological survey data, China's proven oil shale reserves are approximately 972.32 billion tons [6,7]. Therefore, accelerating the efficient development and utilization of oil shale resources has significant political and economic implications for meeting China's energy needs, optimizing its energy structure, and ensuring energy security.

Fluid flow and heat transfer are two closely related physical processes during the pyrolysis of oil shale [8]. Fluid flow not only dictates mass transport within oil shale but also influences the distribution and yield of reaction products. Simultaneously, heat transfer directly affects the rate and efficiency of pyrolysis reactions. Therefore, delving into the mechanisms of fluid flow and heat transfer in oil shale during pyrolysis holds significant importance for optimizing reaction conditions, enhancing oil shale extraction efficiency, and mitigating environmental impacts [9–11]. Current research on heat transfer and fluid flow in oil shale spans several aspects. At the macroscopic scale, researchers utilize porous media models and continuum assumptions to explore mass transfer and heat transfer behaviors during oil shale pyrolysis. These models consider factors such as rock pore structure, porosity, and permeability to evaluate the rate and efficiency of pyrolysis reactions under different conditions. Additionally, through experimental and computational methods, in-depth investigations have been conducted into oil shale's pore structure, pore distribution, and pore connectivity to reveal fluid flow patterns and permeation behavior within rock pores [12–17]. On the other hand, at the microscopic scale, researchers employ simulation methods based on pore network models, such as computational fluid dynamics (CFD) simulations and molecular dynamics simulations [18–22], to simulate fluid flow and heat transfer processes within oil shale pores. These studies provide insights into the impact of pore structure on fluid flow and heat transfer performance from a microscopic perspective, offering new insights into understanding the pyrolysis process of oil shale. Furthermore, some studies focus on utilizing advanced imaging techniques, such as X-ray CT scanning and electron microscopy, to reconstruct and characterize the three-dimensional pore structure of oil shale samples [23–29], aiming to obtain more realistic and accurate pore geometry information for precise simulation and prediction of heat transfer and fluid flow behavior in oil shale. These endeavors provide important experimental and theoretical foundations for a more comprehensive understanding of heat transfer and fluid flow mechanisms during oil shale pyrolysis.

This paper aims to investigate the fluid flow and conjugate heat transfer mechanisms in oil shale during pyrolysis through pore-scale modeling. Utilizing X-ray CT scan images of oil shale samples treated at various pyrolysis temperatures and with different organic matter contents, we convert these images into three-dimensional digital rock cores. Employing CFD methods, we analyze the flow patterns of supercritical CO_2 along the X, Y, and Z directions in various rock samples and study the fluid–solid heat transfer during these flow processes. This research seeks to elucidate the impact of pyrolysis temperature and organic matter content on the fluid heat transfer mechanisms within oil shale. By conducting this study, we aim to provide new perspectives and methodologies for a deeper understanding of the oil shale pyrolysis process. The insights gained will offer a scientific foundation for the development and utilization of oil shale resources, promoting sustainable development in the energy sector.

2. Model Description

2.1. Governing Equations

The Navier–Stokes equations are the basic equations to solver the flow of liquid. When the flow state is steady laminar and the fluid can be regarded as imcompressible Newtonian fluid, the Navier–Stokes equations are as follows:

$$\nabla \cdot \mathbf{u} = 0 \tag{1}$$

$$\nabla \cdot (\mathbf{u} \otimes \mathbf{u}) = -\nabla p + \nu \nabla^2 \mathbf{u} \tag{2}$$

where $\mathbf{u}$ (m/s) is the velocity, p (m^2/s^2) is the kinematic pressure, ν (m^2/s) is kinematic viscosity. The energy equations in the fluid region Ω_f and solid region Ω_s can be written in term of temperature:

$$\nabla \cdot (c_f T_f \mathbf{u}) = \nabla \cdot \kappa_f \nabla T_f \tag{3}$$

$$\nabla \cdot \kappa_s \nabla T_s = 0 \tag{4}$$

where c_f and c_s $(\mathrm{kJ/m^3/K})$ are the fluid and solid heat capacity, respectively, T_f and T_s (K) are the temperatures in the fluid and solid, respectively, κ_f and κ_s $(\mathrm{kW/m/K})$ are the fluid and solid thermal conductivity, respectively.

The micro-continuum approach was used to calculate the heat transfer between fluids and oil shale rocks. The micro-continuum approach is a method that uses volume-averaging of the flow equations over a control volume in the presence of solid material. The heat and mass transport are solved in terms of the volume-averaged properties:

$$\bar{\mathbf{u}} = \frac{1}{V}\int_{V_f} \mathbf{u}\mathrm{d}V \tag{5}$$

$$\bar{p} = \frac{1}{V_f}\int_{V_f} p\mathrm{d}V \tag{6}$$

$$\bar{T_f} = \frac{1}{V_f}\int_{V_f} T_f\mathrm{d}V \tag{7}$$

$$\bar{T_s} = \frac{1}{V_s}\int_{V_s} T_f\mathrm{d}V \tag{8}$$

where V_f and V_s are the volume of fluid and solid in a control volume, respectively, and $V = V_f + V_s$. Then, combined with the Darcy–Stokes–Brinkman equation over the whole region Ω:

$$\nabla \cdot \bar{\mathbf{u}} = 0 \tag{9}$$

$$\frac{1}{\epsilon_f}\nabla(\frac{\mathbf{u}}{\epsilon_f}\otimes\mathbf{u}) = -\nabla\bar{P_f} + \frac{\nu}{\epsilon_f}\nabla^2\bar{\mathbf{u}} - \nu K^{-1}\bar{\mathbf{u}} \tag{10}$$

where K is the permeability of the control volume, $\nu K^{-1}\bar{\mathbf{u}}$ is the Darcy resistance. ϵ_f is the local permeability porosity of fluid. The permeability K is solved by the Kozeny–Carman equation:

$$K^{-1} = \frac{180}{h^2}\frac{(1-\epsilon_f)^2}{\epsilon_f^3} \tag{11}$$

where h is the mesh resolution. More equation details can be found in the literature [30].

2.2. Oil Shale Model and REV Selection

The CT scanner takes a series of 2D images of the rock samples from multiple angles. By stacking the 2D pictures, the 3D CT data of rock samples can be obtained and the fine 3D visualization of rock pore structure can be realized. In the paper, eight sets of open source oil shale CT data [31,32] were used to analyze the effects of pyrolysis temperature and organic matter content on physical properties of oil shale and heat transfer between supercritical CO_2 and oil shale.

The open source data of different pyrolysis temperatures were obtained from the same oil shale sample (i.e., Oil shale) heated from 20 °C to 500 °C by a customized heating furnace under the absolute pressure of 0.1 kPa. During the pyrolysis process, synchrotron radiation real-time CT scanning with a resolution of 2 μm was carried out with the diamond light source I13-2 beam, as shown in Figure 1. For the different organic matter content cases, the CT data of four groups of samples with different organic matter contents (i.e., OM1, OM2, OM3, OM4) were analyzed after pyrolysis at 0.1 KPa and 500 °C. The organic matter content in the four samples is ranked as: OM4 > OM3 > OM2 > OM1. The porosity and size information can be seen in Table 1 and the three-dimensional geometry is shown in Figure 2, respectively.

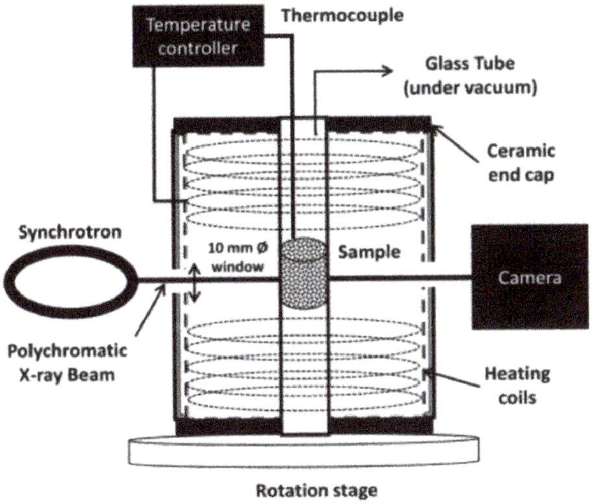

Figure 1. Schematic of CT experimental apparatus [31].

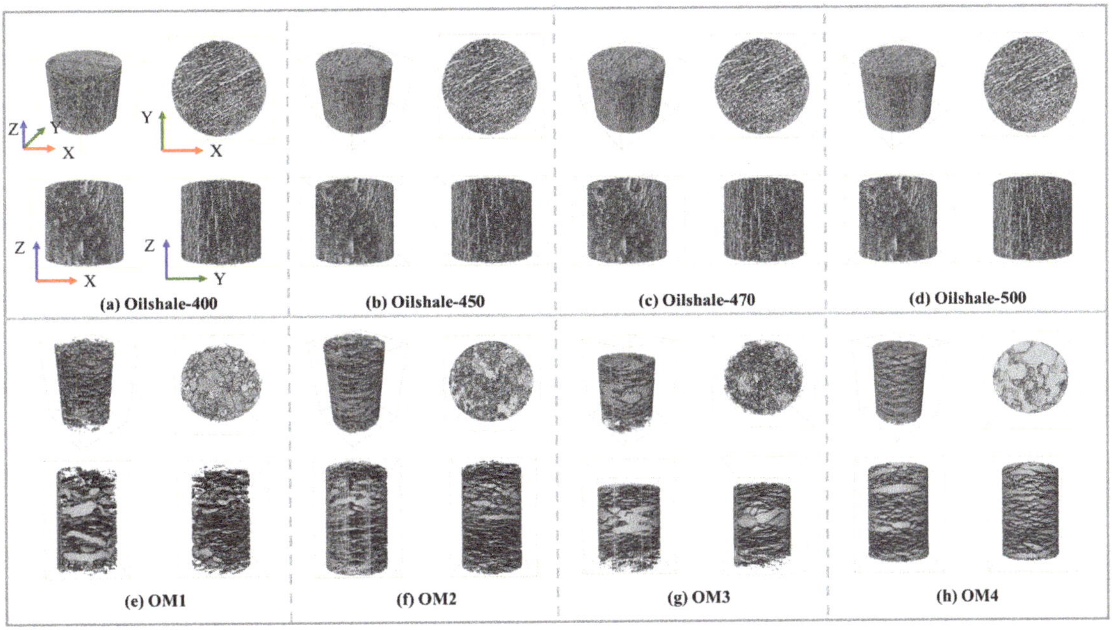

Figure 2. 3D geometry of the eight samples: (**a**) Oilshale-400 °C, (**b**) Oilshale-450 °C, (**c**) Oilshale-470 °C, (**d**) Oil shale-500 °C, (**e**) OM1, (**f**) OM2, (**g**) OM3, (**h**) OM4.

Due to the limitations of computers, it is almost impossible to simulate the entire model region. Moreover, it is not necessary to solve for the entire region, and usually, the Representative Elementary Volume (REV) is used for the calculation. However, it is difficult to find the REV that can fully characterize rock microstructure, especially for oil shale with strong heterogeneity. Based on the performance of the computer, in this study, the cubic consisting of 400 × 400 × 400 voxels was selected for the REV. The specific process of graphics processing and model construction is illustrated in Figure 3. Firstly, the target region is determined. Subsequently, the target region's pores and solid areas are distinguished using a binarized image, thus enabling the construction of the required digital oil shale core model. For ease of result observation, any additional pads and surrounding walls can be trimmed, followed by the threshold setting to obtain pore spaces and solid spaces within the model.

Table 1. Porosity and size of rocks in the CT images.

Name	Width	Height	Number of Images	Porosity
Oilshale-400/450/470/500	1280	1280	1020	21.9%/23.8%/24.1%/24.9%
OM1	1311	1388	1790	12.1%
OM2	1302	1295	1790	19.1%
OM3	1426	1285	1790	23.0%
OM4	1412	1400	1789	29.3%

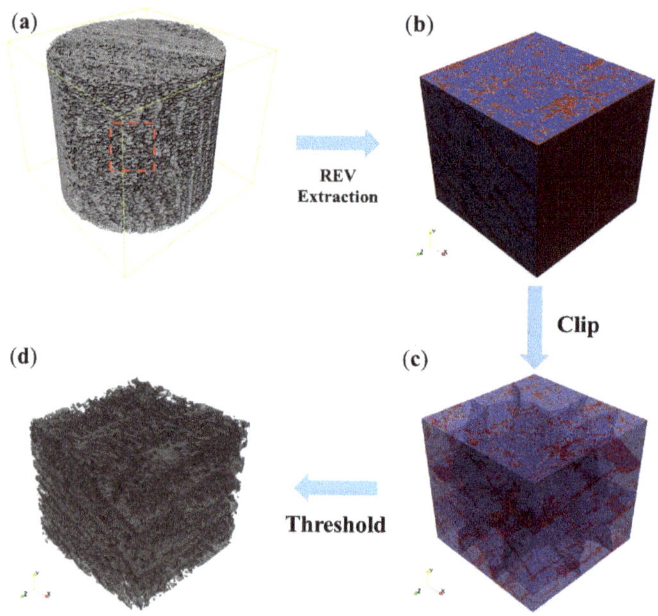

Figure 3. Construct the REV from CT images and image processing procedure: (**a**) binary image of samples, (**b**) extraction of the REV based on binary image, (**c**) structure trimming for convenient display, (**d**) adjust the threshold to obtain solid regions.

Then the porosity of the REV was calculated by GeoChemFoam-5.0 software. GeoChemFoam is an OpenFOAM-based package developed to perform reactive multiphase transport simulation at the micro-scale [30]. For oil shale samples with different pyrolysis temperatures, the central part was chosen as the REV because this region has similar porosity to the whole sample, as shown in Figure 4. The error is 7.17%, 7.61%, 7.26% and 1.12%, respectively, indicating that this calculation method is effective. It has to be mentioned that, the porosity agreement is only the preliminary criteria in the fluid flow and heat transfer studies [33]. Notably, for samples with varying organic matter content, the pyrolysis of organic matter transforms the volume it originally occupied into pore volume, resulting in a highly heterogeneous nature of the rock samples. This high heterogeneity makes it challenging for the selection of REV units to accurately reflect the true physical properties of the rock samples. However, this selection method of the REV does not affect the analysis of the impact patterns of organic matter.

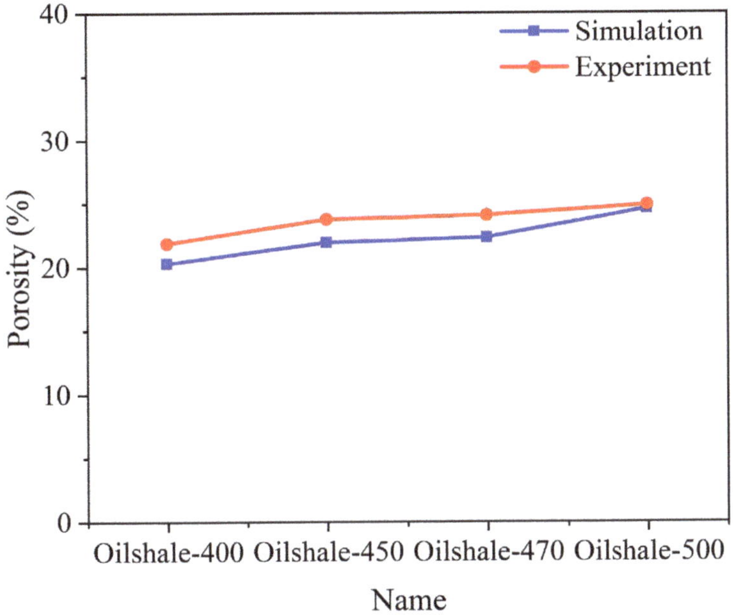

Figure 4. Comparsion of porosity between REV and original sample.

2.3. Simulation Parameters

To reveal the effects of the pyrolysis temperature on the physical properties of the oil shale and the heat transfer between CO_2 and the oil shale, we calculated the porosity, permeability and heat transfer coefficient between CO_2 and oil shale in the X, Y, and Z directions for each rock sample. Because the size of the REV is 400^3 voxels with a resolution of 2 μm, its real physical size is 800 μm^3. The REV was first meshed with a 100^3 cells uniform cartesian mesh, then we selected two-level refinement, which means there are 64 million cells in the whole region and the mesh resolution is also 2 μm. At the same time, we set the porosity in the solid region to 0.0001, which indicates there is only heat conduction within the solid domain and no fluid flow. A buffer of length 10 μm was added to the inlet to facilitate the injection at constant velocity.

CO_2 was injected from the inlet boundary at constant velocity u_{CO_2} and constant temperature $T_{inlet} = T_{CO_2} = 333$ K. The wall boundaries were also constant temperature $T_{walls} = 393$ K and no-flow condition. The outlet boundary was set to no temperature gradient and constant pressure $P = 0$. The specific boundary conditions can be found in Figure 5. Under conditions of 10 MPa pressure and 333 K temperature, the properties of CO_2 were considered constant and specific parameter values can be obtained from the https://webbook.nist.gov/chemistry/ (accessed on 21 May 2024). Numerous studies have demonstrated that, in the case of oil shale, the physical and thermal properties are primarily influenced by many factors such as sand content, mineral composition, temperature, and others [34–37]. The thermal conductivity of rock exhibits anisotropy, with significant variations observed between directions perpendicular and parallel to the bedding plane. The thermal conductivity of the shale samples in the literature [34] is mainly distributed between 1.00 and 2.50 W/m/K. The specific heat capacity of oil shale, calculated to be 1.36 kJ/g/K, was derived using the formula $C = 1.54e^{-8}T^3 - 1.30e^{-5}T^2 + 4.35e^{-3}T + 1.00$ as stated in the literature [37] and the density can be obtained from https://www.aqua-calc.com/page/density-table/substance/shale-coma-and-blank-solid (accessed on 21 May 2024). The specific parameters of the physical and thermal properties are shown in Table 2.

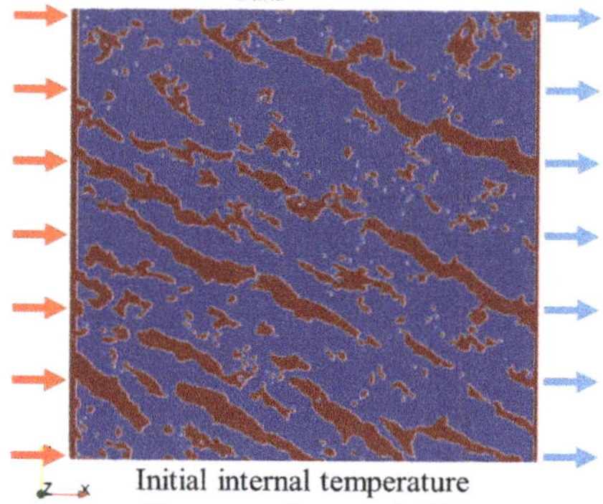

Figure 5. The simulated boundary conditions: the red indicates pore regions, and the blue indicates solid regions.

Table 2. CO_2 and oil shale physical and thermal properties.

Properties	CO_2	Oil Shale
Density (kg/m^3)	290.81	2675.00
Dynamic viscosity (m^2/s)	8.10×10^{-8}	
Specific heat capacity (kJ/g/K)	3.05	1.36
Thermal conductivity (W/m/K)	0.04	1.85

In this study, the *simpleDBSFoam* and *heatTransportFoam* solvers developed by Maes et al. [30] were used to solve the flow and heat transfer equations in GeoChem-Foam software (latest v. 5.0). To mimic laminar flow within the reservoir, we regulated the injection velocity ($u_{CO_2} = 1 \times 10^{-9}$ m^3/s) to ensure that the Reynolds number (Re) remains below 1. The time step was 2×10^{-4} s and the total simulation time was 1.5 s, which proved sufficient to reach steady state temperature.

3. Results and Discussion

3.1. Influence of Pyrolysis Temperature

Velocity fields and streamlines. Figure 6 depicts the pore structure of oil shale treated at different pyrolysis temperatures and streamline of CO_2 flowing along the X, Y, and Z directions. For clarity purposes, the color bar for the velocity field is set from 1×10^{-9} to 1×10^{-2} m/s. However, please note that this does not imply the actual minimum velocity is 1×10^{-9} and the maximum velocity is 1×10^{-2} during the flow process. As the pyrolysis temperature increases, more organic matter in the oil shale samples decomposes to form pores, resulting in an increase in the porosity of the oil shale samples. However, this increase is not significant, making it almost imperceptible from the pore structure, as depicted in Figure 6. Based on the distribution of velocity fields and streamlines in different directions, it is evident that when CO_2 flows through interconnected pore spaces, the flow velocity is relatively high with minimal variation. However, when CO_2 infiltrates into isolated pore spaces from matrix the flow velocity is extremely low. By comparing streamlines of CO_2 flow in the same direction under different pyrolysis temperature conditions, such as the X direction, it is noted that the streamline distribution remains similar regardless of temperature variation. This suggests that for relatively homogeneous oil shale samples, pyrolysis treatment has minimal impact on the original pore structure network of the oil shale, and does not significantly affect the direction of fluid flow. The primary mechanism lies in decomposing organic matter, enlarging the existing pore volume, thereby influencing the velocity of fluid flow. By observing the results for different flow directions under the same pyrolysis temperature conditions, such as in Figure 6a, we find that different directions of CO_2 injection lead to variations in the position of the minimum flow velocity point and changes in the difference in flow velocity. This is due to alterations in the connectivity of pores in different directions and changes in flow resistance.

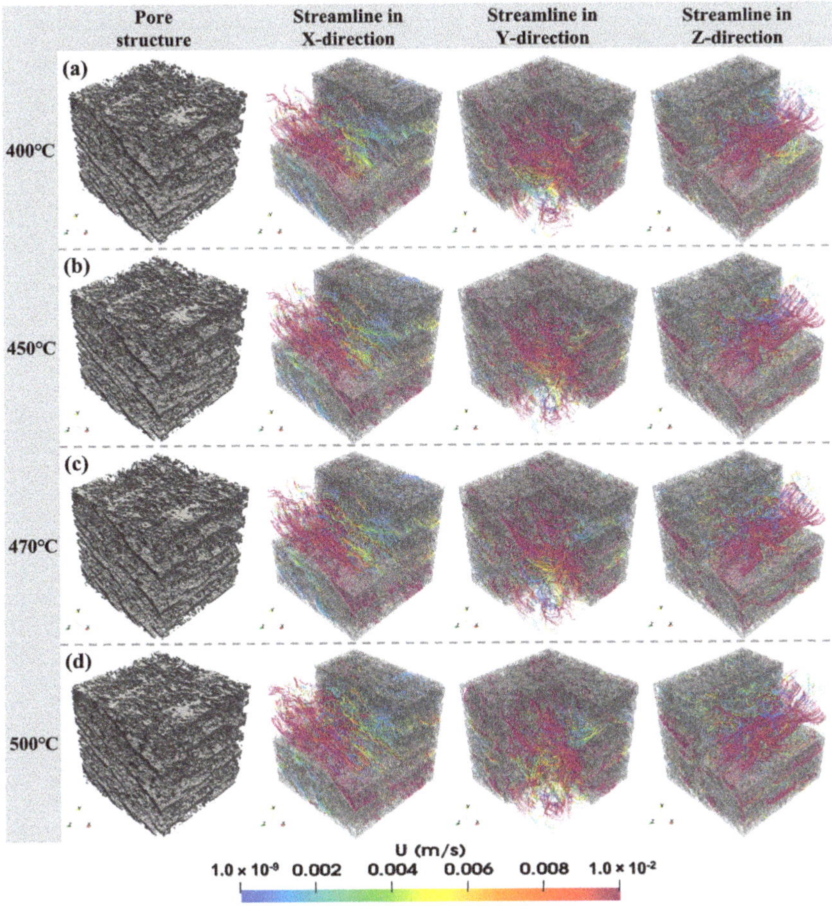

Figure 6. Pore structure of oil shale treated at different pyrolysis temperature and streamlines of CO_2 flowing along the X, Y, and Z directions: (**a**) Oilshale—400, (**b**) Oilshale—450, (**c**) Oilshale—470, (**d**) Oilshale—500.

Permeability. To further investigate the impact of pyrolysis temperature on the fluid flow characteristics in oil shale, the gas permeability of oil shale samples in various directions was calculated under different pyrolysis temperature conditions, as shown in Figure 7. It can be observed that with the increase in pyrolysis temperature, the permeability in all three directions shows a monotonically increasing trend. This is consistent with the principle that the velocity of fluid flowing parallel to the bedding plane of the oil shale is greater than that flowing perpendicular to the bedding plane. When the pyrolysis temperature increases from 400 °C to 500 °C, the permeability in the X, Y, and Z directions increases from 0.541, 0.0395, 0.913 μm^2 to 0.870, 0.135, 1.284 μm^2, respectively, with percentage increases of 60.81 %, 241.77%, and 40.64%, respectively. This indicates that the impact of pyrolysis temperature on the permeability of oil shale is most significant in the direction perpendicular to the bedding plane. It further demonstrates that with the increase in pyrolysis temperature, the size of pores and microfractures in the oil shale samples increases, effectively enlarging the flow path for oil and gas and enhancing their seepage capacity within the matrix.

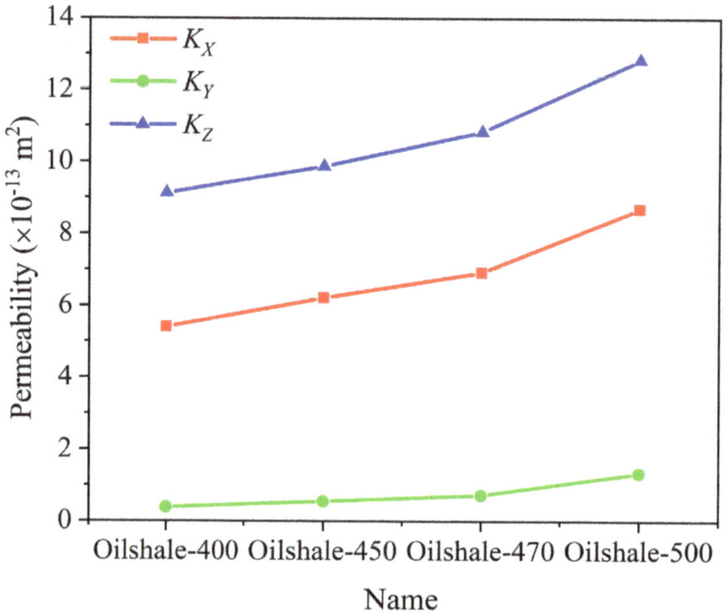

Figure 7. Permeability in the X, Y, and Z directions of oil shale samples treated at different pyrolysis temperature.

Temperature fields. Based on the aforementioned flow field distribution characteristics, the heat transfer during the CO_2 flow in the X, Y and Z directions within the oil shale samples was solved. During the flow of CO_2 in the oil shale samples, the main heat transfer mechanism can be divided into three parts: heat conduction and convection between CO_2 fluids, heat conduction and convection between CO_2 and oil shale, and heat conduction within the solid regions of the oil shale. The temperature fields of the pore regions and the solid regions are shown in Figures 8 and 9, respectively. Since the temperature of the injected CO_2 is lower than that of the rock, the temperature field distribution gradually decreases from the inlet to the outlet along the flow direction. By combining Figures 6 and 8, it can be observed that points with lower temperatures in the temperature field correspond to regions with higher velocities in the velocity field. This indicates that as the flow velocity increases, heat transfer between CO_2 fluids also increases. For the temperature field inside oil shale (as shown in Figure 9), since the heat transfer mechanism within oil shale relies solely on thermal conduction, the heat transfer efficiency is much lower than that of thermal convection. It primarily depends on the position of the fluid–solid interface and the temperature gradient, and is less influenced by the CO_2 flow velocity. Therefore, the temperature distribution within the solid oil shale is relatively uniform. Similar to Figure 6, because the initial organic matter distribution in the oil shale sample is relatively uniform, the pyrolysis process results in minimal changes in the pore space distribution characteristics. Consequently, the temperature field distribution along the same flow direction, as shown in Figures 8a and 9a, exhibits no significant contrast. Therefore, the average temperature and the average temperature difference between CO_2 and oil shale were calculated to further elucidate the flow heat transfer characteristics.

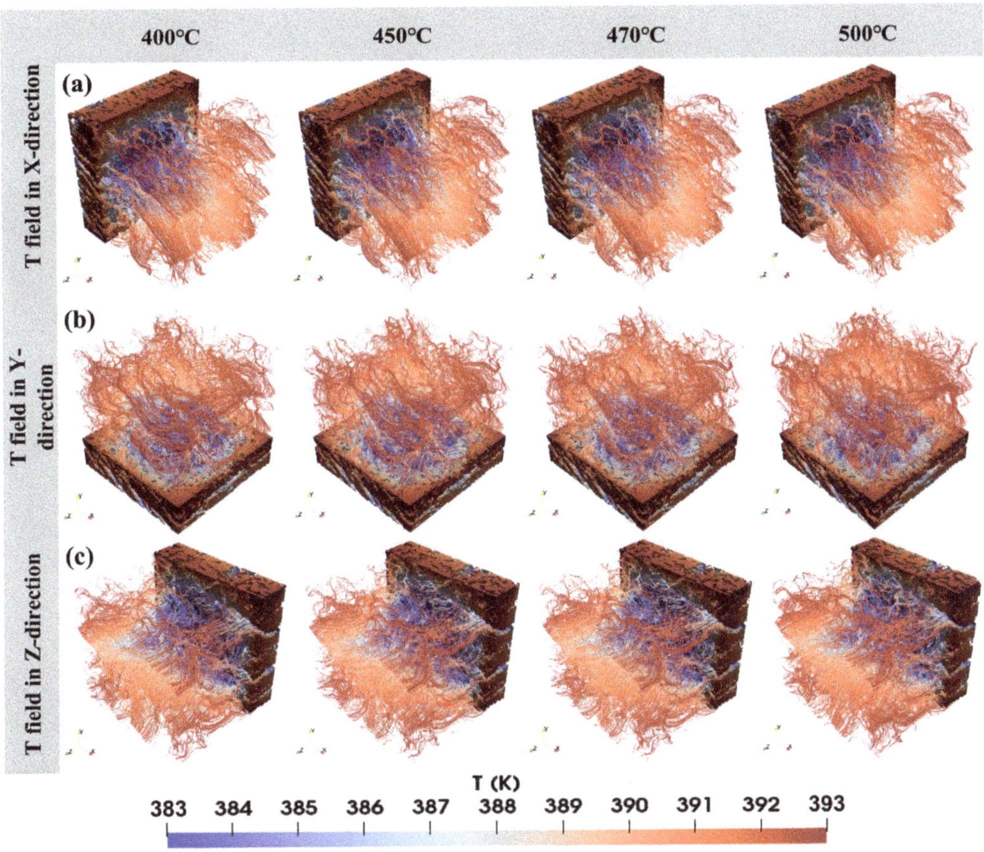

Figure 8. The temperature distribution along streamlines in the pore regions of oil shale treated at different pyrolysis temperature when CO_2 flows along the X, Y, and Z directions, respectively: (**a**) temperature field in X-direction, (**b**) temperature field in Y-direction, (**c**) temperature field in Z-direction.

Average temperature and heat transfer coefficient. Figure 10 illustrates the average temperature of CO_2 fluid and oil shale solid during CO_2 flow in different directions (Figure 10a) and the heat transfer coefficients (Figure 10b). Regardless of the flow direction, the average temperature of the oil shale solid region is always higher than that of the CO_2 fluid, consistent with the heat transfer mechanism between the fluid and solid. As the pyrolysis temperature increases, both the fluid and solid regions exhibit a decreasing trend in average temperature, albeit to a minimal extent, which indicates that under low flow velocity conditions, the pyrolysis temperature has a minor impact on heat transfer between CO_2 and oil shale. For instance, when the pyrolysis temperature increases from 400 °C to 500 °C, the average temperatures in the fluid region of CO_2 flowing along the X, Y, and Z directions decrease from 386.146, 386.959, and 386.681 K to 386.078, 386.908, and 387.364 K, with decreases of 0.018%, 0.013%, and 0.037%, respectively. The average temperatures in the solid region also decrease from 387.648, 388.328, and 387.552 K to 387.364, 388.007, and 387.393 K, respectively, with decreases of 0.073%, 0.083%, and 0.041%. When CO_2 flows vertically to the bedding direction of the oil shale, i.e., the Y direction, the average temperature decrease in the fluid region is minimal compared to the other two flow directions, while the decrease in the solid region is the largest. This is consistent with the results of Figure 7, indicating that the pyrolysis temperature has the greatest impact on the fluid when flowing perpendicular to the bedding direction of the oil shale.

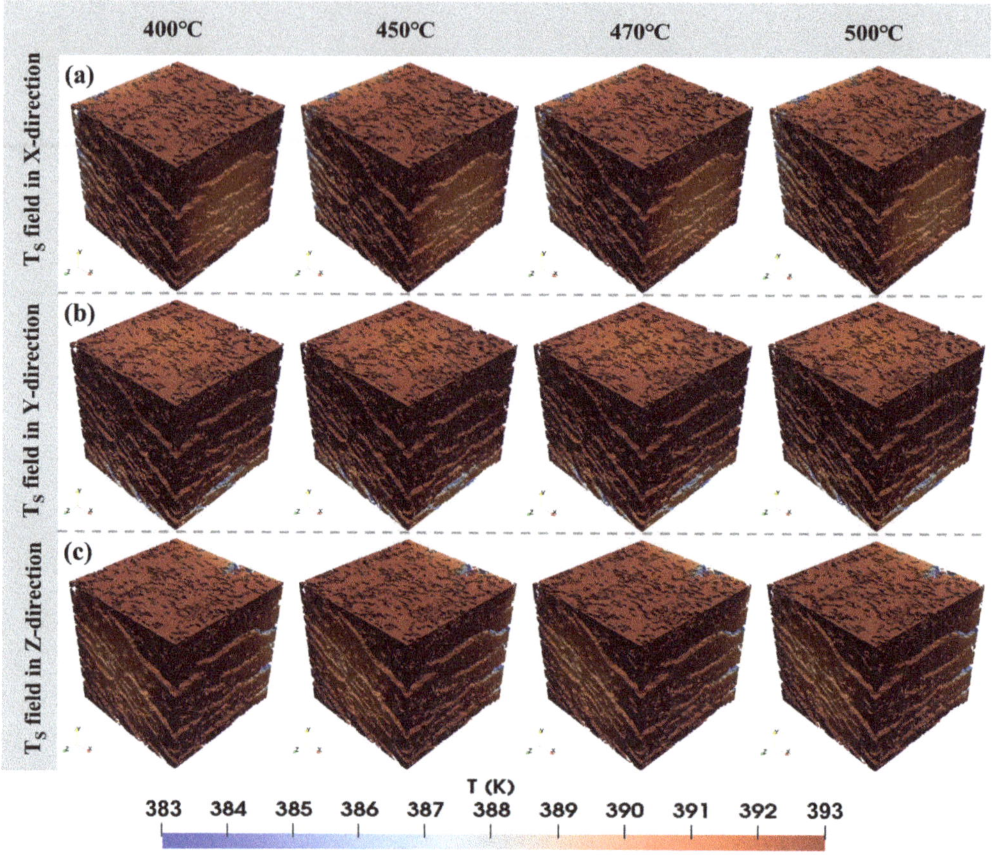

Figure 9. The temperature distribution in the solid regions of oil shale treated at different pyrolysis temperatures when CO_2 flows along the X, Y, and Z directions, respectively: (**a**) temperature field in X-direction, (**b**) temperature field in Y-direction, (**c**) temperature field in Z-direction.

From Figure 10b, it can be seen that when CO_2 flows perpendicular to the bedding direction of the oil shale, the heat transfer coefficient between CO_2 and oil shale increases with rising pyrolysis temperatures, i.e., from 0.0421 to 0.0470 $W/(m^2{\cdot}K)$. This is mainly because pyrolysis creates fractures and pores within the oil shale, primarily forming between the bedding layers, which increases the porosity in the vertical direction and enhances CO_2 permeability and heat transfer. Additionally, the changes in pore structure in the vertical direction facilitate CO_2 convection and diffusion effects, further increasing the heat transfer coefficient.

However, when CO_2 flows parallel to the bedding direction of the oil shale, the heat transfer coefficient decreases or remains nearly unchanged with increasing pyrolysis temperatures, i.e., from 0.0672 to 0.0610 $W/(m^2{\cdot}K)$. This can be attributed to the insulating effect of the bedding layers, which have low thermal conductivity, limiting heat transfer. Additionally, the pyrolysis reactions may be more concentrated within the interlayer gaps, causing localized heat accumulation that is not easily transferred to other areas parallel to the bedding. The heat transfer path parallel to the bedding is longer and more complex, potentially containing thermal resistance that lowers heat transfer efficiency. During pyrolysis, this phenomenon may be more pronounced, resulting in little to no change in the heat transfer coefficient. In summary, the increase in heat transfer coefficient during vertical flow and the decrease or constancy during parallel flow is due to the combined effects of increased porosity, changes in thermal conductivity, enhanced convection and diffusion, as well as the insulating effect of the bedding layers, localized pyrolysis reactions, and the restricted heat transfer path.

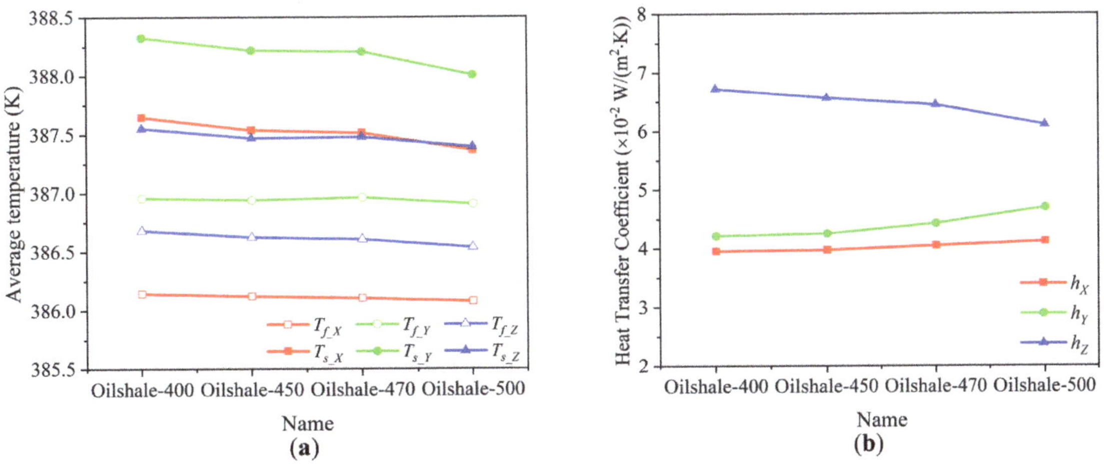

Figure 10. The average temperature and heat transfer coefficient between CO_2 and rock samples when CO_2 flows along the X, Y, and Z directions in oil shale treated at different pyrolysis temperature: (**a**) the average temperature of CO_2 and oil shale samples at steady-state, (**b**) the heat transfer coefficient.

3.2. Influence of Organic Matter Content

Velocity fields and streamlines. Due to the strong heterogeneity of oil shale samples with different organic matter contents, the pore space formed after pyrolysis also exhibits high heterogeneity, as shown by the pore structure in Figure 11. It is important to note that the organic matter content in the original samples increases gradually from OM1 to OM4, which means that after pyrolysis, the porosity of the rock samples should also gradually increase. However, due to the heterogeneity of the samples and the limitations in selecting REVs, the porosity of the selected REVs does not show a gradual increase. Nonetheless, this does not affect the study of the impact of organic matter content on the CO_2 flow and heat transfer mechanisms in oil shale samples, as the pyrolysis of organic matter primarily affects the pore distribution characteristics of the primary pores. The organic matter completely decomposes after pyrolysis, forming large cavities. Some of these cavities are interconnected, enhancing CO_2 flow capability when passing through these pores, while others are isolated, preventing CO_2 fluid from entering. Unlike the previously mentioned oil shale samples treated at different pyrolysis temperatures, the pore distribution characteristics of rock samples with different organic matter contents vary, resulting in significant differences in velocity distribution among the samples, as well as noticeable differences in velocity distribution in different flow directions. For example, comparing the streamlines in Figure 11d when CO_2 flows along the X direction and the Z direction reveals that the velocity of CO_2 flow in the pores in the lower left corner is significantly greater along the Z direction than along the X direction. This indicates that the connectivity of the pores in the Z direction is stronger than in the X direction.

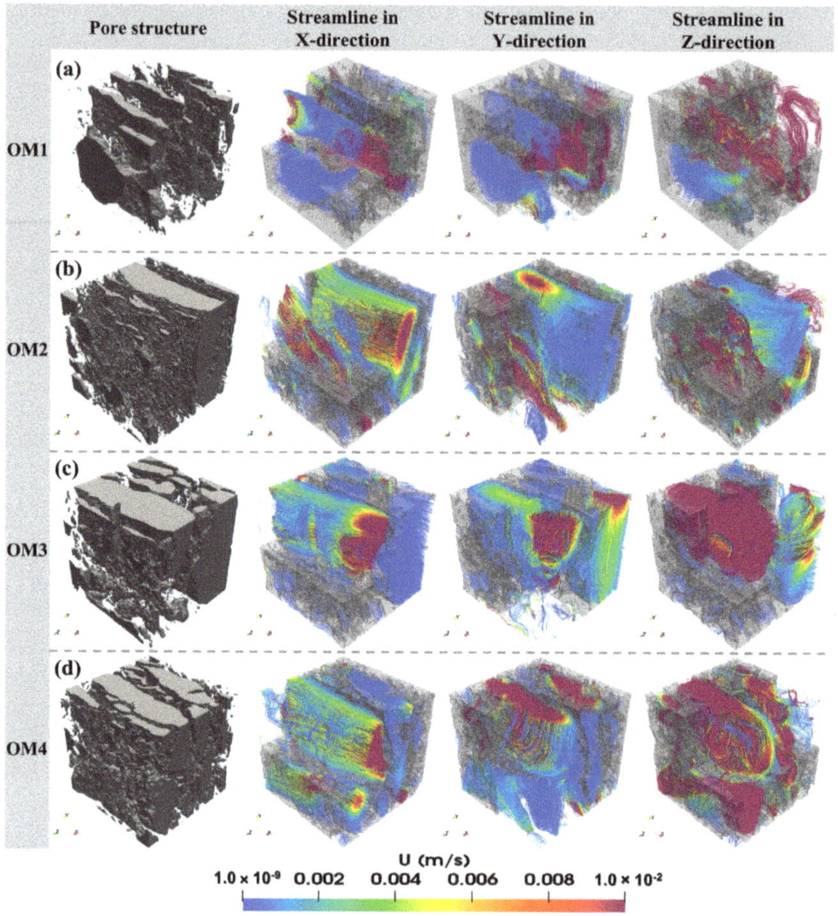

Figure 11. Pore structure of oil shale with different organic matter content treated at 500 °C and streamlines of CO_2 flowing along the X, Y, and Z directions: (**a**) OM1, (**b**) OM2, (**c**) OM3, (**d**) OM4.

Permeability. To further compare the CO_2 flow characteristics in rock samples with different organic matter contents, the permeability of CO_2 flow in three directions was calculated, as shown in Figure 12. Similar to the flow patterns in oil shale samples treated at different pyrolysis temperatures, the permeability of CO_2 flow parallel to the bedding is greater than that perpendicular to the bedding. However, due to the heterogeneity of the rock samples, the changes in permeability are not monotonic. For the same rock sample, the permeability in the three directions may differ slightly, such as in sample OM1 where the permeabilities in the X, Y, and Z directions are 0.896, 0.134, and 0.0003 μm^2, respectively. Conversely, the differences can be significant, as in sample OM2 where the permeabilities in the X, Y, and Z directions are 22.024, 6.669 μm^2, and 8.588 nm^2, respectively, differing by several orders of magnitude. In all four rock samples, the permeability in the direction perpendicular to the bedding, i.e., the Z direction, is extremely low. This indicates that the bedding planes are almost non-contiguous, the original organic matter distribution is relatively isolated, and the pore spaces generated after pyrolysis do not interconnect, thus failing to effectively enhance the permeability of the rock samples.

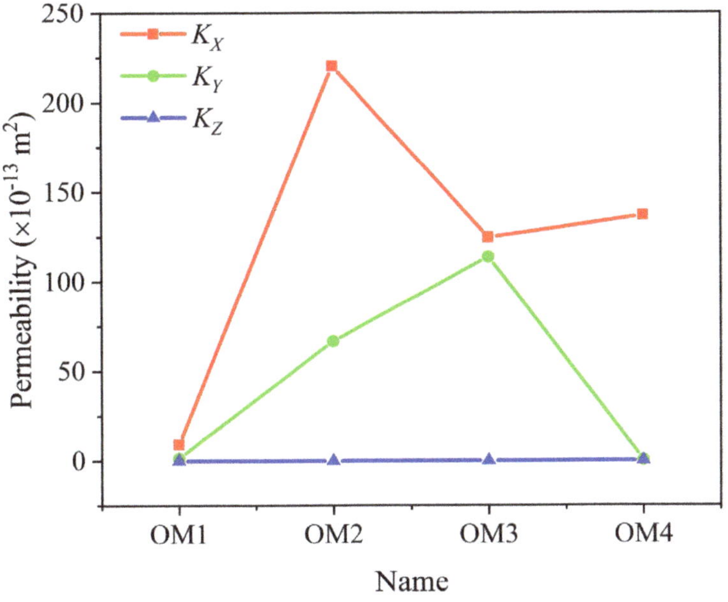

Figure 12. Permeability in the X, Y, and Z directions of oil shale with different organic matter content treated at 500 °C.

Temperature fields. The temperature distribution along streamlines in the pore regions and solid regions of oil shale with different organic matter content treated at 500 °C when CO_2 flows along the X, Y, and Z directions are shown in Figures 13 and 14. From Figure 13a,b, particularly evident in samples OM2, OM3, and OM4, CO_2 flow generates preferential pathways where convective heat transfer between CO_2 molecules is intensified. Consequently, the temperature distribution within the samples exhibits similar characteristics to those observed in these preferential pathways. Due to the poor connectivity between bedding planes in each rock sample, although there are large cavities present in the samples, which facilitates the flow and heat transfer of CO_2. However, when CO_2 flows perpendicular to the bedding direction, it cannot enter these cavities, nor can convective heat transfer occur between them. CO_2 can only rely on heat conduction between CO_2 and the rock (Figure 13c). This results in minimal temperature changes in these cavities, which are almost equal to the initially set temperature. This further illustrates that for the flow and heat transfer problems in rock samples with different organic matter contents, the characteristics of pore structure are crucial factors influencing the fluid flow and heat transfer between fluid–fluid and fluid–solid interfaces. The influence of organic matter content on flow and heat transfer in rock samples essentially stems from the effect of pyrolysis on the characteristics of the original pore network in the rock samples, which is highly correlated with the initial distribution of organic matter and exhibits strong heterogeneity. For instance, if the connectivity of the initial organic matter within the rock specimen is robust, the resultant porosity interconnectivity following the pyrolysis of organic matter will likewise be robust. This can improve fluid flow to some extent, reduce fluid flow resistance, increase fluid flow velocity, and enhance convective heat transfer between fluid–fluid and fluid–solid interfaces, thereby improving heat transfer efficiency. Conversely, even if the organic matter content is high and the organic matter structure is large and intact, if the individual organic matter units are not connected, after pyrolysis, although the porosity of the rock sample can be increased, effective fluid flow channels cannot be formed. Fluids cannot enter, and heat transfer between fluid-fluid and fluid–solid interfaces within the pores can only rely on heat conduction, as shown in Figure 14, which typically has much lower efficiency than convective heat transfer. This results in lower heat transfer efficiency between rock samples. Similar to the previous approach, when heat transfer reaches a steady state, the average temperatures of both the CO_2 and the solid rock samples were calculated, as depicted in Figure 15.

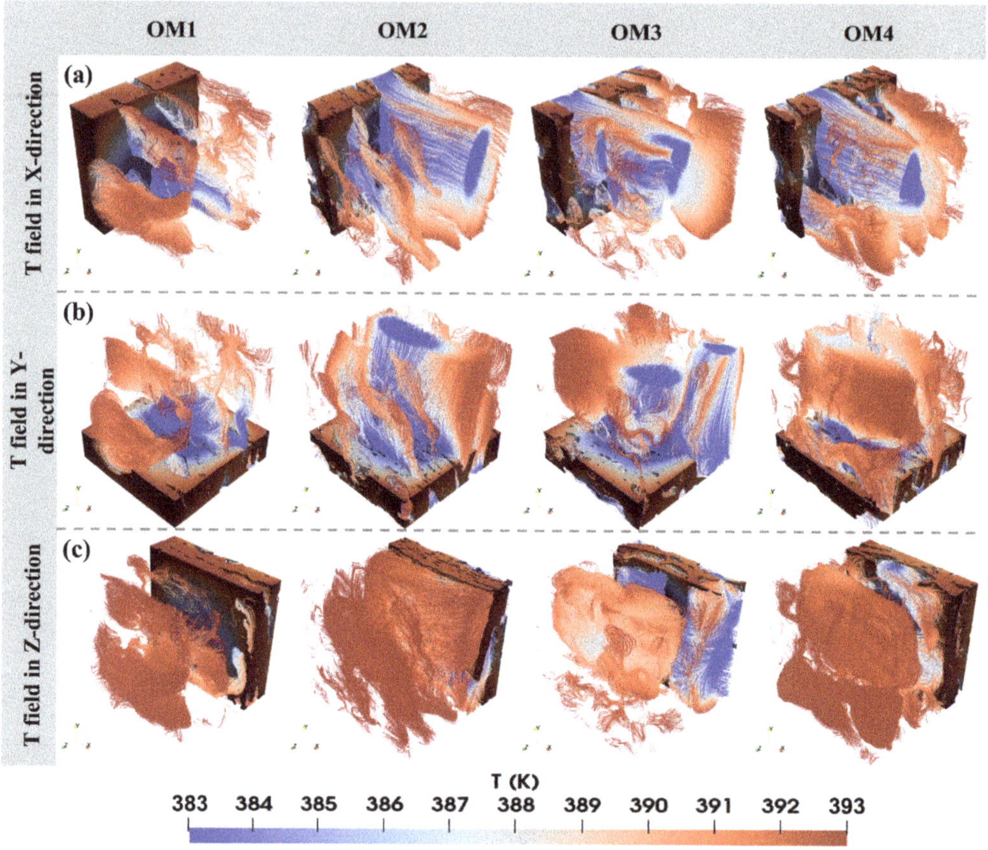

Figure 13. The temperature distribution along streamlines in the pore regions of oil shale with different organic matter content treated at 500 °C when CO_2 flows along the X, Y, and Z directions, respectively, (**a**) temperature field in X-direction, (**b**) temperature field in Y-direction, (**c**) temperature field in Z-direction.

Average temperature and heat transfer coefficient. The average temperatures of CO_2 and the rock solid, as well as the heat transfer coefficients between CO_2 and the rock in rock samples with different organic matter contents, have been calculated, as shown in Figure 15. For CO_2 flow along the Z-direction, the average temperatures of CO_2 in samples OM1 to OM4 are 390.404 K, 386.680 K, 382.145 K, and 388.872 K, respectively, while the average temperatures of the rock solid are 389.660 K, 389.185 K, 388.056 K, and 388.241 K, respectively. Unlike the results shown in Figure 10, when CO_2 flows perpendicular to the bedding planes, the average temperature of the fluid in samples OM1 and OM4 exceeds that of the rock. This indicates that in these two samples, the pyrolysis of organic matter resulted in the formation of numerous isolated pores. These pores are dispersed among the bedding planes and are not interconnected, failing to provide effective channels for fluid flow. Consequently, the CO_2 within these pores cannot engage in convective heat transfer with the flowing CO_2 and relies solely on heat conduction between the rock and CO_2. This results in the average temperature of the CO_2 approaching the initial temperature of the rock, leading to very low heat transfer efficiency.

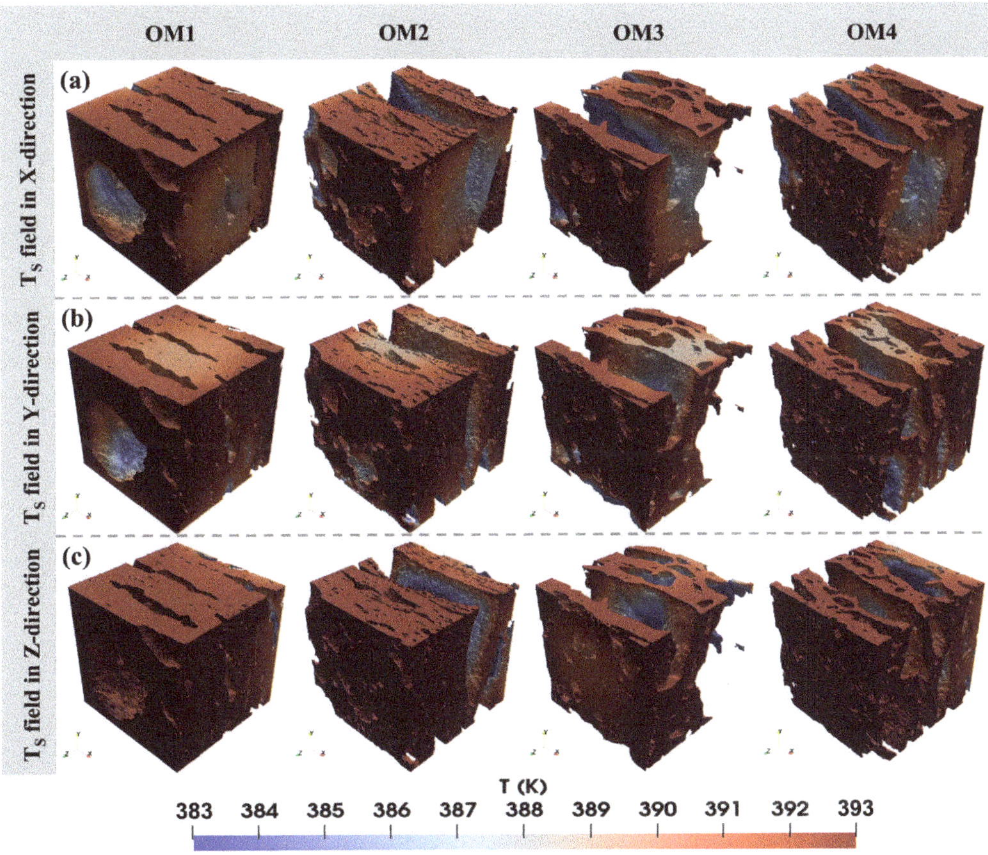

Figure 14. The temperature distribution in the solid regions of oil shale with different organic matter content treated at 500 °C when CO_2 flows along the X, Y, and Z directions, respectively: (**a**) temperature field in X-direction, (**b**) temperature field in Y-direction, (**c**) temperature field in Z-direction.

Figure 15a shows that the average temperature variation of CO_2 in the four rock samples with different organic matter contents is more pronounced than that of the rock solid. This indicates that the changes in the pore structure of the rock caused by the pyrolysis of organic matter have a greater impact on convective heat transfer between fluids than on heat transfer between the fluid and the rock. Furthermore, this suggests that in samples with higher organic matter content, the development and connectivity of the pore network play a significant role in CO_2 flow and its heat transfer efficiency. Additionally, the heat transfer coefficients between CO_2 and the rock were calculated for CO_2 flowing in the X, Y, and Z directions in rock samples with different organic matter contents, as shown in Figure 15b. For different rock samples, it can be observed that the heat transfer coefficient between CO_2 and the rock is almost maximal in the direction perpendicular to the bedding planes. This phenomenon can be attributed to the poor flow capacity of CO_2 in this direction, thereby predominating its thermal transfer with the rock. Conversely, in the other two directions, where the pore space connectivity is better, CO_2

has stronger flow capacity, and convective heat transfer between CO_2 predominates. This results in relatively lower heat transfer between CO_2 and the rock, leading to a decrease in the heat transfer coefficient.

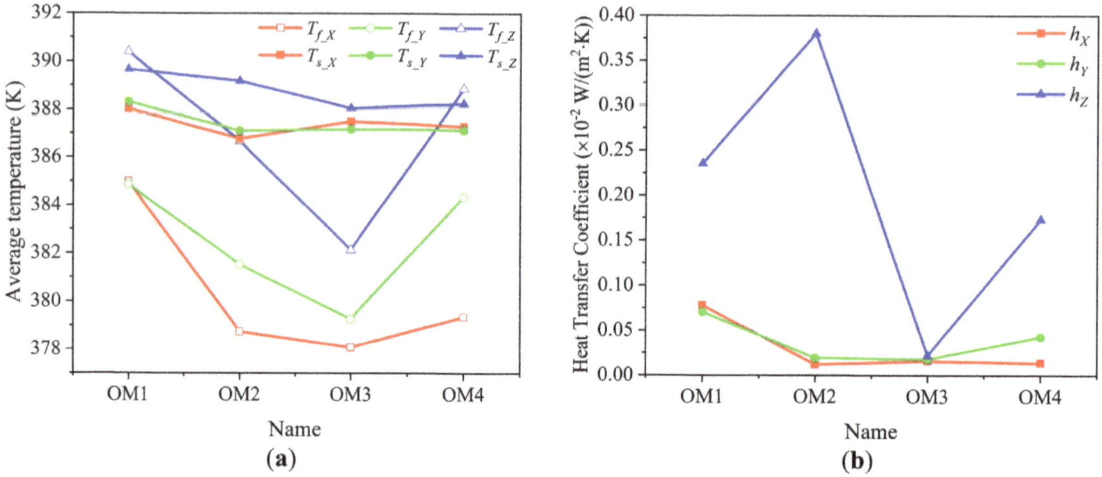

Figure 15. The average temperature and heat transfer coefficient between CO_2 and rock samples when CO_2 flows along the X, Y, and Z directions in oil shale with different organic matter content treated at 500 °C: (**a**) the average temperature of CO_2 and oil shale samples at steady-state, (**b**) the heat transfer coefficient.

4. Conclusions

This study provides an in-depth analysis of the fluid flow and heat transfer characteristics of oil shale samples under different pyrolysis temperatures and organic matter contents. Through investigations into velocity fields, permeability, temperature fields, average temperatures, and heat transfer coefficients, several conclusions can be made:

- The permeability of oil shale samples shows a consistent increase with rising pyrolysis temperatures. While temperature minimally impacts the distribution of velocity and temperature fields, it significantly affects the heat transfer coefficient. Specifically, as temperature increases, the heat transfer coefficient markedly rises perpendicular to the bedding planes but decreases or remains constant parallel to them. This phenomenon is due to changes in the rock's pore structure from organic matter pyrolysis, which influences fluid flow and heat transfer properties.
- Organic matter content significantly influences the fluid flow and heat transfer properties of shale samples. Comparing shale samples with varying organic matter contents after thermal treatment shows that differences in organic matter lead to heterogeneity in pore structure, affecting fluid flow and heat transfer characteristics. After pyrolysis, the complete decomposition of organic matter creates large cavities, altering the native pore distribution. If these cavities are interconnected, they enhance fluid flow capacity and improve convective heat transfer. However, if the cavities are isolated, fluids cannot penetrate despite the increased porosity, and heat transfer relies solely on conduction between the fluid and the solid.
- The influence of organic matter content and pyrolysis temperature on fluid flow and heat transfer in shale is primarily due to the effect of organic matter pyrolysis on the native pore structure. The development and connectivity of the pore network are closely linked to the distribution characteristics of the native organic matter, rather than the quantity of organic matter present.

In summary, our study reveals the significant influence of pyrolysis temperature and organic matter content on the fluid flow and heat transfer characteristics of oil shale samples. These findings provide important theoretical guidance and technical support for the development and utilization of oil shale resources, while also offering valuable insights for future research.

Author Contributions: Conceptualization, Y.S. and D.W.; methodology, B.C. and Y.Z. (Yunpeng Zhang); software, Y.Z. (Yunpeng Zhang); validation, Y.Z. (Yunpeng Zhang) and Y.S.; formal analysis, B.W.; investigation, Y.Z. (Yaochen Zhang); resources, H.W.; data curation, D.W.; writing—original draft preparation, Y.S.; writing—review and editing, B.W., D.W. and Y.S. All authors have read and agreed to the published version of the manuscript.

Funding: This research was funded by the Science Foundation of Key Laboratory of Reservoir Stimulation CNPC (RIPED.CN-2023-JS-432).

Data Availability Statement: The data presented in this study are available on request from the corresponding author.

Conflicts of Interest: The authors declare no conflicts of interest.

References

1. Caineng, Z.; Qun, Z.; Hongyan, W.; Qian, S.; Nan, S.; Zhiming, H.; Chenjun, Z.; Yuping, S.; Meizhu, W.; Dexun, L. Theory and Technology of Unconventional Oil and Gas Exploration and Development Helps China Increase Oil and Gas Reserves and Production. *Oil Forum* **2021**, *40*, 72.
2. Wang, J.; Feng, L.; Steve, M.; Tang, X.; Gail, T.E.; Mikael, H. China's Unconventional Oil: A Review of Its Resources and Outlook for Long-Term Production. *Energy* **2015**, *82*, 31–42. [CrossRef]
3. Zheng, D.; Miska, S.; Ozbayoglu, E.; Zhang, J. Combined Experimental and Well Log Study of Anisotropic Strength of Shale. In Proceedings of the SPE Annual Technical Conference and Exhibition, San Antonio, TX, USA, 16–18 October 2023.
4. Zheng, D.; Ozbayoglu, E.; Miska, S.; Zhang, J. Experimental Study of Anisotropic Strength Properties of Shale. In Proceedings of the 57th U.S. Rock Mechanics/Geomechanics Symposium, Atlanta, GA, USA, 25–28 June 2023.
5. Lei, Y.; Zhijun, J. Global Shale Oil Development and Prospects. *China Pet. Explor.* **2019**, *24*, 553.
6. Hu, S.; Zhao, W.; Hou, L.; Yang, Z.; Zhu, R.; Wu, S.; Bai, B.; Jin, X. Development Potential and Technical Strategy of Continental Shale Oil in China. *Pet. Explor. Dev.* **2020**, *47*, 877–887. [CrossRef]
7. Zhao, W.; Hu, S.; Hou, L.; Yang, T.; Li, X.; Guo, B.; Yang, Z. Types and Resource Potential of Continental Shale Oil in China and Its Boundary with Tight Oil. *Pet. Explor. Dev.* **2020**, *47*, 1–11. [CrossRef]
8. Wang, L.; Yang, D.; Kang, Z. Evolution of Permeability and Mesostructure of Oil Shale Exposed to High-Temperature Water Vapor. *Fuel* **2021**, *290*, 119786. [CrossRef]
9. Knyazeva, A.G.; Maslov, A.L.; Martemyanov, S.M. A Two-Phase Model of Shale Pyrolysis. *Fuel* **2018**, *228*, 132–139. [CrossRef]
10. Skala, D.; Kopsen, H.; Sokić, M.; Neumann, H.J.; Jovanović, J. Modelling and Simulation of Oil Shale Pyrolysis. *Fuel* **1989**, *68*, 168–173. [CrossRef]
11. Wang, G.; Yang, D.; Kang, Z.; Zhao, J.; Lv, Y. Numerical Investigation of the in Situ Oil Shale Pyrolysis Process by Superheated Steam Considering the Anisotropy of the Thermal, Hydraulic, and Mechanical Characteristics of Oil Shale. *Energy Fuels* **2019**, *33*, 12236–12250. [CrossRef]
12. Korba, D.; Li, L. Effects of Pore Scale and Conjugate Heat Transfer on Thermal Convection in Porous Media. *J. Fluid Mech.* **2022**, *944*, A28. [CrossRef]
13. Li, X.; Ma, T.; Liu, J.; Zhang, H.; Wang, Q. Pore-Scale Investigation of Gravity Effects on Phase Change Heat Transfer Characteristics Using Lattice Boltzmann Method. *Appl. Energy* **2018**, *222*, 92–103. [CrossRef]
14. Paknahad, R.; Siavashi, M.; Hosseini, M. Pore-Scale Fluid Flow and Conjugate Heat Transfer Study in High Porosity Voronoi Metal Foams Using Multi-Relaxation-Time Regularized Lattice Boltzmann (MRT-RLB) Method. *Int. Commun. Heat Mass Transf.* **2023**, *141*, 106607. [CrossRef]
15. Rabbani, A.; Baychev, T.G.; Ayatollahi, S.; Jivkov, A.P. Evolution of Pore-Scale Morphology of Oil Shale During Pyrolysis: A Quantitative Analysis. *Transp. Porous Media* **2017**, *119*, 143–162. [CrossRef]
16. Wang, H.; Chen, L.; Qu, Z.; Yin, Y.; Kang, Q.; Yu, B.; Tao, W.Q. Modeling of Multi-Scale Transport Phenomena in Shale Gas Production—A Critical Review. *Appl. Energy* **2020**, *262*, 114575. [CrossRef]
17. Zhang, L.; Li, P.; Chen, J. Pore-Scale Conjugate Heat Transfer of Nanofluids within Fibrous Medium with a Double MRT Lattice Boltzmann Model. *Int. J. Therm. Sci.* **2021**, *163*, 106804. [CrossRef]
18. Kopanidis, A.; Theodorakakos, A.; Gavaises, E.; Bouris, D. 3D Numerical Simulation of Flow and Conjugate Heat Transfer through a Pore Scale Model of High Porosity Open Cell Metal Foam. *Int. J. Heat Mass Transf.* **2010**, *53*, 2539–2550. [CrossRef]
19. Liu, Z.; Yang, Y.; Zhang, Q.; Imani, G.; Zhang, L.; Sun, H.; Zhong, J.; Zhang, K.; Yao, J. Pore-Scale Flow Simulation of CO_2 Sequestration in Deep Shale Based on Thermal-Hydro-Mechanical Coupled Model. *Phys. Fluids* **2024**, *36*, 022005. [CrossRef]

20. Meinicke, S.; Dubil, K.; Wetzel, T.; Dietrich, B. Characterization of Heat Transfer in Consolidated, Highly Porous Media Using a Hybrid-Scale CFD Approach. *Int. J. Heat Mass Transf.* **2020**, *149*, 119201. [CrossRef]
21. Meinicke, S.; Dietrich, B.; Wetzel, T. Pore-scale numerical analysis of hydrodynamics and conjugate heat transfer in solid sponges. In *International Heat Transfer Conference Digital Library*; Begel House Inc.: Danbury, CT, USA, 2018.
22. Meinicke, S.; Wetzel, T.; Dietrich, B. Scale-Resolved CFD Modelling of Single-Phase Hydrodynamics and Conjugate Heat Transfer in Solid Sponges. *Int. J. Heat Mass Transf.* **2017**, *108*, 1207–1219. [CrossRef]
23. Hugo, J.M.; Topin, F.; Tadrist, L.; Brun, E. From Pore Scale Numerical Simulation of Conjugate Heat Transfer in Cellular Material to Effectives Transport Properties of Real Structures. In Proceedings of the 2010 14th International Heat Transfer Conference, Washington, DC, USA, 8–13 August 2010; American Society of Mechanical Engineers Digital Collection: New York, NY, USA, 2011; pp. 931–936.
24. Li, J.; Zhang, P.; Lu, S.; Chen, C.; Xue, H.; Wang, S.; Li, W. Scale-Dependent Nature of Porosity and Pore Size Distribution in Lacustrine Shales: An Investigation by BIB-SEM and X-Ray CT Methods. *J. Earth Sci.* **2019**, *30*, 823–833. [CrossRef]
25. Liu, Z.; Wu, H. Pore-Scale Study on Flow and Heat Transfer in 3D Reconstructed Porous Media Using Micro-Tomography Images. *Appl. Therm. Eng.* **2016**, *100*, 602–610. [CrossRef]
26. Lu, Y.; Lu, T.; Lu, Y.; Wang, B.; Zeng, G.; Zhang, X. The Study on Solving Large Pore Heat Transfer Simulation in Malan Loess Based on Volume Averaging Method Combined with CT Scan Images. *Sustainability* **2023**, *15*, 12389. [CrossRef]
27. Sun, M.; Hu, C.; Zha, L.; Xie, Z.; Yang, L.; Tang, D.; Song, Y.; Zhao, J. Pore-Scale Simulation of Forced Convection Heat Transfer under Turbulent Conditions in Open-Cell Metal Foam. *Chem. Eng. J.* **2020**, *389*, 124427. [CrossRef]
28. Tembely, M.; AlSumaiti, A.M.; Jouini, M.S.; Rahimov, K. The Effect of Heat Transfer and Polymer Concentration on Non-Newtonian Fluid from Pore-Scale Simulation of Rock X-Ray Micro-CT. *Polymers* **2017**, *9*, 509. [CrossRef]
29. Tiwari, P.; Deo, M.; Lin, C.L.; Miller, J.D. Characterization of Oil Shale Pore Structure before and after Pyrolysis by Using X-Ray Micro CT. *Fuel* **2013**, *107*, 547–554. [CrossRef]
30. Maes, J.; Menke, H.P. Geochemfoam: Direct modelling of flow and heat transfer in micro-ct images of porous media. *Heat Mass Transf.* **2022**, *58*, 1937–1947. [CrossRef]
31. Saif, T.; Lin, Q.; Singh, K.; Bijeljic, B.; Blunt, M.J. Dynamic imaging of oil shale pyrolysis using synchrotron X-ray microtomography. *Geophys. Res. Lett.* **2016**, *43*, 6799–6807. [CrossRef]
32. Saif, T.; Lin, Q.; Bijeljic, B.; Blunt, M.J. Microstructural imaging and characterization of oil shale before and after pyrolysis. *Fuel* **2017**, *197*, 562–574. [CrossRef]
33. Liu, J.; Yu, P.; Li, Y.; Wan, C.; Du, D. Numerical simulation on convective heat transfer characteristics in porous media based on the digital rock technology. *Int. J. Heat Mass Transf.* **2022**, *196*, 123323. [CrossRef]
34. Xie, F.; Zhu, C.; Tang, B. An experimental study on anisotropy of thermal conductivity in shale. *Energy Geosci.* **2023**, *4*, 100195. [CrossRef]
35. Cheng, C.; Lin, H.; Jiang, Y.; Feng, L.; Xia, Y.; Mu, C. Thermal conductivity of gas-bearing shale of the Longmaxi Formation in the southern Sichuan. *Pertol. Geol.* **2019**, *41*, 289–294.
36. Chen, C.; Zhu, C.; Tang, B.; Chen, T. Progress in the study of the influencing factors of rock thermal conductivity. *Prog. Geophys.* **2020**, *35*, 2047–2057.
37. Liu, Z.; Ma, H.; Wang, Z.; Guo, Y.; Li, W.; Hou, Z. Experimental study on the thermophysical properties of Jimsar oil shale. *Oil Shale* **2023**, *40*, 194–211. [CrossRef]

processes

MDPI

Article

Numerical Simulation of the Interaction Between Hydraulic Fracture and the Bedding Plane in Shale Formation

Heng Zheng [1,2,3,*], Haibo Wang [1,2], Fengxia Li [1,2] and Ning Li [1,2]

[1] State Key Laboratory of Shale Oil and Gas Enrichment Mechanisms and Effective Development, Beijing 100083, China
[2] State Energy Center for Shale Oil Research and Development, Beijing 102206, China
[3] School of Petroleum Engineering, Yangtze University, Wuhan 430100, China
* Correspondence: hengsmith111@163.com

Abstract: Hydraulic fracturing was the primary technology for shale oil and gas development. Creating a huge stimulated reservoir volume was essential to improving shale oil and gas production. Based on the CDEM, a 3D hydraulic fracturing numerical model was proposed to simulate the interaction between hydraulic fracture and natural fracture. The simulation indicated that net pressure decrease was the main factor affecting fracture propagation in the vertical direction. Under a high stress difference, when the stress difference is over 9.0 MPa, the propagation resistance of rock matrix is higher than that of the bedding plane, thus the bedding plane has a huge opening, and the hydraulic fracture was arrested by the bedding plane. For high formation permeability, increasing the viscosity of the fracturing fluid can effectively improve the fracture height, which can accelerate the hydraulic fracture across the bedding plane. Improving the injection rate also benefitted crossing the bedding plane. This research has great significance in improving the stimulated reservoir volume.

Keywords: bedding plane; minimum horizontal principle stress difference; permeability; hydraulic fracture

1. Introduction

Shale reservoirs have emerged as primary targets for unconventional oil and gas resource development in recent years [1–3]. These reservoirs typically exhibit low porosity, low permeability, and distinctive heterogeneity, characterized by laminated structures and natural fractures. To enhance oil recovery in shale reservoir reconstruction, widespread adoption of the large-scale hydraulic fracturing technique is employed [4,5]. However, studies indicate that geological discontinuities, such as bedding planes, exert complex influences on the fracture morphology.

The bedding planes denote discontinuities developed in formations due to variations in the sedimentary environment. Owing to the intricate sedimentary environments and tectonic movements, occurrences of bedding planes are diverse, and various classification criteria for bedding planes were proposed based on different standards, such as the composition, texture, and structure [6–8]. The composition refers to the mineral components of the bedding planes; the texture refers to the distributions and connections of the organic matter and pores within the bedding planes; and the structure denotes the continuities, shapes, and geometrical relationships of the bedding planes. Among these criteria, structure differences are deemed the most obvious based on the observations of typical shale samples.

Particularly, continuities between the bedding planes and the rock matrix are critical factors affect fracture propagation and should be emphasized in fracturing design [9].

The assessment of hydraulic fracturing under the influence of bedding planes is challenging. A comprehensive understanding of the underlying law of fracture propagation is crucial to optimizing construction parameters and enhancing stimulated reservoir volume (SRV). True tri-axial hydraulic fracturing experiment is a widely adopted approach to investigate fracture propagation in shale [10,11]. Based on the tri-axial hydraulic fracturing experiment, the experimental results indicated that four different fracture geometries was found, namely simple fracture, fishbone-like fracture, fishbone-like fracture with fissure opening, and a multilateral fishbone-like fracture network. The experimental results also indicated that in situ stress, the injection rate, and fluid viscosity had a great effect on fracture geometry. The results showed that low-viscosity fracturing fluid and periodic variable pump rate fracturing are preferable for generating a complex fracture network [12–17]. Various numerical approaches have been employed, including the finite element method, the finite difference method, the phase field method, and the discrete element method (DEM). However, numerical methods based on the continuous medium face challenges in accurately characterizing the heterogeneity of rock samples, such as natural fractures and bedding planes. The dense, discontinuous surface seriously impacts the stability and convergence of the simulation. In contrast, the DEM, based on the discontinuous medium, excels in characterizing rock materials as aggregations of discrete elements, showcasing its superiority in modeling hydraulic fracturing in heterogeneous conditions [18–20]. Specifically, DEM can arbitrarily define natural fractures and bedding planes among particle aggregations without additional computational consumption. Some research proposed fracture propagation models in shale reservoirs based on DEM and discussed fracture behavior in the discontinuous medium. Few studies considered different types of bedding planes and investigated their influences on the shale rocks. Liu et al. conducted the uniaxial compressive tests on rock samples containing bedding planes with different cementation types, namely embedded cementation, oriented cementation, and cracked cementation, and measured the compressive strengths under various conditions [11,21–23]. However, fracture propagation and morphology in shale formations containing different types of bedding planes have rarely been reported, prompting the need for further investigation to fill this blank.

To further understand the hydraulic fracture propagation in shale formation, a 3D numerical model based on CDEM was developed to simulate hydraulic fracturing in shale containing various types of bedding planes. Strategies employed to replicate the mechanical and hydraulic properties of different bedding planes were introduced. Subsequently, fracturing simulations were executed on shale models with different types of bedding planes to scrutinize the impact of bedding plane type on fracture propagation and morphology. Additionally, the effects of multiple influential factors, including bedding plane thickness, scale, permeability, orientation angles, fracturing fluid properties, and injection schemes on fracture propagation were discussed. The comprehensive comparison of the distinct influence of different types of the bedding planes under various conditions can enhance the understanding of fracture propagation in shale reservoirs, and facilitate the optimization of fracturing plans.

2. Mathematical Methods

2.1. Boundary Conditions

As indicated in Figure 1, the whole domain was divided into two parts, $\Omega = \Omega^+ \cup \Omega^-$, and whole domain is bounded by $\Gamma = \Gamma_t \cup \Gamma_u \cup \Gamma_d$. The normal vector n_d^+ indicates the direction of Ω^+ relative to the joint plane.

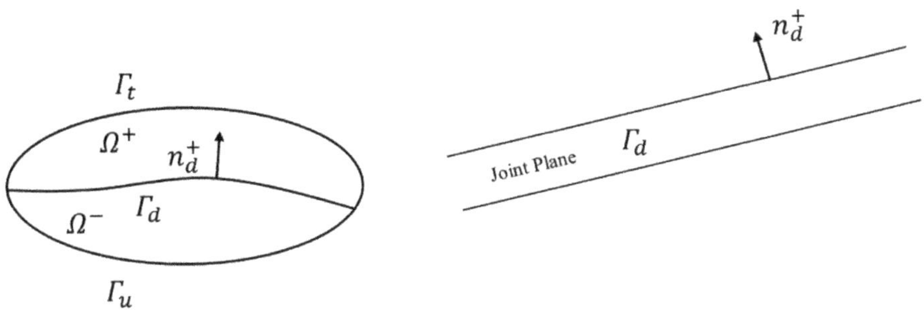

Figure 1. Domain with single joint plane, with jump in the velocity field.

The classical strong forms of strong discontinuity equations are presented in [19] as follows:

$$\rho i = \frac{1}{\rho}\nabla \cdot \sigma + g \tag{1}$$

$$u(x,t) = \overline{u}(x,t) \text{ on } \Gamma_u \tag{2}$$

$$\sigma \cdot \hat{n} = t^* \text{ on } \Gamma_t \tag{3}$$

$$\dot{\sigma}^{+} \cdot n_d^{+} = \dot{\sigma}^{-} \cdot n_d^{-} = \dot{t}_c \text{ on } \Gamma_d \tag{4}$$

where t represents the traction, ρ represents the mass density, u represents the velocity, σ denotes the stress tensor, g represents the body force, and n represents the outward normal vector of the domain.

2.2. Governing Equations

During the simulations, the fracturing fluid was subject to mass and momentum conservation laws. These are expressed in the Lagrangian system as follows. The constitutive stress–strain relationship in the rock has also been given in [20].

$$\sigma^{\alpha\beta} = \frac{1}{3}\sigma^{rr}\delta^{\alpha\beta} + s^{\alpha\beta} = -p\delta^{\alpha\beta} + s^{\alpha\beta} \tag{5}$$

where p represents the pore pressure, $\delta^{\alpha\beta}$ denotes the Kronecker delta, and $\dot{\varepsilon}_p^{\alpha\beta}$ describes the second-order deviatoric stress tensor.

The results of the elastic and plastic strain calculations are as follows.

$$\dot{\varepsilon}^{\alpha\beta} = \frac{1}{2}\left(\frac{\partial v^{\alpha}}{\partial x^{\beta}} + \frac{\partial v^{\beta}}{\partial x^{\alpha}}\right) \tag{6}$$

$$\dot{\varepsilon}^{\alpha\beta} = \dot{\varepsilon}_{\varepsilon}^{\alpha\beta} + \dot{\varepsilon}_{p}^{\alpha\beta} \tag{7}$$

where $\dot{\varepsilon}_{\theta}^{\alpha\beta}$ represents the elastic strain, and $\dot{\varepsilon}_{p}^{\alpha\beta}$ denotes the plastic strain.

The generalized Hook's law was used to determine the stress rate of the tensor [21].

$$\dot{\sigma}_e^{\alpha\beta} = D_{\alpha\beta mn}\dot{\varepsilon}_e^{mn} = 2G\dot{\varepsilon}_{ed}^{\alpha\beta} + K\dot{\varepsilon}_{ev}^{rr}\delta^{\alpha\beta} \tag{8}$$

where K represents the bulk moduli, $D_{\alpha\beta mn}$ represents the standard isotropic elastic tensor of the fourth order, $\dot{\varepsilon}_{ed}^{\alpha\beta}$ represents the rate of deviatoric components of the elastic strain tensor, G represents the shear moduli, and $\dot{\varepsilon}_{ev}^{rr}$ represents the rate of volumetric components of the elastic strain tensor.

This simulation considered the rock as a Drucker–Prager material, with a yield criterion that is represented by a circular cone surface and is given by

$$F(\sigma, c) = \sqrt{J_2(s)} + \eta p(\sigma) - \xi c \tag{9}$$

and

$$\eta = \frac{6 \sin \phi}{\sqrt{3}(3 - \sin \phi)}, \; \xi = \frac{6 \cos \phi}{\sqrt{3}(3 - \sin \phi)}$$

where ϕ and c represent the angle of friction and the cohesion of the material, respectively.

2.3. Damage Evaluation

For the quantitative evaluation of rock damage during hydraulic fracturing, the hydraulic fracture propagation behavior is characterized by the D parameter.

$$D(t) = \frac{4}{3} \pi c_g^3 \int_0^t n'(\varepsilon) \dot{\varepsilon} (1 - D)(t - \tau)^3 d\tau \tag{10}$$

where c_g represents the fracture propagation velocity.

The simulation presented this equation was used to control the growth of damage.

$$\frac{dD^{1/3}}{dt} = \frac{1}{3} c_\xi \varpi^{1/3} \varepsilon^{m/3} \tag{11}$$

where $\varpi = \frac{8\pi k (m+3)^2}{(m+1)(m+2)}$.

A damage threshold in the following form was used to identify the damage condition.

$$F_D\left(\bar{\varepsilon}, \varepsilon_D^0\right) = \bar{\varepsilon} - \varepsilon_D^0 \tag{12}$$

$$\bar{\varepsilon} = \frac{\sigma_{\max}}{K + \frac{4}{3} G} \tag{13}$$

$$\varepsilon_D^0 = (Vk)^{-\frac{1}{m}} \tag{14}$$

where k and m represent the material properties that characterize fracture activation, and V denotes the volume of the SPH particle.

Formation permeability after hydraulic fracturing was quantified as

$$k = k_0 e^{-\beta(\sigma_1 - \alpha p)} \tag{15}$$

where k_0 represents the initial permeability; σ_1 denotes the principle stress; p describes the pore pressure; and α represents the Biot coefficient.

3. Results and Discussion

3.1. Model Description

To further understand the hydraulic fracture propagation when it came across the bedding plane in shale formation, a 3D numerical model was established which considered the leak-off of the fracturing fluid and deformation of fracture surface, and the basic parameters of the proposed model are shown in Table 1. To improve the accuracy of the proposed model, the published experimental results were used to verify the numerical simulation. From the comparison, it was found that one main fracture was created which has three branch fractures, and the simulations also had the same fracture geometry (Figure 2a). For the breakdown pressure and extension pressure, the experiment indicated that the breakdown pressure and extension pressure were 31.4 MPa and 25.3 MPa, and the breakdown pressure and extension pressure were 30.7 MPa and 24.8 MPa (Figure 2b). The main reason for this phenomenon was that the flow resistance resulted in a higher the breakdown pressure and extension pressure than that of the simulation. The error of the breakdown pressure and extension pressure were 2.23% and 1.78%, and the accuracy of the proposed model was verified.

Table 1. Key parameters used in the simulation.

Item	Parameters	Value
Stress field	Vertical stress/MPa	73.6
	Maximum horizontal stress/MPa	68.4
	Minimum horizontal stress/MPa	62.8
Rock matrix mechanical properties	Young's Modulus/GPa	36.5
	Poisson's rate	0.31
	Tensile strength/MPa	6.74
	Shear strength/MPa	7.05
Fracture mechanical properties	Young's Modulus/GPa	30.41
	Poisson's rate	0.32
	Tensile strength/MPa	5.14
	Shear strength/MPa	6.84

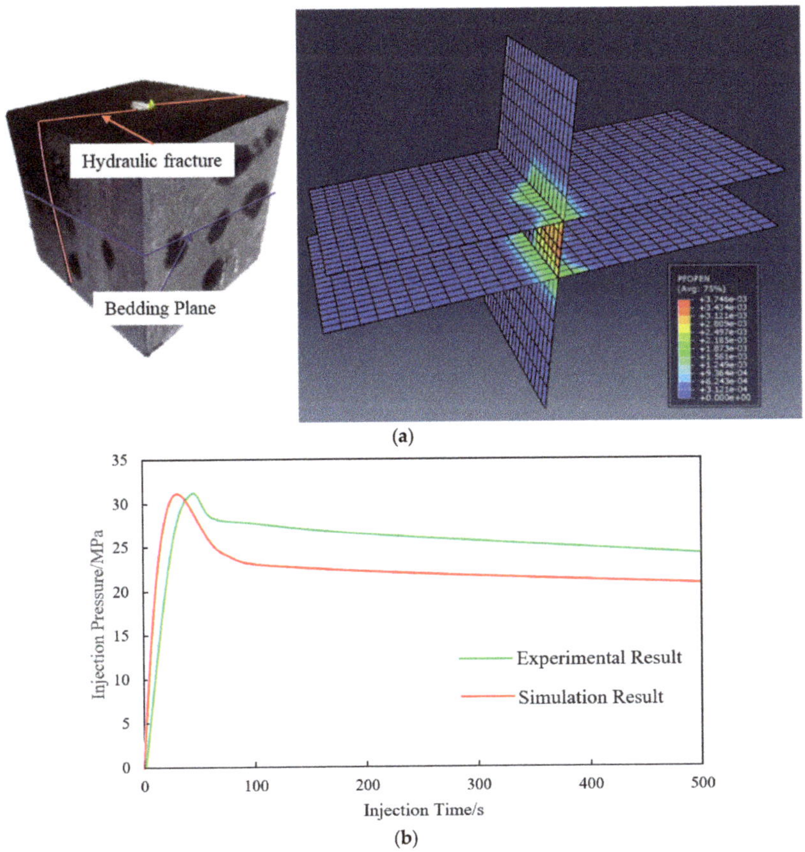

Figure 2. Efficiency verification of the proposed model between experimental result and simulation result. (**a**) Fracture geometry comparison between experimental result (left) and simulation result (right). (**b**) Injection pressure comparison between experimental result and simulation result.

3.2. Simulation and Results

3.2.1. Stress Difference on Fracture Propagation

The minimum horizontal principle stress was the main factor affecting fracture propagation in the vertical direction. To further understand the influence of the minimum horizontal principle stress difference on fracture propagation, different minimum horizontal principle stress differences with 3.0 MPa, 6.0 MPa, 9.0 MPa, 12.0 MPa, and 15.0 MPa were simulated, and the other parameters are listed in Table 1. From the simulation, it was found that the bedding plane was crossed with a length of 1.27 m and bedding plane was activated when the minimum horizontal principle stress difference was 3.0 MPa. After injection, the fracture height was 33.41m. When the stress difference increased to 6.0 MPa, the opening degree of the bedding plane showed a slight increase, increasing to 1.58 m, and the fracture height showed a decrease, decreasing to 30.58 m. When the stress difference rose to 9.0 MPa, great changes took place in the injection of the fracturing fluid. From the simulation, it was found that the opening degree showed an obvious increase as the opening degree of the bedding plane was 4.41 m. Meanwhile, the fracture height showed a huge decrease with the fracture height was 21.42 m. With the minimum horizontal principle stress difference continuously increasing, the fracture height showed a continuous decrease while the opening degree showed an obvious increase when the opening degree of the bedding plane was 5.73 m, the minimum horizontal principle stress difference was 12.0 MPa, and the fracture height decreased to 18.4m. When the minimum horizontal principle stress difference jumped to 15.0 MPa, it was found that the hydraulic fracture was arrested by the bedding plane. With continuous injection, bedding plane activation was completed, thus the fracture height was the distance between two bedding planes, a fracture height of 16.0 m (Figure 3).

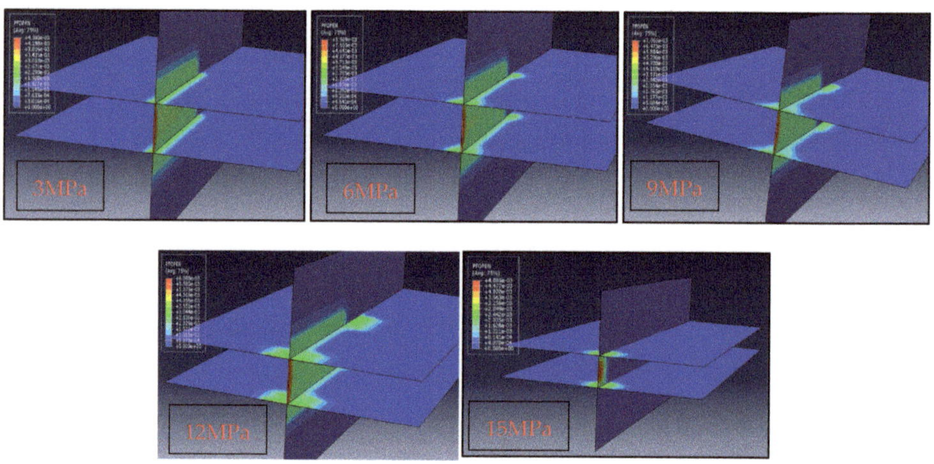

Figure 3. Fracture geometry under different minimum horizontal principle stress differences.

To further understand fracture geometry change during the whole fracturing process, the net pressure and fracture width were quantitatively analyzed. As indicated in Figure 4, it was found that net pressure showed a sharp decrease with the opening of the bedding plane. Compared to net pressure changes under five different scenarios, it was found that the net pressure showed a sharp decrease when the hydraulic fracture came across the bedding plane. It was also found that with the increase in the opening degree of the bedding plane, the net pressure also showed an obvious decrease. The fracture width decreased with the decrease in net pressure. Based on the simulation, it was concluded that the minimum horizontal principle stress difference was the predominant factor controlling fracture geometry. Under a high stress difference, when the stress difference was over 9.0 MPa, fracture propagation resistance showed a huge change during fracture propagation, and propagation resistance in the bedding plane was lower than that of the rock matrix, thus the opening degree of the bedding plane showed a great increase.

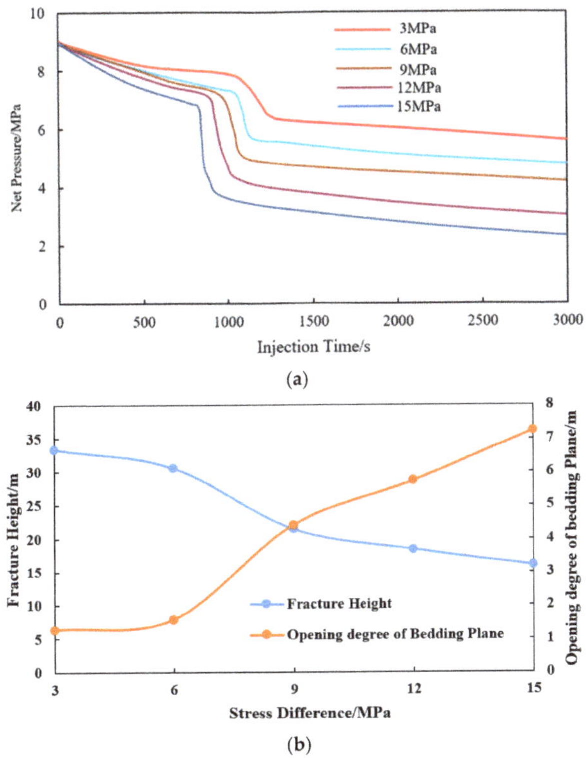

Figure 4. Net pressure and fracture dimension comparison under different minimum horizontal principle stress differences. (**a**) Net pressure change during hydraulic fracturing under different stress differences. (**b**) Fracture dimension comparison under different stress differences.

3.2.2. Permeability of Fracture Propagation

To understand hydraulic fracture propagation under different permeabilities, with a formation permeability of 0.1 mD, 1.0 mD, 10.0 mD and 20.0 mD and the viscosity of the fracturing fluid at 1.0 mPa.s, 10.0 mPa.s and 50 mPa.s, the simulation results were depicted in three groups. In Group I, a formation permeability of 0.1 mD, 1.0 mD, 10.0 mD and 20.0 mD was simulated, with the viscosity of the fracturing fluid at 1.0 mPa.s, and the other parameters as shown in Table 1. From the simulation, it was found that the hydraulic fracture crossed the bedding plane completely when the formation permeability was 0.1 mD and 1.0 mD, the opening degree of the bedding plane was 1.58 m and 2.14 m, and the fracture height decreased to 24.1 m from 30.58 m. However, the opening degree increased to 4.31 m when the formation permeability was increased to 10.0 mPa.s and the fracture height was 18.4 m. With the formation permeability increased to 20.0 mPa.s, it was found that the hydraulic fracture was arrested by the bedding plane when it came across the bedding plane, and the fracture height was 16.0 m (Figure 5).

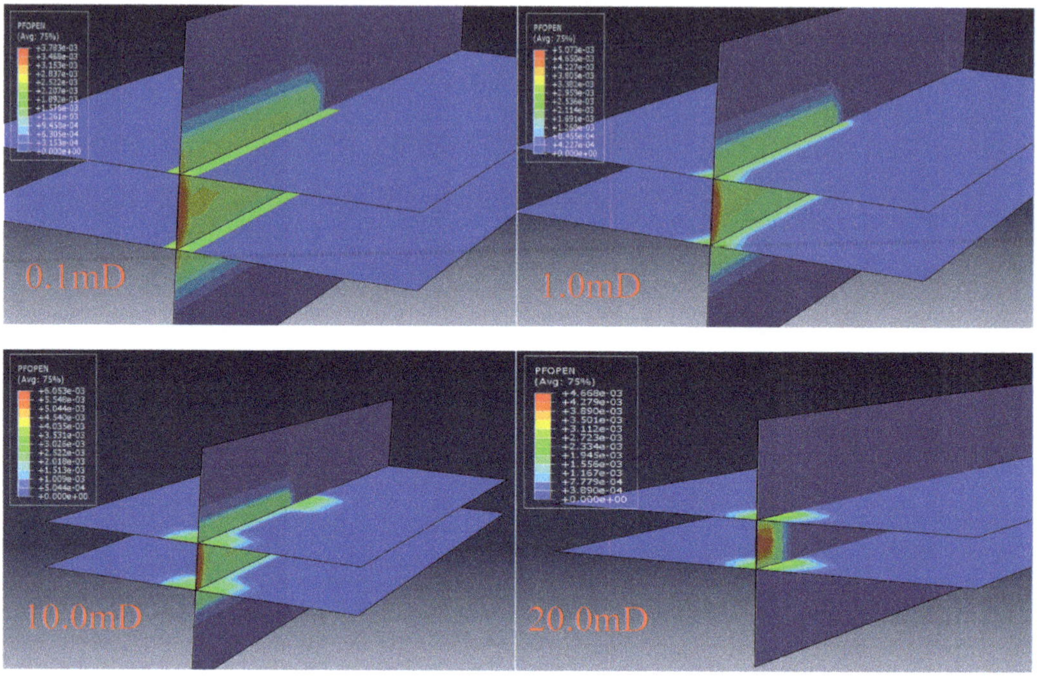

Figure 5. Fracture geometry under different formation permeabilities when the viscosity of the fracturing fluid is 1.0 mPa.s.

For Group II, the experiment was conducted with a formation permeability of 0.1 mD, 1.0 mD, 10.0 mD and 20.0 mD and the viscosity of the fracturing fluid at 10.0 mPa.s. From the simulation, it was found that with the increase in permeability, the opening degree of the bedding plane showed an obvious increase. When the formation permeability was 0.1 mD, 1.0 mD and 10.0 mD, the hydraulic fracture crossed the bedding plane completely, and it was arrested by the bedding plane when the formation permeability increased to 20.0 mD. The simulation results indicated that the fracture height showed a dramatic decrease from 31.2 m to 26.7 m, and it continuously decreased to 19.1 m. When the formation permeability increased to 20.0 mD, the hydraulic fracture height decreased to 16.0 m, and the opening degree of the bedding plane showed a huge increase from 1.24 m with a formation permeability of 0.1 mD to 5.74 m (Figure 6).

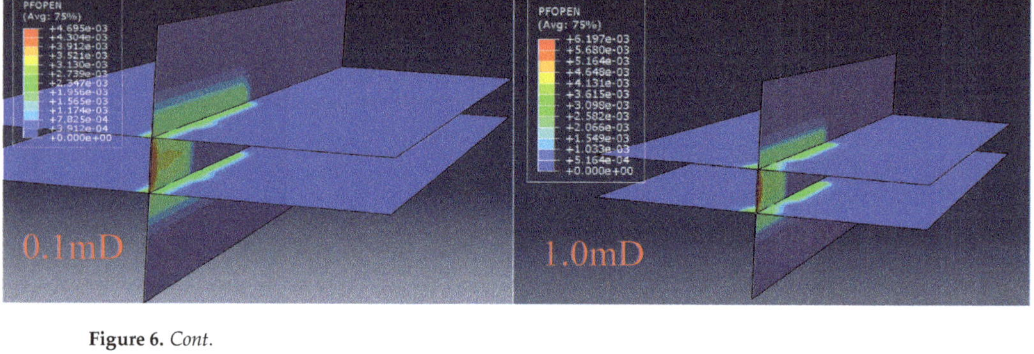

Figure 6. *Cont.*

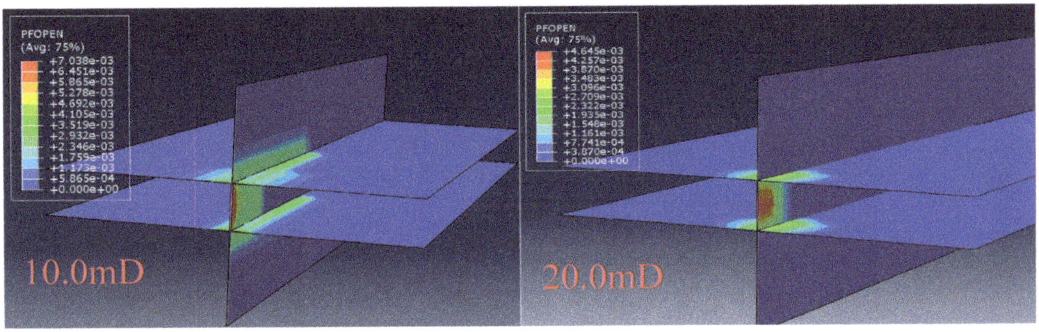

Figure 6. Fracture geometry under different formation permeabilities when the viscosity of the fracturing fluid was 10.0 mPa.s.

For Group III, the experiment was conducted with a formation permeability of 0.1 mD, 1.0 mD, 10.0 mD and 20.0 mD and the viscosity of the fracturing fluid at 50.0 mPa.s. From the simulations, it was found that all hydraulic fractures crossed the bedding planes, and the opening degree of the bedding planes increased with the increase in formation permeability. When the formation permeability was 0.1 mD, 1.0 mD, and 10.0 mD, the opening degree of the bedding planes was 0.8 m, 1.4 m and 2.1 m, and the fracture height was 33.1 m, 28.7 m, and 24.6 m. When the formation permeability increased to 20.0 mD, it was found that the fracture height showed a dramatic decrease; when the fracture height was 18.7 m, the opening degree of the bedding plane increased to 4.6 m (Figure 7).

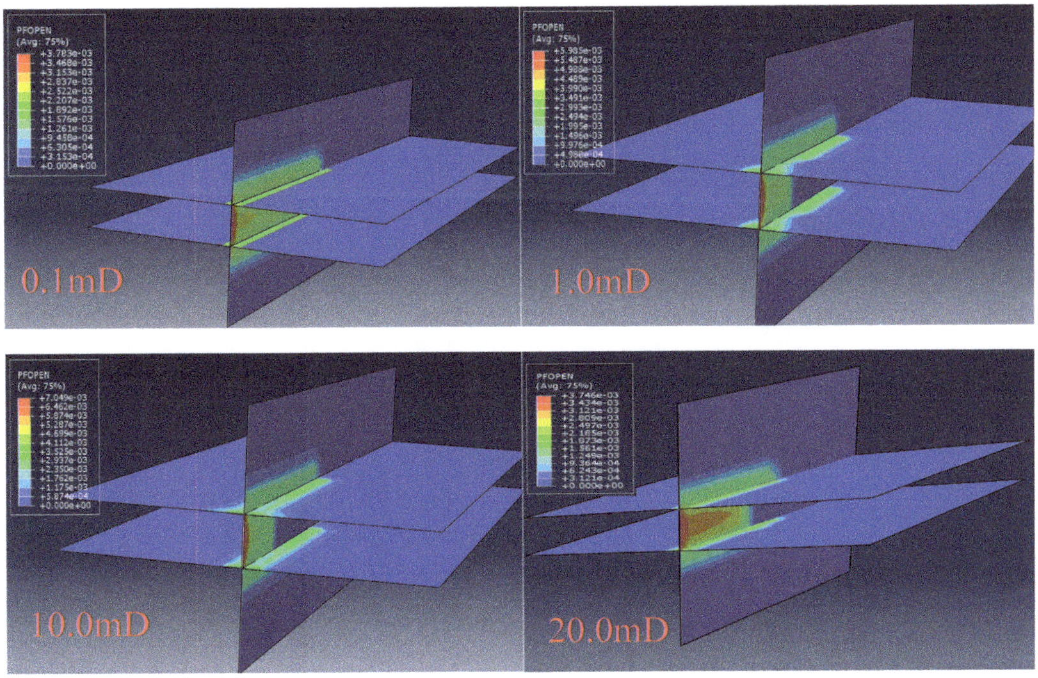

Figure 7. Fracture geometry under different formation permeabilities when the viscosity of the fracturing fluid was 20.0 mPa.s.

To further investigate the pressure change during hydraulic fracturing, the net pressure during fracture propagation was quantitatively analyzed. From Figure 8, it was found that the net pressure increased with the decrease in formation permeability, and decreased with the decrease in the viscosity of the fracturing fluid. The main reason was that a low formation permeability and high viscosity of the fracturing fluid can effectively reduce the seepage of the fracturing fluid from the fracture surface, thus more fracturing fluid can contribute to the propagation of hydraulic fracture. Under a high viscosity of the fracturing fluid and low formation permeability, the hydraulic fracture can cross the bedding plane easily, while it was arrested by the bedding plane under a low viscosity of the fracturing fluid and high formation permeability. It was also found that the net pressure showed a great decrease when the hydraulic fracture interacted with the bedding plane; the main reason was that with the increase in the opening degree of the bedding plane, more fracturing fluid flowed into the bedding plane, thus the net pressure of hydraulic fracture showed a huge decrease.

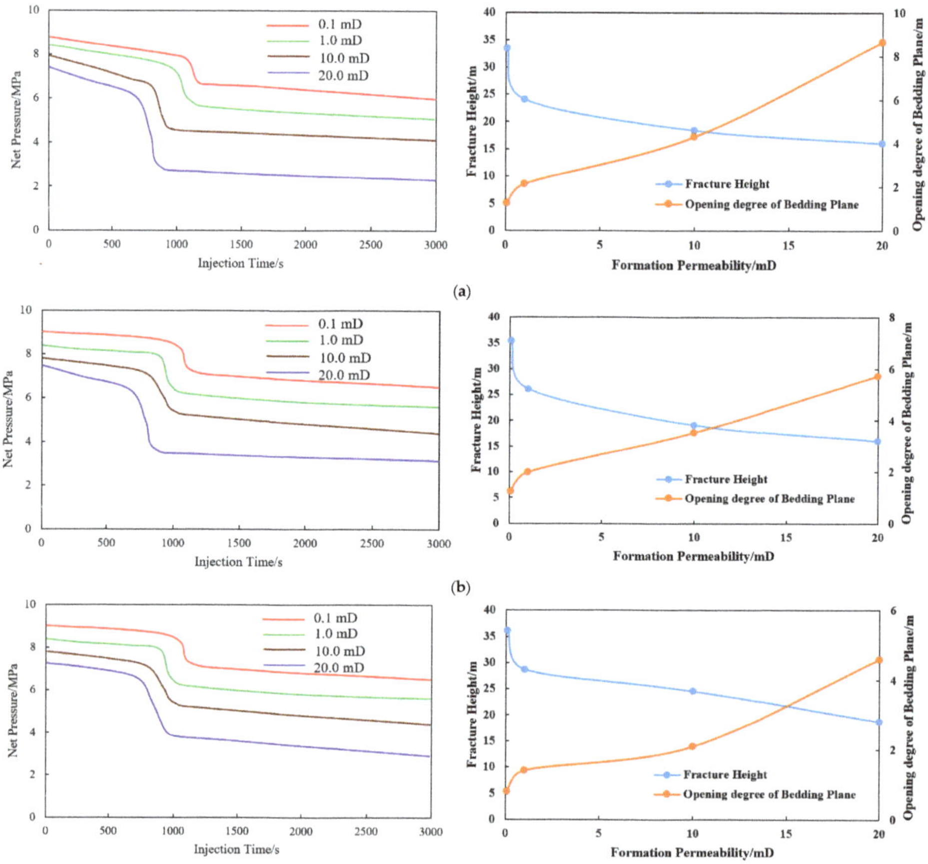

Figure 8. Net pressure change and fracture dimension for different formation permeabilities with different viscosity of the fracturing fluid. (**a**) Net pressure and fracture dimension comparison under different formation permeabilities when the viscosity of the fracturing fluid was 1.0 mPa.s. (**b**) Net pressure and fracture dimension comparison under different formation permeabilities when the viscosity of the fracturing fluid was 10.0 mPa.s. (**c**) Net pressure and fracture dimension comparison under different formation permeabilities when the viscosity of the fracturing fluid was 20.0 mPa.s.

3.2.3. The Injection Rate on Fracture Propagation

To understand the effect of the injection rate on fracture propagation, an injection rate of 2.0 m^3/min, 3.0 m^3/min, 4.0 m^3/min, and 5.0 m^3/min was simulated, and other parameters were as shown in Table 1. From the simulation, it was found that the hydraulic fracture crossed the bedding plane when the injection rate was 4.0 m^3/min and 5.0 m^3/min, while it was arrested by the bedding plane when the injection rate was 2.0 m^3/min and 3.0 m^3/min. When the injection rate was 2.0 m^3/min, the opening degree of the bedding plane was 3.6 m, while it increased to 4.5 m when the injection was 3.0 m^3/min. The main reason was that when the injection rate was 2.0 m^3/min, a lot of fracturing fluid filtered into the shale formation, thus the opening degree of the bedding plane was shorter than when the injection rate was 3.0 m^3/min. When the injection rate was 4.0 m^3/min and 5.0 m^3/min, the net pressure was much higher so it could effectively cross the bedding plane, and the fracture height showed an effective increase of 26.4 m and 31.2 m. The change in net pressure also verified the phenomenon (Figures 9 and 10).

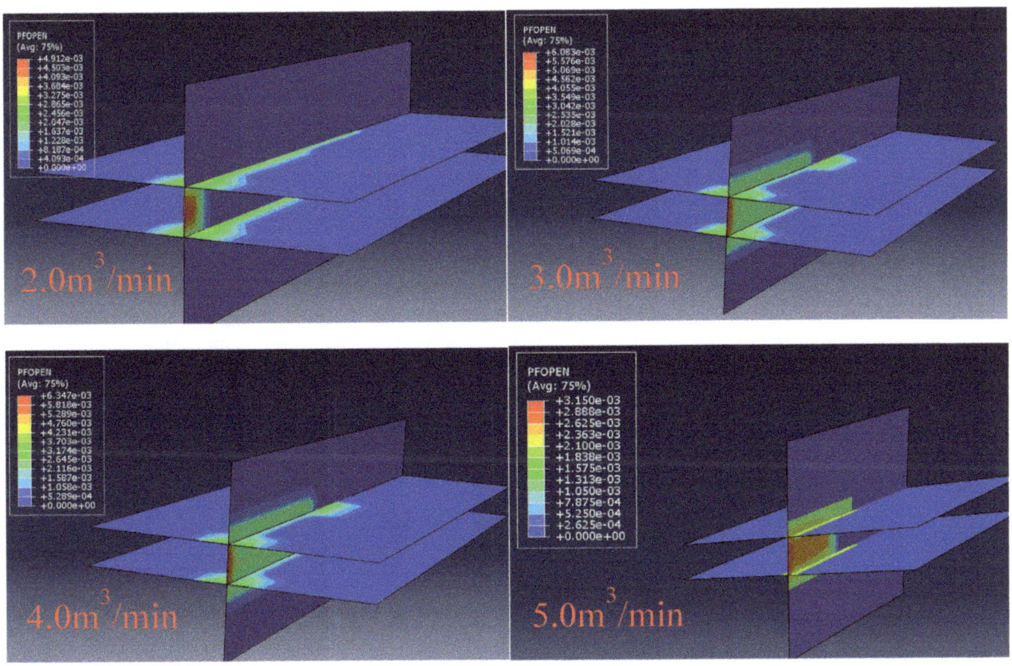

Figure 9. Fracture geometry under different injection rates.

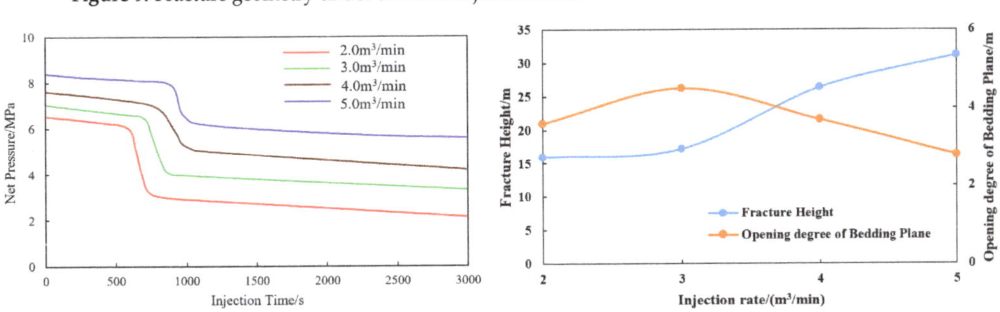

Figure 10. Net pressure and fracture dimension changes under different injection rates.

4. Conclusions

Hydraulic fracturing was the main technology for shale oil and gas development, and the bedding plane was the predominant factor affecting fracture propagation in the vertical direction. Clarifying fracture propagation in shale formation was the basis for deploying the wells for shale oil and gas development. Based on the simulations, the conclusions were shown as follows.

(1) Stress difference, formation permeability, viscosity of the fracturing fluid, and the injection rate are the predominant factors affecting fracture propagation. A high stress difference, formation permeability, and low viscosity of the fracturing fluid, and the injection rate contributed to a higher opening degree of the bedding plane, which arrested the hydraulic fracture by the bedding plane.

(2) When the hydraulic fracture interacted with the bedding plane, the net pressure in the hydraulic fracture decreased. Compared to the hydraulic fracture arrested by the bedding plane when the hydraulic fracture crossed the bedding plane, the net pressure decrease was much higher than that when the hydraulic fracture crossed the bedding plane.

(3) Formation permeability was the main factor affecting fracture propagation in the vertical direction. When the formation permeability was over 10.0 mD, the viscosity of the fracturing fluid was increased and this can promote the hydraulic fracture crossing the bedding plane.

(4) For further research in the future, formation heterogeneity is an important element that should be considered in the simulation, and the effect of temperature on fracture propagation should also be taken into consideration.

Author Contributions: Conceptualization, H.Z., H.W. and F.L.; methodology, H.Z. and H.W.; software, H.Z. and H.W.; validation, H.Z., F.L. and N. L.; formal analysis, H.Z.; investigation, H.Z.; resources, H.Z. and H.W.; data curation, H.Z. and N.L.; writing—original draft preparation, H.Z.; writing—review and editing, H.Z.; visualization, H.Z.; supervision, H.Z.; project administration, H.Z.; funding acquisition, H.Z. All authors have read and agreed to the published version of the manuscript.

Funding: This work was financially supported by the Open fund from State Key Laboratory of Shale Oil and Gas Enrichment Mechanisms and Effective Development (33550000-22-ZC0613-0381).

Data Availability Statement: Data available on requested.

Conflicts of Interest: The authors declare no conflicts of interest.

References

1. Zeng, J.; Wang, X.; Guo, J.; Zeng, F.; Zhang, Q. Composite linear flow model for multi-fractured horizonal wells in tight sand reservoirs with the threshold pressure gradient. *J. Pet. Sci. Eng.* **2018**, *165*, 890–912. [CrossRef]
2. Wang, Z.L.; Lou, Y.; Pan, J.P. China's Oil & gas resources exploration and development and its prospect. *Explor. Prod.* **2017**, *25*, 1–6.
3. Wang, K.; Zhang, H.L.; Zhang, R.H. Characteristics and influencing factors of ultra-deep shale reservoir structural fracture: A case study of Keshen-2 gas field, Tarim Basin. *Acta Petrol. Sin.* **2016**, *37*, 714–727.
4. HongY, Q.; Wu, C.M.; Hu, K. Key Technologies and Practice of Shale Oil Cost-Effective Development in Jimsar Sag, Junggar Basin. *Xinjiang Oil Gas* **2024**, *20*, 15–23.
5. Zeng, L.B.; Ke, S.Z.; Liu, Y. *Research Methods of Low Permeability Oil and Gas Reservoir Fractures*; Petroleum Industry Press: Beijing, China, 2010; pp. 1–187.
6. Olson, J.E.; Laubach, S.E.; Eichhubl, P. Estimating natural fracture producibility in tight gas sandstones, coupling diagenesis with geomechanical modeling. *Lead. Edge* **2010**, *29*, 1494–1499. [CrossRef]
7. Zeng, L.; Liu, W.; Li, J.; Zhu, L.; Weng, J.; Yue, F.; Zu, K. Natural fractures and their influence on shale gas enrichment Sichuan Basin, China. *J. Pet. Sci. Eng.* **2016**, *30*, 1–9. [CrossRef]
8. Liu, W.; Zeng, L.; Zhang, B.; Miao, F.; Lyu, P.; Dong, S. Influence of natural fractures on gas accumulation in the Upper Triassic tight gas sandstones in the northwestern Sichuan Basin, China. *Mar. Pet. Geol.* **2017**, *83*, 60–72. [CrossRef]

9. Luo, Y.; Kong, H.; Xu, K.S. Numerical Simulation for Vertical Propagation Pattern of Hydraulic Fractures in Sand-Shale Interbedded Reservoirs. *Xinjiang Oil Gas* **2023**, *19*, 49–56.
10. Zheng, H.; Liao, R.; Cheng, N.; Shi, S.; Liu, D. Study of mechano-chemical effects on the morphology of hydraulic fractures. *J. Pet. Sci. Eng.* **2021**, *206*, 109031. [CrossRef]
11. Khoei, A.R.; Vahab, M.; Ehsani, H.; Rafieerad, M. XFEM modeling of large plasticity deformation; a convergence study on various blending strategies for weak discontinuities. *Eur. J. Comput. Mech.* **2015**, *24*, 79–106. [CrossRef]
12. Wang, B.; Wang, Q. Numerical Simulation of Three-Dimensional Vertical Fracture Propagation Model under the Influence of Bedding. Xinjiang Oil & Gas. *Xinjiang Oil Gas* **2024**, *20*, 77–88.
13. Tan, P.; Jin, Y.; Han, K.; Hou, B.; Chen, M.; Guo, X.; Gao, J. Analysis of hydraulic fracture initiation and vertical propagation behavior in laminated shale formation. *Fuel* **2017**, *206*, 482–493. [CrossRef]
14. Zhang, H.; Sheng, J.J. Numerical simulation and optimization study of the complex fracture network in naturally fractured reservoirs. *J. Pet. Sci. Eng.* **2020**, *195*, 107726. [CrossRef]
15. Rui, Z.; Guo, T.; Feng, Q.; Qu, Z.; Qi, N.; Gong, F. Influence of gravel on the propagation pattern of hydraulic fracture in the conglomerate reservoir. *J. Pet. Sci. Eng.* **2018**, *165*, 627–639. [CrossRef]
16. Zhang, G.; Sun, S.; Chao, K.; Niu, R.; Liu, B.; Li, Y.; Wang, F. Investigation of the nucleation, propagation and coalescence of hydraulic fractures in conglomerate reservoirs using a coupled fluid loe-DEM approach. *Powder Technol.* **2019**, *354*, 301–313. [CrossRef]
17. Chen, M.; Sun, Y.; Fu, P.; Carrigan, C.R.; Lu, Z.; Tong, C.H.; Buscheck, T.A. Surrogate-based optimization of hydraulic fracture in pre-existing fracture networks. *Comput. Geosci.* **2013**, *58*, 69–79. [CrossRef]
18. Wang, H.Y. Hydraulic fracture propagation in naturally fractures reservoirs: Complex fracture or fracture networks. *J. Nat. Gas Sci. Eng.* **2019**, *68*, 102911. [CrossRef]
19. Wei, Y.S.; Wang, J.L.; Yu, W.; Qi, Y.D.; Miao, J.J.; Yuan, H.; Liu, C.X. A smart productivity evaluation method for shale gas wells based on 3D fractal fracture network model. *Pet. Explor. Dev.* **2021**, *48*, 787–796. [CrossRef]
20. Wan, X.C.; Rasouli, V.; Damjance, B.; Yu, W.; Xie, H.; Li, N.; Rabiei, M.; Miao, J.; Liu, M. Coupling of fracture model with reservoir to simulate shale gas production with complex fractures and nanopores. *J. Pet. Sci. Eng.* **2020**, *193*, 107422. [CrossRef]
21. Sobhaniaraghn, B.; Mansur, W.J.; Peter, F.C. Three-dimensional investigation of multiple stage hydraulic fracturing in unconventional reservoirs. *J. Pet. Sci. Eng.* **2016**, *146*, 1063–1078. [CrossRef]
22. Hanks, C.L.; Lorenz, J.; Teufel, L.; Krumhardt, A.P. Lithologic and structural controls on natural fracture distribution and behavior within the Lisburne Group, northeastern Brooks Range and North Slope subsurface, Alaska. *AAPG Bull.* **1997**, *81*, 1700–1720.
23. Zheng, H.; Pu, C.; Wang, Y.; Sun, C. Experimental and numerical investigation on influence of pore-pressure distribution on multi-fractures propagation in tight sandstone. *Eng. Fract. Mech.* **2020**, *230*, 106993. [CrossRef]

processes

MDPI

Article

Discrimination Model of Interaction Between Hydraulic Fracture and Natural Fracture Based on Energy Balance

Chao Liu [1], Xinggui Yang [2], Wenqi Cao [1], Jin Lin [1], Yuxuan Liu [2,*] and Hang Zhang [2]

[1] Natural Gas Research Institute Branch, Shaanxi Yanchang Petroleum (Group) Co., Ltd., Xi'an 710065, China; lcok101@sina.com (C.L.); cwq4765@163.com (W.C.); linjinn@163.com (J.L.)
[2] State Key Laboratory of Oil and Gas Reservoir Geology and Exploitation, Southwest Petroleum University, Chengdu 610500, China; xg_yang37@163.com (X.Y.); hang-97@163.com (H.Z.)
* Correspondence: liuyx_6@163.com

Abstract: Hydraulic fracturing technology has been extensively applied for the efficient development of unconventional reservoirs. Influenced by geological discontinuities such as naturally fractured weak planes, the complex interaction behaviors between hydraulic fractures and natural fractures significantly challenge the prediction of hydraulic fracture propagation paths. Establishing interaction discrimination models to predict these behaviors proves crucial for characterizing post-stimulation fracture complexity. This study develops a discrimination model for hydraulic-natural fracture interactions based on fracture mechanics theory and energy balance principles. The critical conditions of hydraulic fracture crossing, natural fracture opening, and slippage are quantified, and the interaction propagation behavior of hydraulic fracture and natural fracture under different geological parameters is analyzed. Key findings reveal three interaction modes after fracture intersection: direct hydraulic fracture crossing, natural fracture opening, and natural fracture slippage. However, continuous fluid injection-induced pressure buildup within fractures ultimately drives hydraulic fracture crossing through natural fractures. Threshold effects emerge at specific hydraulic fracture lengths (0.5 m) and fracture toughness values (2 MPa$\cdot$m$^{0.5}$), governing the transition between direct crossing and natural fracture opening behaviors. Horizontal stress difference directly modulates the threshold values of natural fracture cohesion, friction coefficient, and approach angle. These three parameters collectively control the temporal sequence of hydraulic fracture crossing behaviors through natural fractures. The interaction discrimination model established in this study provides theoretical guidance for optimizing fracturing parameter design in fractured reservoirs.

Keywords: hydraulic fracturing; fracture propagation; numerical simulation; interaction discrimination model

1. Introduction

Amid the global transition toward low-carbon energy restructuring, unconventional natural gas resources such as shale gas and tight gas, with their clean and low-carbon characteristics, have emerged as strategic priorities for ensuring energy security and achieving carbon neutrality targets [1]. China possesses vast reserves of unconventional gas resources with extensive geographical distribution, demonstrating enormous exploration and development potential [2]. However, the inherent low porosity and permeability characteristics of unconventional reservoirs render conventional extraction techniques ineffective for obtaining industrial-scale production. Hydraulic fracturing technology has evolved into a pivotal technique for efficient unconventional reservoir development [3]. This method utilizes pressurized fluids to induce rock deformation and fracture initiation, creating artificial fracture networks within reservoirs. Such artificial fractures effectively reduce the seepage distance from matrix fluids to wellbores while enhancing overall reservoir permeability. In homogeneous reservoirs under ideal conditions, hydraulic fractures predominantly propagate perpendicular to the minimum horizontal principal stress direction, forming simple bi-wing fracture geometries [4,5]. Nevertheless, the presence of geological discontinuities such as natural fractures and bedding planes significantly alters hydraulic fracture propagation paths, substantially increasing fracture network complexity [6]. The intricate interaction behaviors between hydraulic and natural fractures, including crossing, slippage, opening, and capture, exhibit strong dependence on multiple factors: in-situ stress regimes, fracture morphology, mechanical properties of weak planes, and fluid viscosity [7]. These complexities necessitate the development of comprehensive discrimination models to elucidate fracture interaction mechanisms.

Significant advancements have been achieved in both experimental and numerical investigations of hydraulic-natural fracture interaction mechanisms. Experimental studies employing true triaxial hydraulic fracturing systems have systematically characterized interaction behaviors under various geological and operational parameters [8]. Key controlling factors identified include natural fracture approach angle, cementation strength, and stress differential [9,10]. Blanton's seminal work demonstrated that hydraulic fractures preferentially cross natural fractures under high-stress differentials and large approach angles, whereas low-stress differentials and small approach angles promote interfacial slippage [11]. Operational parameters such as injection rate, duration, and fracturing fluid viscosity further modulate interaction modes [12,13]. Elevated injection rates and high-viscosity fluids enhance fracture-crossing tendencies, while lower rates and reduced viscosities favor natural fracture activation and complex network development [14,15]. Numerical simulations employing fluid–solid coupling models have replicated interaction processes using diverse computational approaches: finite element method (FEM) [16], extended finite element method (XFEM) [17,18], discrete element method (DEM) [19], and boundary element method (BEM) [20]. These models particularly emphasize how cementation strength, approach angle, stress differential, and fluid viscosity govern fracture propagation patterns [21]. Simulation results confirm that strong cementation, high-stress differentials, and steep approach angles collectively drive direct fracture crossing. Conversely, weak cementation, low-stress differentials, and shallow angles induce fracture capture and bifurcation, generating intricate network geometries [16,22].

Building upon clarified hydraulic-natural fracture interaction mechanisms, researchers have established multi-dimensional theoretical criteria to quantitatively predict fracture propagation paths. Blanton pioneered a criterion incorporating stress intensity factors and approach angles, proposing hydraulic fracture crossing occurs when the fluid pressure at the fracture tip surpasses the combined natural fracture normal stress and rock tensile strength [23]. Warpinski and Teufel developed criteria for determining natural fracture slippage versus opening, simultaneously considering the effects of natural fracture permeability, fluid leak-off on fracture-tip stress fields, and hydraulically induced stress field interactions [24]. Renshaw and Pollard formulated a crossing criterion for orthogonal hydraulic fractures intersecting frictionless interfaces based on linear elastic fracture mechanics [25], subsequently extended by Gu and Weng to non-orthogonal scenarios [26,27]. Sarmadivaleh further refined these models by incorporating cohesive strength and adhesive forces within natural fractures [28]. Recent advancements employ energy balance principles to establish discrimination criteria for post-interaction fracture behaviors. Griffith's foundational concept of surface energy in solid materials established that system-released energy partially converts to surface energy during fracturing [29]. Irwin subsequently defined the energy release rate as the energy liberated per unit area during fracture propagation [30]. Yao and Wang developed an interaction criterion based on Griffith's theory, quantitatively characterizing critical conditions for hydraulic fracture crossing, natural fracture opening, and slippage under far-field stress state [31].

Based on the Griffith energy balance principle and considering the influence of fracture fluid in hydraulic fracture on the stress field, this study establishes an interaction discrimination model between hydraulic fractures and natural fractures. By analyzing energy transformation during fracture interaction, the critical pressure conditions for hydraulic fracture crossing, natural fracture opening, and slippage are derived. The propagation behaviors under different geological parameters are systematically investigated, providing theoretical guidance for optimizing fracturing parameters in naturally fractured reservoirs.

2. Model Establishment

According to the Griffith linear fracture hypothesis, a three-dimensional elliptical linear fracture with a thickness of δ and a half-length of l continuously propagates within an infinite plane, as illustrated in Figure 1.

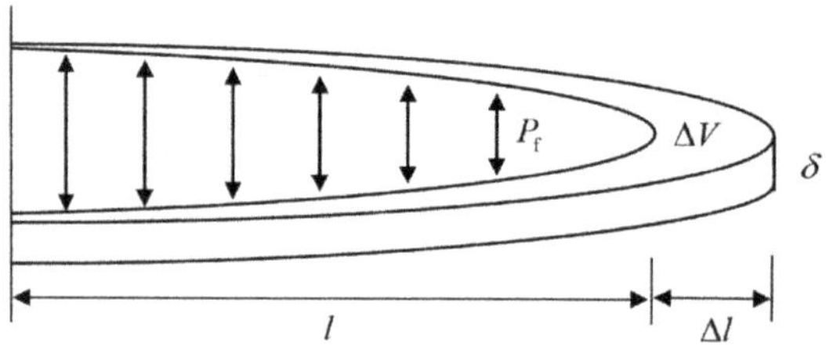

Figure 1. Schematic diagram of linear fractures.

When the fracture propagates a distance of Δl, the fracture width along the length direction can be expressed as follows [32]:

$$w(x) = \frac{4P_f}{E'}\sqrt{l^2 - x^2}\ 0 \le x \le l \tag{1}$$

The incremental fracture volume is given by the following:

$$\Delta V = \frac{\pi \delta P_f l \Delta l}{E'} \tag{2}$$

In the formula, P_f is the fluid pressure inside the fracture, MPa; l is the fracture length, m; E' is the plane-strain Young's modulus, GPa; δ is the linear fracture thickness, m.

When a hydraulic fracture propagates along its path and intersects a natural fracture, the approach angle between the hydraulic fracture and the natural fracture is denoted as θ. The interaction is governed by the maximum horizontal principal stress (σ_H) and the minimum horizontal principal stress (σ_h). A two-dimensional interaction model illustrating this scenario is presented in Figure 2.

Under the combined effects of the in-situ stress field and the fluid stress field within the hydraulic fracture, the stress field around the natural fracture is expressed as follows:

$$\begin{aligned}\sigma_x &= \sigma_H + \frac{K_I}{\sqrt{2\pi r}}\cos\frac{\theta}{2}\left(1 - \sin\frac{\theta}{2}\sin\frac{3\theta}{2}\right)\\ \sigma_y &= \sigma_h + \frac{K_I}{\sqrt{2\pi r}}\cos\frac{\theta}{2}\left(1 + \sin\frac{\theta}{2}\sin\frac{3\theta}{2}\right)\\ \tau_{xy} &= \frac{K_I}{\sqrt{2\pi r}}\sin\frac{\theta}{2}\cos\frac{\theta}{2}\cos\frac{3\theta}{2}\end{aligned} \tag{3}$$

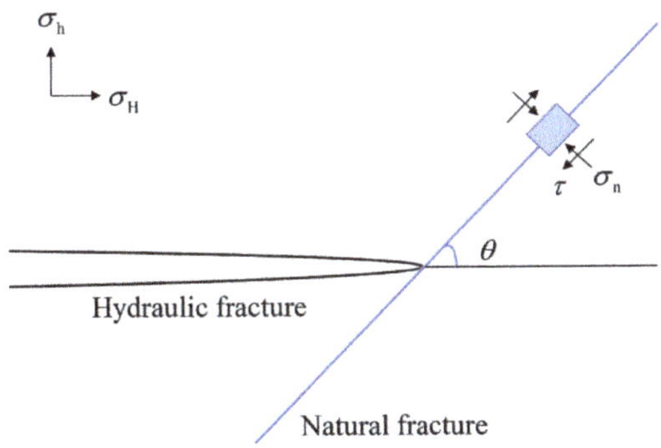

Figure 2. 2D interaction model between hydraulic and natural fractures.

When only the far-field in-situ stress is considered for its effect on the natural fracture, the normal and shear stresses acting on the natural fracture are given by the following:

$$\begin{aligned} \sigma_n &= \frac{\sigma_H+\sigma_h}{2} - \frac{\sigma_H-\sigma_h}{2}\cos 2\theta \\ \tau &= \frac{\sigma_H-\sigma_h}{2}\sin 2\theta \end{aligned} \tag{4}$$

In the formula, K_I represents the mode-I stress intensity factor, MPa·m$^{0.5}$; σ_H is the maximum horizontal principal stress, MPa; σ_h is the minimum horizontal principal stress, MPa; r is the distance from the hydraulic fracture tip to the natural fracture, m; σ_n is the normal stress acting on the natural fracture surface, MPa; τ is the shear stress on the natural fracture surface, MPa.

(1) Critical condition of natural fracture slip

Under the combined influence of the stress concentration at the hydraulic fracture tip and far-field in-situ stresses, natural fractures may experience slip before being intersected by hydraulic fractures. Alternatively, when hydraulic fractures connect with natural fractures, continuous fluid leak-off reduces the pore pressure within natural fractures below the normal stress acting on their surfaces, leading to slip initiation. According to the Mohr–Coulomb criterion, the following can be established:

Conditions for slippage:

$$\tau > k_f(\sigma_n - P_f) + C \tag{5}$$

Critical pressure of natural fracture slip:

$$P_{sp} > \sigma_n - \frac{\tau - C}{k_f} \tag{6}$$

In the formula, k_f represents the friction coefficient of the natural fracture surface (dimensionless). C represents the cohesion of natural fractures, MPa.

(2) Critical condition of hydraulic fracture crossing

According to Griffith's fracture stability theory, during hydraulic fracturing, the injected fluid continuously drives the fracture to propagate forward. Part of the work conducted by the fluid pressure is transformed into the deformation of the rock matrix, which shows as stored elastic energy; part of it is transformed into dissipated surface energy generated by new fractures; and the other part is transformed into heat energy through friction. Accordingly, the principle of conservation of energy is as follows [29]:

$$\Delta W_p = \Delta W_{ela} + 2\gamma\delta\Delta l + Q \tag{7}$$

In the formula, ΔW_{p} represents work conducted by the pressure of the injected fluid, J; ΔW_{ela} represents the elastic energy of rock, J; γ is the surface energy of materials, J/m^2; $2\gamma\delta\Delta l$ represents the surface energy dissipated by the generation of new fractures, J; Q stands for dissipated heat energy, J.

In view of the hydraulic fracture propagation process, the hydraulic fracture is subjected to both fluid pressure and in-situ stress, and the hydraulic fracture is essentially driven by net pressure. When heat dissipation is neglected, Equation (7) is expressed as follows:

$$\Delta W_{\text{net}} = \Delta W_{\text{HF}} - \Delta W_{\text{far}} = \Delta W_{\text{ela}} + 2\gamma\delta\Delta l \tag{8}$$

The work conducted by injecting fluid pressure into hydraulic fracture is the product of fluid pressure and the increased volume of hydraulic fracture, which is expressed as follows:

$$\Delta W_{\text{HF}} = P_{\text{f}}\Delta V \tag{9}$$

In-situ stress work is the product of far-field stress perpendicular to the direction of fracture propagation and the increased volume of hydraulic fracture. This is expressed as follows:

$$\Delta W_{\text{far}} = \sigma_{\text{h}}\Delta V \tag{10}$$

The elastic energy stored in rocks is expressed as follows:

$$\Delta W_{\text{ela}} = \frac{1}{2}P_{\text{net}}\Delta V \tag{11}$$

The net pressure in the hydraulic fracture is the following:

$$P_{\text{net}} = P_{\text{f}} - \sigma_{\text{h}} \tag{12}$$

The net pressure work in the hydraulic fracture is the product of the net pressure and the increased volume of the hydraulic fracture, which is expressed as:

$$\Delta W_{\text{net}} = \Delta W_{\text{HF}} - \Delta W_{\text{far}} = P_{\text{net}}\Delta V \tag{13}$$

The increased volume of hydraulic fracture can be corrected as follows:

$$\Delta V = \frac{\pi\delta P_{\text{net}} l\Delta l}{E'} \tag{14}$$

According to the theory of fracture mechanics, the energy release rate of hydraulic fracture propagation is as follows:

$$G_{\text{HF}} = 2\gamma \tag{15}$$

The hydraulic fracture propagation process belongs to mode-I fracture propagation. The energy release rate can be expressed as follows:

$$G_{\text{HF}} = \frac{K_I^2}{E'} \tag{16}$$

When hydraulic fracture propagates, the mode-I stress intensity factor is equal to the fracture toughness of the rock. At this time, the critical fluid pressure of hydraulic fracture passing through natural fracture can be calculated as follows:

$$P_{\text{cr}} = \sigma_{\text{h}} + \frac{2K_{Ic}}{\sqrt{\pi l}} \tag{17}$$

In the formula, K_{Ic} represents the mode-I fracture toughness of the reservoir rock matrix, MPa·m$^{0.5}$.

(3) Critical condition of natural fracture opening

When the hydraulic fracture communicates with the natural fracture, the fluid enters the natural fracture, and the natural fracture continues to expand under the net pressure. When heat dissipation is neglected, the work conducted by the net pressure in natural fractures can be expressed as follows:

$$\Delta W_{\mathrm{net}}{}' = \Delta W_{\mathrm{NF}} - \Delta W_{\mathrm{far}} = \Delta W_{\mathrm{ela}}{}' + 2\gamma\prime\delta\Delta l \quad (18)$$

The net pressure on a natural fracture is the difference between the fluid pressure in the fracture and the normal stress on the wall, which can be expressed as follows:

$$P_{\mathrm{net}}{}' = P_{\mathrm{f}}{}' - \sigma_{\mathrm{n}} \quad (19)$$

In the formula, $P_{\mathrm{net}}{}'$ represents the net pressure in natural fracture, MPa; $P_{\mathrm{f}}{}'$ represents the fluid pressure in natural fracture, MPa; γ' represents the surface energy of cement in natural fracture, J/m^2.

The stored elastic energy is the following:

$$\Delta W_{\mathrm{ela}}{}' = \frac{1}{2} P_{\mathrm{net}}{}' \Delta V\prime \quad (20)$$

The increased volume of natural fractures is the following:

$$\Delta V\prime = \frac{\pi\delta P_{\mathrm{net}}{}' l \Delta l}{E\prime} \quad (21)$$

According to the theory of fracture mechanics, the energy release rate g of natural fracture propagation is the following:

$$G_{\mathrm{NF}} = 2\gamma' = \frac{(K_{\mathrm{I}}{}')^2}{E\prime} \quad (22)$$

When the mode-I stress intensity factor of the cement in the natural fracture is equal to the fracture toughness, the natural fracture opens, and at this time, the critical minimum fluid pressure of the natural fracture opening can be calculated. This is expressed as follows:

$$\left(P_{\mathrm{op}} - \sigma_{\mathrm{h}}\right)^2 - \frac{1}{2}\left(P_{\mathrm{op}} - \sigma_{\mathrm{n}}\right)^2 = \frac{\eta K_{\mathrm{Ic}}{}^2}{\pi l} \quad (23)$$

$$\eta = \frac{G_{\mathrm{NF}}}{G_{\mathrm{HF}}} = \frac{\gamma\prime}{\gamma} = \frac{(K_{\mathrm{Ic}}{}')^2}{K_{\mathrm{Ic}}{}^2} \quad (24)$$

In the formula, $K_{\mathrm{I}}{}'$ represents the mode-I stress intensity factor of cement in natural fracture, MPa·m$^{0.5}$; $K_{\mathrm{Ic}}{}'$ represents the fracture toughness of cement in natural fracture, MPa·m$^{0.5}$; η represents the bond strength ratio of natural fractures.

3. Numerical Validation

To verify the accuracy of the proposed discrimination model for hydraulic-natural fracture interaction behaviors, a comparative analysis was conducted with Blanton's physical modeling experiments [11]. Given that the cementation strength of natural fractures in physical models is typically significantly lower than the tensile strength of rock matrix, η can be reasonably assumed as zero.

Blanton conducted interaction experiments using gypsum blocks containing prefabricated fractures with varying approach angles (30°, 45°, 60°, 90°) and horizontal stress differentials (minimum horizontal principal stress: 5 MPa or 10 MPa; stress differential: 2–15 MPa). The prefabricated fractures exhibited a frictional coefficient of 0.75 with zero cohesion. The gypsum matrix demonstrated a tensile strength of 3.1 MPa and mode-I fracture toughness of 0.176 MPa·$m^{0.5}$. Fracturing fluid was injected through steel tubing simulating wellbore at a constant flow rate of 0.82 cm^3/s. The hydraulic fracture half-length was maintained at 0.06 m, with fluid viscosity considered negligible under fracture toughness-dominated propagation conditions [33]. Table 1 presents a comparison between Blanton's experimental results and the analytical solution's predictions for fracture interaction behavior. The analytical solution only considers the initial interaction behavior and does not account for subsequent changes induced by continuous fluid injection. The comparison demonstrates that the model established in this study can accurately predict the interaction behavior between hydraulic fractures and natural fractures.

Table 1. Comparison between experimental results and analytical solutions.

Approach Angle (°)	σ_1 (MPa)	σ_3 (MPa)	P_{cr} (MPa)	P_{op} (MPa)	P_{sp} (MPa)	Experimental	Analytical Solution
90	14	5	5.8109	8.7279	13.9904	Crossing	Crossing
60	12	10	10.8109	6.3809	10.3437	Opening	Opening
60	20	5	5.8109	9.6570	7.5775	Crossing	Crossing
60	14	5	5.8109	7.7942	6.5465	Arrested	Crossing
45	20	5	5.8109	8.1041	2.4940	Arrested	Arrested
45	18	5	5.8109	7.6902	2.8282	Arrested	Arrested
45	16	5	5.8109	7.2764	3.1623	Arrested	Arrested
45	14	5	5.8109	6.8625	3.4964	Arrested	Arrested
45	10	5	5.8109	6.0347	4.1647	Opening	Arrested
30	20	5	5.8109	6.5519	0.08895	Arrested	Arrested
30	19	10	10.8109	4.5730	7.0534	Opening	Opening

4. Results and Discussion

The interaction behaviors between hydraulic and natural fractures are governed not only by their mechanical properties but also by critical factors, including horizontal stress difference and approach angles. This section systematically analyzes the effects of five governing parameters on hydraulic-natural fracture interaction mechanisms: hydraulic fracture length, mode-I fracture toughness, frictional properties of natural fractures, horizontal stress difference, and approach angles. The input parameters employed in the computational framework are summarized in Table 2.

Table 2. Basic input parameters.

Parameters	Value	Unit
Minimum horizontal in-situ stress	15	MPa
Horizontal stress difference	5	MPa
Approach angle	75	°
Hydraulic fracture length	0.5	m
mode-I fracture toughness	1.8	MPa·$m^{0.5}$
Friction coefficient of natural fractures	0.3	/
Cohesion of natural fractures	0.2	MPa

4.1. Hydraulic Fracture Length

Figure 3 illustrates the influence of hydraulic fracture length on interaction behaviors. The results demonstrate that hydraulic fracture length exhibits no correlation with the critical pressure required for natural fracture slip. As hydraulic fracture length increases, the critical crossing pressure required for hydraulic fracture crossing through natural fractures progressively decreases, making crossing states more favorable. When hydraulic fracture length is less than 0.5 m, the fluid pressure difference required to satisfy opening conditions under varying η values becomes increasingly pronounced. Conversely, for hydraulic fracture lengths exceeding 0.5 m, variations in η exhibit negligible effects on the minimum fluid pressure required for opening conditions.

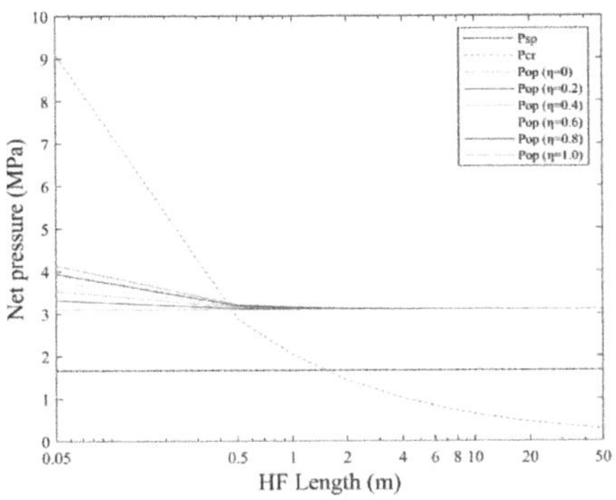

Figure 3. Influence of hydraulic fracture length on interaction behavior.

Hydraulic fracture length exhibits distinct threshold effects on interaction behaviors at 0.5 m and 1.5 m. These thresholds represent critical values for the transition of early-stage propagation behaviors during the interaction between hydraulic fractures and natural fractures. When the length is below 0.5 m, natural fractures undergo slip initiation first, resulting in hydraulic fracture arrest. This induces pressure accumulation within the fracture system, leading to progressive elevation of fluid pressure. Once the internal fluid pressure surpasses the normal stress acting on fracture walls, natural fracture opening occurs, triggering hydraulic fracture reorientation along natural fracture trajectories. For fracture lengths between 0.5 m and 1.5 m, natural fracture slip still causes hydraulic fracture arrest. Subsequent fluid pressure escalation eventually reaches critical crossing conditions, enabling hydraulic fracture crossing through natural fractures. When hydraulic fracture length exceeds 1.5 m, direct crossing behavior dominates post-interaction. The absence of sustained pressure accumulation after crossing prevents the achievement of critical slip conditions for natural fractures. Consequently, interaction mechanisms transition progressively from natural fracture opening to direct hydraulic fracture crossing with increasing fracture length.

4.2. Fracture Toughness

Figure 4 demonstrates the influence of fracture toughness on interaction behaviors. The results reveal that the fracture toughness of reservoir rock exhibits no correlation with the critical pressure for natural fracture slip. Notably, the critical slip pressure consistently remains below the critical opening pressure for natural fractures. Increasing fracture toughness elevates the critical fluid pressure required for hydraulic fracture crossing through natural fractures, thereby reducing the crossing likelihood. When fracture toughness is below 1.5 $\text{MPa}\cdot\text{m}^{0.5}$, variations in η exert negligible influence on the critical fluid pressure for natural fracture opening. Conversely, above 1.5 $\text{MPa}\cdot\text{m}^{0.5}$, distinct critical opening pressures emerge under different η values, demonstrating progressive pressure escalation with increasing η.

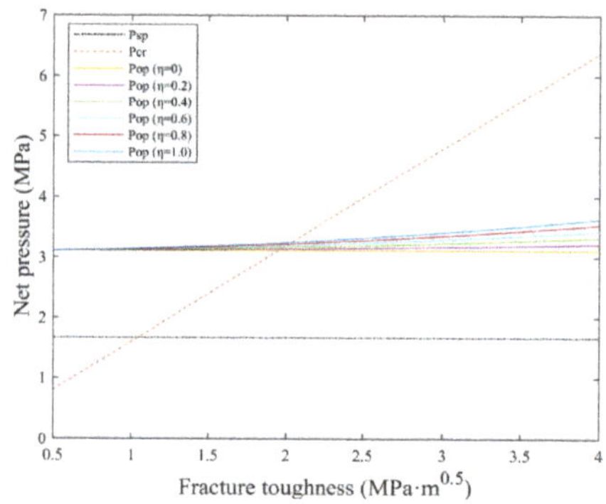

Figure 4. Influence of fracture toughness on interaction behavior.

Fracture toughness manifests threshold effects at 1 MPa·m$^{0.5}$ and 2 MPa·m$^{0.5}$. Below 1 MPa·m$^{0.5}$, direct crossing behavior dominates post-interaction. Between 1–2 MPa·m$^{0.5}$, the system prioritizes slip initiation (lowest critical slip pressure), resulting in hydraulic fracture arrest. Subsequent net pressure buildup within fractures eventually reaches critical crossing thresholds, enabling through-cutting propagation. Above 2 MPa·m$^{0.5}$, initial slip transitions to fracture reorientation along natural fractures when net pressure attains critical opening thresholds. Thus, increasing fracture toughness progressively shifts interaction mechanisms from hydraulic fracture crossing to natural fracture opening dominance.

4.3. Horizontal Stress Difference and Cohesion of Natural Fractures

Figure 5 illustrates the influence of horizontal stress differences and cohesion of natural fractures on interaction behaviors. The results demonstrate that the critical hydraulic fracture crossing pressure shows no dependence on either horizontal stress difference or cohesion, whereas the critical natural fracture opening pressure maintains independence from cohesion but exhibits a positive correlation with horizontal stress difference. With increasing horizontal stress difference and cohesion, the critical natural fracture slip pressure progressively rises, making fracture slip more difficult to initiate. As horizontal stress difference increases, the influence of η variation on the critical fluid pressure for natural fracture opening gradually diminishes, while cohesion changes show negligible effects on η.

Cohesion demonstrates a distinct threshold effect on interaction behaviors. Under low horizontal stress difference conditions ($\leq$5 MPa), the critical natural fracture opening pressure consistently remains lower than the critical hydraulic fracture crossing pressure, prioritizing natural fracture opening over hydraulic fracture crossing. Taking a stress difference of 5 MPa as an example, threshold effects emerge at cohesion values of 0.37 MPa and 0.71 MPa. When cohesion is less than 0.37 MPa, natural fractures initially undergo slip. The subsequent increase in net pressure to the critical opening pressure triggers natural fracture opening, followed by hydraulic fracture crossing upon reaching the critical crossing pressure. For cohesion between 0.37 MPa and 0.71 MPa, natural fractures first open and subsequently slip as net pressure reaches the critical slip condition, with hydraulic fracture crossing occurring at higher net pressures. When cohesion exceeds 0.71 MPa, natural fractures open first, and hydraulic fracture crossing occurs before achieving slip conditions due to pressure release post-crossing. Under high horizontal stress difference conditions ($\geq$10 MPa), only a single cohesion threshold exists. At 15 MPa stress difference, the threshold occurs at 0.4 MPa cohesion. Below 0.4 MPa, the initial natural fracture slip transitions to hydraulic fracture crossing without achieving natural fracture opening conditions. Above 0.4 MPa, the minimized crossing pressure enables direct hydraulic fracture crossing upon reaching critical pressure. The analysis reveals that increasing natural fracture cohesion universally promotes hydraulic fracture crossing through natural fractures. Horizontal stress difference modulates both the sequence of interaction behaviors and the magnitude of cohesion thresholds.

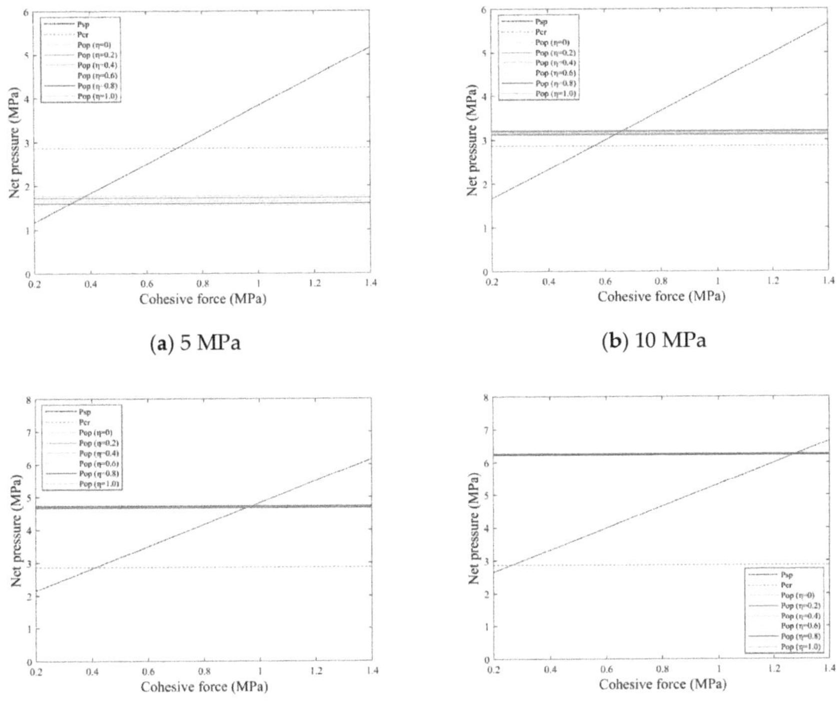

(**a**) 5 MPa (**b**) 10 MPa

(**c**) 15 MPa (**d**) 20 MPa

Figure 5. Influence of natural fracture cohesion on interaction behavior.

4.4. Horizontal Stress Difference and Friction Coefficient of Natural Fractures

Figure 6 illustrates the influence of horizontal stress differences and natural fracture friction coefficients on interaction behaviors. The results indicate that the friction coefficient shows no correlation with either the critical natural fracture opening pressure or critical hydraulic fracture crossing pressure. With increasing friction coefficient, the critical natural fracture slip pressure progressively rises, while the influence of η on the critical opening pressure remains relatively insignificant. As horizontal stress difference increases, both critical opening and slip pressures of natural fractures gradually increase, while the impact of η on critical opening pressure diminishes correspondingly.

Similar to the cohesion effect, the friction coefficient exhibits distinct threshold effects on interaction behaviors under varying horizontal stress difference conditions. Under low horizontal stress difference (≤5 MPa), two friction coefficient thresholds emerge. Taking a 5 MPa stress difference as an example, threshold effects occur at friction coefficients of 0.35 and 0.58. When the friction coefficient is less than 0.35, the interaction sequence progresses from natural fracture slip to opening, followed by hydraulic fracture crossing as net pressure increases. For friction coefficients between 0.35 and 0.58, natural fractures first open, subsequently slip, and finally experience hydraulic fracture crossing. When the friction coefficient exceeds 0.58, natural fractures open initially, followed directly by hydraulic fracture crossing. Under high horizontal stress difference conditions (≥10

MPa), only a single friction coefficient threshold exists. Below this threshold, natural fracture slip precedes hydraulic fracture crossing after reaching critical crossing pressure. Above the threshold, hydraulic fractures penetrate directly upon achieving critical pressure. Increasing natural fracture friction coefficients universally promotes hydraulic fracture crossing through natural fractures. Horizontal stress difference governs both the sequence of interaction behaviors and the magnitude of friction coefficient thresholds.

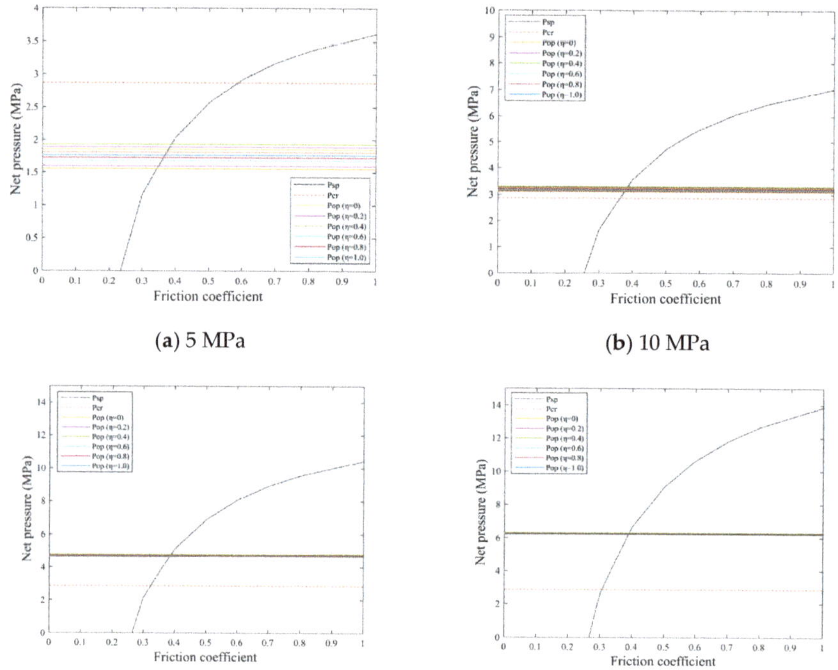

(**a**) 5 MPa (**b**) 10 MPa

(**c**) 15 MPa (**d**) 20 MPa

Figure 6. Influence of friction coefficient of natural fractures on interaction behavior.

4.5. Horizontal Stress Difference and Approach Angle

Figure 7 illustrates the influence of horizontal stress difference and approach angle on interaction behaviors. The results demonstrate that the critical crossing pressure of hydraulic fractures crossing through natural fractures remains independent of both horizontal stress difference and approach angle. With increasing approach angle, the critical opening pressure of natural fractures gradually rises, while the influence of η on this critical opening pressure diminishes progressively. When the approach angle exceeds 70°, the critical slip pressure of natural fractures increases linearly with the approach angle, and its growth rate accelerates with a higher horizontal stress difference. Increasing horizontal stress difference elevates both the critical opening and slip pressures of natural fractures, amplifies the impact of approach angle on these critical pressures, and reduces η's influence on the critical opening pressure.

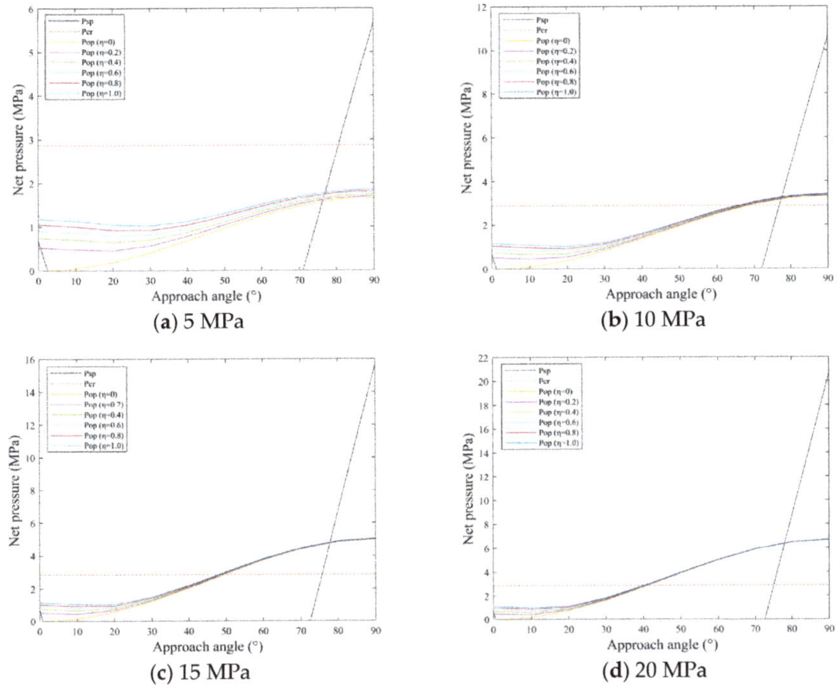

Figure 7. Influence of horizontal stress difference and approach angle on interaction behavior.

Under low horizontal stress difference conditions ($\leq$5 MPa), the critical opening pressure of natural fractures remains consistently lower than the critical crossing pressure of hydraulic fractures. Following the interaction between hydraulic and natural fractures, natural fractures preferentially open, causing hydraulic fractures to reorient and propagate along the natural fracture paths. As hydraulic fractures extend to the tips of natural fractures, pressure gradually accumulates within the fractures. Upon reaching the critical crossing pressure, hydraulic fractures cross through the natural fracture tips and continue propagating along the direction of the maximum horizontal principal stress. At high horizontal stress differences ($\geq$10 MPa), a critical approach angle governs the interaction behavior. When it is below this critical value, the natural fracture opening pressure remains lower than the hydraulic fracture crossing pressure, leading to the fracture opening upon interaction. Crossing occurs only after sufficient pressure accumulation, which enables breakthroughs at the natural fracture tips. Conversely, when the approach angle exceeds this critical value, the natural fracture opening pressure surpasses the crossing threshold, resulting in direct hydraulic fracture crossing through natural fractures. The critical approach angle progressively decreases with increasing horizontal stress difference. For approach angles exceeding 70°, the interaction initiates with natural fracture slip, followed by hydraulic fracture crossing once the critical pressure threshold is achieved through continued pressure accumulation. Consequently, as horizontal stress difference and approach angle increase, the interaction mechanism transitions from a sequential process of natural fracture opening followed by crossing to direct crossing dominance.

5. Conclusions

This study establishes a discriminative model for hydraulic-natural fracture interaction behaviors based on the Griffith energy balance principle, quantifying critical conditions for different interaction behaviors and analyzing the influence of geological factors on fracture interaction behaviors. The principal conclusions are summarized as follows:

(1) Following hydraulic fracture-natural fracture interaction, three distinct behavioral modes may occur: direct hydraulic fracture crossing, natural fracture slip, and natural fracture opening. However, sustained fluid injection induces continuous pressure buildup within the fracture system, ultimately leading to hydraulic fracture crossing through natural fractures in all scenarios.
(2) Natural fracture cohesion and friction coefficient primarily govern the critical conditions for natural fracture slip, while hydraulic fracture length and fracture toughness predominantly influence the critical crossing pressure through natural fractures. Horizontal stress difference and approach angle mainly regulate the critical pressures for hydraulic fracture opening and slip.
(3) Hydraulic fracture length (0.5 m) and fracture toughness (2 MPa$\cdot$m$^{0.5}$) demonstrate significant threshold effects that dictate the transition between direct hydraulic fracture crossing and natural fracture opening behaviors.
(4) Horizontal stress difference directly modulates the threshold values of natural fracture cohesion, friction coefficient, and approach angle. These three parameters collectively control the temporal sequence of hydraulic fracture crossing behaviors through natural fractures.

Author Contributions: Conceptualization, J.L. and Y.L.; Methodology, C.L., W.C. and Y.L.; Validation, X.Y. and H.Z.; Formal analysis, W.C.; Investigation, X.Y.; Data curation, C.L. and H.Z.; Writing—original draft, X.Y. and H.Z.; Writing—review and editing, C.L. and Y.L.; Visualization, W.C. and J.L. All authors have read and agreed to the published version of the manuscript.

Funding: This research received no external funding.

Data Availability Statement: The original contributions presented in this study are included in the article. Further inquiries can be directed to the corresponding author.

Conflicts of Interest: Authors Chao Liu, Wenqi Cao and Jin Lin were employed by the company Shaanxi Yanchang Petroleum (Group) Co., Ltd. The authors declare that they have no known competing financial interests or personal relationships that could have appeared to influence the work reported in this paper.

References

1. Kalam, S.; Afagwu, C.; Al Jaberi, J.; Siddig, O.M.; Tariq, Z.; Mahmoud, M.; Abdulraheem, A. A review on non-aqueous fracturing techniques in unconventional reservoirs. *J. Nat. Gas Sci. Eng.* **2021**, *95*, 104223. [CrossRef]
2. Li, Y.; Zhou, D.; Wang, W.; Jiang, T.; Xue, Z. Development of unconventional gas and technologies adopted in China. *Energy Geosci.* **2020**, *1*, 55–68. [CrossRef]
3. Temizel, C.; Canbaz, C.H.; Palabiyik, Y.; Hosgor, F.B.; Atayev, H.; Ozyurtkan, M.H.; Aydin, H.; Yurukcu, M.; Boppana, N. A review of hydraulic fracturing and latest developments in unconventional reservoirs. In Proceedings of the Offshore Technology Conference, Houston, TX, USA, 2–5 May 2022. [CrossRef]
4. Wu, F.; Li, D.; Fan, X.; Liu, J.; Li, X. Analytical interpretation of hydraulic fracturing initiation pressure and breakdown pressure. *J. Nat. Gas Sci. Eng.* **2020**, *76*, 103185. [CrossRef]
5. Zhang, Y.; Ma, Y.; Hu, Z.; Lei, H.; Bai, L.; Lei, Z.; Zhang, Q. An experimental investigation into the characteristics of hydraulic fracturing and fracture permeability after hydraulic fracturing in granite. *Renew. Energy* **2019**, *140*, 615–624. [CrossRef]
6. Zhang, B.; Liu, J.; Wang, S.G.; Li, S.; Yang, X.; Li, Y.; Zhu, P.; Yang, W. Impact of the distance between pre-existing fracture and wellbore on hydraulic fracture propagation. *J. Nat. Gas Sci. Eng.* **2018**, *57*, 155–165. [CrossRef]
7. Liu, T.; Wei, X.; Liu, X.; Liang, L.; Wang, X.; Chen, J.; Lei, H. A criterion for a hydraulic fracture crossing a natural fracture in toughness dominant regime and viscosity dominant regime. *Eng. Fract. Mech.* **2023**, *289*, 109421. [CrossRef]
8. Zhang, Q.; Zhang, X.; Sun, W. A review of laboratory studies and theoretical analysis for the interaction mode between induced hydraulic fractures and pre-existing fractures. *J. Nat. Gas Sci. Eng.* **2021**, *86*, 103719. [CrossRef]

9. Dehghan, A.N.; Goshtasbi, K.; Ahangari, K.; Jin, Y. Mechanism of fracture initiation and propagation using a tri-axial hydraulic fracturing test system in naturally fractured reservoirs. *Eur. J. Environ. Civ. Eng.* **2016**, *20*, 560–585. [CrossRef]
10. Dehghan, A.N.; Goshtasbi, K.; Ahangari, K.; Jin, Y. The effect of natural fracture dip and strike on hydraulic fracture propagation. *Int. J. Rock Mech. Min. Sci.* **2015**, *75*, 210–215. [CrossRef]
11. Blanton, T.L. An experimental study of interaction between hydraulically induced and pre-existing fractures. In Proceedings of the SPE Unconventional Gas Recovery Symposium, Pittsburgh, PA, USA, 16 May 1982. [CrossRef]
12. Xu, W.; Zhao, J.; Rahman, S.S.; Li, Y.; Yuan, Y. A comprehensive model of a hydraulic fracture interacting with a natural fracture: Analytical and numerical solution. *Rock Mech. Rock Eng.* **2019**, *52*, 1095–1113. [CrossRef]
13. Zhang, W.; Shi, X.; Jiang, S.; Cao, Q.; Wang, F.; Wang, Z.; Ge, Y.; Du, Y. Experimental study of hydraulic fracture initiation and propagation in highly saturated methane-hydrate-bearing sands. *J. Nat. Gas Sci. Eng.* **2020**, *79*, 103338. [CrossRef]
14. Hou, Z.; Cheng, H.; Sun, S.; Chen, J.; Qi, D.; Liu, Z. Fracture propagation and hydraulic Fracturing in different lithologies. *Appl. Geophys.* **2019**, *16*, 243–251. [CrossRef]
15. Yue, Y.; Peng, S.; Liu, Y.; Xu, J. Investigation of acoustic emission response and fracture morphology of rock hydraulic fracturing under true triaxial stress. *Acta Geophys.* **2019**, *67*, 1017–1024. [CrossRef]
16. Guo, J.; Zhao, X.; Zhu, H.; Zhang, X.; Pan, R. Numerical simulation of interaction of hydraulic fracture and natural fracture based on the cohesive zone finite element method. *J. Nat. Gas Sci. Eng.* **2015**, *25*, 180–188. [CrossRef]
17. Wang, W.; Yao, Y. A theoretical model for 3d hydraulic fracture intersecting with natural fracture. *Mech. Res. Commun.* **2022**, *124*, 103961. [CrossRef]
18. Wang, D.; Zhou, F.; Li, Y.; Yu, B.; Martyushev, D.; Liu, X.; Wang, M.; He, C.; Han, D.; Sun, D. Numerical simulation of fracture propagation in russia carbonate reservoirs during refracturing. *Pet. Sci.* **2022**, *19*, 2781–2795. [CrossRef]
19. Cong, Z.; Li, Y.; Tang, J.; Martyushev, D.A.; Hubuqin; Yang, F. Numerical simulation of hydraulic fracture height layer-through propagation based on three-dimensional lattice method. *Eng. Fract. Mech.* **2022**, *264*, 108331. [CrossRef]
20. Zheng, P.; Xia, Y.; Yao, T.; Jiang, X.; Xiao, P.; He, Z.; Zhou, D. Formation mechanisms of hydraulic fracture network based on fracture interaction. *Energy* **2022**, *243*, 123057. [CrossRef]
21. Wang, H.; Xiao, P.; Zhou, D.; Wang, Q. Numerical modelling of interaction of cemented natural fractures and hydraulic fractures. *Sci. Rep.* **2024**, *14*, 32050. [CrossRef]
22. Zhou, J.; Zhang, L.; Pan, Z.; Han, Z. Numerical studies of interactions between hydraulic and natural fractures by smooth joint model. *J. Nat. Gas Sci. Eng.* **2017**, *46*, 592–602. [CrossRef]
23. Blanton, T.L. Propagation of hydraulically and dynamically induced fractures in naturally fractured reservoirs. In Proceedings of the SPE Unconventional Gas Technology Symposium, Louisville, KY, USA, 18–21 May 1986. [CrossRef]
24. Warpinski, N.R.; Teufel, L.W. Influence of geologic discontinuities on hydraulic fracture propagation (includes associated papers 17011 and 17074). *J. Pet. Technol. Pet. Technol.* **1987**, *39*, 209–220. [CrossRef]
25. Renshaw, C.E.; Pollard, D.D. An experimentally verified criterion for propagation across unbounded frictional interfaces in brittle, linear elastic materials. *Int. J. Rock Mech. Min. Sci. Geomech.* **1995**, *32*, 237–249. [CrossRef]
26. Gu, H.; Weng, X. Criterion for fractures crossing frictional interfaces at non-orthogonal angles. In Proceedings of the 44th U.S. Rock Mechanics Symposium and 5th U.S.-Canada Rock Mechanics Symposium, ARMA 2010, Salt Lake City, UT, USA, 37–30 June 2010.
27. Gu, H.; Weng, X.; Lund, J.; Mack, M.; Ganguly, U.; Suarez-Rivera, R. Hydraulic fracture crossing natural fracture at nonorthogonal angles: A criterion and its validation. *Spe Prod. Oper. Prod. Oper.* **2012**, *27*, 20–26. [CrossRef]
28. Sarmadivaleh, M. Experimental and Numerical Study of Interaction of a Pre-Existing Natural Interface and an Induced Hydraulic Fracture. Ph.D. Thesis, Faculty of Science and Engineering, Bentley, Australia, 2012. [CrossRef]
29. Griffith, A.A. The phenomena of rupture and flow in solids. *Philos. Trans. R. Soc. Lond.* **1921**, *221*, 163–198. [CrossRef]
30. Irwin, G.R. Analysis of stresses and strains near the end of a fracture traversing a plate. *J. Appl. Mech.* **1957**, *3*, 361–364. [CrossRef]
31. Yao, Y.; Wang, W.; Keer, L.M. An energy based analytical method to predict the influence of natural fractures on hydraulic fracture propagation. *Eng. Fract. Mech.* **2018**, *189*, 232–245. [CrossRef]
32. Valkó, P.; Economides, M.J. *Hydraulic Fracture Mechanics*; Wiley: Chichester, UK, 1995.
33. Li, M.; Zhang, F.; Wang, S.; Dontsov, E.; Li, P. DEM modeling of simultaneous propagation of multiple hydraulic fractures across different regimes, from toughness- to viscosity-dominated. *Rock Mech. Rock Eng.* **2024**, *57*, 1–23. [CrossRef]

processes

MDPI

Article

Study of the Brittle–Ductile Characteristics and Fracture Propagation Laws of Ultra-Deep Tight Sandy Conglomerate Reservoirs

Xianbo Meng [1], Zixi Jiao [2,*], Haiyan Zhu [2], Peng Zhao [2], Shijie Chen [2], Jun Zhou [3], Hongyu Xian [4] and Yong Wang [1]

1 Sinopec Shengli Oilfield Company, Dongying 257000, China
2 College of Energy, Chengdu University of Technology, Chengdu 610059, China
3 College of Geophysics, Chengdu University of Technology, Chengdu 610059, China
4 College of Environment and Civil Engineering, Chengdu University of Technology, Chengdu 610059, China
* Correspondence: jiaozixi414@163.com

Abstract: Ultra-deep tight sandy conglomerate reservoirs in the Junggar Basin are characterized by vertically alternating lithologies that include mudstone, sandy conglomerate, and sandstone. High in situ stresses and formation temperatures contribute to a brittle–ductile transition process in the reservoir rocks. However, the brittle behavior and ductile hydraulic fracture propagation mechanisms under in situ conditions remain inadequately understood. In this study, ultra-deep core samples were subjected to triaxial compression tests under varying confining pressures and temperatures to simulate different burial depths and evaluate their brittleness. A three-dimensional hydraulic fracture propagation model was developed in ABAQUS 2023 finite element software, incorporating a cohesive zone ductile constitutive model. Numerical simulations were conducted, considering interlayer horizontal stress differences, injection rate, and fracturing fluid viscosity, to systematically analyze the influence of geological and engineering factors on ductile fracture propagation. A fracture length–height competition diagram was constructed to illustrate the propagation mechanisms. The results reveal that high temperatures significantly accelerate the brittle–ductile transition, which occurs at confining pressures between 55 and 65 MPa. Following this transition, failure modes shift from single-shear failure to a multi-localized fracture with bulging deformation. Interlayer horizontal stress differences were found to strongly influence fracture penetration, with larger stress differences hindering vertical growth. Increasing injection rates promoted the uniform distribution of lateral fractures and fracture tip development, while medium- to high-viscosity fracturing fluids enhanced fracture width and vertical stimulation uniformity. These findings provide important insights for optimizing fracturing strategies and expanding the effective stimulation volume in the ultra-deep tight sandy conglomerate reservoirs of the Junggar Basin.

Keywords: ultra-deep tight reservoir; brittle–ductile transition; finite element method; competition propagation; interlayer horizontal stress difference; injection rate; fracturing fluid viscosity

1. Introduction

In the western uplift region of the Junggar Basin, ultra-deep tight sandy conglomerate reservoirs are widely distributed and host abundant oil and gas resources. Hydraulic fracturing is the primary stimulation technique employed for enhancing productivity in these formations [1,2]. Recent exploration results have revealed that these reservoirs consist of vertically alternating layers of sandstone, sandy conglomerate, and mudstone, with pay zones exhibiting both dispersed and continuous distribution patterns. Significant heterogeneities exist among these layers in terms of their physical properties (e.g., porosity and permeability), mechanical properties (e.g., tensile strength and Young's modulus), and interlayer horizontal stress [3,4]. This strong vertical heterogeneity severely limits the vertical propagation of fractures and makes the uniform lateral development of fractures more difficult [4–6]. Additionally, high in situ stresses and high formation temperatures can induce brittle–ductile transition in reservoir rocks, where increased ductility markedly alters hydraulic fracture propagation behavior [6].

The brittleness of rocks plays a critical role in reservoir stimulation because highly brittle rocks are more favorable for fracturing purposes and the development of a larger effective stimulated reservoir volume [7]. Extensive research has been conducted on the brittle–ductile transformation of rocks, revealing that temperature and pressure have significant effects on rock brittleness [8–10]. Following a brittle–ductile transition, the capacity of the rock to sustain macroscopic failure after fracturing undergoes notable changes [7,11]. Ding et al. [12] carried out triaxial compression tests on sandstone and observed that increasing the confining pressure leads to a gradual transition from brittle failure to plastic flow, accompanied by an increase in peak strain. González-Gómez et al. [13] conducted uniaxial compression experiments on heat-treated limestone and found that high-temperature treatment enhanced rock ductility while significantly reducing both strength and elastic modulus. Rybacki et al. [7] performed triaxial compression tests on shale under different temperature conditions and reported that increases in temperature and pressure induce a transition from brittle to semi-brittle behavior, characterized by brittle failure at the micro-scale and ductile deformation at the macro-scale. Xie et al. [14] conducted triaxial compression experiments under in situ stress conditions for rocks at different burial depths, revealing that variations in mineral composition affect brittleness, which tends to increase with depth. Although considerable efforts have been made to investigate the brittle–ductile transformation of deep rocks, few studies have employed ultra-deep core samples while simultaneously considering the coupled effects of temperature and pressure to evaluate the brittle–ductile characteristics of sandy conglomerates. The brittle–ductile behavior of rocks under the in situ geomechanical conditions of the ultra-deep formations in the Junggar Basin remains poorly understood, leading to a lack of theoretical basis for guiding field development [12,15].

In recent years, numerical simulation methods have been widely employed to investigate hydraulic fracture propagation in sandy conglomerate composite reservoirs. Among these, the cohesive zone method based on damage mechanics has become a widely adopted approach for modeling hydraulic fracture propagation [16–20]. For example, Lv et al. [16] developed a two-dimensional hydraulic fracturing model to explore fracture propagation patterns in composite reservoirs and found that high injection rates, combined with low-viscosity fracturing fluids, promote cross-layer fracture propagation. Hou et al. [4] further established a three-dimensional model to analyze fracture growth across lithological interfaces under varying interface strengths and interlayer stress conditions, providing a quantitative assessment of the coupled effects of lithology and geomechanical factors on fracture penetration. However, these studies generally modeled reservoir rock as a purely elastic material, overlooking the significant influence of brittle–ductile behavior on fracture propagation in ultra-deep formations. Therefore, it is imperative to develop a ductile fracture propagation model that more accurately reflects the mechanical and geological features of sandy conglomerate reservoirs in the western Junggar Basin. Such a model would provide important insights into the competitive propagation mechanisms between fracture length and height under different geological and engineering conditions.

To deepen our understanding of brittle–ductile transition characteristics in ultra-deep formations, develop a hydraulic fracturing model suited to regional geological conditions, and elucidate the competitive propagation mechanisms between fracture length and height, this study investigates core samples retrieved from depths greater than 5000 m in the western Junggar Basin. A series of triaxial compression tests were conducted under varying confining pressures and temperatures to simulate different burial environments. Through an analysis of failure modes and stress–strain responses, the brittle–ductile behavior of the ultra-deep tight sandy conglomerate was systematically characterized. Based on the experimental data, a cohesive ductile constitutive model was developed to reflect the geological conditions of the study area. Numerical simulations incorporating both geological and engineering parameters were subsequently carried out, resulting in a diagram that reveals the competitive relationship between fracture length and height. The findings provide valuable guidance for optimizing fracturing design and expanding the effective stimulation range of ultra-deep sandy conglomerate reservoirs in the western Junggar Basin.

2. Geological Characteristics of Tight Sandy Conglomerate Reservoirs

In the tight sandy conglomerate composite reservoirs of the Junggar Basin, mudstone, sandy conglomerate, and sandstone are commonly interbedded. Influenced by in situ stresses and formation temperatures, these rocks exhibit certain brittle–ductile transitions. Taking the Jiangbasi Formation in the Shibei Depression as an example, core observations and logging interpretations (Figure 1) indicate strong vertical heterogeneity within the reservoir. The interbedded distribution of mudstone, sandy conglomerate, and sandstone is irregular, with formation thicknesses ranging from approximately 5 to 20 m. The interlayer horizontal stress difference is typically 2–5 MPa, and the longitudinal architecture consists of connected thin layers (5–10 m) and isolated thick layers (10–20 m).

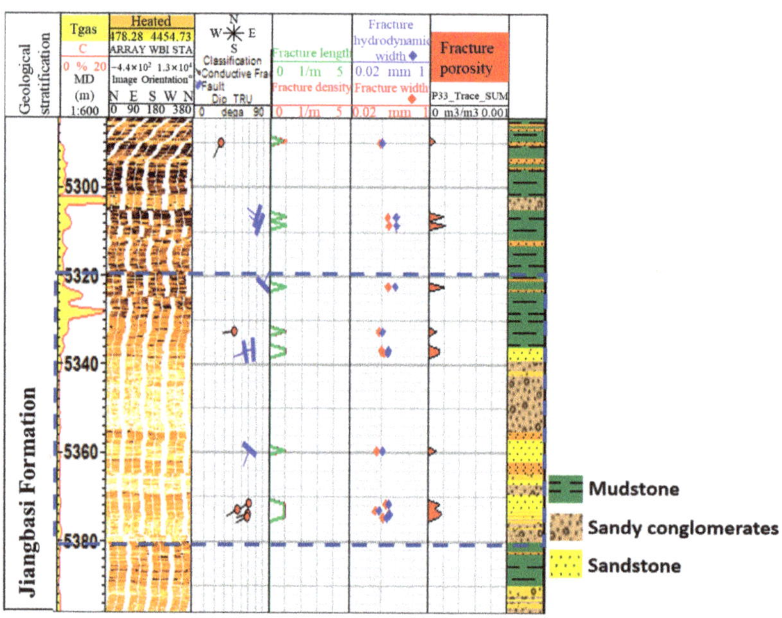

Figure 1. Imaging logging map of dense gravel reservoirs.

For thin single-layer pay zones, the multi-layer fracturing results indicate that vertical heterogeneity, particularly interlayer horizontal stress differences and lithological differences, significantly constrains fracture height growth, thereby limiting cross-layer fracture propagation. In thick single-layer pay zones, when the fluid-driving capacity at the hydraulic fracture tip is insufficient, lateral fracture propagation becomes uneven. Under such conditions, fracturing fluids tend to migrate preferentially along lithological interfaces rather than uniformly penetrating the reservoir, further reducing stimulation efficiency.

3. Triaxial Mechanical Experiments on Ultra-Deep Core Samples

3.1. Materials and Experimental Methodology

The core samples used in this study were obtained from the 5000–5500 m interval of Well X5, located in the western uplift of the Junggar Basin. The cores were processed into standard cylindrical samples measuring 50 mm in height and 25 mm in diameter. To investigate the effects of pressure and temperature on the brittle–ductile mechanical behavior of rocks under ultra-deep reservoir conditions, triaxial compression tests were conducted. The test conditions were designed to simulate reservoir environments at burial depths of 5500 m and 6500 m, with the corresponding experimental parameters summarized in Table 1. Specifically, the sandy conglomerate contains a low gravel content and small particle sizes, exhibiting characteristics similar to sandstone. However, notable differences remain in its physical properties (e.g., porosity and permeability) and mechanical properties (e.g., tensile strength and Young's modulus) compared to typical sandstone.

Table 1. Experimental parameters of samples at different temperatures and circumferential pressures.

Samples	Simulation Depth (m)	Confining Pressure (MPa)	Temperature (°C)
1	5500	55	20
2	6500	65	20
3	5500	55	130
4	6500	65	150

The main experimental procedures were as follows. (1) Prior to testing, each sample was wrapped with a heat-shrinkable film, mounted onto the base of the pressure chamber, and equipped with both axial and circumferential extensometers. The alignment of the loading piston, upper platen, sample, and lower platen was carefully checked to ensure that they were colinear. (2) During the test, the triaxial pressure chamber was sealed, and an initial axial load of 2 MPa was applied to secure the sample. Confining pressure oil was then injected, and the confining pressure was applied. (3) The temperature was raised to the predetermined value using the heating system. (4) Axial loading was applied under strain control at a constant rate of 0.05 mm/min until sample failure. (5) Upon completion of the test, the data were recorded, and the system was shut down.

3.2. Experimental Results

The failure modes of the core samples under different simulated depth conditions are illustrated in Figure 2. Regions outlined or marked in blue indicate the development of shear fractures, while those outlined or marked in red correspond to the formation of tensile fractures.

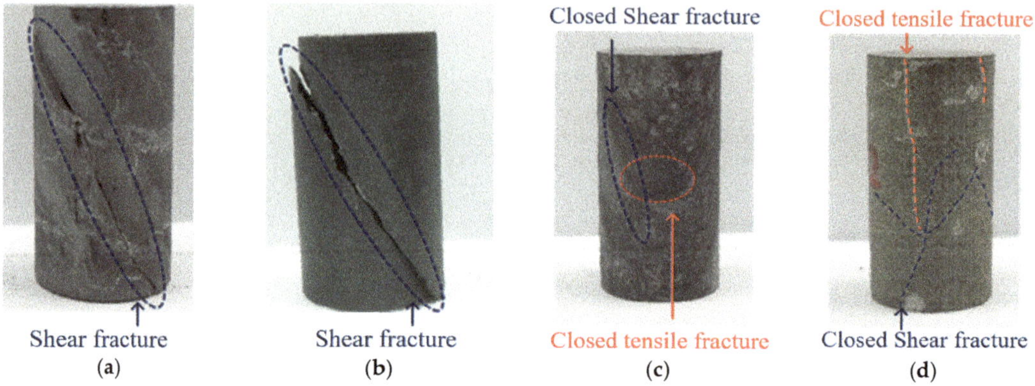

Figure 2. Failure modes of core samples from triaxial compression tests. (**a**) Sample 1. (**b**) Sample 2. (**c**) Sample 3. (**d**) Sample 4.

3.2.1. The Influence of Confining Pressure on the Failure Modes of Ultra-Deep Core Samples

Figure 2a,b shows the failure patterns of core samples under different simulated depth conditions, considering only variations in the stress environment. Under confining pressures of 55 MPa and 65 MPa, the samples predominantly exhibited single shear failure. Apart from the main shear fractures, no significant macroscopic cracks were observed around the samples. After reaching peak stress, the load-bearing capacity of the samples dropped sharply. The applied confining pressure did not prevent the generation of internal micro-fractures, their coalescence into macroscopic fractures, or the eventual formation of a well-defined shear surface. The consistent occurrence of single shear failure under relatively high-stress conditions indicates that the ultra-deep core samples from the study area exhibit high brittleness. As the confining pressure increased, the angle between the fracture surface and the loading axis increased from 23.9° to 27.3°, while the angle of the fracture face decreased from 66.1° to 62.7°. These changes suggest a reduction in overall rock brittleness and an enhancement of ductile behavior [21,22].

3.2.2. The Influence of Confining Pressure and Temperature on the Failure Modes of Ultra-Deep Core Samples

Figure 2c,d shows the failure modes of the core samples under different simulated depths, considering both stress and temperature conditions. At a confining pressure of 55 MPa and a temperature of 130 °C, the sample developed a main shear fracture, accompanied by a tensile fracture in the middle region. The failure mode transitioned from a single localized fracture, indicative of strong brittleness, to a multi-local failure with secondary fractures, suggesting that the high temperature increased the ductility of the rock. Constrained by the confining pressure, the sample was unable to fail rapidly along the main fracture surface. As a result, dislocation slip occurred in some grains, leading to the formation of secondary fractures. At this stage, the angle between the fracture surface and the loading axis was 19.3°, which is smaller than that observed in Figure 2a. This result suggests that when using fracture angle as a brittleness indicator, it is essential to ensure that the sample undergoes single-shear failure. In ultra-deep formations, high temperatures intensify the brittle–ductile transition, and this effect must be evaluated prior to field application. Under combined conditions of 65 MPa confining pressure and 150 °C temperature, no continuous shear fracture was observed on the sample surface after failure. Instead, the failure mode shifted to bulging deformation, accompanied by multiple localized fractures. Compared with Figure 2b, the high temperature significantly intensified the brittle–ductile transition, with the macroscopic failure behavior evolving from localized shear failure to ductile bulging deformation. At this point, the high confining pressure inhibited complete structural instability. Following peak stress, the dominant failure mechanism involved plastic deformation and the dislocation slip of mineral grains. Numerous tensile and shear fractures appeared on the sample surface, yet none penetrated the entire sample.

3.2.3. The Influence of Confining Pressure and Temperature on the Stress–Strain Curves of Ultra-Deep Core Samples

The stress–strain curves of the core samples under different simulated depth conditions are presented in Figure 3. Under ambient temperature conditions at confining pressures of 55 MPa and 65 MPa, the samples predominantly exhibited brittle failure. The axial stress–strain curves displayed a linear trend prior to reaching peak stress, followed by a rapid stress drop after failure. Under high-temperature conditions, when the confining pressure was 55 MPa, the stress–strain curve and failure mode still indicated primarily brittle behavior, although a distinct pre-peak yield stage was observed. Compared to the curve under ambient temperature conditions, the initial portion of the stress–strain response remained nearly identical, but deviations appeared near the peak due to the onset of ductile deformation. At 65 MPa confining pressure and high temperature, both the stress–strain curve and failure mode indicated that the sample underwent a brittle–ductile transition, characterized by a pronounced pre-peak yield stage. After reaching peak stress, the stress decreased at a significantly slower rate, and the energy required for failure markedly increased. Compared to the response observed at 65 MPa under ambient temperature conditions, the combined effects of high temperature and confining pressure led to substantial changes in mechanical behavior, highlighting the influence of the brittle–ductile transition. These results confirm that for rocks in this region, depths between 5500 m and 6500 m, corresponding to confining pressures of 55–65 MPa and gradually increasing reservoir temperatures, represent the critical range for the brittle–ductile transition.

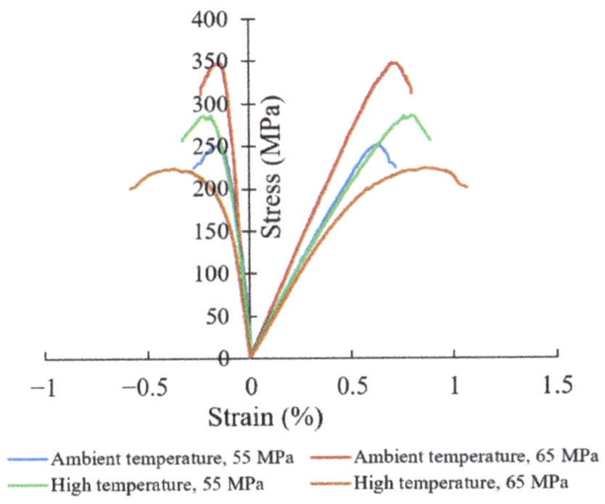

Figure 3. Stress–strain curves of the core samples in triaxial compression tests.

4. Physical Model of Hydraulic Fracturing Process

The hydraulic fracturing process is a complex fluid–solid coupled process involving rock deformation, fracture initiation and propagation, and fluid flow within the fractures. In this study, a three-dimensional hydraulic fracturing model was developed using the cohesive zone method (CZM). The damage factor parameters were calibrated based on the stress–strain curves obtained from triaxial tests on core samples, resulting in a cohesive zone ductile constitutive model that effectively captures the mechanical behavior of sandy conglomerate under regional geological conditions.

4.1. Coupled Flow-Solid Equations

The equilibrium equation for the rock matrix is expressed as follows [23–25]:

$$\int_V \left(\bar{\sigma} - p_w I\right) \delta \varepsilon \mathrm{dV} = \int_S t \cdot \delta v \mathrm{dS} + \int_V f \cdot \delta v \mathrm{dV} \tag{1}$$

where $\bar{\sigma}$ is the effective stress matrix, Pa; p_w is the pore pressure, Pa; I is the unit matrix; $\delta\varepsilon$ is the imaginary strain rate matrix, in s^{-1}; δv is the imaginary velocity matrix, in m/s; t is the surface force matrix, in N/m^2; and f is the body force matrix, in N/m^3.

The continuity equation governing the fluid medium is expressed as follows [4,26]:

$$\frac{1}{J}\frac{\partial}{\partial t}(J\rho_w n_w) + \frac{\partial}{\partial \mathbf{X}}(\rho_w n_w v_w) = 0 \tag{2}$$

where J is the volume change rate of the porous medium; ρ_w is the fluid density, in kg/m^3; n_w is the ratio of the fluid volume to the total volume; $\mathbf{X}$ is the spatial vector, in m; and v_w is the seepage velocity of the fluid, in m/s.

The fluid seepage rate through the rock matrix conforms to Darcy's law [4,26]:

$$v_w = \frac{1}{n_w g \rho_w} k \left(\frac{\partial \rho_w}{\partial x} - \rho_w g\right) \tag{3}$$

where g is the acceleration of gravity, in m/s^2 and k is the permeability matrix of the rock, in m/s.

4.2. Hydraulic Fracture Initiation and Propagation Criteria

In this study, the CZM was employed to simulate the initiation and propagation behavior of hydraulic fractures. The deformation process of the reservoir was divided into two stages: linear elastic behavior prior to fracture initiation and damage evolution following fracture initiation [27,28]. The maximum principal stress criterion was adopted to determine whether a CZM element experienced initial damage. Specifically, when the ratio of the stress in any of the three defined directions to its corresponding critical value exceeds 1, a hydraulic fracture is considered to be initiated [24]:

$$\max\left(\frac{\langle \sigma_n \rangle}{\sigma_n^{max}}, \frac{\tau_s}{\tau_s^{max}}, \frac{\tau_t}{\tau_t^{max}}\right) = 1 \tag{4}$$

where σ_n is the normal stress of the cohesive unit, in MPa; σ_n^{max} is the critical normal stress at the time of cohesive unit failure, in MPa; τ_s and τ_t represent the first tangential and second tangential stresses of the cohesive unit, in MPa; τ_s^{max} and τ_t^{max} represent the critical tangential stresses at the time of cohesive unit failure in the first tangential and second tangential directions, in MPa; $< >$ represents the cohesive unit without damage.

After hydraulic fracture initiation, the stiffness attenuation of the cohesive element was used to characterize its propagation process. The stress expression is as follows [24]:

$$\begin{cases} t_n = \begin{cases} (1-D)\overline{t_n} \\ \overline{t_n} \end{cases} \\ t_s = (1-D)\overline{t_s} \\ t_t = (1-D)\overline{t_t} \end{cases} \tag{5}$$

where t_n, t_s, and t_t represent the actual normal stress and the first and second tangential stresses applied to the cohesive unit, respectively, in Pa; $\bar{t}_n$, $\bar{t}_s$, and $\bar{t}_t$ represent the normal stress and the first and second tangential stresses predicted by the T-S criterion when no damage is produced, respectively, in Pa; D is the damage factor.

The cohesive zone is the area in front of the fracture tip, where micro-fractures are generated and gradually connect to form a macroscopic fracture. The cohesive element uses the bilinear traction-separation (T-S) criterion to determine fracture propagation. Based on the energy balance relationship, the critical displacement of the fracture can be calculated using the following formula [27,28]:

$$\delta_{\mathrm{m}}^{\mathrm{f}} = \frac{2K_{\mathrm{IC}}^2(1-\nu^2)}{E\sigma_{\mathrm{t}}} \tag{6}$$

where $\delta_{\mathrm{m}}^{\mathrm{f}}$ is the displacement of the cohesive unit at complete destruction, in m; K_{IC} is the fracture toughness of the rock, in Pa·m$^{1/2}$; v is the Poisson's ratio; E is the Young's modulus, in Pa; σ_{t} is the tensile strength of the rock, in Pa.

After that, the constitutive curve exhibited linear softening behavior, as shown in Figure 4, and D increased linearly from 0 to 1. When the separation displacement reached its critical value, the interface produced irreversible complete damage, and the interface was detached. The expression of damage factor is as follows [24]:

$$D = \frac{\delta_{\mathrm{m}}^{\mathrm{f}}\left(\delta_{\mathrm{m}}^{\max} - \delta_{\mathrm{m}}^{0}\right)}{\delta_{\mathrm{m}}^{\max}\left(\delta_{\mathrm{m}}^{\mathrm{f}} - \delta_{\mathrm{m}}^{0}\right)} \tag{7}$$

where $\delta_{\mathrm{m}}^{\max}$ is the maximum displacement of the cohesive unit during the loading process, m; δ_{m}^{0} is the displacement of the cohesive unit at the initial damage; when the damage is not generated, D is 0, when the cohesive unit is completely destroyed, D is 1.

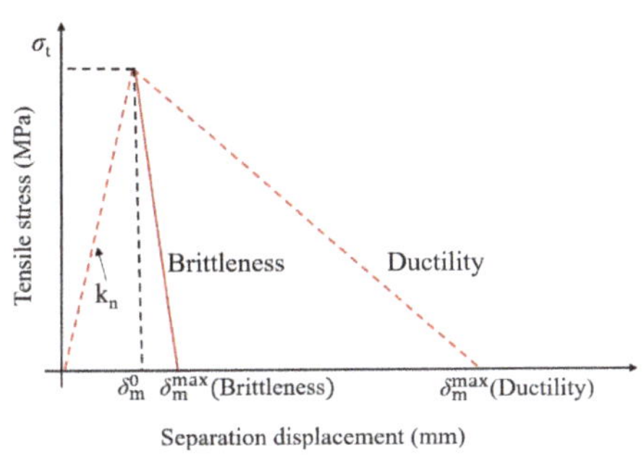

Figure 4. The fracture process and the traction-separation law.

Previous studies have shown that brittle rocks undergo rigid softening after damage, then the stress decreases rapidly. After the brittle–ductile transition, the rock transforms under elastic softening, the rate of stress reduction slows down, and the critical displacement increases [27–29].

4.3. Fluid Flow

During hydraulic fracture propagation, the fracturing fluid enters the fracture, generating a tangential flow that drives the fluid forward along the fracture, as well as a normal flow that leaks off into the surrounding rock matrix.

The tangential flow is governed by the following equation [30,31]:

$$q = -\frac{w^3}{12\mu}\nabla p \tag{8}$$

where q is the flow rate along the fracture propagation direction, in m^3/s; w is the hydraulic fracture width, in m; μ is the fracturing fluid viscosity, in mPa·s; and ∇p is the fluid pressure gradient, in MPa/m.

The normal (leak-off) flow into the rock matrix is governed by the following equation [32]:

$$\begin{cases} q_t = c_t(p_i - p_t) \\ q_b = c_b(p_i - p_b) \end{cases} \tag{9}$$

where q_t and q_b are the flow rate of filtration loss on the upper and lower surfaces of the cohesive element, respectively, in m^3/s; c_t and c_b are the filtration loss coefficients on the upper and lower surfaces of the cohesive element; p_i, p_t, and p_b are the pore pressures at the intermediate node, the upper surface node, and the lower surface node of the cohesive element, respectively, in Pa.

5. Establishment and Validation of Hydraulic Fracturing Models for Tight Sandy Conglomerate Reservoirs

5.1. Calibration of Cohesive Zone Ductile Constitutive Models

A triaxial compression model was developed in ABAQUS 2023 finite element software, based on the mechanical test results of Sample 4, to conduct numerical simulations and calibrate the cohesive zone ductile constitutive model. The Young's modulus was set to 39 GPa, Poisson's ratio to 0.22, and the loading rate to 0.05 mm/min. The critical displacement parameter was adjusted based on the experimental results to ensure that the model transitioned from brittle damage to ductile damage, consistent with the mechanical behavior observed in core samples from the sandy conglomerate reservoirs of the Junggar Basin. The simulated failure mode and the corresponding stress–strain response are presented in Figure 5.

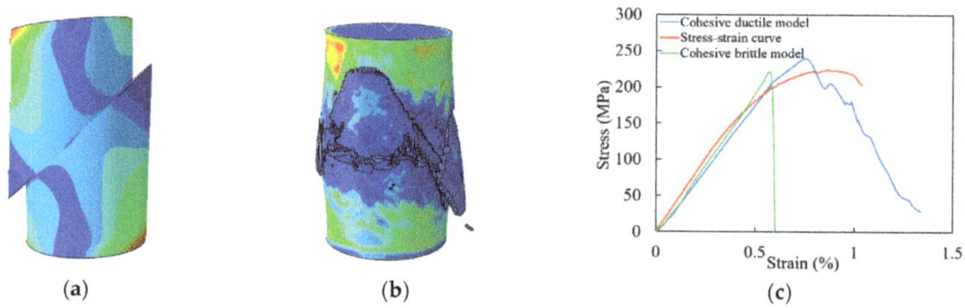

Figure 5. Numerical simulation results of triaxial compression. (**a**) Failure mode of the cohesive brittle constitutive model. (**b**) Failure mode of the cohesive ductile constitutive model. (**c**) Simulated stress–strain curve.

As shown in Figure 5, increasing the critical displacement from 0.001 m to 0.002 m enhances the ductility of the sample. Under the cohesive zone ductile constitutive model, both the simulated failure mode and stress–strain curve closely resemble those observed in high-temperature experimental tests. Therefore, when simulating hydraulic fracture propagation in the tight sandy conglomerate composite reservoirs of the Junggar Basin, the cohesive zone ductile constitutive model provides improved predictive accuracy and better represents the actual mechanical behavior of the reservoir.

5.2. Establishment of Composite Reservoir Models

In this study, a three-dimensional hydraulic fracturing numerical model for tight sandy conglomerate reservoirs in the Junggar Basin was developed using ABAQUS 2023 finite element software. The model incorporates a triaxial in situ stress system represented by σ_h (minimum horizontal principal stress), σ_H (maximum horizontal principal stress), and σ_v (vertical stress). To systematically investigate the controlling mechanisms of geological and engineering factors, including interlayer horizontal stress difference, injection rate, and fracturing fluid viscosity, on fracture height and length propagation, a three-layer model with equal thickness (20 m per layer) and heterogeneous lithology was constructed based on the actual geological data. From top to bottom, the lithological sequence consists of a mudstone layer, a sandy conglomerate layer, and a sandstone layer. Utilizing the geometric symmetry of the reservoir, a semi-model with dimensions of 250 m × 50 m × 60 m was established, as shown in Figure 6. The formation matrix was simulated using C3D8P elements to capture both fluid seepage and mechanical deformation, while hydraulic fracture propagation along predefined paths was modeled using COH3D8P cohesive elements. The injection point was placed at the center of the middle sandy conglomerate layer. The complete model comprised 63,750 elements. Simulation parameters for each lithological layer were assigned based on core mechanical testing, as well as the petrophysical and geomechanical properties interpreted from the well logs, as detailed in Tables 2 and 3.

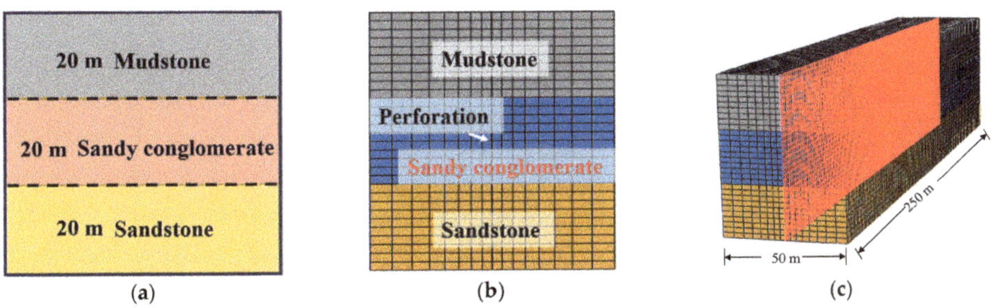

Figure 6. Numerical model of hydraulic fracturing in a tight sandy conglomerate reservoir. (**a**) Schematic of the stratigraphic layering. (**b**) Schematic of the layered numerical model. (**c**) Mesh configuration of the three-dimensional model.

Table 2. Rock matrix model parameters for the different layers.

Property	Mudstone Layer	Sandy Conglomerate Layer	Sandstone Layer
Young's Modulus (GPa)	29	36	26
Poisson's ratio	0.25	0.24	0.28
tensile strength (MPa)	3	4	2
permeability (mD)	0.178	0.31	0.734
porosity (%)	5.3	7.9	7.7
$\sigma_h/\sigma_H/\sigma_v$ (MPa)	55/56/64	51/56/64	55/56/64

Table 3. Defined parameters of the cohesive element model.

Cohesive Element Property	Mudstone Layer	Sandy Conglomerate Layer	Sandstone Layer
tensile strength (MPa)	3	4	2
critical displacement (m)	0.002	0.002	0.002
filtration coefficient (m/(Pa·s))	10^{-14}	10^{-10}	10^{-14}

5.3. Field Data Validation of Hydraulic Fracturing Models

Fracturing operations were conducted in the Baijiantan Formation of the Junggar Basin, with the targeted intervals primarily consisting of sandstone layers. Based on the mechanical testing of core samples and well-logging interpretation, a reservoir model was established with a Young's modulus of 28 GPa, a tensile strength of 3 MPa, and a Poisson's ratio of 0.3. According to the actual field pumping schedule, the injection parameters were set as follows: a flow rate of 12 m^3/min, a fracturing fluid viscosity of 50 mPa·s, and a total fluid volume of 300 m^3.

Numerical simulations were conducted using both the cohesive zone brittle constitutive model and the cohesive zone ductile constitutive model. The corresponding bottomhole pressure–time curves are shown in Figure 7. In the brittle model, the simulated breakdown pressure was 70.2 MPa, and the injection pressure required for stable fracture propagation ranged from 60 to 64 MPa. In contrast, the ductile model yielded a higher breakdown pressure of 80.2 MPa, with a stable propagation pressure range of between 60 and 80 MPa. Field data analysis indicated that the actual breakdown pressure was 78.3 MPa, and the injection pressure for stable fracture propagation ranged from 61.7 to 85.95 MPa.

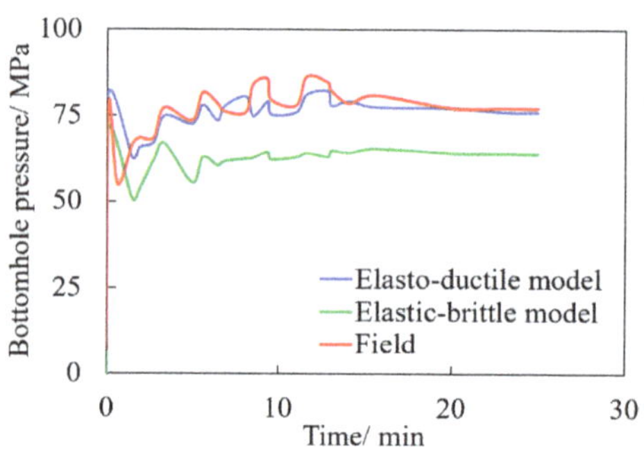

Figure 7. Comparison between the simulated pumping pressure curve and the actual construction pressure curve.

The comparison between the simulated injection pressure curve and the actual field data indicates that the breakdown pressure and the stable propagation pressure, as simulated by the hydraulic fracturing model that was based on the cohesive zone ductile constitutive model, closely align with the measured field values. In contrast, the model based on the cohesive zone's brittle constitutive behavior shows larger deviations. These findings suggest that the cohesive zone ductile model provides higher reliability and better applicability for simulating hydraulic fracturing in this reservoir.

6. Numerical Simulation Results

In this section, the hydraulic fracturing numerical model incorporating the cohesive zone ductile constitutive model, which was established and validated in Section 5, is employed to investigate the ductile fracture propagation behavior of ultra-deep sandy conglomerate composite reservoirs in the Junggar Basin under varying geological and engineering conditions. Based on the actual field construction conditions, the basic simulation parameters were set as follows: an interlayer horizontal stress difference of 4 MPa, an injection rate of 12 m^3/min, a fracturing fluid viscosity of 10 mPa·s, and a total fluid volume of 300 m^3.

6.1. Interlayer Horizontal Stress Difference

After the injection of fracturing fluid, the propagation of hydraulic fractures in both length and height was influenced by in situ stress conditions, resulting in variations in fracture geometry. In composite reservoirs, differences in the minimum horizontal principal stress between layers must be considered when evaluating their impact on fracture propagation. To investigate this effect, the interlayer horizontal stress difference was incrementally varied to 2 MPa, 4 MPa, and 6 MPa. The resulting fracture morphologies under different interlayer stress conditions were obtained after fluid injection, as shown in Figure 8.

During hydraulic fracture propagation, the sandy conglomerate layer, with its higher tensile strength, required a longer period of pressure buildup and higher net pressure within the fracture to initiate and sustain propagation. In the initial stage, fracture initiation occurred within the sandy conglomerate layer. As propagation continued, the hydraulic energy inside the fracture gradually dissipated. In the later stage, a greater volume of fracturing fluid migrated into the underlying sandstone layer, which provided more favorable conditions for fluid flow and fracture propagation. This promoted increased fracture width and enhanced lateral development. Meanwhile, fluid inflow into the sandy conglomerate layer declined, leading to the stagnation of fracture propagation in that zone and the formation of a distinct gap in the middle of the fracture.

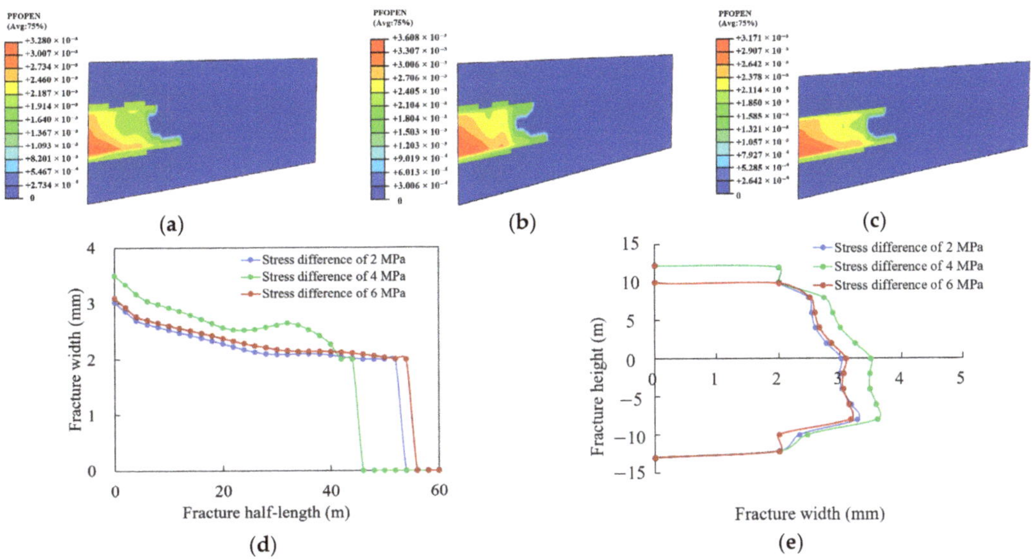

Figure 8. Changes in fracture propagation morphology with interlayer horizontal stress difference. (**a**) Interlayer horizontal stress difference of 2 MPa. (**b**) Interlayer horizontal stress difference of 4 MPa. (**c**) Interlayer horizontal stress difference of 6 MPa. (**d**) Fracture width profile. (**e**) Fracture height profile.

The initiation of a hydraulic fracture requires fluid pressure to exceed both the tensile strength of the rock and the minimum horizontal principal stress. As the interlayer horizontal stress difference increases, the minimum horizontal stress in the barrier layer also increases, enhancing the barrier layer stress effect and making it more difficult for fractures to initiate and propagate across the barrier layer. As shown in Figure 8a, when the interlayer horizontal stress difference was 2 MPa, the hydraulic fracture propagated vertically, penetrating both the upper and lower layers. A larger proportion of the fracturing fluid entered the sandstone layer, which has a lower Young's modulus and higher porosity, resulting in a wider fracture width near the sandstone interface and greater fracture length. When the interlayer horizontal stress difference increased to 4 MPa (Figure 8b), the fracture width increased while the fracture length decreased. At an interlayer horizontal stress difference of 6 MPa (Figure 8c), the enhanced barrier layer stress effect significantly inhibited upward fracture propagation, while the overall fracture length increased. As shown in Figure 8d,e, under the same total fluid volume, increasing the interlayer horizontal stress difference led to a reduction in fracture height, a significant increase in fracture length, and improved lateral fracture development within the central reservoir layer.

Therefore, in tight sandy conglomerate composite reservoirs, increasing the interlayer horizontal stress difference enhances the barrier layer stress effect, thereby restricting cross-layer fracture propagation. However, it also promotes more effective fracture development within the target layer, particularly by increasing the effective fracture length near the fracture tip.

6.2. Injection Rate

After the injection of fracturing fluid, stress concentration occurred at the fracture tip. When the maximum tensile stress generated by the hydraulic energy exceeded the tensile strength of the rock, fracture initiation and propagation occurred. The hydraulic energy available for fracture development varies with the injection rate, thereby influencing the fracture propagation pattern. To investigate this effect, the injection rate was sequentially set to 12 m^3/min, 14 m^3/min, and 16 m^3/min, while maintaining a constant total fluid volume of 300 m^3. The resulting fracture geometries under different injection rate conditions are shown in Figure 9.

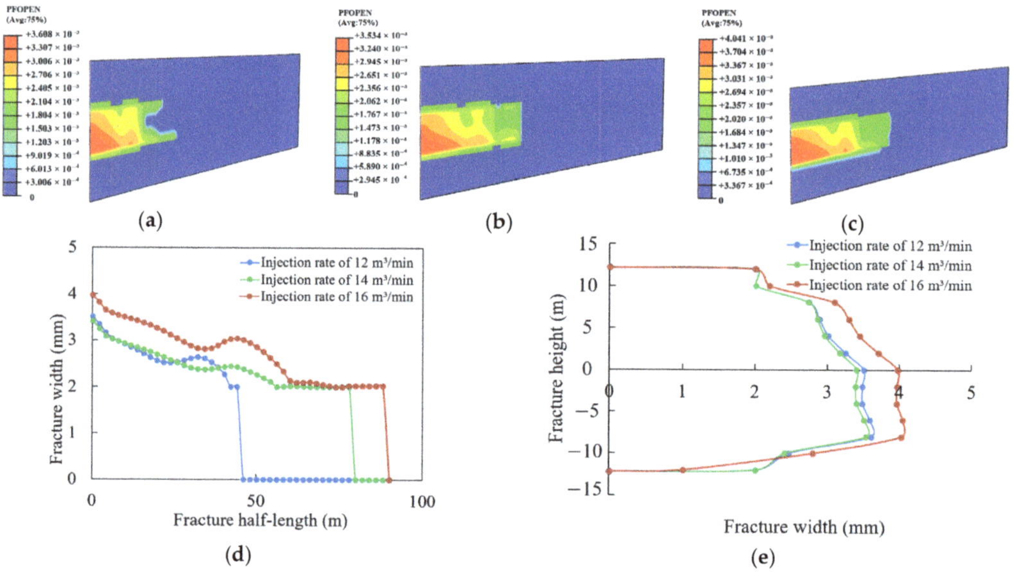

Figure 9. Variation in fracture propagation morphology with injection rate. (**a**) Injection rate of 12 m^3/min. (**b**) Injection rate of 14 m^3/min. (**c**) Injection rate of 16 m^3/min. (**d**) Fracture width profile. (**e**) Fracture height profile.

In tight sandy conglomerate composite reservoirs, variations in lithology and in situ stress result in non-uniform fracture propagation. Fracturing fluid tends to be insufficient at the tip of the central fracture, with a larger portion diverted toward interlayer interfaces.

Near the weaker sandstone layer, the fracture exhibits an increased aperture. As shown in Figure 9a, when the injection rate was 12 m^3/min, nearly half of the leading edge of the central fracture within the sandy conglomerate layer remained unfractured. In the mechanically stronger mudstone layer, cross-layer propagation was uneven and poorly developed. As the injection rate increased from 12 m^3/min to 16 m^3/min (Figure 9b,c), the ability to accumulate net pressure within the fracture improved, enabling the formation of a larger fracture volume within the same injection time. The elevated hydraulic energy facilitated more uniform propagation of the central fracture, significantly increasing the effective fracture length at the tip and moderately improving fracture width. As shown in Figure 9d,e, under a constant total fluid volume, increasing the injection rate notably improved fracture propagation in the central reservoir layer. Although fracture width increased only slightly, the overall fracture volume expanded. Meanwhile, in the direction of fracture propagation, the gradient of decreasing fracture width became less steep, indicating improved uniformity in both central fracture development and cross-layer propagation. Therefore, in "mudstone–sandy conglomerate–sandstone" composite reservoirs, increasing the injection rate effectively mitigates the adverse effects of vertical lithological heterogeneity on the uniform propagation of central fractures and the cross-layer propagation of hydraulic fractures.

6.3. Fracturing Fluid Viscosity

Changes in the viscosity of the fracturing fluid can influence its flow characteristics, alter local fracture friction, and reduce fluid leak-off into the formation, thereby affecting the development of hydraulic fractures. To investigate this effect, the fracturing fluid viscosity was sequentially set to 10 mPa·s, 50 mPa·s, and 100 mPa·s. With a fixed total injection volume of 300 m^3, the resulting fracture propagation patterns under different viscosities were obtained, as shown in Figure 10.

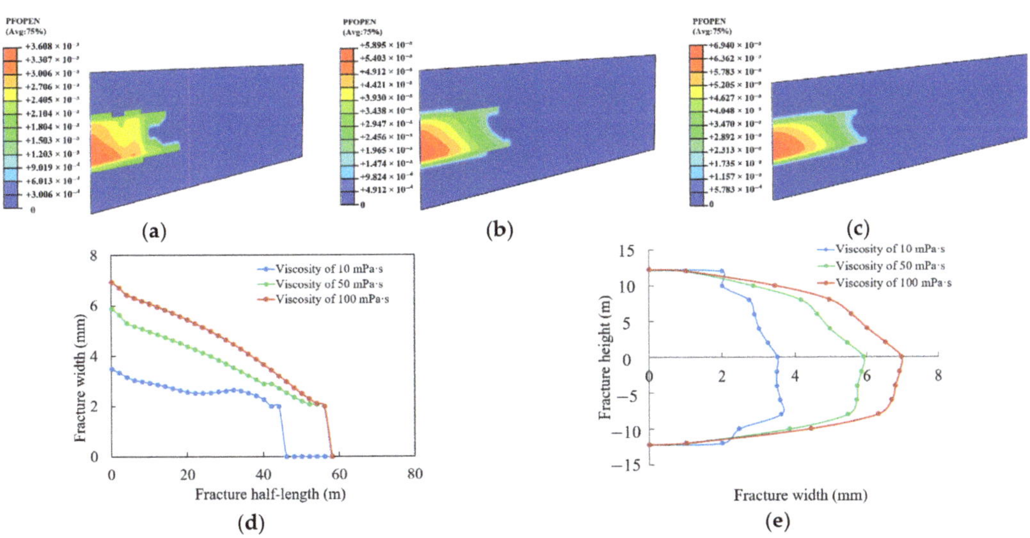

Figure 10. Variation in fracture propagation morphology with fracturing fluid viscosity. (**a**) Viscosity of 10 mPa·s. (**b**) Viscosity of 50 mPa·s. (**c**) Viscosity of 100 mPa·s. (**d**) Fracture width profile. (**e**) Fracture height profile.

As shown in Figure 10a, when the viscosity of the fracturing fluid was 10 mPa·s, vertical lithological heterogeneity within the reservoir significantly influenced the geometry of the hydraulic fracture. Along the fracture length direction, the width of the central fracture exhibited substantial fluctuations, with a low degree of opening at the fracture tip. Vertically, the fracture width was narrower in the upper reservoir and wider in the lower reservoir, indicating a non-uniform distribution. When the viscosity increased to 50 mPa·s (Figure 10b), the higher viscosity fracturing fluid effectively reduced fluid leak-off during the fracturing process. This enhanced the fluid volume available for driving fracture propagation, decreased the pressure drop gradient within the fracture, and led to increases in both overall fracture width and effective fracture length. The variation pattern of fracture width along the fracture length direction transitioned from irregular fluctuations to a more linear decreasing trend. At a viscosity of 100 mPa·s (Figure 10c), the overall fracture width increased further. Vertically, the distribution pattern of fracture width changed from a gradual top-to-bottom decrease to a configuration with maximum width in the middle reservoir and smaller widths in the upper and lower layers. As shown in Figure 10d,e, for tight sandy conglomerate composite reservoirs, high-viscosity fracturing fluids can increase flow resistance within the fracture, significantly enhance fracture width development in the central reservoir, and promote greater fracture length propagation. At the same time, fluid entry into the upper and lower layers is suppressed, thereby improving the uniform propagation of fractures within the target middle layer.

7. Discussion

In tight sandy conglomerate composite reservoirs, fracture propagation is highly non-uniform, due to the combined effects of interlayer horizontal stress differences and vertical lithological heterogeneity. Based on the simulation results presented in Section 6, a fracture length–height competition diagram was developed, as shown in Figure 11. For single pay zone scenarios, optimizing the lateral propagation and uniformity of the central fracture while limiting vertical growth is favorable. In contrast, for multi-layer reservoirs, promoting effective cross-layer fracture propagation to connect multiple intervals can significantly enhance oil and gas recovery. Overall, for multi-layer stimulation, selecting a target interval with low interlayer horizontal stress differences (<4 MPa) and employing medium- to high-viscosity fracturing fluids (10–100 mPa·s) at high injection rates (>14 m^3/min) is conducive to promoting cross-layer fracture propagation and expanding the stimulated reservoir volume. In contrast, in single-layer stimulation, targeting intervals with high interlayer horizontal stress differences (>4 MPa), combined with high viscosity fracturing fluids (50–100 mPa·s) and high injection rates (>14 m^3/min), can improve the lateral propagation balance of fractures within the target layer, increase fracture width and tip opening, and ultimately expand the effective stimulated area.

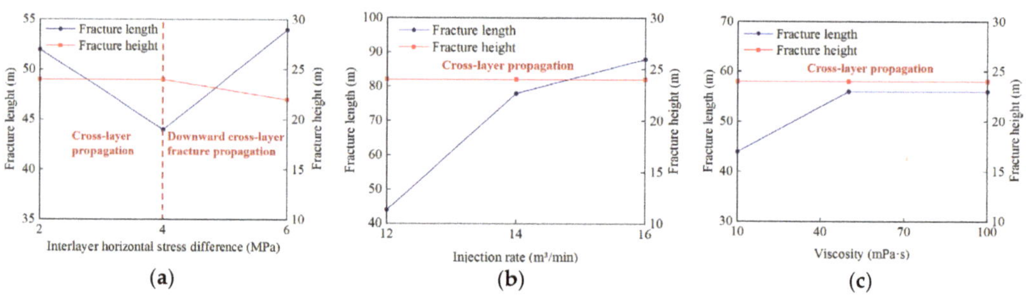

Figure 11. Fracture length–height competition diagram under different influencing factors. (**a**) Interlayer horizontal stress difference. (**b**) Injection rate. (**c**) Fracturing fluid viscosity.

Based on the above findings, a field application was conducted in the Jiangbasi Formation of Well XX6 in the Junggar Basin. The reservoir is characterized by a mudstone–sandy conglomerate–sandstone sequence. Casing fracturing was implemented in three intervals with a total thickness of 11.2 m. The fracturing strategy focused on intervals with low interlayer horizontal stress differences (3–4 MPa), using a high injection rate of 14 m^3/min and alternating injections of medium- and high-viscosity fracturing fluids. This approach facilitated both vertical cross-layer propagation and the relatively uniform lateral propagation of hydraulic fractures. Following the treatment, the well achieved daily gas production of 6581 m^3 and daily water production of 31.96 m^3, indicating effective multi-layer stimulation and sufficient vertical reservoir activation.

8. Conclusions

(1) Core samples from the 5000–5500 m interval of Well X5 in the western uplift of the Junggar Basin were subjected to triaxial compression tests. The results indicate that, when considering both temperature and pressure, the brittle–ductile transition for rocks in this region occurs between 55 MPa (130 °C) and 65 MPa (150 °C). Elevated temperatures significantly accelerate the brittle–ductile transition of the core samples.

(2) After the brittle–ductile transition occurs, the failure mode changes from localized shear failure along a single shear plane to bulging deformation involving multiple localized fractures. The stress–strain curve exhibits a pronounced pre-peak yield stage, and the post-peak stress drop rate is significantly reduced.

(3) Based on the laboratory experiment results, a cohesive ductile constitutive model tailored to regional geological characteristics was developed, and a ductile hydraulic fracturing model for ultra-deep sandy conglomerate reservoirs was established. A fracture length–height competition diagram was constructed to analyze fracture propagation behavior. The results indicate that vertical heterogeneity within the reservoir leads to unbalanced lateral fracture propagation. Fracturing fluid tends to migrate toward interlayer interfaces, and fracture development is more pronounced near sandstone layers with weaker mechanical properties. Increasing the injection rate and employing medium- to high-viscosity fracturing fluids effectively enhances the lateral uniformity of fracture propagation within the target layer.

Author Contributions: Methodology, X.M.; Software, H.Z. and J.Z.; Data curation, S.C. and Y.W.; Writing—original draft, Z.J., P.Z. and H.X.; Writing—review & editing, Z.J., P.Z. and H.X. All authors have read and agreed to the published version of the manuscript.

Funding: This work was financially supported by the National Key Research and Development Program of China "Key Technical Index System and Standardization of the Whole Process of CO_2 Fracturing and Storage of Large Oil and Gas Reservoirs" (No. 2023YFF0614102), the transverse project of Sinopec Xinjiang Xinchun Petroleum Development Co., Ltd. "Research on crack initiation and propagation of ultra-deep fracturing" (No. 30203569-23-ZC0613-0012), and the Natural Science Foundation of Sichuan Province (No. 2025ZNSFSC0371).

Data Availability Statement: The original contributions presented in the study are included in the article; further inquiries can be directed to the corresponding author.

Conflicts of Interest: Authors Xianbo Meng and Yong Wang were employed by the company Sinopec Shengli Oilfield Company. The remaining authors declare that the research was conducted in the absence of any commercial or financial relationships that could be construed as a potential conflict of interest. The authors declare that this study received funding from the National Key Research and Development Program of China, the transverse project of Sinopec Xinjiang Xinchun Petroleum Development Co., Ltd. and the Natural Science Foundation of Sichuan Province. The funder had the following involvement with the study: The funder provided financial support and technical assistance during the research process.

References

1. Liu, X.; Wang, X.; Zhao, B.; Xiong, J.; Liu, L. Propagation of Hydraulic Fractures and Fracability Evaluation of Sandy Conglomerate Reservoirs. *Xinjiang Pet. Geol.* **2023**, *44*, 169–177.
2. Zan, L.; Wang, S.; Zhang, Z.; Sheng, L.; Xing, H. Research Status of Glutenite Reservoir. *J. Yangtze Univ. (Nat. Sci. Ed.)* **2011**, *8*, 63–66.
3. Zeng, F.; Guo, J.; Chen, M. Research and Application of Fracturing Technology for Large Thick Glutenite Reservoir. *J. Oil Gas Techno.* **2011**, *33*, 123–126.
4. Cui, Z.; Hou, B.; Fu, S.; Lv, J. Fractures Cross-Layer Propagation Characteristics of Integrated Fracturing in Shale Oil Tight Reservoir. *Fault-Block Oil Gas Field* **2022**, *29*, 111–117.
5. Yan, L.; Chen, F.; Wang, Z.; Yan, Y.; Cao, J.; Wang, S. Challenges and Technical Countermeasures for Effective Development of Shale Oil in China. *Pet. Drill. Tech.* **2024**, *48*, 63–69.
6. Yang, Z.; Hou, L.; Lin, S.; Luo, X.; Zhang, L.; Wu, S.; Cui, J. Geologic Characteristics and Exploration Potential of Tight Oil and Shale Oil in Lucaogou Formation in Jimsar Sag. *China Pet. Explor.* **2018**, *23*, 76–85.
7. Rybacki, E.; Meier, T.; Dresen, G. What Controls the Mechanical Properties of Shale Rocks?—Part II: Brittleness. *J. Pet. Sci. Eng.* **2016**, *144*, 39–58. [CrossRef]
8. Zhang, G.; Wu, S.; Guo, P.; Zhang, S. Mechanical deformation, acoustic emission characteristics, and microcrack development in porous sandstone during the brittle–ductile transition. *Rock Mech. Rock Eng.* **2023**, *56*, 9099–9120. [CrossRef]
9. Zhang, S.; Abdallah, K.; Li, L.; Hamdi, E.; Liu, J. Multiphase flow controlled by synergistic injection-production pressure: Enabling CO_2 geo-sequestration with additional gas recovery from vertically heterogeneous depleted shale reservoirs. *Fuel* **2025**, *398*, 135591. [CrossRef]
10. Ranjith, P.; Viete, D.; Chen, B.; Perera, M. Transformation plasticity and the effect of temperature on the mechanical behaviour of Hawkesbury sandstone at atmospheric pressure. *Eng. Geol.* **2012**, *151*, 120–127.
11. Tarasov, B.; Potvin, Y. Universal Criteria for Rock Brittleness Estimation Under Triaxial Compression. *Int. J. Rock Mech. Min. Sci.* **2013**, *59*, 57–69. [CrossRef]
12. Ding, R.; Zhang, F.; Wang, Y.; Li, X.; Lin, B. Brittleness of Deep Sandstone in Central Junggar Basin and Its Influence on Fracture Modes of Rocks. *Pet. Geol. Recovery Effic.* **2024**, *31*, 54–62.
13. González-Gómez, W.; Quintana, P.; May-Pat, A.; Avilés, F.; May-Crespo, J.; Alvarado-Gil, J. Thermal Effects on the Physical Properties of Limestones from the Yucatan Peninsula. *Int. J. Rock Mech. Min. Sci.* **2015**, *75*, 182–189. [CrossRef]
14. Xie, H.; Gao, M.; Fu, C.; Lu, Y.; Yang, M.; Hu, J.; Yang, B. Mechanical Behavior of Brittle-Ductile Transition in Rocks at Different Depths. *J. China Coal Soc.* **2021**, *46*, 701–715.
15. Zhang, P.; Mishra, B.; Heasley, K. Experimental Investigation on the Influence of High Pressure and High Temperature on the Mechanical Properties of Deep Reservoir Rocks. *Rock Mech. Rock Eng.* **2015**, *48*, 2197–2211. [CrossRef]
16. Lv, Z.; Pan, L.; Hao, L.; Zou, N.; Zou, Z. Numerical Simulation of Factors Influencing Hydraulic Fracture Propagation in Sandstone-Mudstone Interbedded Reservoirs. *Xinjiang Pet. Geol.* **2023**, *44*, 729–738.
17. Pan, R.; Zhang, G. The Influence of Fracturing Energy Anisotropy on Hydraulic Fracturing Path in Layered Rocks. *Chin. J. Rock Mech. Eng.* **2018**, *37*, 2309–2318.
18. Lan, T.; Hu, B.; Dong, W.; Zhang, X. Numerical Simulation Research of Hydraulic Fracture Extension Laws in Coal and Sandstone Interbed. *China Energy Environ. Prot.* **2018**, *40*, 38–44,49.
19. Zhao, H.; Wang, X.; Liu, Z.; Yan, Y.; Yang, H. Investigation on the Hydraulic Fracture Propagation of Multilayers-Commingled Fracturing in Coal Measures. *J. Pet. Sci. Eng.* **2018**, *167*, 774–784. [CrossRef]
20. Guo, J.; Luo, B.; Lu, C.; Lai, J.; Ren, J. Numerical Investigation of Hydraulic Fracture Propagation in a Layered Reservoir Using the Cohesive Zone Method. *Eng. Fract. Mech.* **2017**, *186*, 195–207. [CrossRef]
21. Zhou, H.; Meng, F.; Zhang, C.; Xu, R.; Lu, J. Quantitative Evaluation of Rock Brittleness Based on Stress-Strain Curve. *Chin. J. Rock Mech. Eng.* **2014**, *33*, 1114–1122.
22. Xia, Y.; Li, L.; Tang, C.; Ma, S.; Li, M. Rock Brittleness Evaluation Based on Stress Dropping Rate After Peak Stress and Energy Ratio. *Chin. J. Rock Mech. Eng.* **2016**, *35*, 1141–1154.
23. Li, M.; Zhou, F.; Huang, G.; Wang, B.; Hu, X.; Li, H. A Finite Element Simulation Method for Multi-Fracture Propagation in Horizontal Wells Based on Fluid Pipe Element. *J. China Univ. Pet. (Ed. Nat. Sci.)* **2022**, *46*, 105–112.
24. Wang, C.; Zhang, H.; Zhao, J.; Wang, T. Study on Hydraulic Fracture Propagation in Elastoplastic Reservoir. *China Pet. Mach.* **2022**, *50*, 80–88.
25. Jiao, Z.; Zhang, A.; Du, L.; Yang, Y.; Fan, H. Numerical Study of the Effect of Perforation Friction and Engineering Parameters on Multicluster Fracturing in Horizontal Wells. *Geofluids* **2021**, *2021*, 18. [CrossRef]

26. Yang, Y.; Jiao, Z.; Du, L.; Fan, H. Numerical Simulation of Shale Reservoir Fluid-Driven Fracture Network Morphology Based on Global CZM. *Front. Earth Sci.* **2021**, *9*, 775446. [CrossRef]
27. Sarris, E.; Papanastasiou, P. The Influence of the Cohesive Process Zone in Hydraulic Fracturing Modelling. *Int. J. Fract.* **2011**, *167*, 33–45. [CrossRef]
28. Sarris, E.; Papanastasiou, P. Modeling of Hydraulic Fracturing in a Poroelastic Cohesive Formation. *Int. J. Geomech.* **2011**, *12*, 160–167. [CrossRef]
29. Wang, H. Poro-elasto-plastic Modeling of Complex Hydraulic Fracture Propagation: Simultaneous Multi-Fracturing and Producing Well Interference. *Acta Mech.* **2016**, *227*, 507–525. [CrossRef]
30. Guo, J.; Lu, Q.; Zhu, H.; Wang, Y.; Ma, L. Perforating Cluster Space Optimization Method of Horizontal Well Multi-Stage Fracturing in Extremely Thick Unconventional Gas Reservoir. *J. Nat. Gas Sci. Eng.* **2015**, *26*, 1648–1662. [CrossRef]
31. Wang, B.; Zhou, F.; Wang, D.; Liang, T.; Yuan, L.; Hu, J. Numerical Simulation on Near-Wellbore Temporary Plugging and Diverting During Refracturing Using XFEM-Based CZM. *J. Nat. Gas Sci. Eng.* **2018**, *55*, 368–381. [CrossRef]
32. Yang, Q.; Li, B.; Shao, M.; Zhang, H. Production Performance of Perforation Clusters During Multistage Fracturing in Shale Gas Reservoirs. *ACS Omega* **2021**, *6*, 26231–26238. [CrossRef] [PubMed]

processes

MDPI

Article

Numerical Simulation of Stress Disturbance Mechanism Caused by Hydraulic Fracturing of Shale Formation

Yinghui Zhu [1,2], Heng Zheng [1,*], Yi Liao [2] and Ruiquan Liao [1]

[1] School of Petroleum Engineering, Yangtze University, Wuhan 434025, China; liaoruiquan@263.net (R.L.)
[2] Research Institute of CNOOC Shenzhen Branch, Shenzhen 518052, China
[*] Correspondence: hengsmith111@163.com

Abstract: Characterizing changes in rock properties is essential for the hydraulic fracture and re-fracture parameter optimization of shale formations. This paper proposed a hydraulic fracturing model to investigate the changes in rock properties during hydraulic fracturing using SPH, and the changes in the stress field and rock properties were quantitatively characterized. The simulation results indicated that the minimum horizontal principal stress increased by 10 MPa~15 MPa during fracture propagation, which is the main reason for the uneven propagation in multi-fracture propagation. Affected by the stress disturbance, the stimulated area was divided into four parts based on the changes in Young's modulus and permeability; the more seriously the stress disturbance was affected, the higher the permeability of the stimulated zone was, and the smaller the stimulated zone was. Meanwhile, a zone with reduced permeability appeared due to the compression effect caused by the high injection pressure, and this increased with the increase in stress disturbance. The main reason for this was that strain formed because of the compression effect from the high injection pressure. The higher the stress disturbance, the higher the accumulated strain. This new model provides a new method for fracture parameter optimization, which also provides a foundation for the re-fracture parameter optimization of shale formations.

Keywords: stress disturbance; smoothed-particle hydrodynamics; fracture propagation; numerical simulation

1. Introduction

Creating a complex fracture network is the key element for shale gas development because of the low permeability of shale formations [1,2]. This is an effective technology to improve the permeability of shale formations, but hydraulic fracturing also causes micro-earthquakes due to rock strain or stress disturbance. Hydraulic fracturing has become essential in recent decades [3]. During hydraulic fracturing, a large volume of fracturing fluid is injected into the shale formation. This leads to the compressional deformation of rocks due to the high net pressure inside the fractures. This compressional deformation is the main cause of stress disturbance during hydraulic fracturing. For the re-fracturing and hydraulic fracturing of adjacent wells, understanding the mechanism of stress perturbation and changes in rock properties is critical.

To determine the mechanism of hydraulic fracture distribution, numerical models of hydraulic fractures based on FEM, DEM, and FEM-DEM have been proposed, and the behavior of hydraulic fracture propagation has been determined [4,5]. It has clearly been shown that the horizontal principal stress difference is the parameter controlling crack propagation in multi-crack propagation simulations. This study suggests that increasing the cluster space and injection rate can cause the uniform propagation of multi-stage fractures. Furthermore, it has been observed that the pore pressure can control fracture propagation [6,7]. The hydraulic fractures tended to deflect to the high-pore-pressure areas while fracturing. In addition, the main factor limiting the propagation of intermediate fractures was the deflection of the maximum principal stress in the horizontal direction. The simulations showed that the spreading resistance increased by 6–8 MPa as a result of the deflection of maximum principal stress in the horizontal direction [8]. Meanwhile, natural fractures are also an important element in fracture networks. A series of numerical models have been developed to observe how hydraulic fractures propagate under varying principal horizontal stresses and angles of approach in order to investigate the relationship between natural and hydraulic fractures in fractured formations. The results showed that hydraulic fracturing occurred in the presence of natural fractures, and the following three types of fractures emerged: crossing, arresting, and terminating [9–12]. On the other hand, high principal stress in the horizontal direction and the approach angles inhibited the propagation of fractures [13,14]. The area with horizontal principal stress and angle of approach can be split into three parts based on how hydraulic fracture propagation occurs. The bedding surface interfered with the propagation of the hydraulic fracture in the vertical direction. In addition, the minimum principal stress in the horizontal direction and the stratigraphic thickness were found to be the most important geologic parameters affecting the fracture height. On the other hand, a large horizontal principal stress difference and great stratigraphic thickness may restrict the fracturing height in a single formation [15,16]. Additionally, thickening the fracturing fluid and boosting the injection rate have been shown to cause an increase in fracture height. The main goal of this study was to examine the factors contributing to surpassing the low fracture limit in shale formations, and it was found that high net pressures resulting from high injection rates and fracturing fluid viscosities play a significant role [17,18]. Experimental simulations to directly observe fracture extension were also conducted. The effects of stress, the injection parameters, and the rock mechanics on fracture extension were investigated using CT scans, and the results of the simulation were in good agreement with the experimental data.

As noted above, the study of hydraulic fracture growth in shale involves factors that influence the creation of an intricate fracture network. Understanding changes in the stress field and the rock properties of shale formations is of importance for significantly improving the productivity of the shale oil and gas industries via widespread re-fracturing. The currently used numerical model is inadequate to effectively describe the changes in the stress field and rock properties after hydraulic fracturing. Smoothed-particle hydrodynamics (SPH) is better suited to calculating the strain and stress fields. To investigate the stress perturbation mechanism and changes in rock properties during hydraulic fracturing, this work presents a new simulation method based on SPH to characterize the changes in the stress field and rock properties after hydraulic fracturing. In this simulation, the rock matrix is characterized by its particles, which effectively solves the problem of the predetermined fracture propagation path in the hydraulic fracturing procedure, resulting in a variety of extended fracture paths and the better characterization of stress field variations. Additionally, permeability is introduced to characterize fracture propagation and the creation of a high-permeability flow channel during the hydraulic fracturing process. Thus, the extension of hydraulic fractures can be effectively described by the changes in permeability.

2. Mathematical Methods

2.1. Boundary Conditions

As indicated in Figure 1, the whole domain was divided into two parts, $\Omega = \Omega^+ \cup \Omega^-$, and the whole domain was bounded by $\Gamma = \Gamma_t \cup \Gamma_u \cup \Gamma_d$. The normal vector n_d^+ indicates the direction of Ω^+ relative to the joint plane.

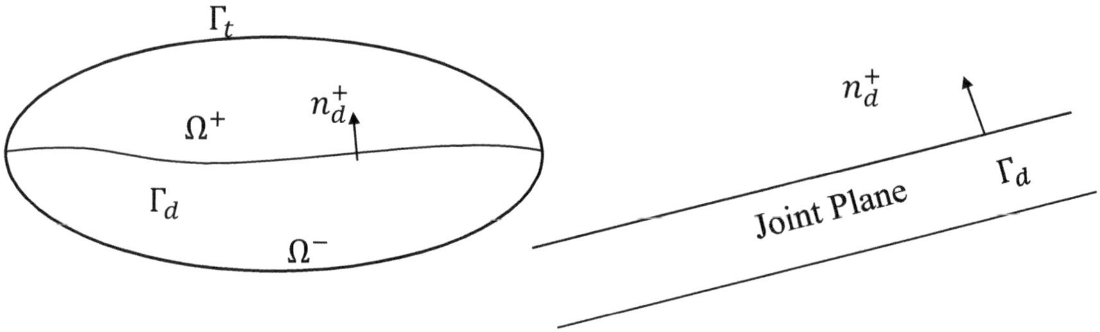

Figure 1. Domain with single joint plane, with a jump in the velocity field.

The classical strong forms of strong discontinuity equations are presented in [19] as follows:

$$\rho\dot{u} = \frac{1}{\rho}\nabla\sigma + g \tag{1}$$

$$u(x,t) = \overline{u}(x,t) \ on \ \Gamma_u \tag{2}$$

$$\sigma \cdot \hat{n} = t^{*} \ on \ \Gamma_t \tag{3}$$

$$\dot{\sigma}^{+} \cdot n_d^{+} = \dot{\sigma}^{-} \cdot n_d^{-} = \dot{t}_c \ on \ \Gamma_d \tag{4}$$

where t represents the traction, ρ represents the mass density, u represents the velocity, σ denotes the stress tensor, g represents the body force, and n represents the outward normal vector of the domain.

2.2. Governing Equations

During the simulations, the fracturing fluid was subject to mass and momentum conservation laws. These are expressed in the Lagrangian system as follows. The constitutive stress–strain relationship in rocks is also given in [20].

$$\sigma^{\alpha\beta} = \frac{1}{3}\sigma^{rr}\delta^{\alpha\beta} + s^{\alpha\beta} = -p\delta^{\alpha\beta} + s^{\alpha\beta} \tag{5}$$

In the above equation, p represents the pore pressure, $\delta^{\alpha\beta}$ denotes the Kronecker delta, and $s^{\alpha\beta}$ describes the second-order deviatoric stress tensor.

The results of the elastic and plastic strain calculations are as follows:

$$\dot{\varepsilon}^{\alpha\beta} = \frac{1}{2}\left(\frac{\partial v^{\alpha}}{\partial x^{\beta}} + \frac{\partial v^{\beta}}{\partial x^{\alpha}}\right) \tag{6}$$

$$\dot{\varepsilon}^{\alpha\beta} = \dot{\varepsilon}_e^{\alpha\beta} + \dot{\varepsilon}_p^{\alpha\beta} \tag{7}$$

where $\dot{\varepsilon}_e^{\alpha\beta}$ represents the elastic strain, and $\dot{\varepsilon}_p^{\alpha\beta}$ denotes the plastic strain.

The generalized Hook's law was used to determine the stress rate of the tensor [21]:

$$\dot{\sigma}_e^{\alpha\beta} = D_{\alpha\beta mn}\dot{\varepsilon}_e^{mn} = 2G\dot{\varepsilon}_{ed}^{\alpha\beta} + K\dot{\varepsilon}_{ev}^{rr}\delta^{\alpha\beta} \tag{8}$$

where K represents the bulk moduli, $D_{\alpha\beta mn}$ represents the standard isotropic elastic tensor of the fourth order, $\dot{\varepsilon}_{ed}^{\alpha\beta}$ represents the rate of deviatoric components of the elastic strain tensor, G represents the shear moduli, and $\dot{\varepsilon}_{ev}^{rr}$ represents the rate of volumetric components of the elastic strain tensor.

This simulation considered the rock as a Drucker–Prager material, with a yield criterion that is represented by a circular cone surface and is given by

$$F(\sigma, c) = \sqrt{J_2(s)} + \eta p(\sigma) - \xi c \tag{9}$$

and

$$\eta = \frac{6sin\phi}{\sqrt{3}(3 - sin\phi)}, \xi = \frac{6cos\phi}{\sqrt{3}(3 - cos\phi)}$$

where ϕ and c represent the angle of friction and the cohesion of the material, respectively.

2.3. Damage Evaluation

For the quantitative evaluation of rock damage during hydraulic fracturing, the hydraulic fracture propagation behavior is characterized by the D parameter:

$$D = \frac{4}{3}\pi c_g^3 \int_0^t n'(\varepsilon)\dot{\varepsilon}(1 - D)(t - \tau)^3 d\tau \tag{10}$$

where c_g represents the fracture propagation velocity.

The simulation presented in this equation was used to control the growth of damage:

$$\frac{dD^{1/3}}{dt} = \frac{1}{3} c_g \overline{w}^{1/3} \varepsilon^{m/3} \tag{11}$$

where $\overline{w} = \frac{8\pi k(m+3)^2}{(m+1)(m+2)}$.

A damage threshold in the following form was used to identify the damage condition:

$$F_D(\bar{\varepsilon}, \varepsilon_D^0) = \bar{\varepsilon} - \varepsilon_D^0 \tag{12}$$

$$\bar{\varepsilon} = \frac{\sigma_{max}}{K + 4/3G} \tag{13}$$

$$\varepsilon_D^0 = (VK)^{-1/m} \tag{14}$$

where K and m represent the material properties that characterize fracture activation, and V denotes the volume of the SPH particle.

Formation permeability after hydraulic fracturing was quantified as

$$k = k_0 e^{-\beta(\sigma_1 - \alpha p)} \tag{15}$$

where k_0 represents the initial permeability; σ_1 denotes the principal stress; p describes the pore pressure; and α represents the Biot coefficient.

3. Simulation Results

3.1. Verification

In comparison with Zheng's experiments [22], the effectiveness of the proposed model was verified. The rock specimens used were 30 cm × 30 cm × 30 cm. The center of the rock sample was the injection point. The simulated accumulated displacement was found to be similar to the results of the experiments. As illustrated in Figure 2, the established model proved to be effective, with a maximum accumulated displacement error of 5.7% for 4#DDS and an average error of 3.2% for the other three DDSs, which were compared to the experimental and simulation findings.

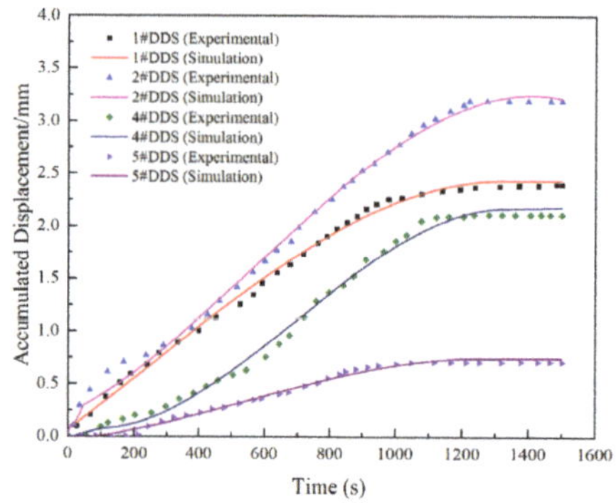

Figure 2. Comparison between experimental and simulation results.

3.2. Model Description

To study the changes in stress distribution and rock properties during hydraulic fracturing, numerical models were used to simulate the stress field and rock properties under different stimulated volumes. The numerical model size was 120 m × 120 m, and the injection points were located at the fracture opening. The principal stress distribution is shown in Figure 3a, and the calculation procedure is presented in Figure 3b. The stress field was under observation during the simulations, and the boundary conditions are listed in Table 1. The stress boundary conditions comprised the maximum and minimum horizontal principal stresses, the vertical stress, and the pore pressure. The generated fracture was characterized by permeability (Equation (15)), and the initial value was 0.35 mD, which was determined from its distribution. The calculation of the proposed hydraulic fracturing model used in the simulations is depicted in Figure 3b.

Table 1. Parameters of the hydraulic fracturing simulation.

Item	Parameters	Unit	Value
Rock mechanical properties	Young's modulus	GPa	35.6
	Poisson's rate	Dimensionless	0.34
	Tensile strength	MPa	8.42
	Shear strength	MPa	8.61
	Permeability	mD	0.35
Injection parameters	Injection rate	m^3/min	4.0
	Viscosity of fracturing fluid	mPa.s	20.0
Stress field	Vertical stress	MPa	87.5
	Maximum horizontal principal stress	MPa	78.6
	Minimum horizontal principal stress	MPa	68.6
	Pore pressure	MPa	23.8

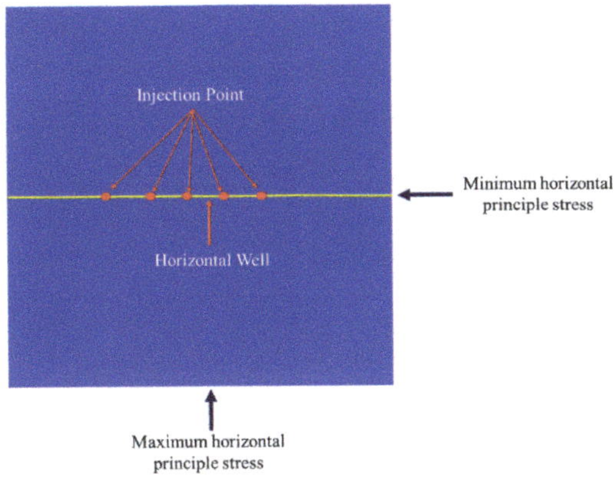

(a)

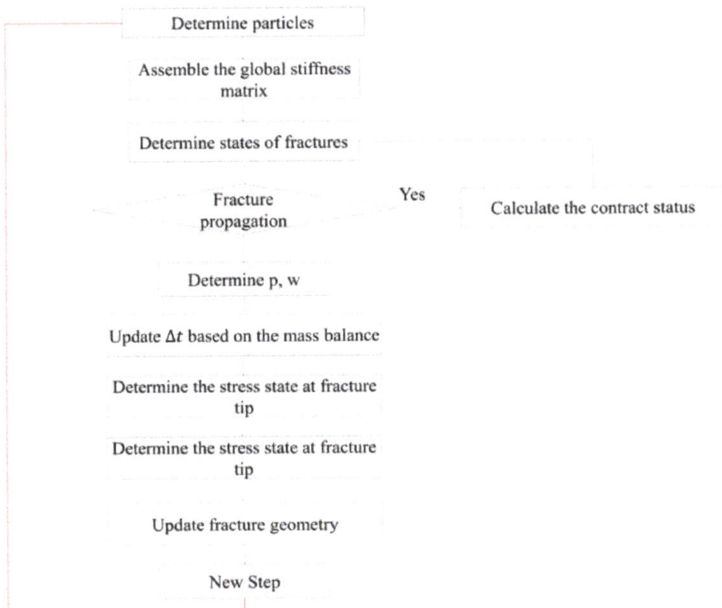

(b)

Figure 3. (**a**) Sketch of hydraulic fracturing. (**b**) Calculation of the proposed hydraulic fracturing model. Proposed hydraulic fracturing model description and calculation procedure.

3.3. Results

Figure 4 depicts the distribution of hydraulic fractures under different clusters, with a cluster space of 15 m. The simulation results indicate that the hydraulic fractures were asymmetric in distribution, and the central fractures were much shorter than the adjacent fractures. Meanwhile, the fracture geometry was also very different. The main reason for this was that the horizontal principal stress experienced great changes during hydraulic fracturing. The whole simulation lasted for 1.4 h using the Windows 11 system. The UMAT program was developed, which was embedded into ABAQUS to study the propagation of fractures.

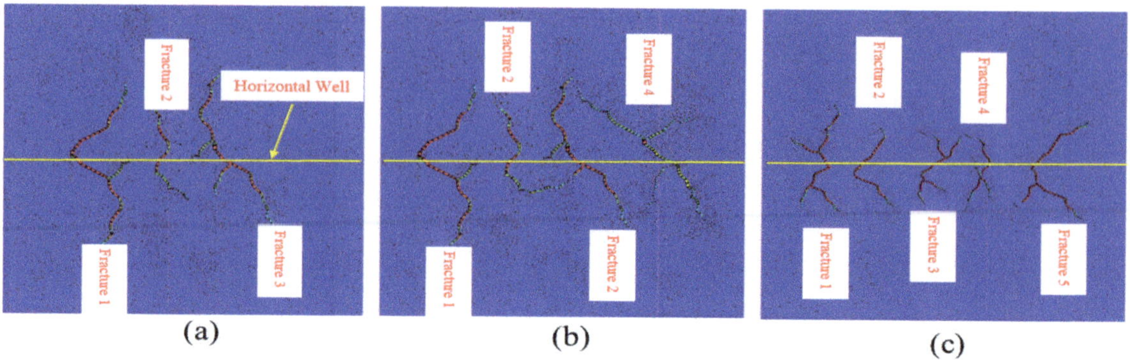

Figure 4. Hydraulic fracture distribution under different clusters. (**a**) three clusters; (**b**) Four clusters; (**c**) five clusters.

To help define the changes in this stress, Figure 5 shows the minimum principal stress in the horizontal direction along the hydraulic fracture. From the simulations, it can be seen that the minimum principal stress in the horizontal direction increased greatly during hydraulic fracture propagation. From Figure 5a, it can be seen that the minimum principal stress in the horizontal direction increased to 76.96 MPa when three cracks were formed, and those of the other two cracks were 75.31 MPa and 76.27 MPa. The average minimum horizontal principal stresses were 70.39 MPa, 74.27 MPa, and 71.32 MPa, which was the main reason for the limitation of the middle crack. Figure 5b shows that the lowest value of minimum principal stress in the horizontal direction was at the injection point, and this increased with hydraulic fracture propagation. After hydraulic fracturing, the highest value of minimum principal stress in the horizontal direction was 84.39 MPa for fracture 3, and its average increase in the horizontal direction was 8.27 MPa. The minimum principal stress in the horizontal direction increased by 45.34% on average, as shown in Figure 5a. Figure 5c shows that the minimum principal stress in the horizontal direction was 81.62 MPa for fracture 1 after hydraulic fracturing, and this increased to 85.61 MPa and 86.34 MPa for fractures 2 and 3, respectively. Meanwhile, the average increase in minimum principal stress in the horizontal direction increased to 11.37 MPa, which increased by 99.82%. In addition, it can also be found that the minimum principal stress in the horizontal direction had a funnel shape. This indicates that the stress disturbance area increased with the number of fractures.

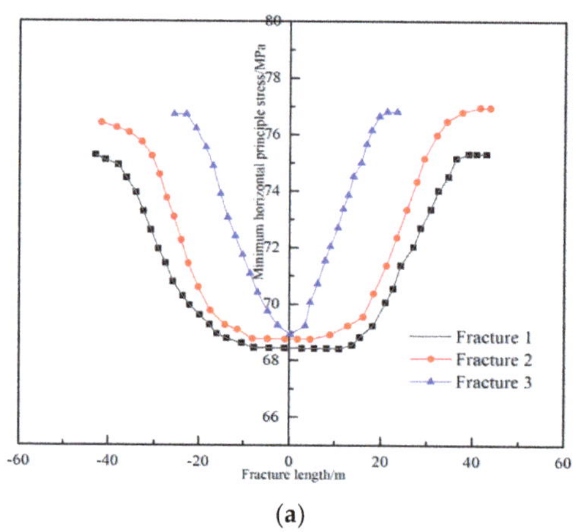

Figure 5. *Cont.*

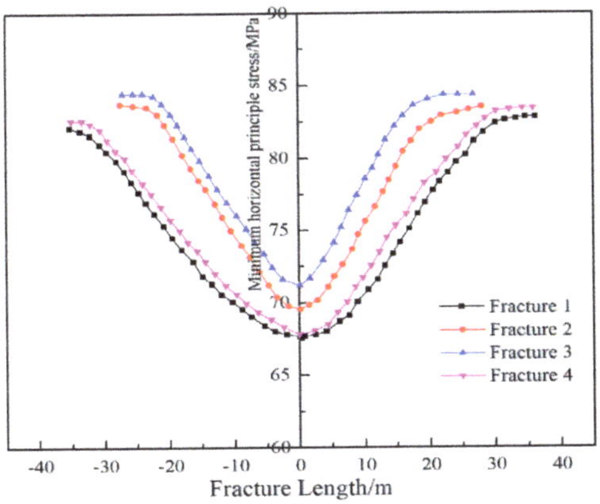

(b)

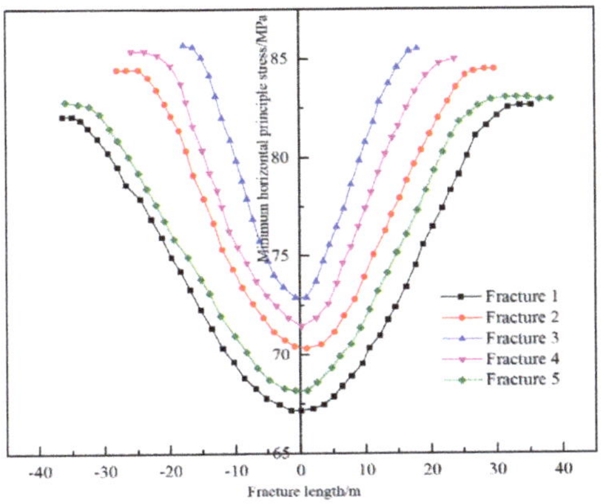

(c)

Figure 5. (**a**) The minimum horizontal stress distribution along three hydraulic fractures. (**b**) The minimum horizontal stress distribution along four hydraulic fractures. (**c**) The minimum horizontal stress distribution along five hydraulic fractures.

The physical properties of the shale formation after hydraulic fracturing are plotted in Figures 5–7 to further characterize the changes in rock properties after hydraulic fracturing. Three fractures were generated post-hydraulic fracturing, and the Young's modulus of the rock and the permeability distribution of the shale are illustrated in Figure 6. From the simulations, the original rock mechanical properties and permeability are shown in Table 1. It is clear from Figure 6a that the Young's modulus of shale decreased following hydraulic fracturing. The fractured zone can be divided into four parts according to the degree of the Young's modulus reduction. Zone I: The Young's modulus experienced a large decrease, which varied from 15.0 GPa to 20.0 GPa. The formation permeability greatly increased in this area. Zone II: The Young's modulus grew from 20.0 GPa to 25.0 GPa. In this zone, the degree of stress perturbation was slightly reduced compared with that in zone I, and the changes in Young's modulus and permeability were smaller than those in zone I. For zones III and IV, the degree of stress perturbation was obviously reduced, and the Young's modulus in these areas was slightly reduced. From Figure 6b, it can be seen that the permeability in zones I and II obviously increased, while the permeability in zone III slightly decreased. The main reason for this was that the fracture surface suffered compression deformation under high pressure inside the fracture, resulting in damage to rocks in the deep formation. The porosity of the rocks under compression deformation decreased, and the local permeability of the rocks slightly decreased due to the characteristics of porous media in the rock. According to the experimental findings, a shear failure occurred in the rocks in the deep formation under compressive deformation, so the permeability in zone IV slightly increased. When four fractures occurred or fractures were made by hydraulic action, the changes in the Young's modulus and permeability showed the same trend (Figures 7 and 8). As the number of fractures increased, the area with reduced permeability also increased. The main reason for this physical phenomenon was that as the number of fractures increased, compression deformation caused by the high injection pressure became more serious, and thus, the area with reduced permeability increased. It should also be noted that the area with decreased permeability enlarged as more fractures developed. This accounts for the strain generated in the stimulated area during hydraulic fracturing, and this strain was increased with the increase in fracture degree in the central zone. Thus, as more fractures developed, the area of zone III largely increased.

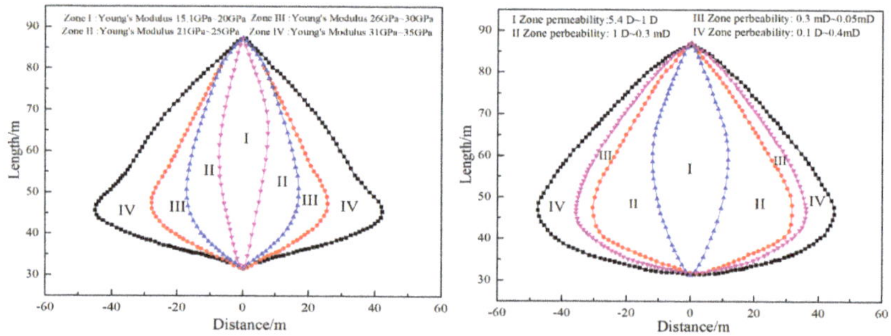

Figure 6. Young's modulus (**a**) and permeability distribution (**b**) of three fractures.

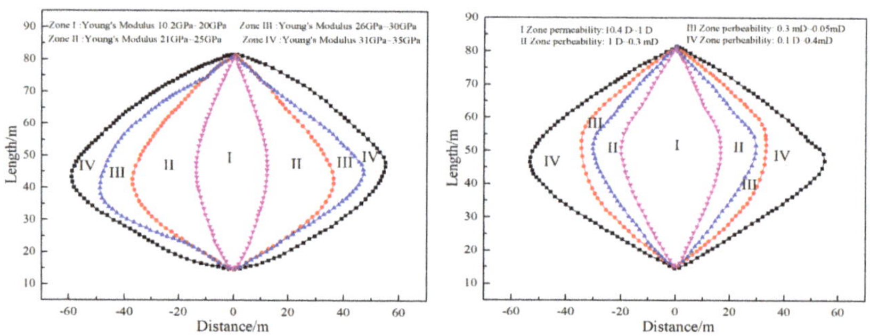

Figure 7. Young's modulus (**a**) and permeability distribution (**b**) of four fractures.

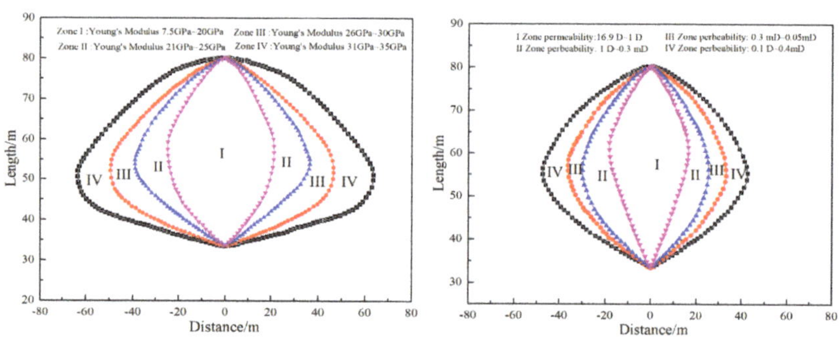

Figure 8. Young's modulus (**a**) and permeability distribution (**b**) of five fractures.

4. Discussion

The effects of the stress field, cluster space, and fracturing fluid viscosity on the rock properties were simulated to further clarify the changes in the zone with reduced permeability.

4.1. Stress Field

The horizontal principal stress difference was the main factor influencing hydraulic fracture distribution. In the simulations, stress fields with horizontal principal stress differences of 15 MPa, 12 MPa, 9 MPa, and 6 MPa were simulated. The cluster space was 15 m, four fractures were created in this process, and the other parameters were the same as those given in Table 1. It can be seen that the change in Young's modulus varied from −53.75 m to 55.92 m, and the permeability varied from −43.28 m to 43.64 m, as shown in Figure 9a. When the stress difference between two horizontal principal stresses increased to 9 MPa, the change in Young's modulus decreased from 109.67 m to 87.5 m, and the permeability decreased from 86.92 m to 64.57 m. The change in Young's modulus decreased continuously to 79.4 m, and the permeability decreased to 61.3 m when the stress difference reached 15 MPa. Although the horizontal principal stress differential significantly reduced the stimulated region, the highest value of permeability was obviously larger. The main reason for this is that the generated fractures were confined to a smaller area as the horizontal principal stress difference increased. The central area was stimulated more thoroughly, so the Young's modulus degraded more, and the permeability improved more under the same injection conditions. The strain in the center area also increased due to the restriction of fractures, which increased with the horizontal principal stress difference. Meanwhile, that of the whole stimulated area obviously decreased due to the fact that the created fractures were restricted to a smaller area under a high horizontal principal stress differential, while the strain of the other area was comparatively smaller.

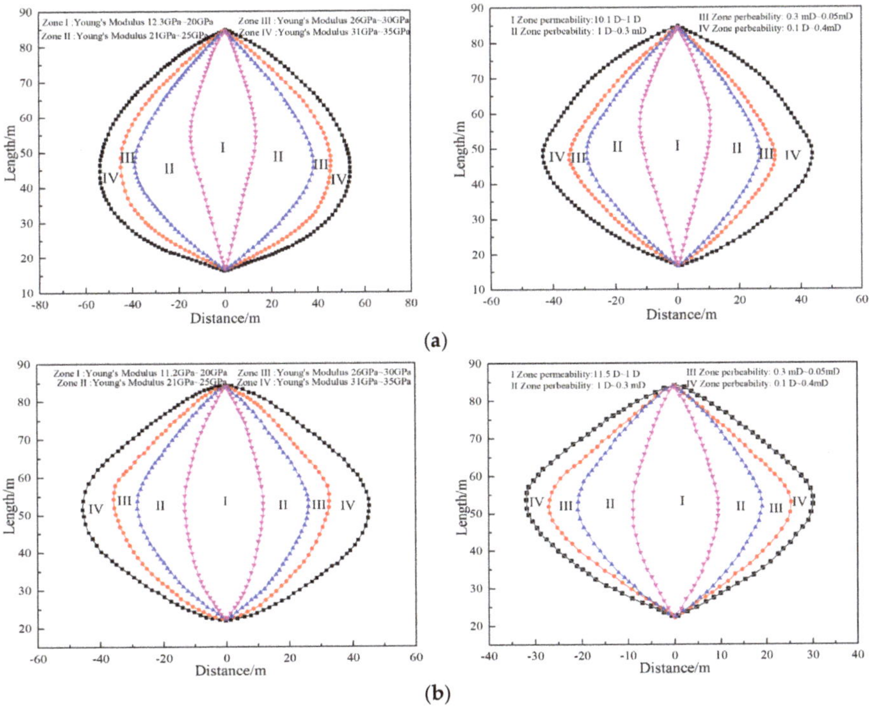

Figure 9. *Cont.*

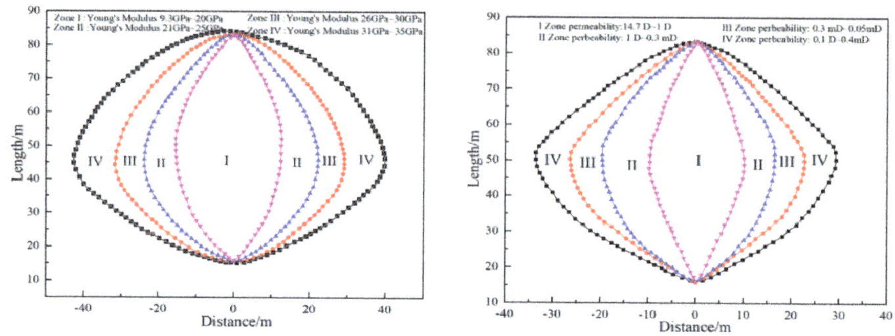

(c)

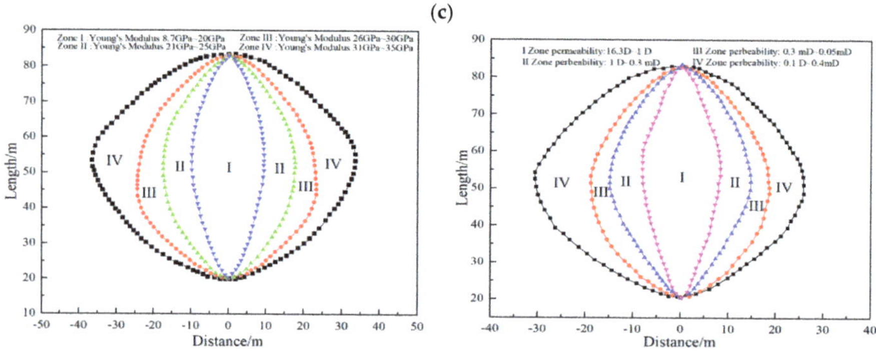

(d)

Figure 9. (**a**) Young's modulus and permeability distribution under the horizontal principal stress difference of 6 MPa. (**b**) Young's modulus and permeability distribution under the horizontal principal stress difference of 9 MPa. (**c**) Young's modulus and permeability distribution under the horizontal principal stress difference of 12 MPa. (**d**) Young's modulus and permeability distribution under the horizontal principal stress difference of 15 MPa.

4.2. Cluster Space

The other factor influencing hydraulic fracture distribution is cluster space, according to the existing research. Hydraulic fracture propagation with different cluster spaces was simulated to further investigate the changes in rock properties during hydraulic fracturing. In these simulations, the other parameters were the same as shown in Table 1, and cluster spaces of 8 m, 12 m, 16 m, and 20 m were simulated. From the simulations, the Young's modulus decreased to below 20 GPa in zone I, and the lowest value was 6.2 GPa, with the highest permeability of 11.6 D, as shown in Figure 10. In addition, the area in which the Young's modulus changed was larger than the area in which the permeability changed. The main reason for this was that the permeability of a shale formation can be obviously improved only when the current Young's modulus decreases to the threshold. When the cluster space increased to 12 m, the lowest Young's modulus was 7.6 GPa in zone I, and the highest permeability was 11.2 D. When the cluster space continuously increased to 20 m, the lowest Young's modulus increased to 13.8 GPa, while the highest permeability decreased to 8.2 D. Based on the simulations, it can be concluded that decreasing the cluster space can dramatically increase the number of hydraulic fractures formed. The highest permeability of the stimulated area also showed the same tendency. The width of the stimulated area decreased by 49.1%, while the highest permeability increase was 41.5% compared with that when the cluster space was 8 m. Therefore, the changes in Young's modulus and permeability also decreased, and the peak permeability decreased with an increase in cluster space. In addition, the formation strain also decreased with an increase in cluster space, so the area with reduced permeability decreased.

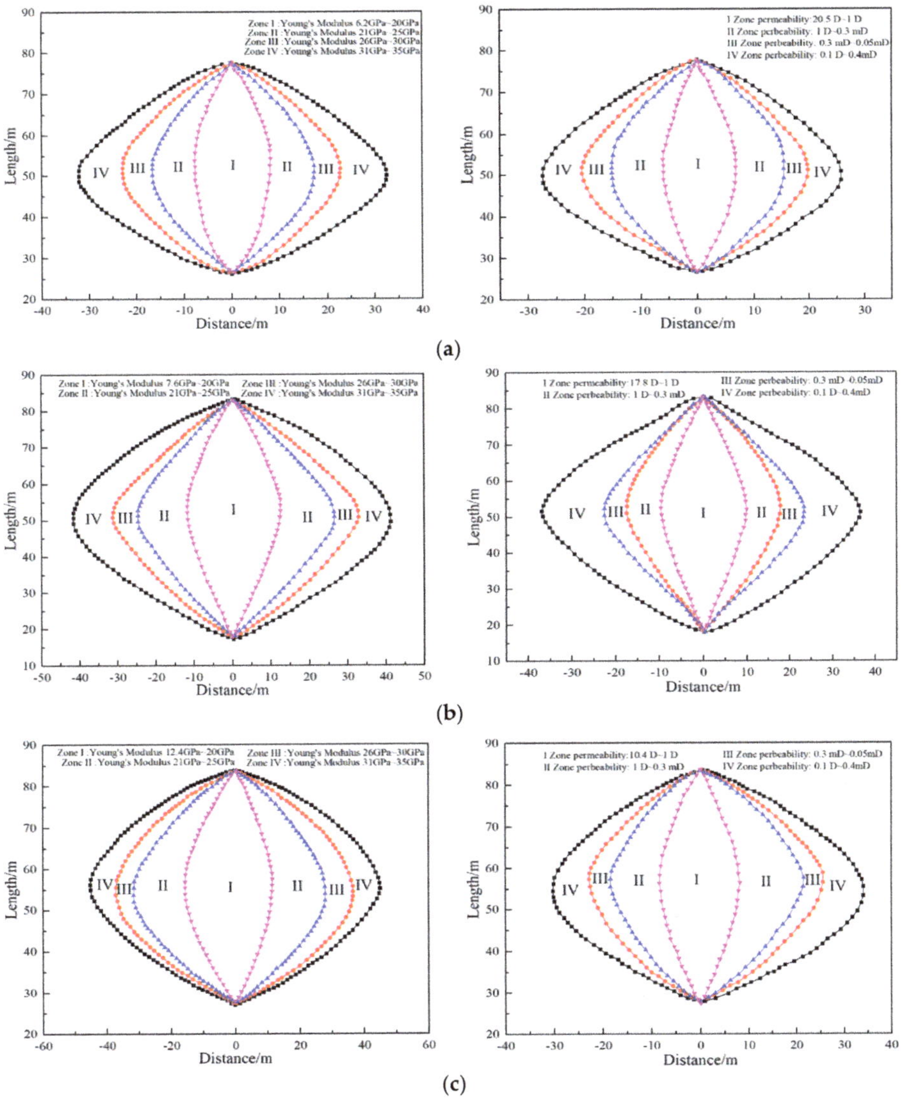

Figure 10. *Cont.*

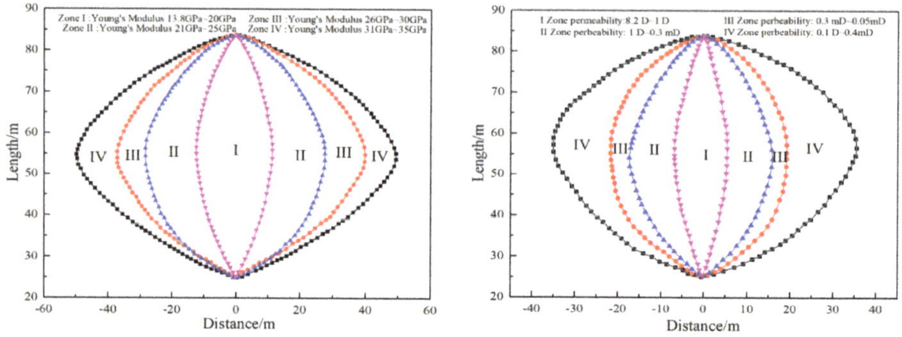

(d)

Figure 10. (**a**) Young's modulus and permeability distribution under the cluster space of 8 m. (**b**) Young's modulus and permeability distribution under the cluster space of 12 m. (**c**) Young's modulus and permeability distribution under the cluster space of 16 m. (**d**) Young's modulus and permeability distribution under the cluster space of 20 m.

4.3. Viscosity of Fracturing Fluid

The viscosity of the fracturing fluid was the most important technical factor influencing the fracture geometry during hydraulic fracturing. We simulated different viscosities (1.0 mPa.s, 10.0 mPa.s, 50.0 mPa.s, and 100.0 mPa.s) and characterized the changes in formation properties after hydraulic fracturing to investigate the effect of fracturing fluid viscosity on the formation properties. The Young's modulus and permeability are shown in Figure 11a–d. The simulation results show that the total stimulated area decreased with increasing fracturing fluid viscosity. Under a viscosity of 1.0 mPa.s, the width of the area with changed permeability was 76.4 m, while it decreased to 71.3 m when the fracturing fluid viscosity increased to 10.0 mPa.s. A large change also took place when the fracturing fluid viscosity climbed to 50.0 mPa.s, and the width of the area with changed permeability was 60.3 m. As the width of the area with changed permeability decreased, the peak permeability largely increased from 9.0 D to 18.1 D when the fracturing fluid viscosity increased from 1.0 mPa.s to 100.0 mPa.s. This indicated that high fracturing fluid viscosity was beneficial for creating high-permeability fractures, while low fracturing fluid viscosity was helpful for creating a complex fracture network. The most important reason why high fracturing fluid viscosity contributed to higher hydraulic fracture permeability was because the net pressure generated by high viscosity was much higher than that caused by low viscosity. In addition, the area with reduced permeability was also larger because the high net pressure was also a source of high stress.

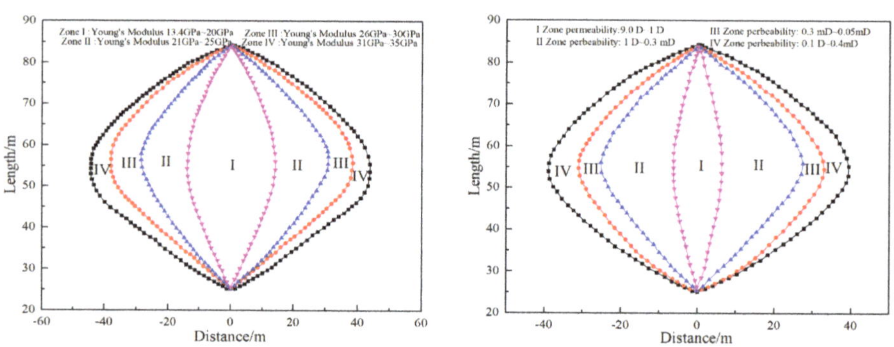

(a)

Figure 11. *Cont.*

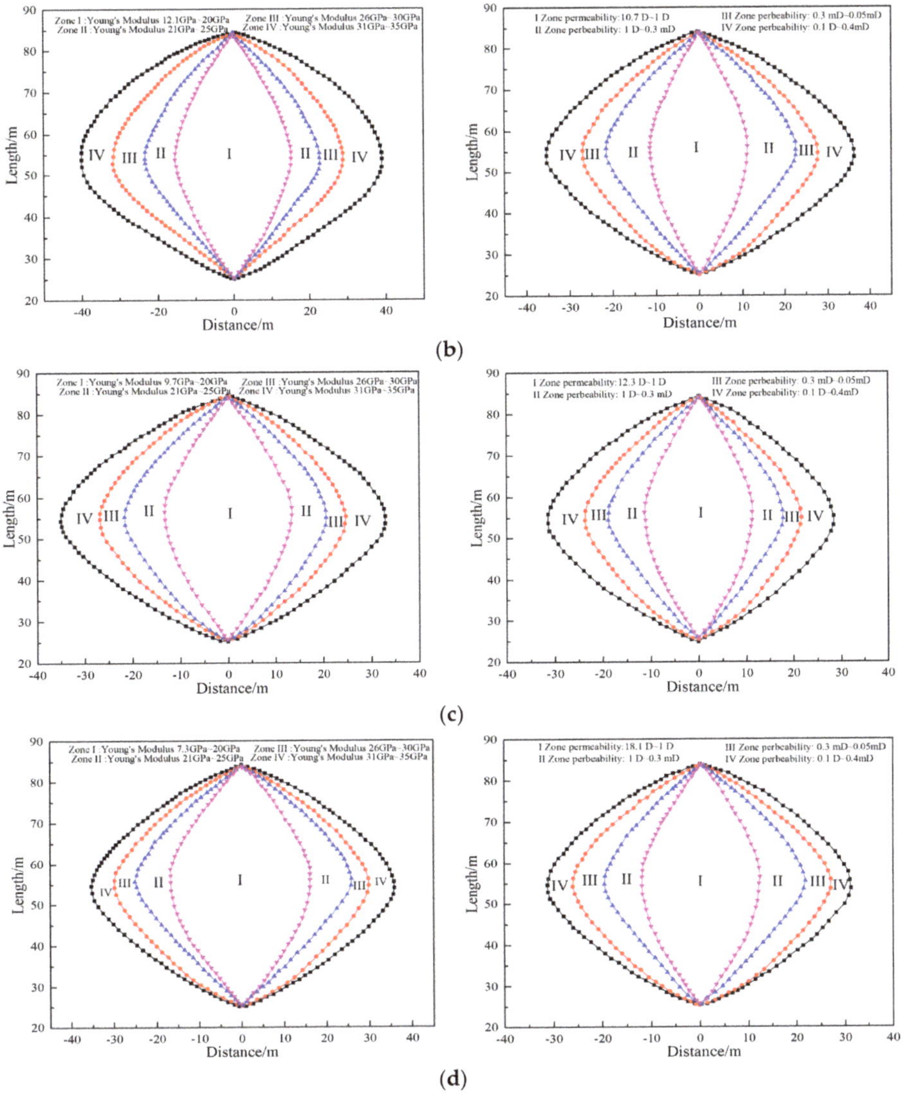

Figure 11. (**a**) Young's modulus and permeability distribution when viscosity of fracturing fluid was 1.0 mPa.s. (**b**) Young's modulus and permeability distribution when fracturing fluid viscosity was 10.0 mPa.s. (**c**) Young's modulus and permeability distribution when fracturing fluid viscosity was 50.0 mPa.s. (**d**) Young's modulus and permeability distribution when fracturing fluid viscosity was 100.0 mPa.s.

5. Conclusions

Stress disturbance is an important factor affecting the stimulated volume during hydraulic fracturing, and knowledge of the stress disturbance mechanism is critical for the optimization of hydraulic fracture parameters. In this paper, a numerical model was developed to describe the alterations in Young's modulus and the permeability distribution following hydraulic fracturing. The findings are summarized as follows.

(1) Increasing the minimum principal stress in the horizontal direction was found to be the main factor influencing the fracture propagation behavior in multi-stage hydraulic

fracturing. During fracture propagation, the minimum principal stress in the horizontal direction increased from 10 MPa to 15 MPa, and the maximum value of minimum principal stress in the horizontal direction was located at the fracture tip, which largely restricted fracture propagation.

(2) The whole stimulated area can be divided into four parts. In the parts which were severely affected by stress disturbance, the Young's modulus severely decreased, and the permeability of the formation greatly increased. A zone with reduced permeability formed in the stimulated area due to the stress caused by the high injection pressure.

(3) The main reason for the permeability decrease in the stimulated zone during hydraulic fracturing was formation compression. For rocks, the porosity and local permeability decreased under compression deformation. The compression effect on reservoir rocks increased with the heterogeneity of the reservoir stress and fracturing fluid viscosity, whereas it was negative in the cluster space.

Author Contributions: Conceptualization, Y.Z. and H.Z.; methodology, Y.Z. and H.Z.; formal analysis, Y.Z., H.Z., Y.L. and R.L.; investigation, Y.Z., H.Z., Y.L. and R.L.; resources, H.Z.; data curation, Y.Z., H.Z., Y.L. and R.L.; writing—original draft preparation, Y.Z. and H.Z.; writing—review and editing, Y.Z. and H.Z.; supervision, H.Z.; project administration, H.Z., Y.L. and R.L.; funding acquisition, R.L. All authors have read and agreed to the published version of the manuscript.

Funding: This work was supported by National Natural Science Foundation of China [Grant Number: [62173049].

Data Availability Statement: The authors confirm that the data supporting the findings of this study are available within the article.

Conflicts of Interest: The authors declare no conflict of interest.

References

1. Taheri-Shakib, J.; Ghaderi, A.; Hosseini, S.; Hashemi, A. Debonding and coalescence in the interaction between hydraulic and natural fracture: Accounting for the effect of leak-off. *J. Nat. Gas Sci. Eng.* **2016**, *36*, 454–462. [CrossRef]
2. Shi, S.; Zhuo, R.; Cheng, L.; Xiang, Y.; Ma, X.; Wang, T. Fracture Characteristics and Distribution in Slant Core from Conglomerate Hydraulic Fracturing Test Site (CHFTS) in Junggar Basin, Northwest China. *Processes* **2022**, *10*, 1646. [CrossRef]
3. Wei, C.; Zhang, B.; Li, S.; Fan, Z.; Li, C. Interaction between Hydraulic Fracture and Pre-Existing Fracture under Pulse Hydraulic Fracturing. *SPE Prod. Oper.* **2021**, *36*, 553–571. [CrossRef]
4. Liu, B.H.; Jin, Y.; Chen, M. Influence of vugs in fractured-vuggy carbonate reservoirs on hydraulic fracture propagation based on laboratory experiments. *J. Struct. Geol.* **2019**, *124*, 143–150. [CrossRef]
5. Shakib, J.T.; Akhgarian, E.; Ghaderi, A. The effect of hydraulic fracture characteristics on production rate in thermal EOR methods. *Fuel* **2015**, *141*, 226–235. [CrossRef]
6. Liu, X.; Qu, Z.; Guo, T.; Sun, Y.; Wang, Z.; Bakhshi, E. Numerical simulation of non-planar fracture propagation in multi-cluster fracturing with natural fractures based on Lattice methods. *Eng. Fract. Mech.* **2019**, *220*, 106625. [CrossRef]
7. Guo, J.; Lu, Q.; Zhu, H.; Wang, Y.; Ma, L. Perforating cluster space optimization method of horizontal well multi-stage fracturing in extremely thick unconventional gas reservoir. *J. Nat. Gas Sci. Eng.* **2015**, *26*, 1648–1662. [CrossRef]
8. Yin, P.F.; Yang, S.Q.; Gao, F.; Tian, W.L.; Zeng, W. Numerical investigation on hydraulic fracture propagation and multi-perforation fracturing for horizontal well in Longmaxi shale reservoir. *Theor. Appl. Fract. Mech.* **2023**, *125*, 103921. [CrossRef]
9. Tang, J.; Wu, K.; Li, Y.; Hu, X.; Liu, Q.; Ehlig-Economides, C. Numerical investigation of the interactions between hydraulic fracture and bedding planes with non-orthogonal approach angle. *Eng. Fract. Mech.* **2018**, *200*, 1–16. [CrossRef]
10. Wang, S.; Li, H.; Li, D. Numerical Simulation of Hydraulic Fracture Propagation in Coal Seams with Discontinuous Natural Fracture Networks. *Processes* **2018**, *6*, 113. [CrossRef]
11. Tan, P.; Chen, G.; Wang, Q.; Zhao, Q.; Chen, Z.; Xiang, D.; Xu, C.; Feng, X.; Zhai, W.; Yang, Z.; et al. Simulation of hydraulic fracture initiation and propagation for glutenite formations based on discrete element method. *Front. Earth Sci.* **2023**, *10*, 1116492. [CrossRef]
12. Zheng, H.; Pu, C.S.; Sun, C. Study on the interaction between hydraulic fracture and natural fracture based on extended finite element method. *Eng. Fract. Mech.* **2020**, *230*, 106981. [CrossRef]
13. Cao, M.; Hirose, S.; Sharma, M.M. Factors controlling the formation of complex fracture networks in naturally fractured geothermal reservoirs. *J. Pet. Sci. Eng.* **2022**, *208*, 109642. [CrossRef]
14. Liu, Y.; Zheng, X.; Peng, X.; Zhang, Y.; Chen, H.; He, J. Influence of natural fractures on propagation of hydraulic fractures in tight reservoirs during hydraulic fracturing. *Mar. Pet. Geol.* **2022**, *138*, 105505. [CrossRef]

15. Wang, S.; Li, Z.; Yuan, R.; Li, G.; Li, D. A shear hardening model for cohesive element method and its application in modeling shear hydraulic fractures in fractured reservoirs. *J. Nat. Gas Sci. Eng.* **2020**, *83*, 103580. [CrossRef]
16. Jiang, M.; Wei, Y.; Li, T.; Dong, K. Influence of Horizontal Multi-Bedding on Hydraulic Fracture Propagation in Shale Reservoirs. *Processes* **2023**, *11*, 2846. [CrossRef]
17. Chang, X.; Qiu, G.; Li, J.; Guo, Y. Research on the impact of pre-existing geological fractures on hydraulic fracturing in high in situ stress environments. *Energy Sci. Eng.* **2024**, *12*, 934–948. [CrossRef]
18. Zhai, M.; Wang, D.; Li, L.; Zhang, Z.; Zhang, L.; Huang, B.; Li, A.; Zhang, Q. Investigation on the mechanism of hydraulic fracture propagation and fracture geometry control in tight heterogeneous glutenites. *Energy Explor. Exploit.* **2022**, *40*, 246–278. [CrossRef]
19. Zhao, X.L.; Huang, B.X.; Grasselli, G. Numerical Investigation of the Fracturing Effect Induced by Disturbing Stress of Hydrofracturing. *Front. Earth Sci.* **2021**, *9*, 751626. [CrossRef]
20. Zhu, D.; Zhang, L.; Song, X.; Lian, H.; Niu, D. Propagation mechanism of the hydraulic fracture in layered-fractured-plastic formations. *Int. J. Fract.* **2023**, *241*, 189–210. [CrossRef]
21. Zheng, Y.; He, R.; Huang, L.; Bai, Y.; Wang, C.; Chen, W.; Wang, W. Exploring the effect of engineering parameters on the penetration of hydraulic fractures through bedding planes in different propagation regimes. *Comput. Geotech.* **2022**, *146*, 104736. [CrossRef]
22. Tang, Y.; Zheng, H.; Xiang, H.; Nie, X.; Liao, R. Experimental Simulation on the Stress Disturbance Mechanism Caused by Hydraulic Fracturing on the Mechanical Properties of Shale Formation. *Processes* **2023**, *11*, 2931. [CrossRef]

processes

MDPI

Article

Study on the Stress Distribution Characteristics of Rock in the Bottomhole and the Influence Laws of Various Parameters Under the Impact of a Liquid Nitrogen Jet

Bo Xiao [1], Chengzheng Cai [2,3,*], Yanan Gao [2,4,*], Bo Wang [3], Yinrong Feng [3], Jiacheng Li [3] and Zengxin Zou [5]

[1] Sinopec Key Laboratory of Drilling Completion and Fracturing of Shale Oil and Gas, Sinopec Research Institute of Petroleum Engineering Co., Ltd., Beijing 102206, China
[2] YunLong Lake Laboratory of Deep Underground Science and Engineering, Xuzhou 221006, China
[3] State Key Laboratory of Intelligent Construction and Healthy Operation & Maintenance of Deep Underground Engineering, China University of Mining and Technology, Xuzhou 221116, China
[4] Carbon Neutrality Institute, China University of Mining and Technology, Xuzhou 221116, China
[5] School of Low-Carbon Energy and Power Engineering, China University of Mining and Technology, Xuzhou 221116, China
* Correspondence: caicz@cumt.edu.cn (C.C.); yngao@cumt.edu.cn (Y.G.)

Abstract: This study presents research on the stress distribution characteristics of rock in the bottomhole and the influence laws of various parameters under the impact of liquid nitrogen jet. A multi-field coupled numerical model considering transient flow field, conjugate heat transfer, and nonlinear solid deformation was established to investigate the damage-induced fracturing mechanism of rock under liquid nitrogen jet. The study compares the impact effects of liquid nitrogen jet and water jet on rock and analyzes the variations in the stress field under different parameters. Due to its extremely low temperature, the liquid nitrogen jet creates a strong thermal stress gradient in a short time, significantly increasing the maximum principal stress and Mises stress in the rock compared to a water jet. Solid parameters, particularly the confining pressure and elastic modulus of the rock, have a more significant impact on stress distribution, while fluid parameters such as outlet pressure and fluid temperature have a smaller and more volatile effect. An increase in confining pressure inhibits tensile failure in the rock, while a higher elastic modulus enhances both tensile and shear failure. The initial rock temperature significantly affects the stress distribution, with optimal tensile failure observed at intermediate temperatures. The liquid nitrogen jet achieves a higher maximum velocity and overflow velocity than the water jet, contributing to more effective rock fracturing. The results provide a theoretical basis for the optimization of liquid nitrogen jet drilling parameters, which can help improve drilling efficiency.

Keywords: liquid nitrogen jet; flow field; stress distribution; parameter analysis

1. Introduction

Since the Industrial Revolution, the extensive use of fossil fuels and the resulting greenhouse gas emissions have exacerbated global warming. In response, China has committed to reaching the peak of CO_2 emissions by 2030 and achieving carbon neutrality by 2060 [1]. Geothermal energy, as a clean energy source, is abundant in China but has low exploitation efficiency [2]. Exploiting the geothermal resource of hot, dry rock is a crucial objective for optimizing the energy structure, conserving energy, and reducing emissions; however, it presents significant challenges. Hydraulic fracturing is the primary technique used, but it faces several issues [3]. Hydraulic fracturing technology requires the consumption of vast amounts of freshwater resources, which are often difficult to recycle and reuse, resulting in the wastage of water resources. A single well typically consumes millions of liters of water, equivalent to the daily water usage of thousands of people.

The fracturing fluids used in the hydraulic fracturing process contain a large number of chemical additives, and these chemicals have the potential to contaminate soil and groundwater resources. Once wastewater leaks, it can pollute drinking water sources, posing a threat to the ecological environment and residents' health. Additionally, the cost of treating wastewater generated by hydraulic fracturing is high. Due to the high concentration of chemicals in this wastewater, treatment is extremely difficult and might even be beyond the capabilities of large-scale wastewater treatment plants. Therefore, researchers have proposed waterless fracturing methods [4–9], with liquid nitrogen fracturing standing out due to its ability to generate thermal stress, which promotes crack propagation in reservoirs at extremely low temperatures. Liquid nitrogen is readily available and cost-effective, making the liquid nitrogen fracturing technique promising. For deep reservoirs, liquid nitrogen jet-assisted fracturing has been proposed to reduce fracture initiation pressure and enhance rock-breaking efficiency [10]. Hot dry rock refers to deep underground rocks characterized by high temperatures, strong abrasiveness, and low permeability. Therefore, efficiently extracting heat from hot dry rock reservoirs requires efficient drilling and completion engineering. Currently, hydraulic-assisted drilling is the most used technique for reservoir breaking and well building. The working principle involves injecting water as the drilling fluid into the target reservoir to alter its stress field by increasing water pressure and expanding the breaking area [11].

With increasing concern for environmental protection and water resource conservation, waterless drilling and fracturing methods have begun to attract attention. Among these, liquid nitrogen jetting, as a novel and efficient method for rock breaking and hole formation, shows promising application prospects in assisting drilling acceleration and jet fracturing. Liquid nitrogen demonstrates significant advantages in enhancing rock cracking and breaking compared to traditional methods. For example, it has been applied in unconventional reservoir fracturing in the United States [12,13]. McDaniel et al. [12] indicated that when liquid nitrogen contacts high-temperature reservoirs, it creates a strong cold impact, causing the generation and propagation of fractures on the rock surface. This indicated that the injection of low-temperature liquid nitrogen into the well bottom through pipelines is feasible and operationally viable. Finnie et al. [14] focused on the influence of transient cooling on fracture initiation and expansion and found that cracks are more likely to form at pore edges under thermal impact. Thereby, it was confirmed that the use of rapid cooling methods could accelerate the fracturing of rocks. Ren et al. [15] observed a significant increase in permeability, along with changes in the internal strength and structural characteristics of coal rock, by cold impacting the coal rock with liquid nitrogen. Kim et al. [16] further demonstrated that rapid cooling of a hot rock alters thermal distribution, with significant tensile stresses appearing on the outer surface of the sample, contributing to the formation of microcracks. This indicated that the liquid nitrogen cooling method was able to weaken the mechanical strength of the rock and alter its crack distribution state. Cai et al. [17] employed nuclear magnetic resonance to analyze changes in the mechanical parameters of sandstone, shale, and coal rock after liquid nitrogen cooling. They revealed that liquid nitrogen not only induced cracks on the surface of the rock samples but also facilitated the propagation of pre-existing cracks, with permeability enhancement positively correlated with water content. This indicated that under the effect of rapid cooling with liquid nitrogen, significant changes occurred in the pore structure within the rock. Cha et al. [18] confirmed the feasibility of low-temperature fracturing in the reservoir by developing a liquid nitrogen fracturing test device. In addition, they observed that the pore structure within the rock is influenced by the temperature gradient created by liquid nitrogen, and the main fracture direction in low-temperature fracturing does not necessarily align with the principal stress direction. It can be seen that when liquid nitrogen is used for stratum rock breaking and drilling, the low-temperature cracking effect produced by liquid nitrogen helps to improve the efficiency of rock fragmentation. However, the analysis of the rock fracture mechanism was mainly conducted from the perspective of temperature changes, without addressing the aspect of stress distribution.

In addition, the aforementioned experimental studies primarily focused on the cooling effect of stationary liquid nitrogen and were not able to accurately reflect the cooling and cracking effect of high-speed flowing liquid nitrogen.

The liquid nitrogen jet, which combined the dual mechanisms of high-speed impact and low-temperature fracturing, exhibited significantly different rock-breaking characteristics from traditional water jets. Cai et al. [19,20] explored the impact of a liquid nitrogen jet on reservoir fracturing through numerical simulation and compared it to a water jet. The results indicated that abrasive particles in the liquid nitrogen jet achieved better acceleration and that the rock experienced more severe damage due to thermal stress under the low temperature of liquid nitrogen. The feasibility of forming a liquid nitrogen jet under downhole conditions was demonstrated by this research. Zhang et al. [21] established a coupled model to study the effects of liquid nitrogen cooling on coal bodies and discovered that low temperatures significantly increased the permeability of coal rock, with damage and deformation intensifying closer to the borehole. Huang et al. [22] investigated the effect of nozzle parameters on particle velocity in liquid nitrogen abrasive jets, while Li et al. [23] proposed utilizing liquid nitrogen vaporization to fracture shale reservoirs as a solution to water resource limitations and environmental concerns. The feasibility of forming a liquid nitrogen jet under downhole conditions was demonstrated by this research. Zhai et al. [24,25] experimentally verified that the time and cycle of liquid nitrogen freezing increase with higher effective and total porosity of coal. Yao et al. [26] employed TOUGH2-EGS to simulate liquid nitrogen fracturing for enhanced production and found that the high-pressure injection of low-temperature fluid significantly improves reservoir permeability. Cai et al. [27,28] analyzed the flow field characteristics of liquid nitrogen jets through laboratory experiments and demonstrated their superior impact performance compared to water jets. Experiments by Wu [29] on coal fracturing with a liquid nitrogen jet also showed that the volume of crushed coal under the impact of a liquid nitrogen jet was greater than that under the impact of the water jet. The experimental results indicated that the combined effects of low-temperature fracturing and the high-speed impact of the liquid nitrogen jet effectively enhanced the efficiency of rock fragmentation. Numerical simulations by Zhang et al. [30,31] revealed that the thermal stress generated by temperature gradients reduced rock fracture pressure and promoted crack formation. They further simulated hot dry rock reservoirs with a liquid nitrogen abrasive jet and conducted experiments. They concluded that the combined effects of low temperature, high pressure, and abrasives in the liquid nitrogen abrasive jet caused severe damage to high-temperature rock, forming intricate pore networks [32,33]. The research findings revealed the effect of thermal stress on rock fracturing during the impact of liquid nitrogen jets, further elucidating the differences in rock-breaking mechanisms between liquid nitrogen jets and water jets. Yang et al. [34] conducted a series of gas fracturing tests on granite samples treated with a liquid nitrogen abrasive jet. They demonstrated that the rock samples treated with the liquid nitrogen abrasive jet experienced significant changes in fracture pressure. After the coal body was treated with the liquid nitrogen abrasive jet, staggered fracture networks gradually formed on the coal surface, improving reservoir permeability. Liu et al. [35] proposed a novel water–liquid nitrogen (W-LN_2) jet method and revealed concrete's crushing rules and crack characteristics at macro and micro levels. Their results indicated that a W-LN_2 jet increases crushing pit width by 139.25%, depth by 21.65%, and volume by 59.68% compared to a pure water jet (PWJ). Wang et al. [36] designed an experimental platform to study the surface heat transfer induced by the liquid nitrogen jet and calculated the temperature-dependent surface heat transfer coefficient (T-SHTC) using an inverse heat conduction theory. They found that factors like LN_2 mass flow, jet distance, and nitrogen phase influence SHTC, with jet distance being more significant. Dai et al. [37] conducted experiments to study rock failure under cutting and jet impingement. The results showed that jet impact, especially LN_2 with its cryogenic properties and low viscosity, induced thermal stress and improved rock-breaking efficiency more than the water jet. The cutting force reduced by 43.96% with the water jet and by

52.53% with the LN_2 jet. Cai et al. [38] investigated the impact of a liquid nitrogen jet (LNJ) and liquid nitrogen immersion (LNI) cooling on high-temperature granite. The results showed that mechanical properties deteriorated more with increasing temperature and cooling cycles for both methods. However, LNI's effect weakened with cycles, while LNJ's deteriorated significantly. This indicated that there were significant differences in the fracturing mechanisms between liquid nitrogen jet cooling and immersion cooling. Li et al. [39] experimentally investigated the transient cooling of rock surfaces using liquid nitrogen (LN_2) jets. The results showed sharp thermal gradients and tensile thermal stresses, with a maximum tensile stress of 5 MPa, potentially damaging the rock structure and reducing its strength. The aforementioned research findings primarily focused on the flow field characteristics and differences in rock-breaking effects between liquid nitrogen jets and water jets, with limited exploration of the underlying mechanical mechanisms involved.

In summary, current research on liquid nitrogen jets has primarily concentrated on aspects such as jet flow field characteristics and rock fragmentation features. However, the differences in flow field characteristics between liquid nitrogen jets and water jets result in distinct stress distributions on the rock surface, thereby influencing the stress state of the rock at the wellbore bottom and leading to variations in rock-breaking effects. Therefore, to better elucidate the rock-breaking mechanism of liquid nitrogen jets and subsequently identify the key factors affecting the impact effectiveness of liquid nitrogen jets, it is necessary to analyze the impact characteristics and key influencing factors of liquid nitrogen jets from the perspective of stress distribution. This analysis should not merely focus on the impact pressure of the jet on the rock surface but should shift its emphasis to the stress distribution on and within the rock surface. This study aims to utilize the finite element method to construct numerical models of impacting flow fields in the bottomhole under liquid nitrogen and water jets and to compare the distribution characteristics of impact velocity, pressure, temperature, and overflow velocities under both jets. In addition, the stress fields of rock under the impact of both jets are simulated and analyzed to explore the differences in their effects on the rock stress field in the bottomhole, providing theoretical support for understanding the differences between liquid nitrogen and water jets in rock-fracturing effects. In addition, this study investigates the influence of various parameters on the maximum principal stress and Mises stress of rock, aiming to offer a scientific basis for the optimized design of liquid nitrogen jet parameters.

2. Numerical Model

2.1. Problem Description and Geometric Model

The high-pressure jet drilling technique, which utilizes jet impact to fracture rock, offers multiple advantages, including a low cost, high efficiency, and environmental friendliness. This technique primarily employs high-pressure water jets to achieve the desired impact. However, due to their low-temperature properties, research reveals that liquid nitrogen jets can induce thermal stress in the rock, enhancing fracturing efficiency. Therefore, this study develops a model using water and liquid nitrogen as impact fluids to investigate their effects. Given that the impact of a liquid nitrogen jet on rock is a complex multi-field coupling process involving transient flow fields, conjugate heat transfer, and solid nonlinear deformation, this study focuses on the coupling of thermal stress and jet impact pressure to reveal the underlying mechanisms of rock fracturing by liquid nitrogen jets.

A geometric model is established to thoroughly analyze the fluid velocity, pressure, and temperature fields of high-temperature rock during the impact of water and liquid nitrogen jets in the bottomhole, as shown in Figure 1. The model consists of the solid and fluid domains, where the fluid domain encompasses the region inside the nozzle and the impact area of the wellbore annulus. The space between the nozzle outlet and the bottomhole is the underground region, while the space between the drill rod and rock mass is considered the annular region. The solid domain primarily represents the rock surrounding the wellbore, including the surrounding rocks and rock masses in the bottomhole. The nozzle design features a symmetrical structure composed of conical and cylindrical sections. The distance from the nozzle outlet to the bottomhole is the jet distance. Based on gas drilling technology, during jet drilling operations, the high-pressure jet enters the nozzle and underground region through the nozzle inlet and accelerates to form a high-speed jet at the outlet. This high-speed jet significantly impacts the rock in the bottomhole and then flows out through the annular region. This study defines the jet centerline and its extension line as axial, while the direction perpendicular to the jet centerline is referred to as radial. The specific parameters of the model are listed in Table 1.

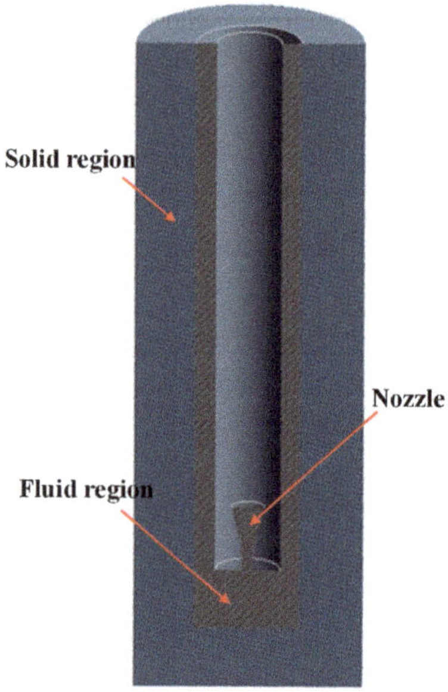

Figure 1. The model for the flow and stress fields in the bottomhole.

Table 1. Parameters of the model.

Wellhole Height	Borehole Diameter	Drill Pipe Diameter	Rock Mass Thickness	Total Model Height	Nozzle Jet Distance	Nozzle Outlet Diameter
300 mm	50.8 mm	31.8 mm	30 mm	330 mm	30 mm	6 mm

2.2. Mathematical Model

Both liquid nitrogen and water jets impacting high-temperature rock in the bottomhole involve fluid flow and convective heat transfer. The mathematical model is related to fluid viscosity, flow velocity, density, thermal conductivity, specific heat capacity, and other factors.

(1) Fluid region

The liquid nitrogen jet process involves heat transfer and fluid compressibility. Therefore, the energy equation must also be solved in addition to the mass conservation and momentum equations. During the impact of the liquid nitrogen jet, the pressure and temperature in the flow field undergo rapid changes. These sudden changes in pressure and temperature affect the physical properties of liquid nitrogen and nitrogen gas, influencing the characteristics of the flow field. A more precise model for calculating the physical properties of liquid nitrogen and nitrogen gas is employed during the calculations to enhance computational accuracy.

Given the high Reynolds number of liquid nitrogen jets, the turbulence model provides high accuracy and stability for fully developed turbulent flow at high Reynolds numbers.

$$\frac{\partial(\rho k)}{\partial t}+div(\rho Uk)=div\left[\left(\mu+\frac{\mu_r}{\sigma_k}\right)\cdot gradk\right]-\rho\varepsilon+\mu_t P_G \quad (1)$$

$$\frac{\partial(\rho\varepsilon)}{\partial t}+div(\rho Uk)=div\left[\left(\mu+\frac{\mu_t}{\sigma_\varepsilon}\right)\cdot grad\varepsilon\right]-\rho C_2\frac{\varepsilon^2}{k}+\mu_t C_1\frac{\varepsilon}{k}P_G \quad (2)$$

where k and ε are the turbulent kinetic energy and dissipation rate, respectively; σ_k, σ_ε, C_1, and C_2, are constants, and their values are generally taken as σ_k = 1.0, σ_ε = 1.3, C_1 = 1.44, and C_2 = 1.92 in numerical calculations; and μ_t is the turbulent viscosity. μ_t and P_G can be expressed as follows:

$$\mu_t=\rho C_\mu\frac{k^2}{\varepsilon},P_G=2S_{ij}\cdot S_{ij} \quad (3)$$

The continuity equation can be expressed as

$$\frac{\partial\rho_f}{\partial_t}+\frac{\partial\left(\rho_f u_i\right)}{\partial x_i}=0 \quad (4)$$

The momentum conservation equation can be expressed as

$$\frac{\partial\left(\rho_f u_i\right)}{\partial t}+\frac{\partial\left(\rho_f u_i u_j\right)}{\partial x_j}=-\frac{\partial p}{\partial x_i}+\frac{\partial}{\partial x_i}(\mu\frac{\partial u_i}{\partial x_j})+S_i \quad (5)$$

The energy conservation equation can be expressed as

$$\frac{\partial\left(\rho_f T_f\right)}{\partial t}+\frac{\partial\left(\rho_f u_i\right)}{\partial x_i}=S_T+\frac{\partial}{\partial x_i}\left(\frac{k_h}{c_p}\frac{\partial T_f}{\partial x_i}\right) \quad (6)$$

where t is the time; ρ is the density; u is the velocity component; x is the displacement component; μ is the hydrodynamic viscosity of fluid; k_h is the fluid thermal conductivity; c_p is the fluid specific heat; and s_i and s_T are the source terms of the momentum and energy conservation equations, respectively.

Heat exchange occurs between the fluid and the rock, and conjugate heat transfer is employed to calculate heat transfer at the fluid–solid interface. Since the heat fluxes at the interface are identical, the governing equation for fluid thermal conductivity can be derived as follows:

$$\begin{cases} qw|_{solid}=qw|_{fluid} \\ -k_s\left(\frac{\partial T_s}{\partial n}\right)_w\Big|_{solid}=k_h\left(T_W-T_{adj}\right)\Big|_{fluid} \end{cases} \quad (7)$$

where q_w is the heat flux at the flow–solid surface; k_s is the heat transfer coefficient of the fluid at the wall surface; and T_w and T_{adj} are the rock temperature at the flow–solid coupling surface and the near-wall fluid temperature, respectively.

(2) Solid region

As heat transfer occurs in the rock in the solid region, solving the heat transfer equation for the solid region is necessary.

$$k_s\nabla^2 T_s=c_s\rho_s\frac{\partial T_s}{\partial t} \quad (8)$$

where k_s is the heat transfer coefficient of the wall surface to the fluid; T_s is the temperature of the rock; ρ_s is the density of the rock; and c_s is the specific heat capacity of the rock.

The impact pressure from the jet affects the stress field within the rock. In addition, due to heat transfer in the solid region, varying temperatures influence the thermal stress state within the rock. It is necessary to solve the physical, equilibrium, and compatibility equations for thermal stress calculations in the solid region. The governing equations are as follows:

Physical equation:

$$\varepsilon_{ij} = \frac{1}{2G}\left(\sigma_{ij} - \frac{\nu}{1+\nu}\sigma_{kk}\delta_{ij}\right) + \alpha\Delta T_s\delta_{ij} \tag{9}$$

Equilibrium equation:

$$\left(\frac{E\nu}{(1+\nu)(1-2\nu)} + G\right)\nabla^2 e - \frac{\alpha E}{1-2\nu}\nabla^2 T_S = 0 \tag{10}$$

Compatibility equation:

$$\begin{cases} \nabla^2\sigma_i + \frac{1}{1+\nu}\frac{\partial^2\sigma_{ii}}{\partial x_i^2} = 1\alpha E\left(\frac{1}{1-\nu}\nabla^2 T_s + \frac{1}{1+\nu}\frac{\partial^2 T_s}{\partial x_i^2}\right) \\ \nabla^2\tau_{ij} + \frac{1}{1+\nu}\frac{\partial^2\sigma_{ii}}{\partial x_i \partial x_{ij}} = \left(\frac{\alpha E}{1+\nu}\frac{\partial^2 T_s}{\partial x_i \partial x_j}\right) \end{cases} \tag{11}$$

where k_s is the heat transfer coefficient of the wall surface to the fluid; T_s is the temperature of the rock; ρ_s is the density of the rock; c_s is the specific heat capacity of the rock; E is the modulus of elasticity; G is the shear modulus; ν is the Poisson's ratio of the rock; α is the thermal expansion coefficient; ε_{ij} is the strain in the rock; σ_{ij} is the rock stress; ν is the Poisson's ratio of the rock; σ_{kk} is the sum of the total stresses; δ_{ij} is the Kroenke's symbol (its value is specified as 1 for $i = j$ and 0 for $i \neq j$); α is the thermal expansion coefficient; ΔT_s is the temperature difference in the rock; F_i is the volumetric force in the i-direction; and u_i is the displacement in the i-direction.

2.3. Mesh Division and Boundary Conditions

Figure 2 indicates that the geometric model in Figure 1 is meshed using triangular elements. The mesh is refined at this interface to enhance computational convergence and accuracy to facilitate the convergence of results at the fluid–solid coupling interface.

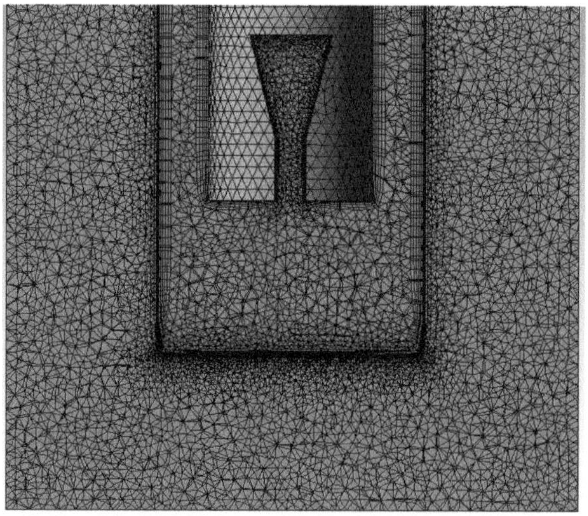

Figure 2. Mesh division.

The nozzle inlet in the simulation is designated as the pressure inlet boundary, while the wellhole annular outlet is set as the pressure outlet boundary. The interface between the fluid and solid regions is defined as the fluid–solid conjugate boundary, ensuring that the temperatures and heat fluxes of the fluid and solid remain consistent at this boundary while the remaining areas are set as wall boundaries. During the impact of the liquid nitrogen jet, a significant temperature difference between the liquid nitrogen and the deep hot rock results in a strong cold impact effect on the rock, causing energy exchange. Therefore, activating the energy equation and considering heat transfer boundary conditions is necessary under this condition. In addition, under high-speed jet conditions, the low-temperature liquid nitrogen entering from the nozzle generates turbulence in the flow field at the bottomhole; thus, the k-ε model is used for turbulence calculations. Since the deformation of the solid region is not considered, the boundary at the bottomhole in the solid domain is set as a fixed constraint to limit its degrees of freedom. The fluid–solid contact surface is the coupling surface, consistent with the fluid region, while the side boundaries are adiabatic, with zero heat flux.

2.4. Model Parameter Settings and Calculation Process

The inlet temperature, inlet pressure, outlet pressure, initial fluid temperature, initial rock temperature, fluid density, fluid viscosity, thermal conductivity, and specific heat capacity for both fluids (water and liquid nitrogen) are defined. The baseline calculation conditions are detailed in Table 2.

Table 2. Parameters for the fluid region.

Parameter Settings	Liquid Nitrogen Jet	Water Jet
Inlet pressure (MPa)	45	45
Outlet pressure (MPa)	25	25
Initial fluid temperature (K)	110	298.15
Initial rock temperature (°C)	150	150
Density (kg/m^3)	806.08	998.2
Viscosity (Pa·s)	1.6065×10^{-4}	10.03×10^{-4}
Thermal conductivity (W/m·K)	0.14581	0.6
Specific heat capacity (kJ/kJ·K)	2.0415	4.182

The density, specific heat capacity, heat conductivity coefficient, elastic modulus, and Poisson's ratio for the solid region are defined as listed in Table 3.

Table 3. Parameters for the solid region.

Density	Specific Heat Capacity	Heat Conductivity Coefficient	Elastic Modulus	Poisson's Ratio
2300 kg/m^3	760 J/(kg·K)	2.5 W/(m·K)	30 Gpa	0.25

3. Characteristics of the Impact Flow Field in the Bottomhole

The flow and heat transfer characteristics in the bottomhole were analyzed under baseline conditions, with an inlet pressure of 45 MPa and an outlet pressure of 25 MPa. Since the flow field structure stabilizes after 10 s, choosing any time after 10 s is sufficient; the simulation time in this section is set to 10 s. The significant temperature difference between the water jet and the liquid nitrogen jet results in noticeable variations in physical parameters during heat transfer with the surrounding rock.

3.1. Velocity Field

A curve depicting velocity variation along the axis was derived to compare the velocity changes along the axial direction of water and liquid nitrogen jets, as depicted in Figure 3. At the right end of the figure, specifically at the nozzle inlet, the velocities of both jets are nearly identical. As the nozzle diameter gradually narrows, the jet velocities increase, forming an acceleration segment and reaching their peak at the nozzle outlet. Then, the jets are expelled at a constant maximum velocity over a distance, creating a constant velocity segment. As the jets approach the bottomhole, their velocity declines rapidly, forming an impact segment. The lengths of these three segments for the water and liquid nitrogen jets are nearly the same. However, in the constant velocity segment, the velocity of the water jet is approximately 20 m/s slower than that of the liquid nitrogen jet. This difference can be attributed to the fact that, under identical jet pressure differences and nozzle sizes, the dynamic viscosity of liquid nitrogen is lower than that of water, resulting in relatively less kinetic energy dissipation due to viscous forces during the jetting process of the liquid nitrogen jet.

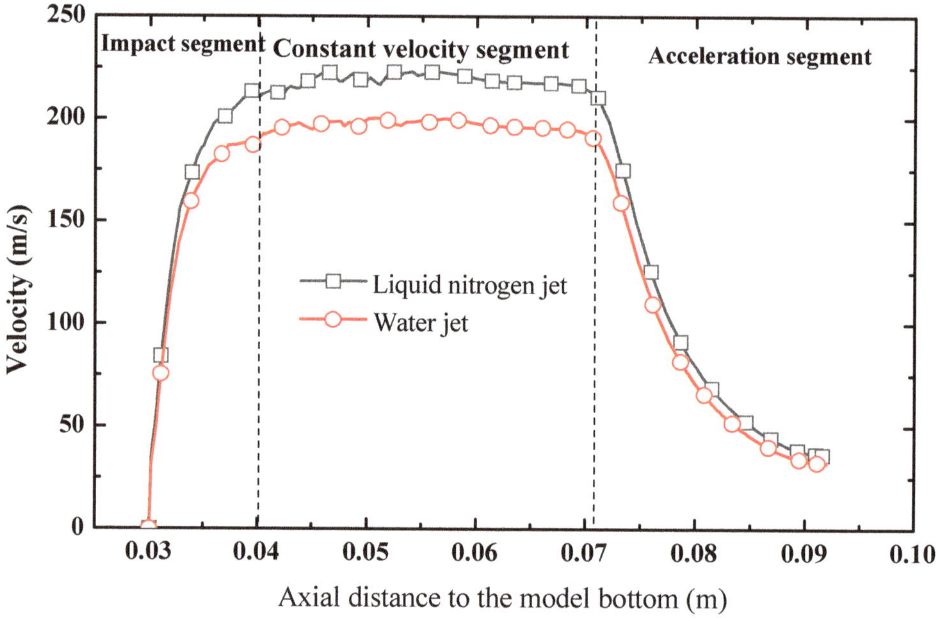

Figure 3. Curves of velocities along the axis.

3.2. Pressure Field

Figure 4 indicates that the pressure distributions created in the bottomhole by the water jet and liquid nitrogen jet exhibit similar characteristics. The pressure in the central region of the bottomhole is the highest, reaching 43 MPa, primarily due to the direct impact of the high-speed jets on the bottomhole. Then, the pressure gradually decreases in the radial direction, reaching a minimum of 23.5 MPa. However, the pressure near the edges, close to the surrounding rock, rises again to about 37.5 MPa, which results from the tangential velocity of the jets impacting the surrounding rock. The fluid pressure distribution in different regions of the bottomhole is uneven, leading to a varied pressure impact on the rock in the bottomhole. Specifically, the pressure at the center of the bottomhole is nearly twice that at its edge. This distribution pattern, where the center undergoes high compression while the edges experience lower pressure, significantly compresses the rock at the bottomhole center and facilitates its fracturing. The pressure distribution characteristics generated by the water jet and liquid nitrogen jet in the bottomhole play a significant role in rock fracturing, aiding in rock breakage and providing favorable conditions for subsequent drilling operations.

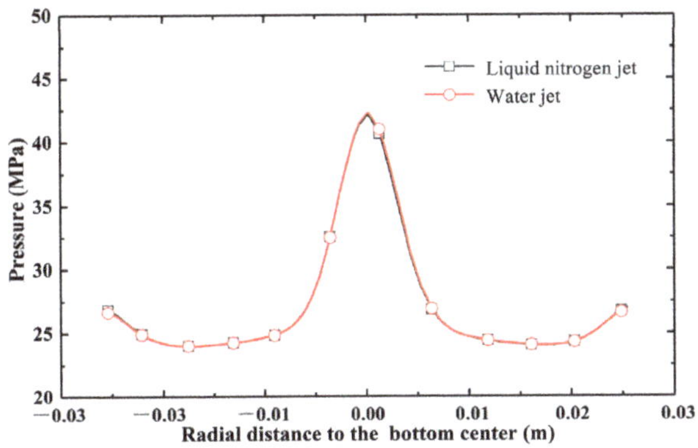

Figure 4. Curve of pressures in the bottomhole.

3.3. Overflow Velocity Field

After impacting the bottomhole, the jets generate a high-velocity overflow along the bottomhole, exerting significant shear effects on the rock within the bottomhole and enhancing the convective heat transfer between the jets and the rock. Considering the tangential velocity approximately 3 mm above the bottomhole as the subject of analysis, this study compares the tangential velocity characteristics in the bottomhole for both jets, and the velocity variations along the diameter of the bottomhole are depicted in Figure 5. The analysis reveals that the tangential velocity of both jets shows a symmetrical distribution, characterized by an acceleration segment where it rapidly rises from 0 m/s, a fluctuation segment with a peak velocity, and a deceleration segment where it quickly drops to 0 m/s near the surrounding rock at both ends. At the same location, the tangential velocity of the liquid nitrogen jet is significantly higher than that of the water jet. For instance, at a distance of 15 mm from the bottomhole center, the tangential velocity of the liquid nitrogen jet surpasses that of the water jet by approximately 6 m/s, marking an increase of nearly 20%, which indicates a more pronounced trend of velocity increase. The shear effect of the jet within the bottomhole directly affects its heat transfer with the rock. Variations in tangential velocity across the acceleration, fluctuation, and deceleration segments lead to significant differences in convective heat transfer along these paths, resulting in uneven thermal stress distribution and exacerbating the rock fracturing effect. Specifically, the lower tangential velocity of the water jet compared to the liquid nitrogen jet results in a higher convective heat transfer rate for the liquid nitrogen jet, leading to a more significant impact on thermal stress in the rock and a more pronounced fracturing effect.

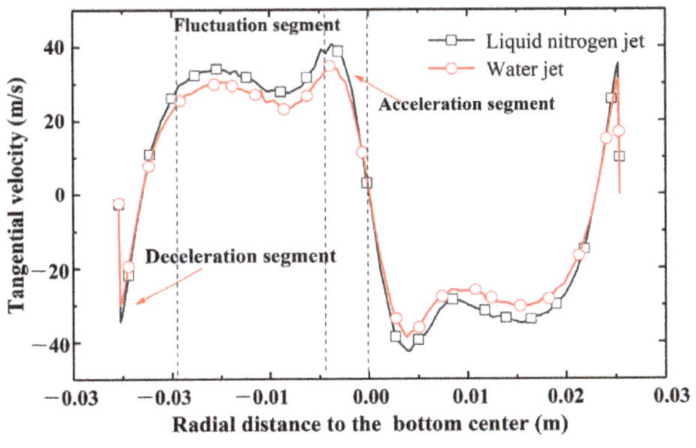

Figure 5. Curve of tangential velocity in the bottomhole.

3.4. Thermal Stress

Given that the water jet's temperature is significantly higher than that of the liquid nitrogen jet and its temperature gradient is smaller, its thermal stress effects on the rock are less pronounced than those caused by the liquid nitrogen jet. This study analyzes the liquid nitrogen jet to more clearly observe its unique effects. Figure 6 shows that this study examines rock layers at depths of 2 and 10 mm. Under thermal stress, the variation in the maximum principal stress over time at these two depths exhibits a trend similar to that observed under thermomechanical coupling. The maximum principal stress within the rock at a 2 mm depth rapidly increases from 2 to 10 s, then decelerates after 10 s, and slightly decreases at 120 s. In contrast, the maximum principal stress within the rock at 10 mm depth demonstrates a continuous upward trend, although slower. This shows that thermal stress significantly affects the rock layer at 2 mm depth. Comparatively, previous findings indicate that under the impact of the liquid nitrogen jet, the maximum principal stress of the rock at 2 mm depth is nearly negative at 1 s under thermomechanical coupling. This indicates that the rock is initially in compression, which transitions to tension after 10 s, aligning with the thermal stress trend. In the rock layer at 2 mm depth, the maximum principal stress at the jet center changes by about 9.627 MPa from 1 to 10 s under thermomechanical coupling, with approximately 8.168 MPa of this change attributed to thermal stress, accounting for 84.8% of the total change magnitude. This underscores that thermal stress plays a significant supportive role in the tensile failure of the rock. Figure 7 further illustrates the changes in Mises stress of rocks at different depths under thermal stress. Specifically, for the rock sample at 2 mm depth, the Mises stress at the center reaches a minimum at 1 s, then gradually increases in the radial direction. At 10 s, the trend rapidly changes so that the stress at the center reaches its maximum value, then gradually decreases along the radial direction. During this period, the Mises stress at the center increases by 5.516 MPa, while under thermomechanical coupling, this increase reaches 9.62 MPa, marking an increase of about 40%. This indicates a significant contribution of thermal stress to Mises stress, reinforcing its pivotal role in the shear failure of the rock.

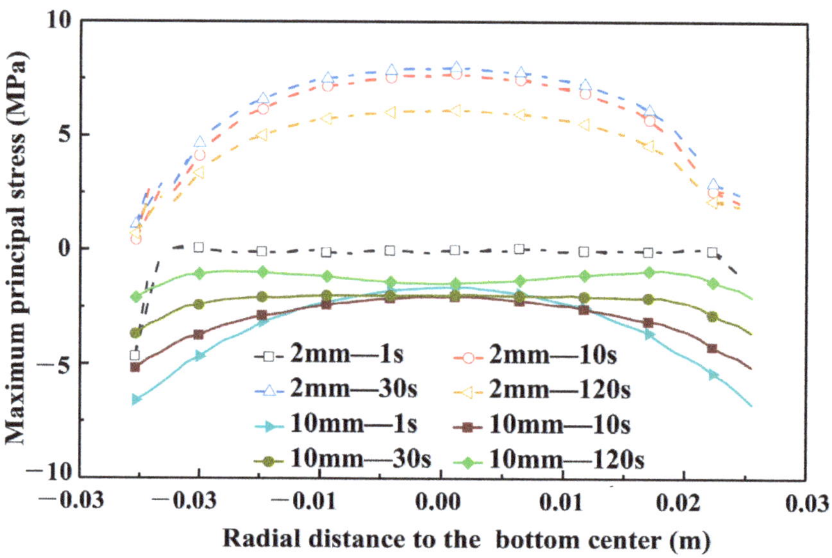

Figure 6. Maximum principal stress distribution of liquid nitrogen jet in radial direction in the bottomhole under the effect of temperature.

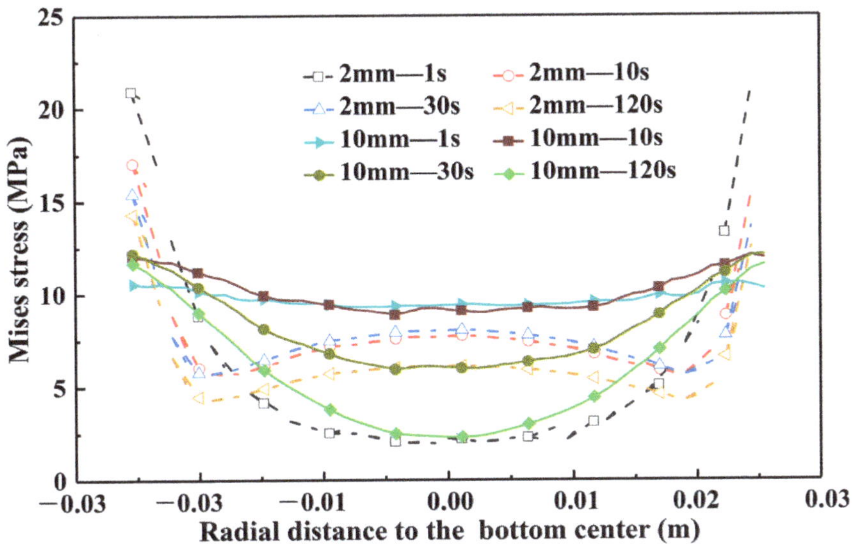

Figure 7. Mises stress distribution of liquid nitrogen jet in radial direction in the bottomhole under the effect of temperature.

4. Analysis of the Influence of Parameters on Stress Distribution in the Rock in the Bottomhole Under the Impact of a Liquid Nitrogen Jet

4.1. Parameter Selection

Previous research indicates that the stress field in rock is influenced by impact pressure and thermal stress under the effect of the liquid nitrogen jet. According to the governing equations, four rock parameters, confining pressure, elastic modulus, Poisson's ratio, and rock temperature, affect the internal stress field within the rock. The impact pressure of the jet is directly influenced by jet velocity, which is related to the inlet and outlet pressures of the jet. Therefore, it is necessary to investigate the influence of rock confining pressure, elastic modulus, Poisson's ratio, and rock temperature, as well as the inlet pressure and temperature of the fluid, on rock stress. The studied parameter ranges may not fully cover all possible conditions encountered in real-world applications. For example, rock properties and fluid conditions can vary widely in different geological formations and drilling operations. However, the conclusions drawn from this range could provide valuable insights into the stress distribution characteristics of rock under liquid nitrogen jet impact.

4.2. Influence of Solid Parameters on Stress Distribution

4.2.1. Influence of Rock Confining Pressure on Stress Distribution

The stress distributions of the rock at the bottomhole under different confining pressure conditions are obtained by adjusting and recalculating the rock confining pressure in the basic numerical example, as illustrated in Figure 8. As the confining pressure increases, both the maximum principal stress and Mises stress in the rock exhibit a trend of gradual decrease, with the magnitude of change at the center point of the bottomhole being relatively small. Specifically, when the rock confining pressure increases from 5 MPa to 12.5 MPa, the peak maximum principal stress at 2 mm depth decreases from 30.09 MPa to 9.86 MPa. This indicates that an increase of 7.5 MPa in confining pressure reduces 20.23 MPa in maximum principal stress, with a change magnitude ratio of approximately 270% between the confining pressure and the maximum principal stress. This indicates that rock confining pressure significantly affects the maximum principal stress at 2 mm depth. At 10 mm depth, the peak maximum principal stress decreases from −1.71 to −10.42 MPa, with a change magnitude ratio of approximately 139%. A positive value of maximum principal stress indicates a tensile state, while a negative value indicates a compressive state. The reduction in maximum principal stress indicates a gradual decrease in tensile stress, showing a clear trend of transitioning toward compressive stress. Since rock has greater compression resistance than tension, tensile failure predominates, aligning with the damage region assessment methods discussed in Chapter 4. The red dashed line in Figure 8a indicates the tensile failure limit of granite, which is 8 MPa. As confining pressure increases, the range of maximum principal stress exceeding the tensile failure limit decreases gradually, from 19.417 mm at 5 MPa to 8.897 mm at 12.5 MPa, a reduction of 121.6%. This shows that a higher confining pressure inhibits tensile failure, while a lower confining pressure promotes tensile failure under the impact of the liquid nitrogen jet. The changes in Mises stress show that the peak stress of the rocks at 2 mm and 10 mm depths does not vary significantly with the confining pressure. At 10 mm depth, the change in the peak Mises stress is only 4.6 MPa, indicating that confining pressure has a smaller effect on Mises stress than on maximum principal stress. This further demonstrates that variations in confining pressure have a limited impact on the shear failure of rock under the influence of the liquid nitrogen jet.

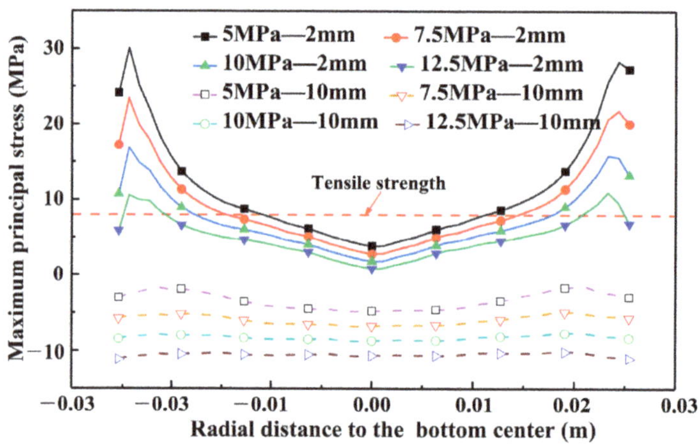

(**a**) Maximum principal stress.

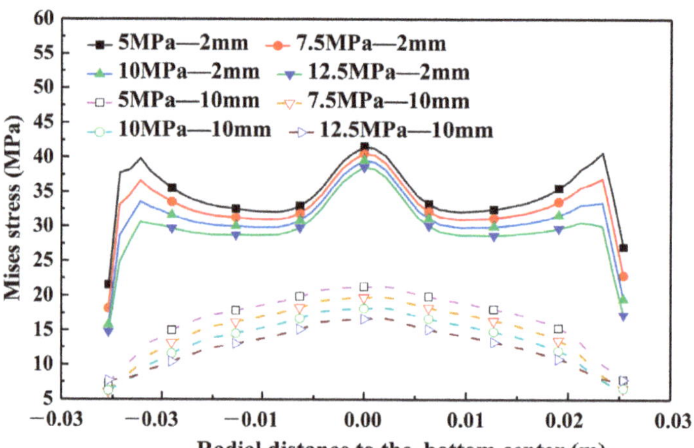

(**b**) Mises stress

Figure 8. Influence of rock confining pressure on stress distribution.

4.2.2. Influence of Elastic Modulus and Poisson's Ratio of the Rock on Stress Distribution

The stress distributions of the rock at the bottomhole under different parameter conditions are obtained by adjusting the elastic modulus and Poisson's ratio of the rock in the basic numerical example, as illustrated in Figures 9 and 10. The elastic modulus has a more pronounced effect on the changes in the rock stress field, which is reflected in larger stress variations, whereas the influence of Poisson's ratio is smaller. This difference is particularly significant at 2 mm depth.

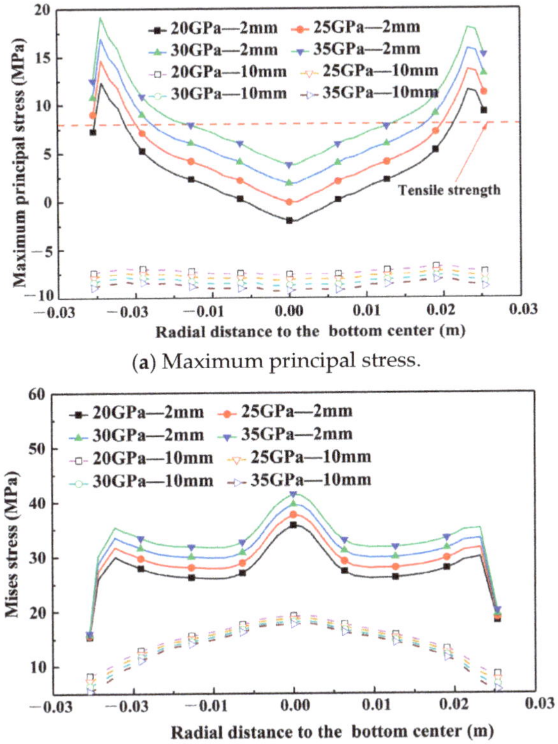

(**a**) Maximum principal stress.

(**b**) Mises stress

Figure 9. Influence of elastic modulus on stress distribution.

Figure 9a demonstrates that with the impact center point in the bottomhole used as the study object, the maximum principal stress at a depth of 2 mm increases significantly from −2.026 MPa to 3.787 GPa when the elastic modulus rises from 20 to 35 GPa, representing a 289% increase. The red dashed line in the figure indicates a tensile failure limit of 8 MPa for the rock. As the elastic modulus increases, the tensile failure zone expands from 8.793 to 16.242 mm, showing an 84% increase. With each 5 GPa increase in the elastic modulus, the maximum principal stress increases further. Similarly, the stress at the center point at a depth of 10 mm gradually rises with the elastic modulus, representing a 22% increase. This indicates that rocks with higher elastic modulus values are more prone to tensile failure under the impact of a liquid nitrogen jet, leading to increased tensile stress.

Figure 9b reveals that the Mises stress at the impact center point increases by 16.3% at a 2 mm depth and decreases by 8.5% at a 10 mm depth. This further demonstrates that a larger elastic modulus enhances the shear failure of shallow rocks under the impact of a liquid nitrogen jet while suppressing shear failure in deeper rocks.

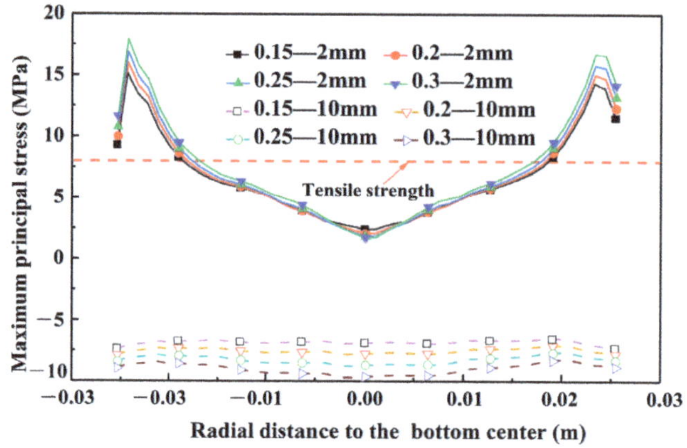

(**a**) Maximum principal stress

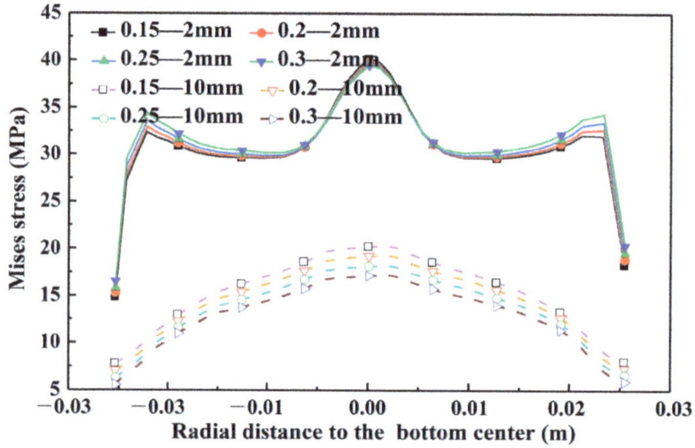

(**b**) Mises stress

Figure 10. Influence of Poisson's ratio on stress distribution.

Figure 10 shows that the maximum principal stress and Mises stress at the center point at a depth of 2 mm only increase slightly, while the tensile failure zone expands from 10.92 mm at a Poisson's ratio of 0.15 to 13.067 mm at a Poisson's ratio of 0.3. The change in rock stress at the center point at the 10 mm depth is more pronounced than that at the 2 mm depth, with the maximum principal stress increasing from −7.76 to −9.64 MPa, indicating increased compressive stress. This shows that at a larger Poisson's ratio, the liquid nitrogen jet has minimal impact on the maximum principal stress of shallow rocks, with only a slight effect on the expansion of the damage zone. In addition, deeper rocks with higher Poisson's ratios experience greater compressive stress under the impact of the liquid nitrogen jet, which is unfavorable for rock failure.

4.2.3. Influence of Initial Rock Temperature on Stress Distribution

Figure 11 indicates that the trends in Mises stress and maximum principal stress with increasing temperature vary by depth. At a depth of 2 mm, the maximum principal stress at the impact center point increases from −0.144 MPa at 50 °C to 1.849 MPa at 150 °C, then decreases to 0.621 MPa as the temperature continues to rise to 200 °C, with the tensile failure zone reaching its largest extent at 150 °C. In contrast, at a depth of 10 mm, the maximum principal stress at the impact center point increases from −5.399 MPa at 50 °C to −10.768 MPa at 200 °C, showing a 99.46% increase. The change in Mises stress at a 2 mm depth exhibits a similar trend of initial increase followed by a decrease, while at 10 mm depth, it gradually decreases with rising temperature.

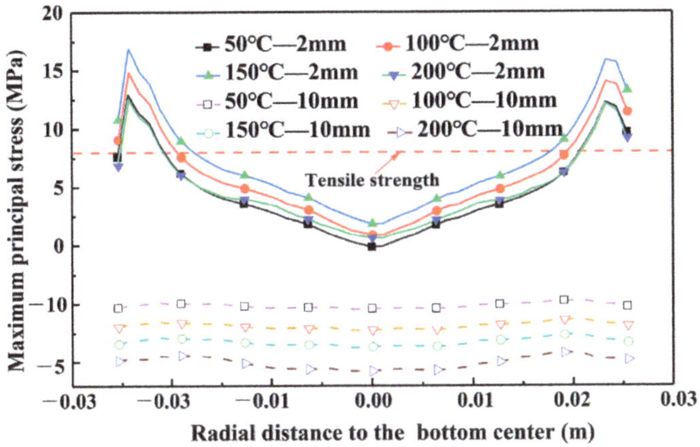

(**a**) Maximum principal stress

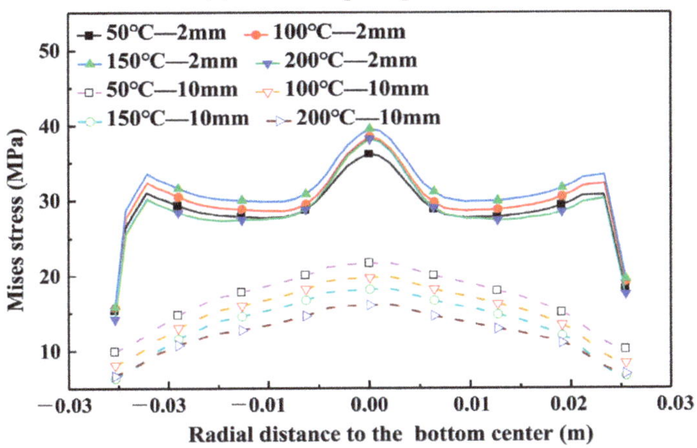

(**b**) Mises stress

Figure 11. Influence of initial rock temperature on stress distribution.

4.3. *Influence of Fluid Parameters on Stress Distribution*

4.3.1. Influence of Outlet Pressure on Stress Distribution

Figure 12 demonstrates that unlike the trends observed with solid parameters, the changes in maximum principal stress under varying outlet pressures are insignificant. At outlet pressures of 35 and 40 MPa, the maximum principal stress of the rock at a depth of 2 mm decreases as the pressure increases, while the changes at a depth of 10 mm are minimal. The maximum principal stress range at the impact center point is 2.855 MPa at a 2 mm depth and 1.286 MPa at a 10 mm depth. The tensile failure zone does not reach its limit before 40 MPa but at 45 MPa, with the damage zone increasing from 6.008 to 7.067 mm as the pressure rises. This indicates that variation in outlet pressure between 40 and 45 MPa significantly affects the expansion of the tensile failure zone.

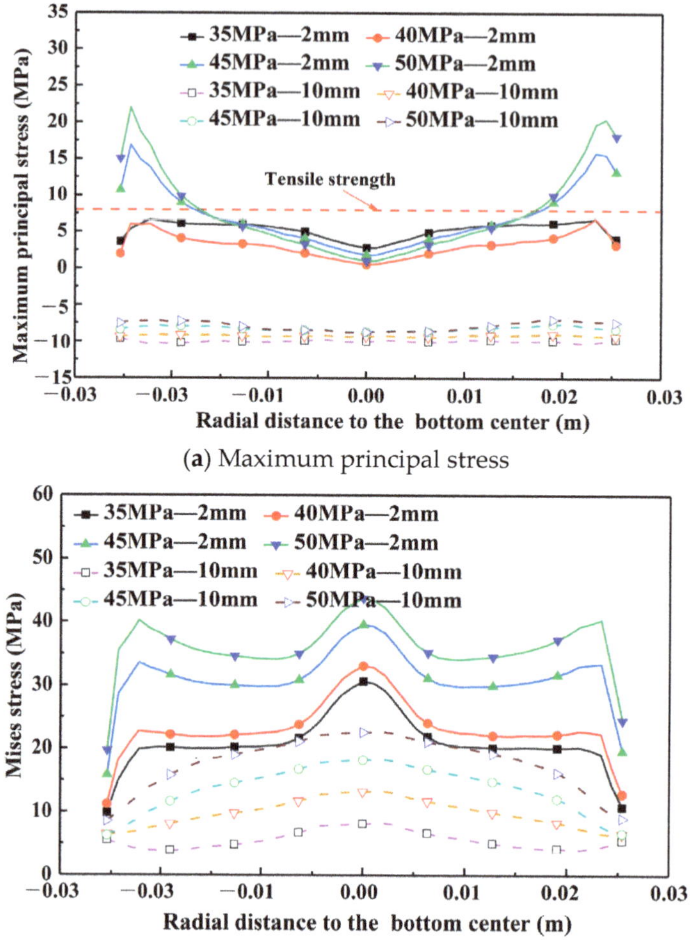

(**a**) Maximum principal stress

(**b**) Mises stress

Figure 12. Influence of outlet pressure on stress distribution.

The changes in Mises stress are significant. For instance, at the impact center point of the jets, it increases from 21.805 to 43.711 MPa at a 2 mm depth, with an increase of 19.906 MPa, while at a 10 mm depth, it increases from 8.1024 to 22.513 MPa, representing an increase of 14.411 MPa. When the outlet pressure increases by 5 MPa, the most significant change in Mises stress occurs between 40 and 45 MPa, indicating that this pressure range has the greatest effect on the Mises stress and shear failure of the rock. This is consistent with the outlet pressure range that most influences the expansion of the tensile failure zone.

4.3.2. Influence of Fluid Temperature on Stress Distribution

Figure 13 shows that the maximum principal stress distribution does not change significantly with increasing temperature. It maintains a pattern of compression at the center and tension at both ends, generated by stress concentrations at the impact center and the two ends of the bottomhole. Similarly, Mises stress also does not change significantly with increasing temperature. Stress concentration occurs near the bottomhole plane, with a trend of decreasing radially and then increasing at both ends.

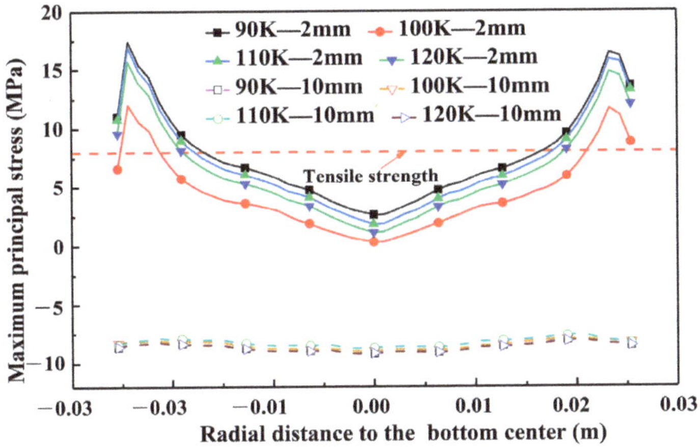

(**a**) Maximum principal stress

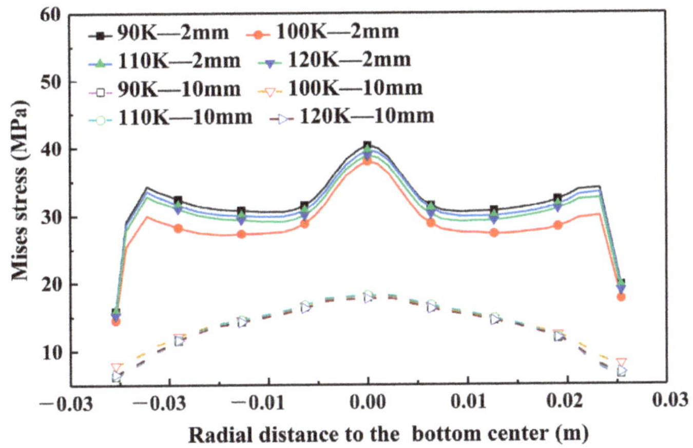

(**b**) Mises stress

Figure 13. Influence of fluid temperature on stress distribution.

The maximum principal and Mises stresses remain essentially unchanged at a depth of 10 mm. However, at a depth of 2 mm, they reach their maximum at 90 K, decrease at 100 K, increase at 110 K, and decrease again at 120 K, indicating an unstable trend with significant fluctuations. The tensile failure zone exhibits a similar trend to the stress changes, with damage zones measuring 7.067, 2.833, 6.01, and 4.96 mm as fluid temperature increases. The volatility in the influence of fluid temperature on stress distribution can be due to a small temperature change gradient. In this case, the temperature change between adjacent test groups is 10 K, while in the previous study on the influence of rock temperature, the temperature change between adjacent groups was 50 °C, leading to a more distinct pattern of change.

5. Advantages of Drilling with Liquid Nitrogen

Based on the findings of this study on the stress distribution characteristics of rock in the bottomhole and the influence laws of various parameters under the impact of a liquid nitrogen jet, the potential economic benefits and environmental impacts of implementing optimized liquid nitrogen jet drilling techniques can be analyzed as follows:

5.1. Potential Economic Benefits

This study indicates that liquid nitrogen jet drilling can enhance rock-breaking efficiency due to the thermal stress created by its extremely low temperature. This could lead to faster drilling rates, reducing the time and cost associated with drilling operations. Compared to traditional hydraulic drilling, liquid nitrogen jet drilling does not require large volumes of water, which could reduce energy consumption associated with water transportation and treatment. The ability of liquid nitrogen to effectively create and propagate fractures within the reservoir rock can lead to improved reservoir permeability and production capacity, potentially extending the life of the reservoir and increasing its economic value.

5.2. Comparison to Existing Technologies

The economic benefits can be quantified through a cost–benefit analysis that compares the initial investment, operational costs, and potential revenue gains from increased production between liquid nitrogen jet drilling and existing technologies such as hydraulic drilling. The environmental impact, particularly related to energy use and greenhouse gas emissions, can be assessed by comparing the life cycle energy consumption and carbon footprint of liquid nitrogen jet drilling with those of traditional methods. This would involve considering the energy required for producing, transporting, and utilizing liquid nitrogen, as well as the impacts of reduced water use.

5.3. Environmental Impacts

One of the key environmental benefits is the reduced reliance on water, which is particularly important in water-scarce regions. This can alleviate the pressure on local water resources and reduce the potential for water contamination associated with hydraulic fracturing. The use of liquid nitrogen may result in lower greenhouse gas emissions compared to hydraulic drilling, primarily due to the elimination of large-scale water usage and associated energy consumption. However, a complete life cycle assessment would need to consider the emissions associated with liquid nitrogen production and transportation.

6. Conclusions

(1) Due to lower kinetic viscosity, the liquid nitrogen jet experiences less energy dissipation and achieves a significantly higher maximum velocity than the water jet under the same conditions, with an axial velocity about 20 m/s higher.

(2) Both water and liquid nitrogen jets have high-pressure regions concentrated at the bottomhole center, with comparable pressure magnitudes and impact forces. The pressure at the center is twice that at the ends, causing significant rock compression.

(3) Both jets exhibit similar tangential velocity distributions with high-velocity overflow at the bottomhole center, but the liquid nitrogen jet's maximum velocity is 20% higher. This overflow boosts heat exchange efficiency, which is crucial for thermal rock fracturing.

(4) The rate of stress change in rock under the liquid nitrogen jet is significantly higher than that under the water jet, especially within the first 10 s. At 2 mm depth, the liquid nitrogen jet induces tensile stress, making it more effective at causing tensile failure and promoting rock fracturing.

(5) The Mises stress from the water jet is concentrated at the impact center, while the liquid nitrogen jet shows a broader stress distribution, leading to a larger rock fracturing region.

Author Contributions: Methodology, C.C. and B.X.; conceptualization, C.C.; investigation, B.W. and Y.F.; data curation, Z.Z. and J.L.; writing—original draft, C.C. and B.X.; writing—review and editing, B.W. and Y.G.; funding acquisition, Y.G. All authors have read and agreed to the published version of the manuscript.

Funding: This research was supported by the Yunlong Lake Laboratory of Deep Underground Science and Engineering Project (No. 104024001) and Sinopec Key Laboratory of Shale Oil and Gas Drilling and Hydraulic Fracturing (No. 36650000-23-ZC0609-0025).

Data Availability Statement: Data will be made available on request.

Conflicts of Interest: Author Bo Xiao was employed by the company Sinopec Research Institute of Petroleum Engineering Co., Ltd. The remaining authors declare that the research was conducted in the absence of any commercial or financial relationships that could be construed as a potential conflict of interest.

References

1. Li, Y.; Wang, H.; Wang, Y.; Zhao, J. Background and routs of carbon neutrality and its nature-derived thermal solutions. *Huadian Technol.* **2021**, *43*, 5–14.
2. Wang, G.; Zhang, W.; Liang, J.; Lin, W.; Liu, Z.; Wang, W. Evaluation of Geothermal Resources Potential in China. *Acta Geosci. Sin.* **2017**, *38*, 449–450.
3. Lei, H. Coupling Mechanics with Thermal-Hydrodynamic Processes for Heat Development in Enhanced Geothermal Systems (EGS). Doctoral Thesis, Jilin University, Changchun, China, 2014.
4. Rauenzahn, R.M.; Tester, J.W. Rock failure mechanisms of flame-jet thermal spallation drilling-theory and experimental testing. *Int. J. Rock Mech. Min. Sci. Geomech. Abstr.* **1989**, *26*, 381–399. [CrossRef]
5. Hu, X.; Song, X.; Li, G.; Shen, Z.; Lyu, Z.; Shi, Y. Shape factor of the flake-like particle in thermal spallation and its effects on settling and transport behavior in drilling annulus. *Powder Technol.* **2018**, *335*, 211–221. [CrossRef]
6. Wang, H.; Li, G.; Shen, Z. A Feasibility Analysis on Shale Gas Exploitation with Supercritical Carbon Dioxide. *Energy Sources Part A Recover. Util. Environ. Eff.* **2012**, *34*, 1426–1435. [CrossRef]
7. Wang, H.; Li, G.; Shen, Z.; Tian, S.; Sun, B.; He, Z.; Lu, P. Experiment on rock breaking with supercritical carbon dioxide jet. *J. Pet. Sci. Eng.* **2015**, *127*, 305–310. [CrossRef]
8. Zhao, X.; Huang, B.; Xu, J. Experimental investigation on the characteristics of fractures initiation and propagation for gas fracturing by using air as fracturing fluid under true triaxial stresses. *Fuel* **2018**, *236*, 1496–1504. [CrossRef]
9. Wang, L.; Yao, B.; Cha, M.; Alqahtani, N.B.; Patterson, T.W.; Kneafsey, T.J.; Miskimins, J.L.; Yin, X.; Wu, Y.-S. Waterless fracturing technologies for unconventional reservoirs-opportunities for liquid nitrogen. *J. Nat. Gas Sci. Eng.* **2016**, *35*, 160–174. [CrossRef]
10. Kim, K.M.; Kemeny, J. Effect of thermal shock and rapid unloading on mechanical rock properties. In Proceedings of the 43rd US Rock Mechanics Symposium & 4th US-Canada Rock Mechanics Symposium, Asheville, NC, USA, 28 June–1 July 2009; American Rock Mechanics Association: Seattle, WA, USA, 2009.
11. Brown, D.W.; Duchane, D.V. Scientific progress on the Fenton Hill HDR project since 1983. *Geothermics* **1999**, *28*, 591–601. [CrossRef]
12. McDaniel, B.W.; Grundmann, S.R.; Kendrick, W.D.; Wilson, D.R.; Jordan, S.W. Field applications of cryogenic nitrogen as a hydraulic fracturing fluid. In Proceedings of the SPE Annual Technical Conference and Exhibition, San Antonio, TX, USA, 5–8 October 1997; Society of Petroleum Engineers: Richardson, TX, USA, 1997.
13. Grundmann, S.R.; Rodvelt, G.D.; Dials, G.A. Cryogenic nitrogen as a hydraulic fracturing fluid in the Devonian Shale. In Proceedings of the SPE Eastern Regional Meeting, Pittsburgh, PA, USA, 9–11 November 1998; Society of Petroleum Engineers: Richardson, TX, USA, 1998.
14. Finnie, I.; Cooper, G.A.; Berlie, J. Fracture propagation in rock by transient cooling. *Int. J. Rock Mech. Min. Sci. Geomech. Abstr.* **1979**, *16*, 11–21. [CrossRef]
15. Ren, S.; Fan, Z.; Zhang, L. Mechanisms and Experimental Study of Thermal-shock Effect on Coal-rock Using Liquid Nitrogen. *Chin. J. Rock Mech. Eng.* **2013**, *32*, 3790–3794.
16. Kim, K.; Kemeny, J.; Nickerson, M. Effect of Rapid Thermal Cooling on Mechanical Rock Properties. *Rock Mech. Rock Eng.* **2014**, *47*, 2005–2019. [CrossRef]
17. Cai, C.; Li, G.; Huang, Z.; Tian, S.; Shen, Z.; Fu, X. Experiment of coal damage due to super-cooling with liquid nitrogen. *J. Nat. Gas Sci. Eng.* **2015**, *22*, 42–48. [CrossRef]
18. Cha, M.; Yin, X.; Kneafsey, T.; Johanson, B.; Alqahtani, N.; Miskimins, J.; Patterson, T.; Wu, Y.-S. Cryogenic fracturing for reservoir stimulation—Laboratory studies. *J. Pet. Sci. Eng.* **2014**, *124*, 436–450. [CrossRef]
19. Cai, C.; Huang, Z.; Li, G.; Gao, F. Particle velocity distributions of abrasive liquid nitrogen jet and parametric sensitivity analysis. *J. Nat. Gas Sci. Eng.* **2015**, *27*, 1657–1666. [CrossRef]
20. Cai, C.; Huang, Z.; Li, G.; Gao, F.; Wei, J.; Li, R. Feasibility of reservoir fracturing stimulation with liquid nitrogen jet. *J. Pet. Sci. Eng.* **2016**, *144*, 59–65. [CrossRef]
21. Zhang, C.; Wang, L.; Zhao, Q.; Li, W. Permeability evolution model and numerical analysis of coupled coal deformation, failure and liquid nitrogen cooling. *J. Hebei Univ. Sci. Technol.* **2015**, *36*, 90–99.
22. Huang, Z.; Cai, C.; Li, G.; Tian, S. Flow behavior and particle acceleration effect of abrasive liquid nitrogen jet. *J. China Univ. Pet. (Ed. Nat. Sci.)* **2016**, *40*, 80–86.
23. Li, Z.; Xu, H.; Zhang, C. Liquid nitrogen gasification fracturing technology for shale gas development. *J. Pet. Sci. Eng.* **2016**, *138*, 253–256. [CrossRef]
24. Zhai, C.; Qin, L.; Liu, S.; Xu, J.; Tang, Z.; Wu, S. Pore Structure in Coal: Pore Evolution after Cryogenic Freezing with Cyclic Liquid Nitrogen Injection and Its Implication on Coalbed Methane Extraction. *Energy Fuels* **2016**, *30*, 6009–6020. [CrossRef]

25. Zhai, C.; Wu, S.; Liu, S.; Qin, L.; Xu, J. Experimental study on coal pore structure deterioration under freeze–thaw cycles. *Environ. Earth Sci.* **2017**, *76*, 1–12. [CrossRef]
26. Yao, B.; Wang, L.; Yin, X.; Wu, Y.-S. Numerical modeling of cryogenic fracturing process on laboratory-scale Niobrara shale samples. *J. Nat. Gas Sci. Eng.* **2017**, *48*, 169–177. [CrossRef]
27. Cai, C.; Gao, F.; Huang, Z.; Yang, Y. Numerical simulation on the flow field characteristics and impact capability of liquid nitrogen jet. *Energy Explor. Exploit.* **2018**, *36*, 989–1005. [CrossRef]
28. Cai, C.; Yang, Y.; Liu, J.; Gao, F.; Gao, Y.; Zhang, Z. Downhole Transient Flow Field and Heat Transfer Characteristics During Drilling with Liquid Nitrogen Jet. *J. Energy Resour. Technol.* **2018**, *140*, 12290212. [CrossRef]
29. Wu, X.; Huang, Z.; Li, G.; Li, R.; Yan, P.; Deng, X.; Mu, K.; Dai, X. Experiment on coal breaking with cryogenic nitrogen jet. *J. Pet. Sci. Eng.* **2018**, *169*, 405–415. [CrossRef]
30. Zhang, S.; Huang, Z.; Li, G.; Wu, X.; Peng, C.; Zhang, W. Numerical analysis of transient conjugate heat transfer and thermal stress distribution in geothermal drilling with high-pressure liquid nitrogen jet. *Appl. Therm. Eng.* **2018**, *129*, 1348–1357. [CrossRef]
31. Zhang, S.; Huang, Z.; Wang, H.; Zhang, H.; Zhang, C.; Xiong, C. Thermal characteristics analysis with local thermal non-equilibrium model during liquid nitrogen jet fracturing for HDR reservoirs. *Appl. Therm. Eng.* **2018**, *143*, 482–492. [CrossRef]
32. Zhang, S.; Huang, Z.; Huang, P.; Wu, X.; Xiong, C.; Zhang, C. Numerical and experimental analysis of hot dry rock fracturing stimulation with high-pressure abrasive liquid nitrogen jet. *J. Pet. Sci. Eng.* **2018**, *163*, 156–165. [CrossRef]
33. Zhang, S.; Huang, Z.; Wang, H.; Li, G.; Sepehrnoori, K.; Wu, X.; Hong, C. Experimental study on the rock-breaking characteristics of abrasive liquid nitrogen jet for hot dry rock. *J. Pet. Sci. Eng.* **2019**, *181*, 106166. [CrossRef]
34. Yang, R.; Hong, C.; Huang, Z.; Song, X.; Zhang, S.; Wen, H. Coal breakage using abrasive liquid nitrogen jet and its implications for coalbed methane recovery. *Appl. Energy* **2019**, *253*, 113485. [CrossRef]
35. Liu, J.; Li, J.; Long, W.; Yuting, C.; Jin, Y. Failure analysis of water liquid nitrogen cyclic jet impacting concrete. *Int. J. Mech. Sci.* **2022**, *235*, 107714. [CrossRef]
36. Wang, Y.; Dai, M.; Liu, K.; Liu, J.; Han, L.; Liu, H. Research on surface heat transfer mechanism of liquid nitrogen jet cooling in cryogenic machining. *Appl. Therm. Eng.* **2020**, *179*, 115607. [CrossRef]
37. Dai, X.; Huang, Z.; Wu, X.; Shi, H.; Xiong, C. Failure Analysis of High-Temperature Granite Under the Joint Action of Cutting and Liquid Nitrogen Jet Impingement. *Rock Mech. Rock Eng.* **2021**, *54*, 6249–6264. [CrossRef]
38. Cai, C.; Wang, B.; Huang, Z.; Yue, W.; Wang, H.; Gao, Y.; Feng, Y.; Yang, Y.; Guo, C. Comparison with the effect of the cyclic liquid nitrogen jet and liquid nitrogen immersion cooling on the deterioration of mechanical properties of high temperature rocks. *Geoenergy Sci. Eng.* **2024**, *243*, 213295. [CrossRef]
39. Li, R.; Yan, Y.; Huang, Z. Thermal and mechanical analysis of LN2 jet impinging on rock surface. *Appl. Therm. Eng.* **2020**, *178*, 115581. [CrossRef]

processes

MDPI

Article

Numerical Investigation of Complex Hydraulic Fracture Propagation in Shale Formation

Heng Zheng [1,2,3,*], **Fengxia Li** [1,2] **and Di Wang** [1,2]

[1] State Key Laboratory of Shale Oil and Gas Enrichment Mechanisms and Effective Development, Beijing 100083, China
[2] State Energy Center for Shale Oil Research and Development, Beijing 100010, China
[3] School of Petroleum Engineering, Yangtze University, Wuhan 434000, China
* Correspondence: hengsmith111@163.com

Abstract: Due to the high flow resistance of shale oil and gas, creating artificial flow channels with high conductivity in shale formation was the main challenge for the development of shale oil and gas resources. To further understand the fracture propagation mechanism in shale formation, this paper proposed a global cohesive element method to simulate the hydraulic fracture propagation behavior in which natural fractures were distributed randomly, and the fracture geometry was quantitatively analyzed. From the simulation, it can be found that the horizontal stress difference was the determining factor affecting the generation of a complex fracture network. The simulation indicated that a low horizontal stress difference was beneficial for improving the stimulated volume. When the stress difference was below 5.0 MPa, numerous branch fractures were created which was the foundation of a complex fracture network. High injection rates with low-viscosity fracturing fluid were helpful for creating a complex fracture network, while high-viscosity fracturing fluid limited the fracture fluid flow into the deep formation.

Keywords: natural crack; hydraulic fracture; complex fracture network; global cohesive element method

1. Introduction

Due to the abundance of oil and gas resources in shale formations, shale oil and gas resources have become the most promising energy sources to support industrial production. Limited by the low permeability and porosity of shale formations, hydraulic fracturing has become the key technology for the effective development of shale oil and gas [1–3].

During the recent decades, many experiments and simulation results were conducted to understand the mechanism of complex fracture network formation [4]. The experiments and simulations indicated that horizontal stress difference was the main factor affecting the interaction between a natural fracture and a hydraulic fracture [5,6]. With experimental research, it was demonstrated that when HFs interact with NFs, different mechanical properties and approach angles may lead to different interaction results. The laboratory can only study a fracture geometry with limited factors [7–9]. In addition, it was difficult to observe the fracture network extension in these experiments because the scale of core samples was too small, which was far from the reservoir scale. Recently, many simulation models were established based on the FEM, DEM, and DDM. During the simulation, the fracture propagation trajectory was pre-determined, it was difficult to simulate the interaction between a hydraulic fracture with multiple natural fractures [10,11]. Based on previous research, it is indicated that three different styles were displayed when the

https://doi.org/10.3390/pr12122630

hydraulic fracture interacted with a natural fracture, namely crossing, arrest, and offset. The fracture geometry in interaction with a natural fracture was comprehensively affected by the horizontal stress difference, approach angle, injection rate, and viscosity of fracturing fluid [12,13]. To overcome the barrier of horizontal stress differences on complex fracture networks, some experiments and simulations were conducted to investigate the cluster space, injection rate, and viscosity of fracturing fluid on fracture propagation [14–16]. The simulations showed that cluster space was the main factor affecting the fracture propagation, increasing the injection rate can release the "stress shadow" on fracture propagation under certain cluster spaces. It also showed that temporary plugging and diverting fracturing was an effective method to increase the number of branch fractures during multi-staged hydraulic fracturing [17,18].

To further investigate hydraulic fracture propagation in shale formations abundant in natural fractures, this paper proposed a global cohesive method to simulate and analyze the complex fracture network. Compared with previous research on fracture propagation behavior using CZM, the method proposed in this paper has the advantages as follows. In previous research, most studies focused on single fractures and the interaction of single fractures under the influence of geological and engineering factors, which only represent local correlations within the reservoir. At the same time, the path of fracture propagation should be strictly planned. The cohesive elements were inserted in the whole reservoir model randomly, the distribution of weak surfaces was generally more complicated. The randomly occurring and distributed weak surfaces in the model were closer to the real reservoir, which can more accurately simulate the extension of hydraulic fractures under multi-weak surfaces. During the simulations, the effects of mechanical rock properties, ground stress fields, and injection parameters on fracture network generation were analyzed. This provided a new method to optimize the hydraulic fracture processing parameters during the exploration and development of shale oil and gas.

2. Governing Equation

The physical processes of hydraulic fracturing can be divided into four processes: (1) fracture initiation; (2) Fracturing fluid flows in the fracture network, including tangential flow in the fracture and normal flow in the rock matrix; (3) extension of natural fractures; (4) Fracturing fluid causes deformation of rock around fractures.

2.1. Stress Equilibrium Equation

In the simulation, the principle of effective stress was used to describe the coupling between the stress field of the rock matrix and fluid pressure, as shown below [19]:

$$\sigma_{total} = \sigma\prime + \alpha P_w \delta \tag{1}$$

where σ_{total} is the total stress; $\sigma\prime$ is the tensor of the effective stress; α is the Biot coefficient; P_w is the pore pressure; and δ is the Kronecker constant.

During fracture propagation, the rock surface deformation caused by the fracturing fluid injection was described by the virtual work principle, which was defined as follows [20]:

$$\int (\overline{\sigma} - p_w I) : \delta\varepsilon dV = \int_S t \cdot \delta v dS + \int_V f \cdot \delta v dV \tag{2}$$

where $\overline{\sigma}'$ and ε' are effective stress and virtual deformation rate, respectively. I is the unit matrix, and t and f are surface traction per unit area and body force per unit volume, respectively. In this equation, the displacement is a node variable, discretized by the Lagrange formula.

2.2. Fracturing Fluid Flow Equation

For the fracturing fluid flow in the hydraulic fracture, the cubic law was introduced to calculate the fracture width under a certain viscosity of fracturing fluid which is shown below as Equation (3):

$$q_f = -\frac{w^3}{12\mu}\nabla p_f \tag{3}$$

∇P_f is the pressure gradient along the cohesive element, μ is the fluid viscosity, and w is the fracture opening width.

In hydraulic fracturing, the leak-off between the fracturing fluid and rock matrix was the main factor affecting fracture propagation. In this simulation, the Carter leak-off model was introduced, which fully coupled the pore elasticity method to calculate the leakage state of the fracture surface when the fracturing fluid flowed through the fracture surface [21], as shown below:

$$q_t = c_t(p_i - p_t) \tag{4}$$

$$q_b = c_b(p_i - p_b) \tag{5}$$

where q_t and q_b are the typical flow flux crossing into the cohesive element's top and bottom surfaces, respectively. P_t and P_b are the pore pressures on the top and bottom surfaces, respectively, while P_i is the fluid pressure within the cohesive element gap. C_t and C_b are the fluid leak-off parameters.

2.3. Fracture Propagation Criterion

In the simulation, it was assumed that rock failure followed the linear elastic fracture mechanics characteristic; thus, the traction-separation criterion was used to describe rock failure during hydraulic fracturing. The traction-separation criterion is shown as Figure 1.

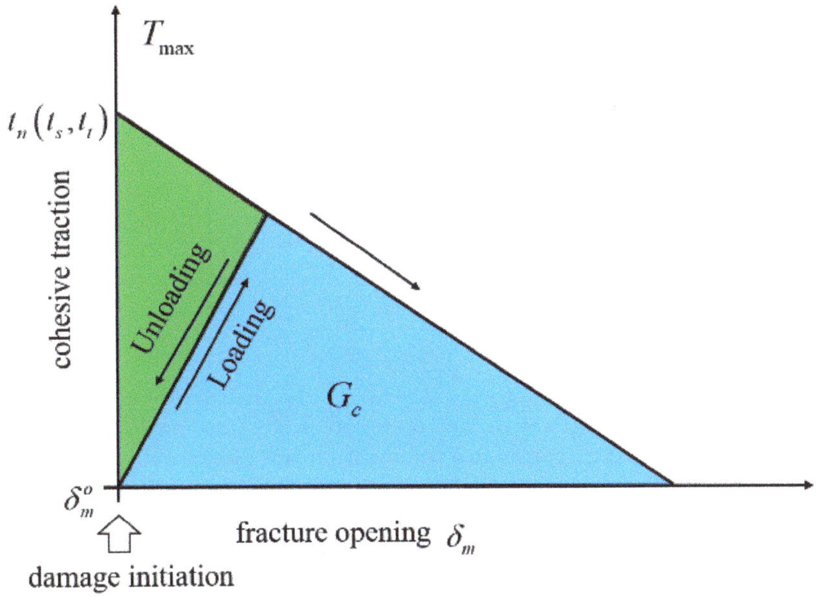

Figure 1. Traction-separation relationship for fracture initiation.

In the simulation, the second nominal principal stress criterion was introduced to describe the damage caused by fracturing fluid injection. In the simulation, both the tensile failure and shear failure were considered in the criterion [22], as follows:

$$\left\{\frac{t_n}{t_n^0}\right\}^2 + \left\{\frac{t_s}{t_s^0}\right\}^2 = 1 \tag{6}$$

where t_s^0 and t_n^0 are the normal and shear strengths of an undamaged cohesive element, respectively; the symbol stands for the Macaulay bracket, and t_n and t_s stand for the normal and shear stress components, respectively.

The normal and shear stress components can be defined as follows:

$$t_n = \begin{cases} (1-D)T_n, T_n \geq 0 \\ T_n, T_n < 0 \end{cases} \tag{7}$$

$$t_s = (1-D)T_s \tag{8}$$

$$t_t = (1-D)T_t \tag{9}$$

To characterize the damage caused by fracturing fluid injection, the scalar damage variable D was used to describe the entire fracturing process based on the mixed-mode energy theory. The area under the traction-separation curve represents the breaking energy. The damage variable for linear softening was shown as follows [23]:

$$D = \frac{U_m^f(U_m^{\max} - U_m^0)}{U_m^{\max}(U_m^f - U_m^0)} \tag{10}$$

where U_m^0 is the effective displacement during the damage evolution; U_m^f is the displacement when the material fails; and $U_m^{\max}$ is the maximum effective displacement during the loading.

For the fracture propagation, the Benzeggagh-Kenane (B-K) criterion was used to describe the fracture propagation, and the Benzeggagh-Kenane (B-K) criterion was shown as follows [24]:

$$G_n^C + \left(G_s^C - G_n^C\right)\left\{\frac{G_s}{G_T}\right\}^\eta = G^C \tag{11}$$

G_c is the calculated total critical energy release rate due to mixed-mode failure; G_{nc} and G_{sc} are the critical energy release rates due to Mode I and Mode II fractures, respectively. The energy release rate of Mode I and Mode II cracks, G_n and G_s, are constants that depends on the type of material.

2.4. Verification

To verify the effectiveness of the proposed model, a true tri-axial hydraulic fracturing experiment was conducted with maximum and minimum horizontal principle stresses of 28.4 MPa and 21.4 MPa and vertical stress of 32.4 MPa. The Young modulus and Poisson rate of the prepared rock sample were 35.7 GPa and 0.32, the injection rate was 15 mL/s, and the viscosity of the fracturing fluid was 1.0 mPa.s. Figure 2a indicates the fracture geometry after the hydraulic fracturing experiment, and Figure 2b shows the numerical simulation result. From the comparison, it can be found that the fracture created during the experiment had a slight curve, which was the same in the numerical simulation. For the injection pressure, the breakdown pressures of the experiment and numerical simulation were 35.4 MPa and 34.8 MPa, with an error of 1.69%, and the average extension pressures of the experiment and numerical simulation were 21.6 MPa and 22.4 MPa, with an error of 3.7%. From the comparison, the effectiveness of the proposed model is verified.

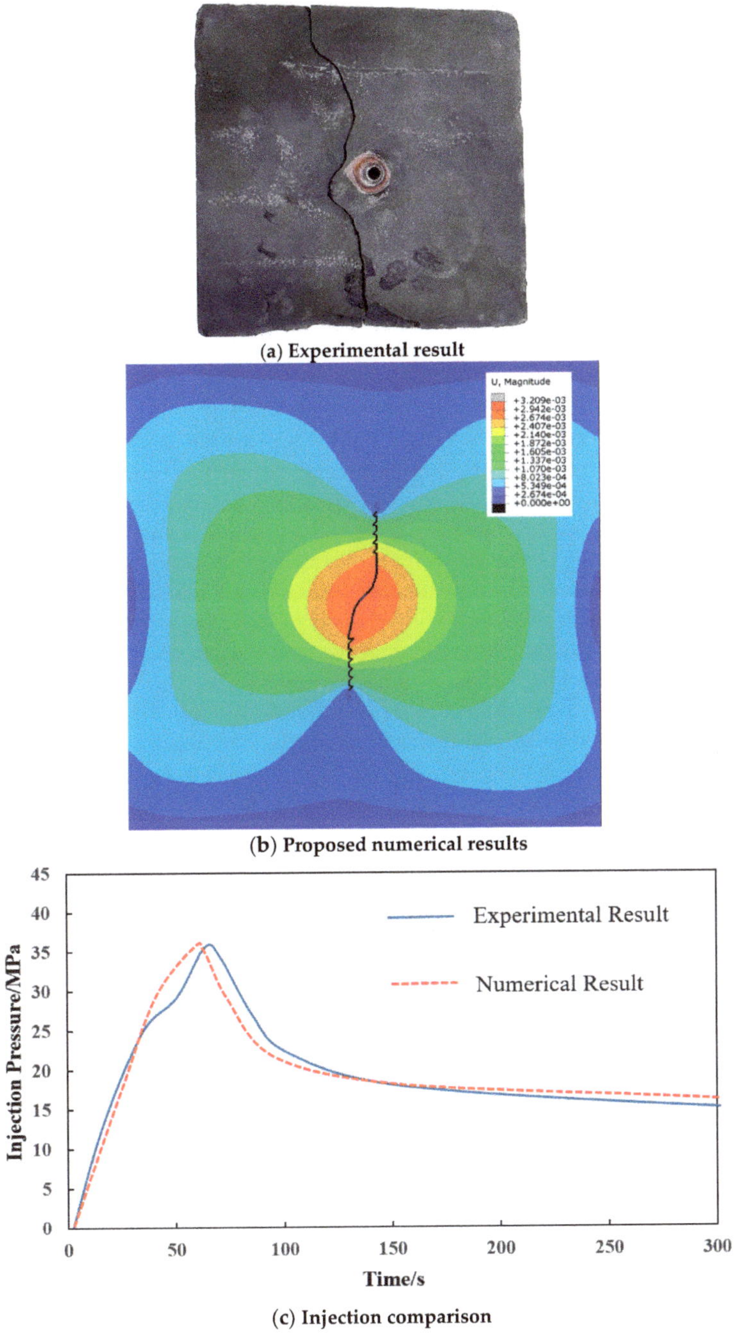

Figure 2. (**a**) Experimental numerical results. (**b**) Proposed model. (**c**) Injection comparison. Simulation results comparison between the experimental results and the proposed model.

3. Simulation and Discussion

3.1. Model Description

In the simulation, a 2D model with a length of 400 m and a width of 200 m was established, and the formation was discretized using the CPE4P (four-node pore pressure-stress) element, and the COH2D4P element was intersected between the CPE4P elements to describe the natural fracture present in the formation. The boundary of the established model was set as a fixed boundary because the stimulated area was much smaller than that of the whole model area, thus the displacement and stress on the boundary remained stable. In the model, the natural fracture angles were 30° and 150°, the initial natural fracture length was 0.5 m, the fracture density was 0.4 m, and the perforation direction was aligned with the direction of maximum horizontal stress. In the model, the minimum horizontal principal stress was kept stable, and the maximum horizontal principal stress was changed, thus the extension pressure of the hydraulic fracture did not have huge fluctuations. The other parameters were obtained from the Jimusar shale oilfield.

3.2. Discussion

3.2.1. Injection Rate of Fracturing Fluid on Fracture Propagation

To understand the effect of injection rate on fracture propagation in shale formations, different pumping rates of 8 m^3/min, 10 m^3/min, and 12 m^3/min were simulated to investigate the fracture propagation. The other parameters in the modeling were shown in Table 1. Figure 3 indicates the fracture geometry under different pumping rates. From the simulation, it can be found that the hydraulic fracture propagation was not planar due to natural fractures. During hydraulic fracture propagation, the fracture propagation path was determined by two competing forces: one was the tendency to propagate along the weak plane generated by the natural fracture set, and the other was the tendency to propagate along the direction of the maximum principal stress. Generally speaking, the direction of natural fractures was inconsistent with the direction of in situ principal stress. Fracture propagation must balance these two forces, minimize fracturing energy, and find the path of least resistance, which was the main reason causing the zigzag path during hydraulic fracture propagation. It can be seen that the geometric shape of the crack was similar to the shape of an "elbow", and the trajectory of the two hydraulic fracture wings was different. The deflection direction of the two hydraulic fracture wings on both sides was inclined to the *X*-axis. When the pumping rate increased to 10 m^3/min, branch fractures were created, and the upper fracture propagated more easily with the increase in pumping rate. Meanwhile, it should be noted that the length of the created fractures had an obvious increase due to the amount of fracturing fluid also experiencing an increase, which was shown in Figure 4a. When the injection rate increased from 8 m^3/min to 10 m^3/min, it can be found that the total fracture length increased by 139.98%, from 105.3 m to 252.7 m, while it only increased by 40.12% when the injection rate increased from 10 m^3/min to 12 m^3/min. The breakdown pressure and extension pressure under different injection rates is shown in Figure 4b. From the simulation results, it can be concluded that breakdown pressures increased with the increase in pumping rate. When the injection rate was 8 m^3/min, the breakdown pressure and average extension pressure were 44.6 MPa and 24.7 MPa, respectively. When the injection rate was 10 m^3/min, the breakdown pressure and average extension pressure were 52.3 MPa and 35.2 MPa, respectively. When the injection rate was 12 m^3/min, the breakdown pressure was close to that when the injection rate was 10 m^3/min, while the extension pressure increased to 43.7 MPa. The main reason for this phenomenon was that as the injection rate increased, the number of branch fractures also increased, leading to a noticeable increase in breakdown pressure between 8 m^3/min and 10 m^3/min. However, when the injection rate increased to 12 m^3/min, the number of branch fractures stabilized, resulting in little change in breakdown pressure. Furthermore, the tortuosity of the created fractures increased with the injection rate, directly causing an increase in extension pressure.

Table 1. Input parameters of the numerical simulation.

Parameter	Value
Injection rate	8 m^3/min
Young's modulus	35 GPa
Poisson's ratio	0.35
The minimum horizontal stress	36 MPa
The maximum horizontal stress	41 MPa
Fluid leak-off parameter	1×10^{-6} $m^3/Pa{\cdot}s$
Fluid viscosity	10 mPa·s
Formation porosity	0.15
Size of the study area	400 × 200 m

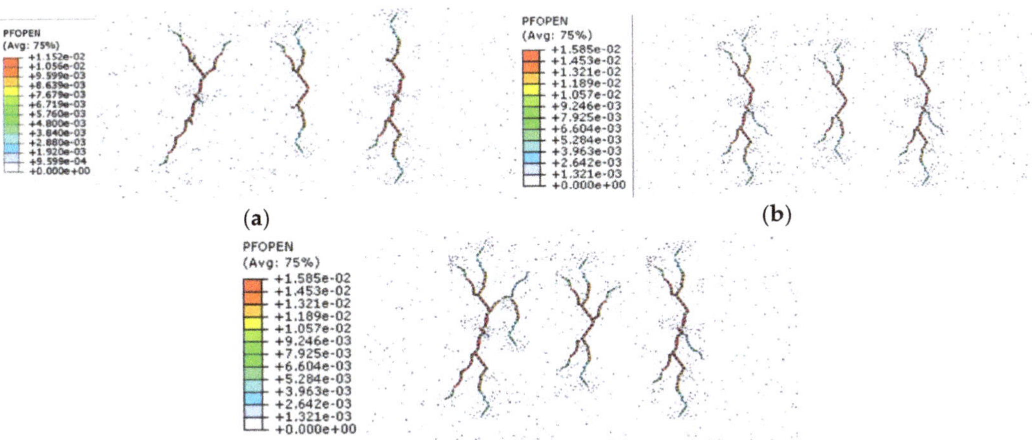

(a) (b)

(c)

Figure 3. (**a**) 8.0 $m^3{\cdot}min^{-1}$. (**b**) 10.0 $m^3{\cdot}min^{-1}$. (**c**) 12.0 $m^3{\cdot}min^{-1}$. Distribution of hydraulic fracture under different injection rate.

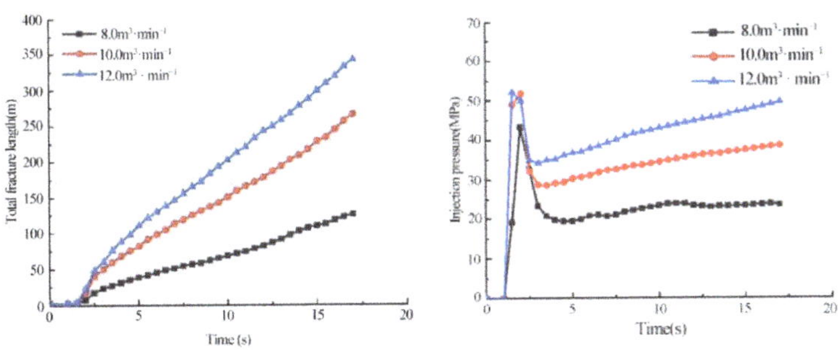

(a) (b)

Figure 4. (**a**) Comparison of the total length of fractures. (**b**) Injection pressure comparison. Changes of total fracture length and fracture pressure at different injection rate.

3.2.2. Influence of Viscosity of Fracturing Fluid on Fracture Propagation

To understand the effect of viscosity of fracturing fluid on fracture propagation in shale formations, different viscosities with 10 mPa·s, 50 mPa·s, and 100 mPa·s were simulated. To investigate the fracture propagation, the injection rate was 10 m^3/min, the other parameters were as shown in Table 1. The simulation results (Figure 5) indicated that when the fracturing fluid viscosity increased from 10 mPa·s to 50 mPa·s, the total fracture length experienced a slight increase, and the tortuosity of the created fracture decreased obviously, the expansion path of the viscosity with 50 mPa·s tended to follow a straight line. When the viscosity continuously increased to 100 mPa·s, the tortuosity of the created fracture decreased further. It also can be found that the number of branch fractures decreased with the increase in viscosity of fracturing fluid. Figure 6 indicates that the total fracture length of the created fractures showed slight difference, and the breakdown pressure and extension pressure showed the same tendency. The main reason accounting for this was that under a low viscosity of fracturing fluid, the flow resistance was comparatively low, thus the fracturing fluid flowed much easier into the natural fracture, hence the number of branch fractures increased. With the increase in viscosity, the hydraulic fracture created crossed the natural fracture much easier, resulting in a straight fracture. From the simulation, when the fracturing fluid viscosity exceeded 50 mPa·s, the viscosity dominated the propagation direction of the fracture, allowing the hydraulic fracture to pass through the natural fracture. Meanwhile, although the extension pressure increased with the increase in tortuosity of the created fracture, the flow resistance of the created fracture also increased with higher fracturing fluid viscosity. Thus, while the straight fractures were created under high fracturing fluid viscosity, the extension pressure did not show an obvious decrease.

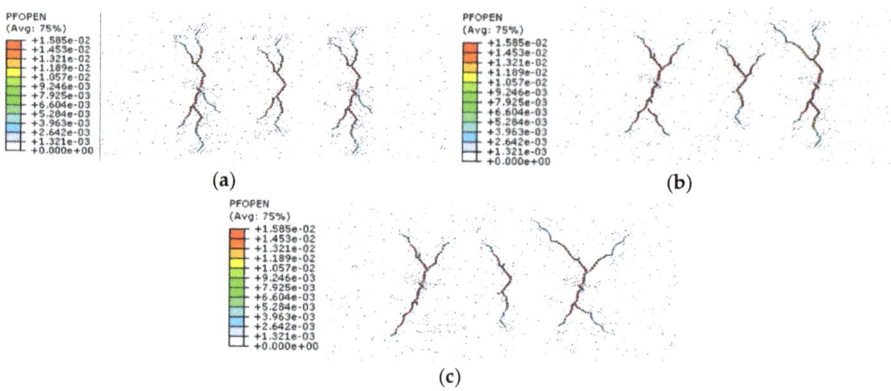

Figure 5. (**a**) 10 mPa·s. (**b**) 50 mPa·s. (**c**) 100 mPa·s. Distribution of hydraulic fracture under different fracturing fluid viscosity.

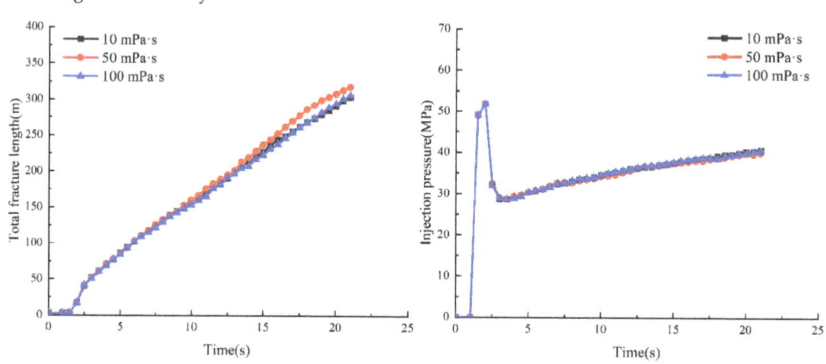

Figure 6. (**a**) Comparison of the total length of cracks. (**b**) Rupture pressure comparison chart. Changes of total fracture length and injection pressure under different fracturing fluid viscosity.

3.2.3. Influence of Elastic Modulus on Fracture Propagation

The elastic modulus of reservoir rock is an essential factor controlling fracture cracking and expansion. To further understand the effect of elastic modulus on fracture propagation, the elastic moduli of 20 GPa, 30 GPa, and 40 GPa were calculated. From the simulation, it was found that with an increase in elastic modulus, the number of branch fractures increased, the tortuosity of the created fractures increased significantly, and the total fracture length also increased. The simulation results (Figure 7) showed that the higher the elastic modulus was, the more complex the hydraulic fracture propagation was, and the easier it was to connect the natural fracture. As for the breakdown pressure and extension pressure, it can be found that both the breakdown pressure and extension pressure had a significant increase (Figure 8). The main reason accounting for this was that with the increase in elastic modulus, the strength difference between the rock matrix and natural fractures also increases, thus the interaction between hydraulic fractures and natural fractures changed from crossing to offset, thus more branch fractures were created when the hydraulic fracture propagated along the natural fracture under high elastic modulus. Meanwhile, with the increase in elastic modulus, the energy consumption for hydraulic propagation also increased, thus both the breakdown pressure and extension pressure increased.

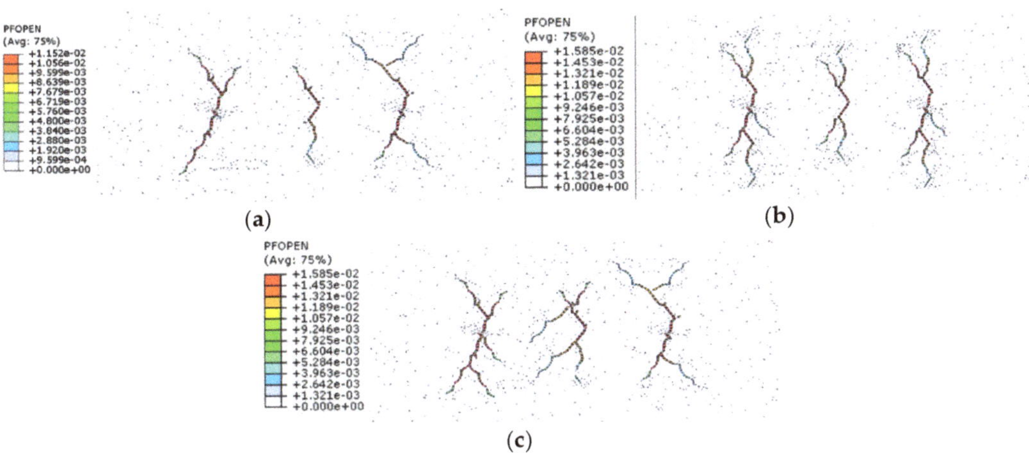

Figure 7. (**a**) 20 GPa. (**b**) 30 GPa. (**c**) 40 GPa. Distribution of hydraulic fracture under different elastic modulus.

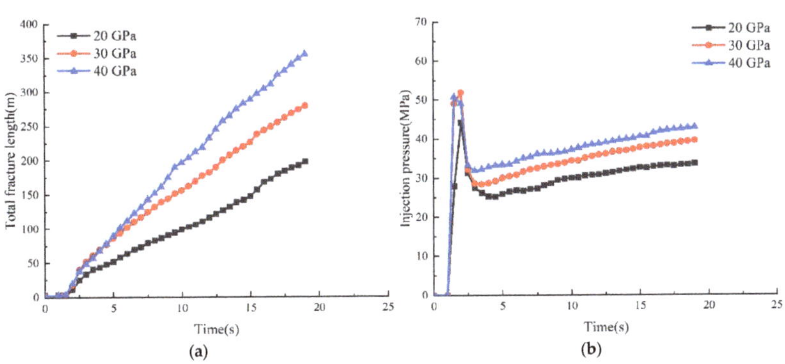

Figure 8. (**a**) Comparison of the total length of cracks. (**b**) Rupture pressure comparison chart. Changes of total fracture length and injection pressure under different elastic modulus.

3.2.4. Influence of Horizontal Stress Differences on Fracture Propagation

In order to study the influence of horizontal stress differences on fracture propagation, the stress differences of 5 MPa, 10 MPa, and 15 MPa were simulated. In the simulation, to avoid the change of the minimum horizontal stress difference on fracture propagation, only the maximum horizontal stress difference was changed to obtain the target stress difference. The simulation (Figure 9) indicated that with an increase in horizontal stress difference, the number of branch fractures significantly increased, and the total length of the created fractures increased. Meanwhile, it also can be found that with an increase in horizontal stress difference, the fracture tended to propagate in a straight line, thus the tortuosity decreased. Under high stress differences (10 MPa and 15 MPa), it can be found that the direction of maximum horizontal stress was the propagation direction, while the propagation direction increased significantly when the stress difference was 5 MPa. The main reason was that under low stress differences, the simulated formation was comparatively uniform, thus many energy release channels were created when the fracturing fluid was injected, and the number of activated natural fractures also increased, so the number of branch fracture increased, and a hydraulic fracture with high tortuosity was created (Figure 10). With the increase of stress difference, the direction of maximum horizontal stress was the propagation direction, thus the number of branch fractures decreased, the number of activated natural fractures also decreased, the number of branch fracture also followed the same tendency.

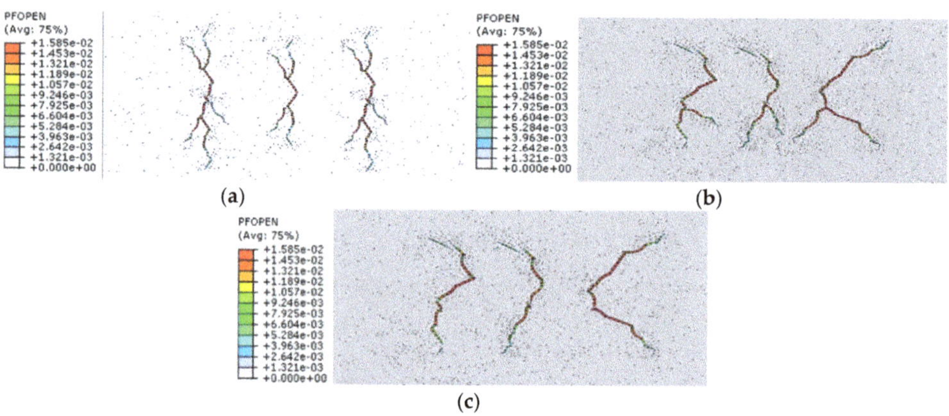

Figure 9. (**a**) 5 MPa. (**b**) 10 MPa. (**c**) 15 MPa. Distribution of hydraulic fractures under different stress differences.

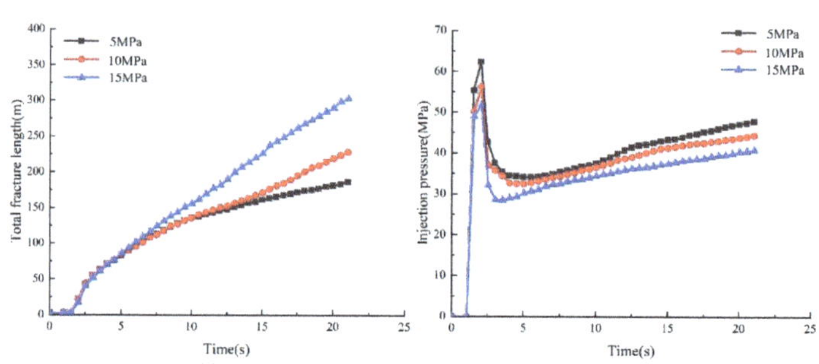

Figure 10. (**a**) Comparison of the total length of cracks. (**b**) Rupture pressure comparison chart. Changes in total fracture length and fracture pressure under different horizontal stress differences.

3.2.5. Influence of Cluster Space on Fracture Propagation

Cluster space was also an important factor affecting fracture propagation. To understand the effect of cluster space on fracture network creation, cluster spaces with 20 m, 40 m, and 60 m were simulated. From the simulation, it can be concluded that with an increase in cluster space, the number of branch fractures decreased, while the length of middle fractures increased (Figure 11). The main reason was that the "stress shadow" caused between fractures was more serious with the decrease in cluster space. When the cluster space was 20 m, more natural fractures were activated, thus more branch fractures were created during the hydraulic fracture propagation. While the "stress shadow" can also limit the propagation of middle fractures. The main reason was that the propagation resistance had an increase at the tip of the middle fractures, the length of the middle fractures was shorter than that of adjacent fractures. With the increase in cluster space, the "stress shadow" on fracture propagation was weakened, thus the number of activated natural fractures decreased, thus the branch fractures decreased, the tortuosity also decreased (Figure 12).

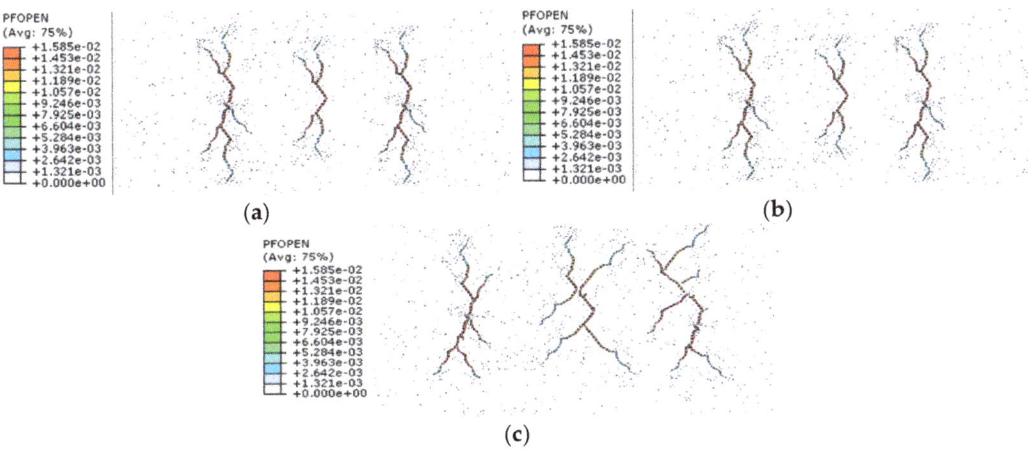

Figure 11. (**a**) 20 m. (**b**) 40 m. (**c**) 60 m. Distribution of hydraulic fractures under different cluster spaces.

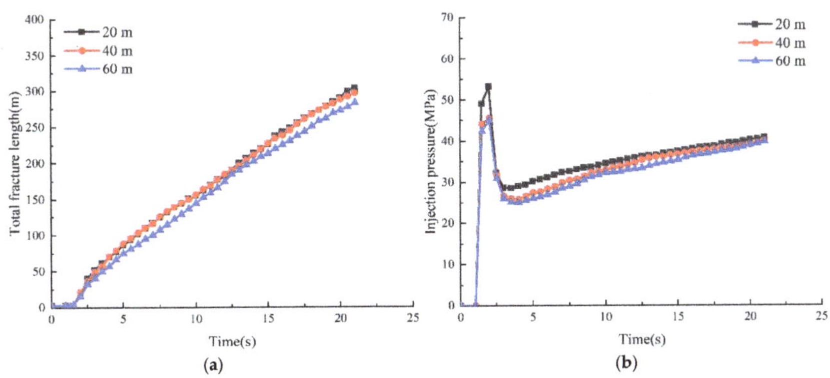

Figure 12. (**a**) Comparison of the total length of cracks. (**b**) Rupture pressure comparison chart. Changes in total fracture length and fracture pressure under different cluster spaces.

4. Conclusions

Hydraulic fracture has become the most important technology for shale oil and gas development. To further understand hydraulic fracture distribution in shale formations, a 2D hydraulic fracturing model was established which considered the natural fracture effect on fracture propagation. It was established based on the global cohesive element method and the influence of stress field, injection parameters and rock properties on fracture propagation were simulated. The conclusions were shown as follows:

(1) Natural fracture and stress field were the predominant factors affecting the hydraulic fracture propagation. Under low horizontal stress difference, the hydraulic fracture propagation was diverse, thus many branch fractures were created, and the tortuosity of hydraulic fractures also increased due to the increase in the activated natural fracture.

(2) Low viscosity of the fracturing fluid and high injection rates were helpful for creating complex fracture networks. The main reason was that low flow resistance of the low-viscosity fracturing fluid made the natural fractures to be activated more, thus the hydraulic fracture had a difference in propagation trajectory. High injection rate can provide high net pressure which was helpful in activating the natural fracture.

(3) With the increase in elastic modulus, the number of branch fractures increased, and the tortuosity of the created fractures increased significantly. The main reason was that the high difference in strength between the rock matrix and the natural fracture can change the interaction between a hydraulic fracture and a natural fracture from crossing to offset, which activated the natural fracture more, thus more branch fractures were created.

Author Contributions: Conceptualization, H.Z. and F.L.; methodology, H.Z.; software, H.Z.; validation, H.Z., F.L. and W.D; formal analysis, H.Z.; investigation, H.Z.; resources, H.Z.; data curation, H.Z. and D.W.; writing—original draft preparation, H.Z.; writing—review and editing, H.Z.; visualization, H.Z.; supervision, H.Z.; project administration, H.Z.; funding acquisition, H.Z. All authors have read and agreed to the published version of the manuscript.

Funding: This work was financially supported by the Open fund from State Key Laboratory of Shale Oil and Gas Enrichment Mechanisms and Effective Development (33550000-22-ZC0613-0381).

Data Availability Statement: Data available on requested.

Conflicts of Interest: The authors declare no conflict of interest.

References

1. Zeng, J.; Wang, X.; Guo, J.; Zeng, F.; Zhang, Q. Composite linear flow model for multi-fractured horizonal wells in tight sand reservoirs with the threshold pressure gradient. *J. Pet. Sci. Eng.* **2018**, *165*, 890–912. [CrossRef]
2. Wang, Z.L.; Lou, Y.; Pan, J.P. China's Oil & gas resources exploration and development and its prospect. *Explor. Prod.* **2017**, *25*, 1–6.
3. Wang, K.; Zhang, H.L.; Zhang, R.H.; Wang, J.P.; Dai, J.S.; Yang, X.J. Characteristics and influencing factors of ultra-deep shale reservoir structural fracture: A case study of Keshen-2 gas field, Tarim Basin. *Acta Petrol. Sin.* **2016**, *37*, 714–727.
4. Yang, T.; Zhang, G.S.; Liang, K.; Zheng, M.; Guo, B.J. The exploration of global shale gas and forecast of the development tendency in China. *Strateg. Study CAE* **2012**, *14*, 64–68.
5. Zeng, L.B.; Ke, S.Z.; Liu, Y. *Research Methods of Low Permeability Oil and Gas Reservoir Fractures*; Petroleum Industry Press: Beijing, China, 2010; pp. 1–187.
6. Olson, J.E.; Laubach, S.E.; Eichhubl, P. Estimating natural fracture producibility in tight gas sandstones, coupling diagenesis with geomechanical modeling. *Lead. Edge* **2010**, *29*, 1494–1499. [CrossRef]
7. Zeng, L.; Liu, W.; Li, J.; Zhu, L.; Weng, J.; Yue, F.; Zu, K. Natural fractures and their influence on shale gas enrichment Sichuan Basin, China. *J. Pet. Sci. Eng.* **2016**, *30*, 1–9. [CrossRef]
8. Liu, W.; Zeng, L.; Zhang, B.; Miao, F.; Lyu, P.; Dong, S. Influence of natural fractures on gas accumulation in the Upper Triassic tight gas sandstones in the northwestern Sichuan Basin, China. *Mar. Pet. Geol.* **2017**, *83*, 60–72. [CrossRef]

9. Fisher, M.K.; Wright, C.A.; Davidson, B.M.; Steinsberger, N.P.; Buckler, W.S.; Goodwin, A.; Fielder, E.O. Integrating fracture mapping technologies to optimize stimulations in the Barnett shale. In Proceedings of the SPE Annual Technical Conference and Exhibition, San Antonio, TX, USA, 29 September–2 October 2002; SPE-77411.
10. Zheng, H.; Liao, R.; Cheng, N.; Shi, S.; Liu, D. Study of mechano-chemical effects on the morphology of hydraulic fractures. *J. Pet. Sci. Eng.* **2021**, *206*, 109031. [CrossRef]
11. Maxwell, S.C.; Urbanci, K.T.; Steinsberger, N.P.; Zinno, R. Microseismic imaging of hydraulci fracture complexity in the barnett shale. In Proceedings of the SPE Annual Technical Conference and Exhibition, San Antonio, TX, USA, 29 September–2 October 2002; SPE-77440.
12. Perkins, T.K.; Kern, L.R. Width of hydraulic fractures. *J. Pet. Technol.* **1961**, *13*, 937–949. [CrossRef]
13. Gu, H.; Weng, X. Criterion for fractures crossing frictional interfaces at nonorthogonal angles. In Proceedings of the 44th US Rock Mechanics Symposium and 5th US-Canada Rock Mechanics Symposium, Salt Lake City, UT, USA, 27–30 June 2010; ARMA-10-198.
14. Tan, P.; Jin, Y.; Han, K.; Hou, B.; Chen, M.; Guo, X.; Gao, J. Analysis of hydraulic fracture initiation and vertical propagation behavior in laminated shale formation. *Fuel* **2017**, *206*, 482–493. [CrossRef]
15. Zhang, H.; Sheng, J.J. Numerical simulation and optimization study of the complex fracture network in naturally fractured reservoirs. *J. Pet. Sci. Eng.* **2020**, *195*, 107726. [CrossRef]
16. Rui, Z.; Guo, T.; Feng, Q.; Qu, Z.; Qi, N.; Gong, F. Influence of gravel on the propagation pattern of hydraulic fracture in the conglomerate reservoir. *J. Pet. Sci. Eng.* **2018**, *165*, 627–639. [CrossRef]
17. Zhang, G.; Sun, S.; Chao, K.; Niu, R.; Liu, B.; Li, Y.; Wang, F. Investigation of the nucleation, propagation and coalescence of hydraulic fractures in conglomerate reservoirs using a coupled fluid loe-DEM approach. *Powder Technol.* **2019**, *354*, 301–313. [CrossRef]
18. Chen, M.; Sun, Y.; Fu, P.; Carrigan, C.R.; Lu, Z.; Tong, C.H.; Buscheck, T.A. Surrogate-based optimization of hydraulic fracture in pre-existing fracture networks. *Comput. Geosci.* **2013**, *58*, 69–79. [CrossRef]
19. Wang, H.Y. Hydraulic fracture propagation in naturally fractures reservoirs: Complex fracture or fracture networks. *J. Nat. Gas Sci. Eng.* **2019**, *68*, 102911. [CrossRef]
20. Wei, Y.S.; Wang, J.L.; Yu, W.; Qi, Y.D.; Miao, J.J.; Yuan, H.; Liu, C.X. A smart productivity evaluation method for shale gas wells based on 3D fractal fracture network model. *Pet. Explor. Dev.* **2021**, *48*, 787–796. [CrossRef]
21. Wan, X.C.; Rasouli, V.; Damjance, B.; Yu, W.; Xie, H.; Li, N.; Rabiei, M.; Miao, J.; Liu, M. Coupling of fracture model with reservoir to simulate shale gas production with complex fractures and nanopores. *J. Pet. Sci. Eng.* **2020**, *193*, 107422. [CrossRef]
22. Sobhaniaraghn, B.; Mansur, W.J.; Peter, F.C. Three-dimensional investigation of multiple stage hydraulic fracturing in unconventional reservoirs. *J. Pet. Sci. Eng.* **2016**, *146*, 1063–1078. [CrossRef]
23. Khoei, A.R.; Vahab, M.; Ehsani, H.; Rafieerad, M. XFEM modeling of large plasticity deformation; a convergence study on various blending strategies for weak discontinuities. *Eur. J. Comput. Mech.* **2015**, *24*, 79–106. [CrossRef]
24. Hanks, C.L.; Lorenz, J.; Teufel, L.; Krumhardt, A.P. Lithologic and structural controls on natural fracture distribution and behavior within the Lisburne Group, northeastern Brooks Range and North Slope subsurface, Alaska. *AAPG Bull.* **1997**, *81*, 1700–1720.

processes

MDPI

Article

Numerical Simulation of Particle Migration and Settlement in Hydraulic Fractures Using the Multiphase Particle-in-Cell Method

Youshi Jiang [1,*], Zhibin He [2], Shuxia Jiang [2], Mouxiang Cai [1], Fujian Liu [3] and Ying Yuan [1]

1 State Key Laboratory of Oil and Gas Reservoir Geology and Exploitation, Southwest Petroleum University, Chengdu 610500, China; cmx.20020103@outlook.com (M.C.); yuanying6927@outlook.com (Y.Y.)
2 Exploration and Development Research Institute, Zhongyuan Oilfield Branch, China Petroleum and Chemical Corporation, Puyang 457001, China; hezhibin240549@163.com (Z.H.); jiangshuxia2024@163.com (S.J.)
3 CCDC Well Intervention Company, Chengdu 610052, China; liufujian_sc@cnpc.com.cn
* Correspondence: jiangyswpu@gmail.com

Abstract: Solid–liquid two-phase flow often occurs when pumping proppant or temporary plugging agents into hydraulically fractured wells. The final distribution of these injected particles in the fracture has an important influence on the well productivity after hydraulic fracturing. This paper focuses on simulating and analyzing particle migration within slug injection hydraulic fractures in the Sulige gas reservoir. In this study, a particle migration and settlement model in hydraulic fractures is established based on the Multiphase Particle-in-Cell (MP-PIC) method, allowing for effective simulation of particle migration and settlement in fractures. This model is validated by the results of particle-pumping experiments. The influences of fluid viscosity, injection rate, particle density, particle diameter, and particle concentration on the distribution of particles are studied. The results indicate that keeping the viscosity of the particle-carrying liquid above 50 mPa·s is necessary. It is recommended to keep the liquid viscosity above 200 mPa·s so that the particles can move farther in the fractures. For pulse fracturing, a lower flow rate leads to a more dispersed distribution of particles, but for temporary plugging with particles, a lower flow rate can lead to a decrease in particle concentration and reduce the success rate of temporary plugging. Low particle density can lead to more dispersed particles, but the amount of particle settlement will be less, so from the perspective of pulse fracturing, it is recommended that the particle density should not be lower than 2200 kg/m^3. Similarly, the particle size should not be too large for pulse fracturing, and the initial particle concentration should be maintained above 18%.

Keywords: particle migration; hydraulic fracture; Euler–Lagrange method; multiphase particle-in-cell

1. Introduction

Hydraulic fracturing has become an important method for oil and gas extraction and production enhancement. The primary purpose is to create a high-permeability propped fracture within the reservoir. Hydraulic fracturing is not only applied in oil extraction but also plays a significant role in the exploration and development of oil and gas reservoirs. This technique effectively reduces exploration risks and liberates inefficient reservoirs while providing valuable reservoir information. It enhances the efficiency of hydrocarbon discovery and evaluation. Additionally, hydraulic fracturing enables the efficient development of low-permeability oil and gas reservoirs, maintains stable production in conventional oil fields, and supports the adjustment and enhancement of mature oil fields. It also facilitates the development and production increase of specialized types of oil and gas reservoirs. After hydraulic fracturing, the productivity of the well is primarily determined by the effective fracture length. Therefore, the placement of proppants within the fractures is crucial for increasing and stabilizing oil and gas production. Increasing the placement distance of proppants can be achieved by modifying the viscosity of the fracturing fluid or adjusting the density of the proppants. However, due to current field conditions and cost constraints, proppant particles often settle in the near-well area before reaching the target distance, resulting in a short effective support distance and suboptimal development outcomes. Therefore, investigating the efficient settling of proppant particles is essential for improving the conductivity of hydraulic fractures and enhancing post-fracturing production. It is necessary to study the effective migration and settling patterns of particles within hydraulic fractures.

The migration and settlement of particles in liquids is a typical solid–liquid two-phase mixed flow with interaction forces and complex mechanical properties in the process. As early as the 19th century, particle migration and settlement were studied systematically. For the particle settlement problem, Stokes [1] first derived the hydrodynamic formula for particle settlement and obtained the famous Stokes formula, whose application is related to the Reynolds number. Since then, Schiller [2], Yourcenar et al. [3], Lapple et al. [4], and Kim et al. [5] studied and obtained formulae for proppant settlement at different Reynolds numbers. In 1923, Faxen et al. [6] investigated the migration of particles between two parallel plates, considering two special cases. One of the cases was particles in the middle, and the other one was particles at a quarter position on one side of the plate. Kern et al. [7] first proposed the concept of "equilibrium height" by systematically studying the migration of particles in low-viscosity fluids. Clifton et al. [8] proposed a dimensionless proppant transport model, but the model did not consider the influence of particle position in the fracture. The model strongly relied on the Stokes formula without considering the influence of sand-carrying fracturing fluid density on the vertical flow of fluid. Guo et al. [9] used slug injection for particle delivery and calculated it in combination with the definition of equilibrium height. Still, the model did not consider the longitudinal flow field distribution of fluid in the fracture. Wang et al. [10] took into account the velocity of fluid in two directions in the fracture. They not only utilized the solid–liquid two-phase migration equation but also considered the change in fracture width to analyze the particle migration combined with the Green formula. Although such studies provide an important basis for understanding the behavior of particle migration and settlement, they generally ignore the convective transport of particles. Therefore, Barree et al. [11,12], Abdurrahman [13], and other scholars carried out small laboratory experiments of particle convective transport, which provided them with a profound understanding of particle convective transport characteristics. Malhotra [14] and Wen [15] used a large visual single-plate device to simulate fracturing in artificial fractures, analyzing the influence of various factors such as fluid viscosity, fracture width, and sand ratio on the particle migration and settlement laws and establishing sensitivity analysis of different influencing factors [16]. Xiao et al. [17] designed three particle size combination methods to explore the effects of step-by-step injection of particles with different combined particle sizes (small particle size first, large particle size later) and mixed injection methods for particle migration.

Compared with physical experiments, numerical simulation technology has low cost and diverse model forms and boundary conditions, which can achieve more complex conditions. Tehrani et al. [18] proposed a transport model that considers gravity settlement and convective transport of particles and used the Lax–Wendroff method to solve the problem jointly. However, due to the extremely complex numerical solution, it was not applied in practice but laid a foundation for subsequent theoretical research. With the development of research, the numerical simulation of particle migration and settlement was mainly completed using the Euler–Euler and Euler–Lagrange methods. The former treats the solid phase and liquid phase as a continuous medium and adopts the Euler method to describe the motion process of the fluid and solid particles [19–23]. This method requires little calculation and is often used to simulate the migration process of low-concentration solid particles, but it cannot consider the motion and action of each particle. It is often difficult to obtain accurate simulations when dealing with uneven distributions of real particles and drastic changes in particle concentration. The Euler–Lagrange method [24–26] can solve these problems well. Therefore, it is necessary to establish a set of numerical simulation models of particle migration with high accuracy for complex solid–liquid two-phase mixed flow.

The multiphase flow point grid model (MP-PIC) is a simulation technique based on the Euler–Lagrange framework, which analyzes the fluid as a continuous phase while treating the particles as discrete phases. The dynamic properties of the liquid phase are described by the N-S governing equations of volume average. In contrast, the particle phase is simulated by tracking the motion ofeach particle. This approach uses the concept of numerical particles to characterize large populations of particles with consistent properties, which significantly reduces the number of particles that need to be tracked. The model does not directly deal with collisions between particles, but it uses the particle stress model to prevent excessive agglomeration of particles. This method can prevent the particle volume fraction from exceeding the physical limit, thereby indirectly ensuring the realism of the simulation results.

Snider [27] first introduced a three-dimensional, incompressible, multiphase particle-in-cell (MP-PIC) method in 2000. This approach used a continuum model to solve the governing equations for the fluid phase while employing a Lagrangian framework to handle the governing equations for the particle phase. This method ensured computational accuracy and efficiency by cleverly mapping particle properties onto an Eulerian grid and subsequently applying the computed stress tensor back to the particle positions. Over time, the MP-PIC method has been increasingly used to solve solid–liquid two-phase flow problems in devices such as fluidized bed reactors, risers, and bubbling beds. Discretizing particles efficiently simulates the behavior of particles and gas–solid interactions, thereby enhancing computational efficiency and accuracy. Subsequently, researchers like Zeng [28], Mao [29], and Suri [30] coupled the MP-PIC method with fracture propagation models such as the Perkins–Kern–Nordgren (PKN) and Kristianovich–Geertsma–De Klerk–Carter (KGD) models. These coupled numerical models were employed to study proppant transport in CO_2 fracturing and channel fracturing. With the development of unconventional hydrocarbon reservoirs, hydraulic fracturing plays a crucial role in shale development. Siddhamshetty and colleagues, focusing on shale gas development, utilized the MP-PIC method to design a multi-scale proppant injection scheme to maximize shale gas production. In 2024, Wen and others [31] combined the displacement discontinuity method with the MP-PIC method. These methods are used for fracture deformation and propagation simulations, fluid flow, and proppant transport, respectively. Their model was employed to investigate the fracturing process in a layered reservoir.

In summary, the MP-PIC method has been widely used to simulate the particle transport process in fracture propagation, providing a powerful tool for understanding complex fluid–solid coupling mechanisms. However, slug injection is also applied during the hydraulic fracturing process for proppant pumping in field operations. The impact of slug injection on particle transport is not yet fully understood. This paper focuses on studying the influence of slug injection on particle transport under the fracturing operation parameters of the Sulige gas reservoir in Nei Mongol, providing guidance for future development.

In this study, a numerical model of particle migration and settlement in hydraulic fractures was established based on the MP-PIC method, with the effects of some key parameters on the particle distributions being studied. In Section 2, the governing equations of solid–liquid two-phase flow are shown, and the boundary conditions are determined. In Section 3, our model is validated, followed by a sensitivity analysis to determine the influences of liquid viscosity, injection rate, particle density, particle diameter, and particle concentration on the particle distribution. The main conclusions are summarized in Section 4.

2. Methodology

This paper focuses on simulating and analyzing particle migration within slug injection hydraulic fractures in the Sulige gas reservoir. A numerical model based on the MP-PIC method was developed to study particle migration and settlement in hydraulic fractures. The model incorporates commonly used parameters such as liquid viscosity, injection rate, particle density, particle diameter, and particle concentration to simulate the fundamental patterns of particle transport under slug injection conditions. Consequently, this model does not account for the heterogeneity of natural fractures or the coupling of multiple physical fields, such as fracture deformation, chemical reactions, and heat transfer.

In this study, in order to better observe the distribution of particles during migration, the fracture was regarded as a slab with a length of L_x , a height of H_s, and a width of W_s (see Figure 1). The model was meshed by a regular hexahedron structure. Six speed entrances were distributed on the left side of the model, each with a height of l_h and a width of l_w. The constant pressure outlet was on the right side of the model. In the simulation, the particles and fluid enter the fracture model through the speed entrance at a flow rate of u_s, and the outlet pressure is constant P_e.

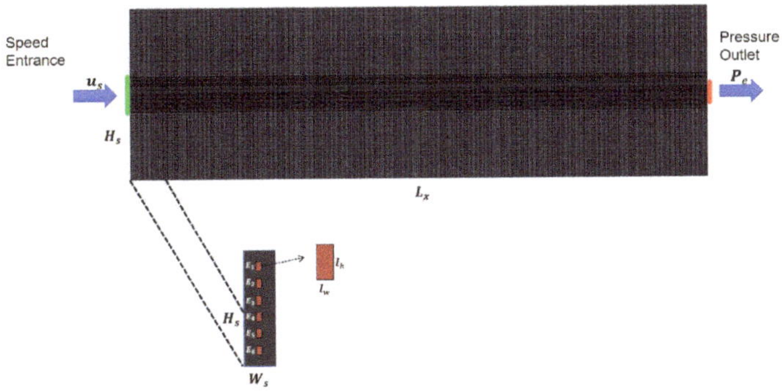

Figure 1. Schematic diagram of the fracture physical model.

2.1. Govering Equations

(1) Governing equation of the liquid phase

The volume-average N-S governing equations describe the flow of the liquid phase. Considering the interaction between the liquid and solid phase, the influence of pressure gradient and gravity, the volume mean momentum equation of the liquid phase is established. The volume-averaged continuity equation [32] and volume mean momentum equation for the liquid phase are shown in Equations (1) and (2):

$$\frac{\partial(\varepsilon_l \rho_l)}{\partial t} + \nabla \cdot (\varepsilon_l \rho_l \vec{u}_l) = 0 \quad (1)$$

$$\frac{\partial(\varepsilon_l\rho_l\vec{u}_l)}{\partial t}+\nabla\cdot(\varepsilon_l\rho_l\vec{u}_l\vec{u}_l)=-\nabla p+\nabla\cdot(\varepsilon_l\vec{\tau}_l)+\varepsilon_l\rho_l g-\vec{F} \tag{2}$$

where ε_l denotes the liquid phase volume fraction; ρ_l represents the density of the liquid phase in kg/m^3; $\vec{u}_l$ is the liquid velocity in m/s; t denotes the time in s; $(\nabla\cdot)$ represents the divergence operator; ∇p is the pressure gradient in Pa/m; $\vec{\tau}_l$ represents the liquid stress tensor in Pa; g is gravitational acceleration in m/s^2 ; $\vec{F}$ denotes the liquid and solid phase momentum exchange rate per unit volume in $kg/(m^2\cdot s^2)$.

The liquid and solid phase momentum exchange rate per unit volume $\vec{F}$ in $kg/(m^2{\cdot}s^2)$ [33] can be calculated according to the following equation:

$$\vec{F}=\iiint fm_p\left\{D_P(\vec{u}_l-\vec{u}_p)-\frac{\nabla p}{\rho_p}\right\}dm_p d\vec{u}_p \tag{3}$$

where f represents the particle distribution function; D_P represents the particle drag function; m_p is the particle mass in kg; ρ_p represents the density of the solid phase in kg/m^3; u_p is the solid velocity in m/s.

In order to ensure the conservation of mass and volume, the sum of phase volume fractions at any point in the flow field is 1.

$$\varepsilon_l+\varepsilon_p=1 \tag{4}$$

where ε_p denotes the solid phase volume fraction.

(2) Governing equation of the solid phase

The distribution of solid particles in space is described by a probability distribution function $f(x,u_l,\rho_p,V_p,t)$ [34], which represents the probability of finding a particle with velocity u_l, density ρ_p, and volume V_p at location x at time t. The equation for the change in f with time t is as follows:

$$\frac{\partial f}{\partial t}+\nabla\cdot(f\vec{u}_p)+\nabla\cdot(f\vec{a})=0 \tag{5}$$

$$\vec{u}_p=\frac{dx}{dt} \tag{6}$$

$$\vec{a}=\frac{d\vec{u}_p}{dt}=D_P(\vec{u}_l-\vec{u}_p)-\frac{\nabla p}{\rho_p}+g-\frac{\nabla\vec{\tau}_p}{\varepsilon_p\rho_p} \tag{7}$$

where x denotes the position of any point in space in m; $\vec{a}$ represents the acceleration of the particles in m/s^2. u_p and $\vec{a}$ can be obtained using Equations (6) and (7), respectively [35].

In the MP-PIC method, the collision between particles is regarded as the stress gradient of the particle phase. The particle stress gradient in liquid–solid flow is interpolated into discrete particles by calculating in Euler mesh. Therefore, the normal stress of the particles from dilute phase transport to dense phase accumulation can be calculated via interpolation of the particle stress tensor formula, as follows [36]:

$$\vec{\tau}_p=\frac{p_s\varepsilon_p^{\alpha}}{\max[(\varepsilon_{cp}-\varepsilon_p),\beta(1-\varepsilon_p)]} \tag{8}$$

where p_s represents the pressure constant in Pa; α denotes coefficients with values between 2 and 5, where the default is 3; ε_p represents the maximum particle stack volume fraction; max represents the maximum of all numbers in parentheses; β is a constant, taken as 10^{-8}.

(3) Drag model

Drag force refers to the resistance acting on the surface of an object in the process of fluid flow. For the migration of particles in fracture, drag force is one of the main forces caused by the relative flow between fluid and particles. By describing the drag force exerted by fluid on particles, the drag model directly affects the velocity, trajectory, and distribution characteristics of particles in multiphase flow. Selecting a suitable drag model can improve the precision of numerical simulation. The drag force $\vec{F}_p$ in $(\mathrm{kg} \cdot \mathrm{m})/\mathrm{s}$ on a particle can be expressed as shown in Equation (9) [32]:

$$\vec{F}_p = \mathrm{m}_p D_P(\vec{u}_l - \vec{u}_p) \tag{9}$$

Classical drag models include the Ergun model [37], Wen-Yu model [38], Wen-Yu/Ergun model, and Turton model. The Wen-Yu/Ergun model is a hybrid model composed of the Ergun model and Wen-Yu model, which can solve discontinuous problems. In this study, the maximum bulk fraction of the particle phase is 60%, and considering that the system in this study has a transition region from the dilute phase to the dense phase, the Wen-Yu/Ergun model is selected.

$$D_p = \begin{cases} D_{W-Y} = \frac{3}{4} C_D \frac{(1-\varepsilon_P)\rho_l|\vec{u}_l - \vec{u}_p|}{d_p}(1-\varepsilon_P)^{-2.65} & \varepsilon_P \leq 0.75\varepsilon_{cp} \\ D_{Trans} = \frac{\varepsilon_P - 0.75\varepsilon_{cp}}{0.85\varepsilon_{cp} - 0.75\varepsilon_{cp}}(D_{p,Ergun} - D_{p,W-Y}) + D_{p,W-Y} & 0.75\varepsilon_{cp} < \varepsilon_P \leq 0.85\varepsilon_{cp} \\ D_{\mathrm{Ergun}} = \frac{180\varepsilon_P \mu \vec{u}_l}{(1-\varepsilon_P)d_p^2} + \frac{2\rho_l|\vec{u}_l - \vec{u}_p|}{d_p} & \varepsilon_P > 0.85\varepsilon_{cp} \end{cases} \tag{10}$$

$$C_D = \begin{cases} \frac{24}{Re_p}\left(1 + 0.15 Re_p^{0.687}\right) & Re_p < 1000 \\ 0.44 & Re_p > 1000 \end{cases} \tag{11}$$

$$Re_p = \frac{\varepsilon_l \rho_l |\vec{u}_l - \vec{u}_p| d_p}{\mu_l} \tag{12}$$

where D_{W-Y}, D_{Trans}, D_{Ergun} are different models selected for the proportion of different solid volume fractions to the maximum particle stack volume fraction; C_D represents the drag coefficient; d_p represents the particle diameter in m; $D_{p,Ergun}$ and $D_{p,W-Y}$ are the drag coefficients of the particles from different models; Re_p denotes the Reynolds number; μ_l is the fluid viscosity in Pa·s; C_D and Re_p can be obtained using Equations (11) and (12), respectively.

2.2. Boundary Conditions

The left boundary of the model in this paper is assumed to be the inlet (see Figure 1). The initial velocities of the liquid phase and the particle phase at the inlet are assumed to be constants, as well as the volume fraction of the particle phase. No-slip walls are assumed at the top and bottom boundaries.

In order to ensure that the simulation is consistent with the actual particle injection process inside hydraulic fractures, it is necessary to use the similarity criterion to convert the real parameters of hydraulic fracturing into simulation parameters to ensure that the flow regime of this numerical simulation is the same as those real parameters. In this paper, Equations (13) and (14) are established according to the similarity criterion [39]:

$$Q_s = \frac{Q_f}{2} \times \frac{H_s \times W_s}{H_f \times W_f} \times 1000 \tag{13}$$

$$u_s = \frac{60 \times Q_s}{1000 \times n \times S} \tag{14}$$

where Q_s is the stimulation injection rate in L/min; Q_f is field construction injection rate in L/min; H_s, W_s are the height and width of the physical model in m, respectively; H_f, W_f are the height and width of the actual fracture in m, respectively; n denotes the number of speed entrances; S denotes the area of a single speed entrance in m^2.

3. Results and Discussion

3.1. Model Validation

In this section, the numerical method is validated with the physical experiment of particle migration and settlement by Ma et al. [40]. The injection parameters of this numerical simulation are set to be the same as those used by that experiment, and the slug injection of particles is adopted.

Ma et al. conducted simulations using a large-scale visualized fracture flow experimental setup. The geometry of the visualized flat fracture was 4 m in length, 0.6 m in height, and 6 mm in width. The inlet consisted of 3 to 12 adjustable perforations, each with a diameter of 15 mm, to simulate the process of particles entering the fracture through wellbore perforations. During the experiments, two storage tanks were used to hold particles and liquid, and the slug injection process was achieved by alternately switching between the tanks. Different experimental conditions were employed to observe the particle transport patterns, and high-definition cameras were used to record the process.

With an injection rate of 15 L/min, the particle concentration distributions of the numerical simulation and physical experiment are shown in Figure 2. The particles move forward within the fracture in the form of slugs. The different colors represent different particle concentrations; the red area has a higher particle concentration, and the blue area has a lower particle concentration. As shown in Figure 2, the particle distribution obtained via numerical simulation is the same as in the experiment. Therefore, the numerical method proposed in this paper can effectively simulate particle migration in the fracture.

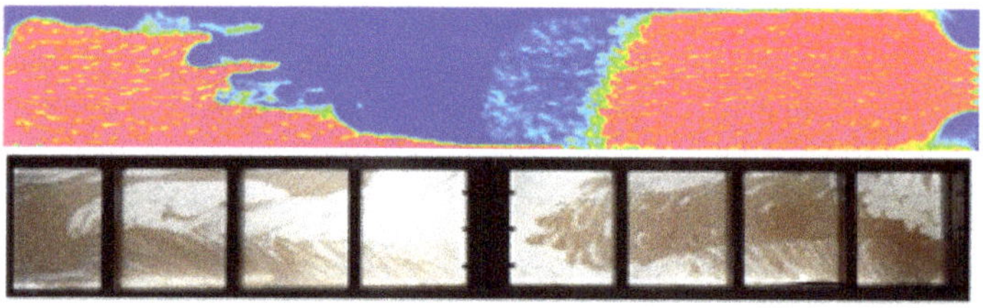

Figure 2. Distribution of the particle distribution at an injection rate of 15 L/min.

3.2. Simulation and Analysis

(1) Basic example and results

In this paper, particle migration inside the hydraulic fracture is simulated on the basis of the basic parameters listed in Table 1 with slug injection being adopted for particle transport. Figure 3 shows the distribution of particle concentrations in this basic example. It can be seen that after the particles enter the fracture from the left entrance, they gradually disperse during the migration from left to right, and the overall particle concentration gradually decreases in the meantime. In Section 3.2, an index parameter called the concentrated degree of particles is used to characterize the dispersion of particles during migration. Due to the adoption of the slug injection method, the larger the concentrated degree of the particles, the less dispersed the particles are during particle migration.

The concentrated degree of particles is calculated based on the area occupied by the particles in the T region in Figure 4. The T region is surrounded by the black line in Figure 4. The concentrated degree of particles is equal to the ratio of the area not occupied by the particles in the T region to the area of the entire T region.

Table 1. Basic parameters.

Parameter Type	Value	Unit
Length of the model	4	m
Height of the model	1.2	m
Width of the model	0.01	m
Injection rate	48	L/min
Liquid viscosity	200	mPa·s
Particle diameter	20/40	mesh
Particle concentration	12	%
Particle density	2650	kg/m^3

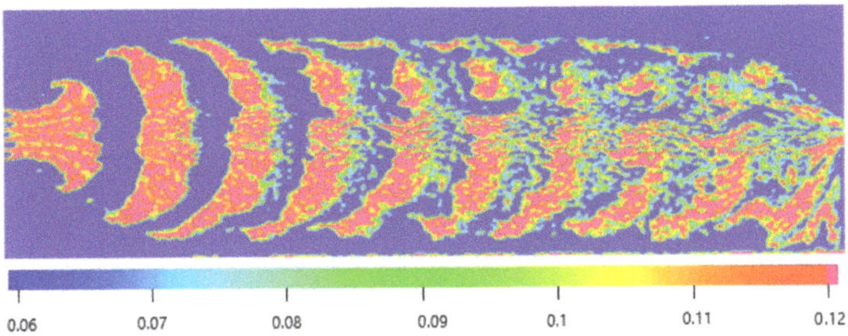

Figure 3. The distribution of particle concentrations. The different colors represent different particle concentrations; the legend on the bottom shows that the red area has a higher particle concentration, and the blue area has a lower particle concentration.

Figure 4. Calculation of the concentration of particles.

(2) Sensitivity analysis

(1) Liquid viscosity

As shown in Figure 5, when the viscosity is equal to 50 mPa·s, the concentration of the particles is at a minimum, mainly because a large number of particles have settled at a low viscosity. When the viscosity increases from 100 to 200 mPa·s, the concentration of the particles decreases due to the combination of liquid impact and particle settling. When the viscosity rises from 200 to 300 mPa·s, the concentration of the particles remains unchanged. This stability occurs because the forces exerted by the fluid on the particles and the inter-particle forces reach an equilibrium state under high-viscosity conditions, maintaining a stable concentration of the particles.

Figure 6 shows that as the liquid viscosity increases, the concentration of the particles first increases, then decreases, and finally stabilizes. When the viscosity is equal to 50 mPa·s, the concentration of the particles is minimal, mainly because a large number of particles have settled. When the viscosity increases from 100 to 300 mPa·s, the concentration of the particles gradually decreases, indicating that the higher the viscosity, the more the particles disperse. Therefore, for channel fracturing, a higher viscosity is conducive to forming better channels. On the other hand, for temporary plugging with particles, the higher the viscosity is, the higher the particle concentration that can be maintained, and the success rate of temporary plugging treatment can be improved.

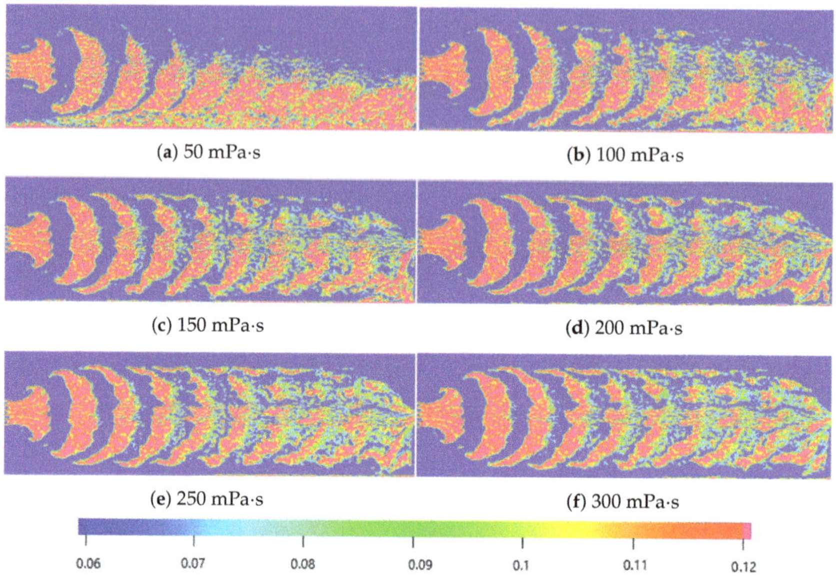

Figure 5. Particle distributions under different liquid viscosities of 50 mPa·s, 100 mPa·s, 150 mPa·s, 200 mPa·s, 250 mPa·s and 300 mPa·s.

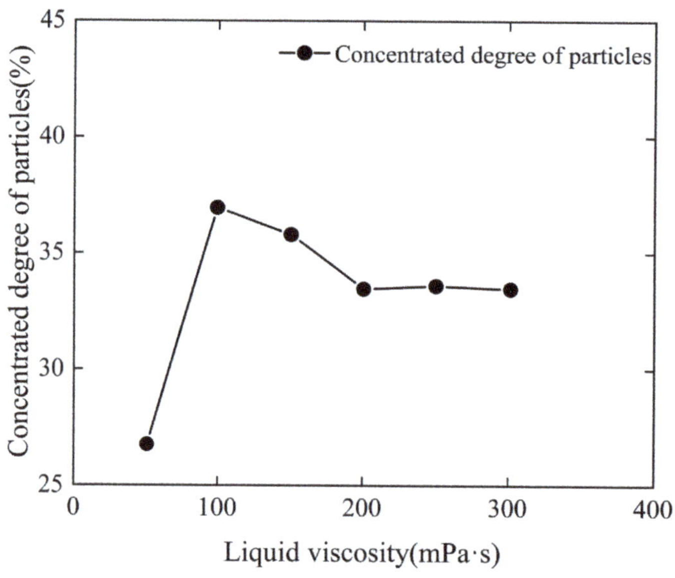

Figure 6. The influence of liquid viscosity on particle concentration.

(2) Injection rate

Figure 7 shows that when the injection rate is low, the particles are more easily dispersed by the liquid, leading to back-and-forth particle aggregation. This particle aggregation can easily cause fracture plugging and reduce the fracture's diversion capability. Additionally, a higher injection rate results in fewer particles settling. Figure 8 also shows that with the increase in injection rate, the concentration of the particles gradually increases, reflecting a gradual decrease in the degree of dispersion. Therefore, a high injection rate is conducive to the stability of channel fracturing and the stability of particle concentration for temporary plugging.

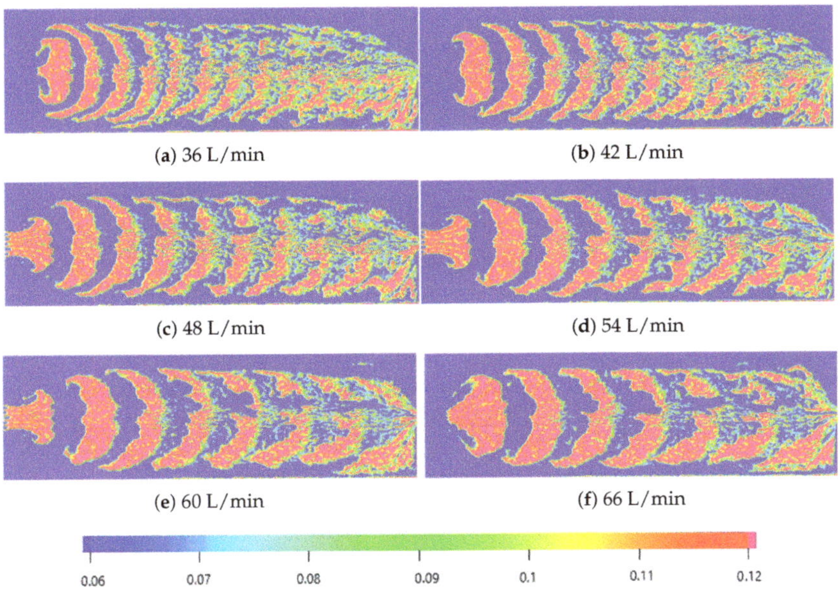

Figure 7. Particle distributions under different injection rates of 36 L/min, 42 L/min, 48 L/min, 54 L/min, 60 L/min and 66 L/min.

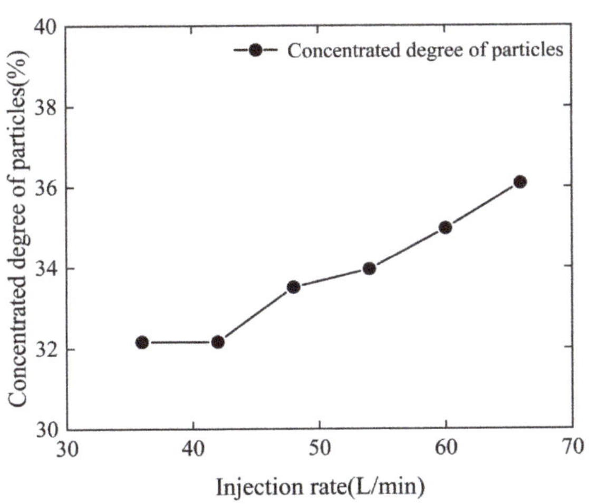

Figure 8. The influence of injection rate on particle concentration.

(3) Particle density

Figure 9 shows that particle density significantly affects both the degree of dispersion and particle settlement during particle migration. The higher the particle density, the more particles settle to the bottom of the fracture. On the other hand, when the particle density is less than 2200 kg/m^3, the particle distribution is obviously more dispersed than when the particle density is greater than 2200 kg/m^3. Figure 10 also shows this phenomenon. When the particle density is above 2200 kg/m^3, the concentration of the particles obviously increases with the increase in particle density. In general, the higher the particle density, the lower the degree of particle dispersion, but more particles settle at the same time, so it is necessary to consider these two trends comprehensively when optimizing the particle density.

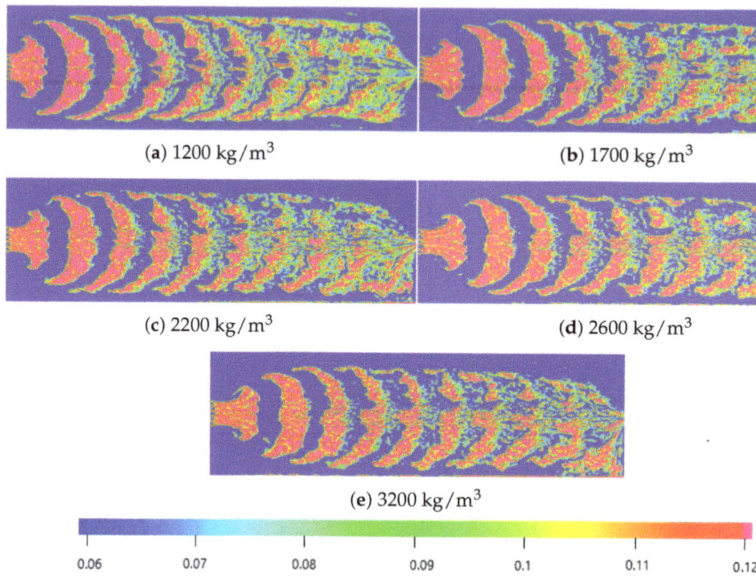

Figure 9. Particle distributions under different particle densities of 1200 kg/m^3, 1700 kg/m^3, 2200 kg/m^3, 2600 kg/m^3 and 3200 kg/m^3.

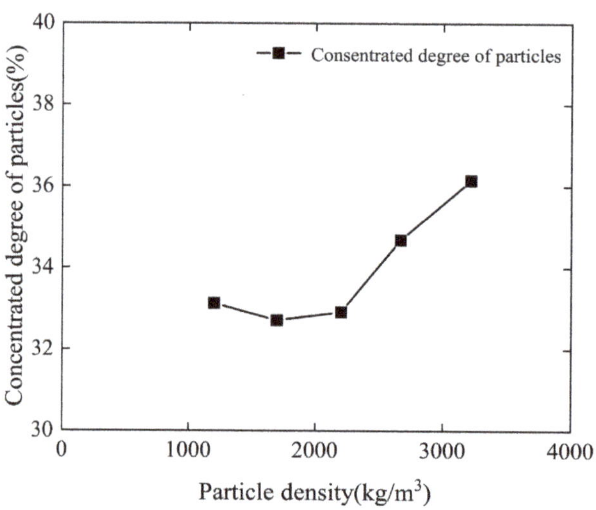

Figure 10. The influence of particle density on particle concentration.

(4) Particle diameter

As can be seen from Figures 11 and 12, the larger the particle size, the more dispersed the particles are. The effects of particle dispersion can be analyzed from two perspectives. Based on the conventional theory of hydraulic fracturing, particle dispersion can help particles to better prop fractures. However, for channel fracturing or temporary plugging, particle dispersion results in worse channels or lower particle concentrations, especially after the particles have migrated over a long distance inside the hydraulic fracture.

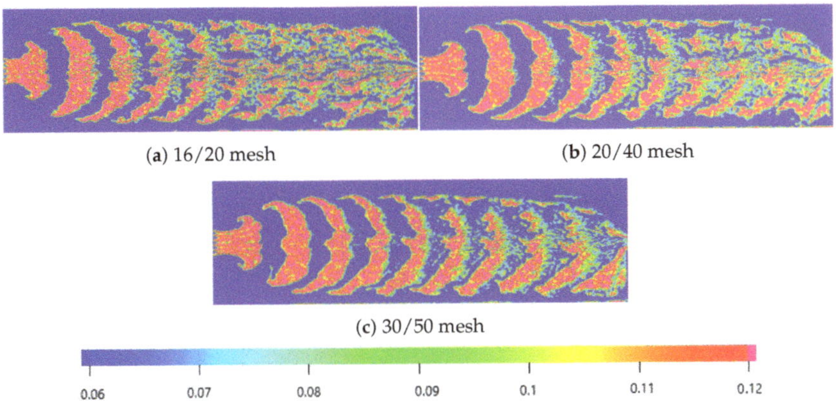

Figure 11. Particle distributions under different particle diameters of 16/20 mesh, 20/40 mesh and 30/50 mesh.

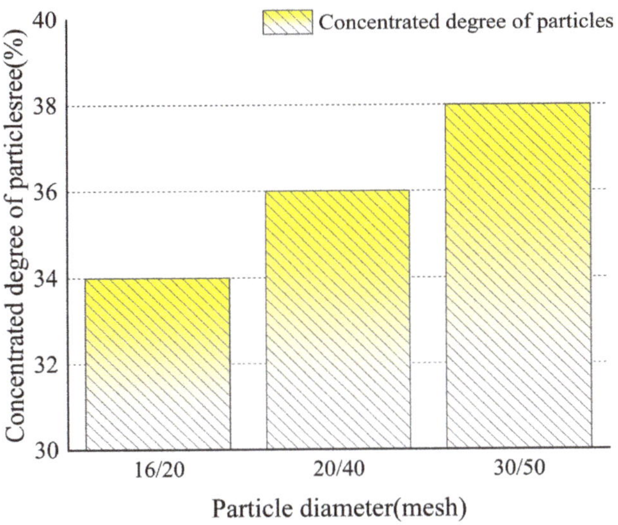

Figure 12. The influence of particle diameter on particle concentration.

(5) Initial particle concentration

In this paper, particle concentration is defined as the volume fraction of the solid phase. As shown in Figure 13, when the initial particle concentration is low, the particles are more easily dispersed by the liquid. On the contrary, when the initial particle concentration is high, the particles appear more concentrated, and the average particle concentration is also higher. Figure 14 indicates that when the initial particle concentration is between 5% and 15%, the concentration of the particles increases slowly. As the concentration rises from 15% to 18%, the concentration increase rapidly. When the initial particle concentration exceeds 18%, the rate of increase slows down again. This behavior is attributed to the fact that particle collisions and aggregation are initially accelerated as the initial particle concentration increases. However, as the particles become concentrated, the repulsive forces between particles become more pronounced, leading to an initial slow increase in concentration, followed by rapid acceleration and finally a deceleration.

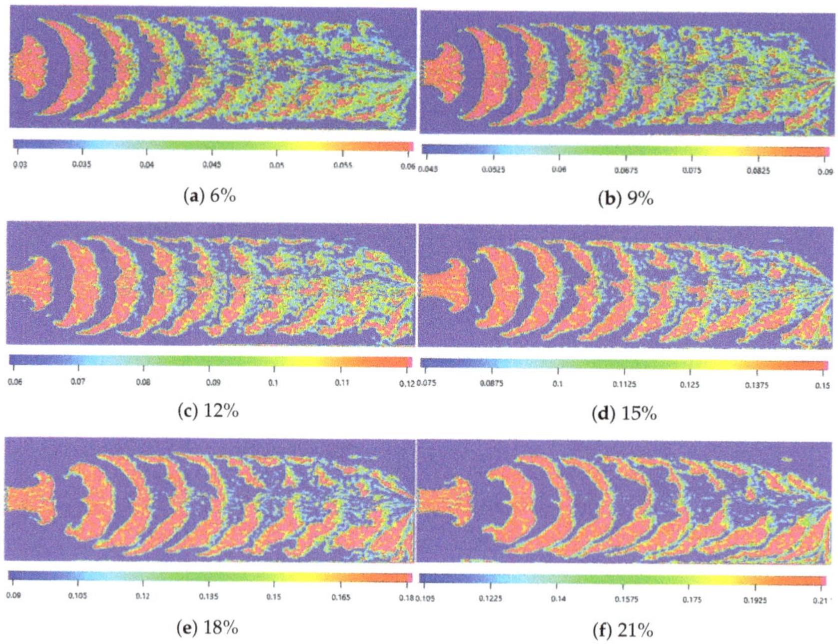

Figure 13. Particle distributions under different particle concentrations (volume fractions) of 6%, 9%, 12%, 15%, 18% and 21%.

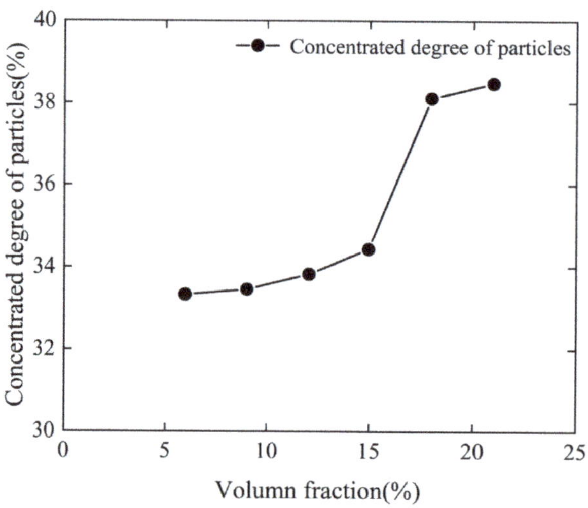

Figure 14. The influence of particle concentration (volume fraction) on overall particle concentration.

3.3. Discussion

A higher viscosity is conducive to the formation of better channels and temporary plugging. Increased viscosity enables the maintenance of higher particle concentrations and improves the success rate of temporary plugging treatment. The higher injection rate is conducive to channel fracturing stability and particle concentration stability for temporary plugging. The concentration of the particles obviously increases with increases in particle density. Larger particle sizes result in greater particle dispersion. Moreover, elevated initial particle concentration tends to promote particle concentration. Consequently, in fracturing construction, it is essential to ensure that the particle density and initial concentration are not excessively low and that the particle size is not too large.

The above studies have demonstrated the effectiveness of the Euler–Lagrange method in simulating solid–liquid two-phase flow (particle migration and settlement). The Euler–Lagrange method is superior to the Euler–Euler method in calculation accuracy. In addition, because it can accurately capture the migration behavior of each particle, the Euler–Lagrange method can obtain better simulation results when simulating pulse fracturing, temporary plugging with particles, and the effects of vortices on particle migration and settlement. On the other hand, the Euler–Lagrange method also has disadvantages in the simulation cost. The high computational cost makes it difficult to apply it to the numerical simulation of particle migration and settlement at the field scale. Although the problem of high computational cost may be alleviated to a certain extent after a substantial increase in computing power in the future, the research areas with more realistic potential for the Euler–Lagrange method are still the particle migration and settlement laws and the temporary plugging mechanisms with particles at the decimeter to meter scales.

4. Conclusions

In this study, we established the numerical model for the migration and settling of particles in hydraulic fractures using the MP-PIC method. The impact of critical parameters on particle distribution was examined, and the model's accuracy was validated by comparing it with experimental results from Ma et al. The patterns of slug injection particle migration in the Sulige gas reservoir can be summarized as follows:

(1) In terms of liquid viscosity, keeping the viscosity of the particle-carrying liquid above 50 mPa·s is necessary. we recommend keeping the liquid viscosity above 200 mPa·s so that the particles can move farther in the fractures.
(2) For pulse fracturing, a lower flow rate leads to a more dispersed distribution of particles, but for temporary plugging with particles, a lower flow rate can lead to a decrease in particle concentration and reduce the success rate of temporary plugging.
(3) Low particle density can lead to more dispersed particles, but the amount of particle settlement will be less, so from the perspective of pulse fracturing, it is recommended that the particle density should not be lower than 2200 kg/m^3.
(4) Similarly, the particle size should not be too large for pulse fracturing, and the initial particle concentration should be maintained above 18%.

Author Contributions: Conceptualization, Y.J., Z.H., S.J. and F.L.; methodology, Y.J., Z.H., S.J. and F.L.; validation, Z.H. and S.J.; formal analysis, Y.J., Z.H., S.J. and F.L.; investigation, Y.J., Z.H., S.J., M.C. and Y.Y.; resources, Y.J., Z.H. and S.J.; data curation, Y.J., Z.H., S.J. and F.L.; writing—original draft preparation, Y.J., Z.H., S.J., M.C. and Y.Y.; writing—review and editing, Y.J., Z.H., S.J., M.C. and Y.Y.; visualization, Y.J., Z.H. and S.J.; supervision, Y.J.; project administration, Y.J.; funding acquisition, Y.J. All authors have read and agreed to the published version of the manuscript.

Funding: This work was supported by the National Natural Science Foundation of China (No. 52104038) and the National Natural Science Foundation of China (U23B20156).

Data Availability Statement: The data presented in this study are available upon request from the corresponding author.

Conflicts of Interest: Authors Zhibin He and Shuxia Jiang were employed by the China Petroleum and Chemical Corporation Zhongyuan Oilfield Branch Exploration and Development Research Institute. Fujian Liu was employed by CCDC Well Intervention Company. The remaining authors declare that the research was conducted in the absence of any commercial or financial relationships that could be construed as a potential conflict of interest.

References

1. Stokes, G.G. On the effect of the internal friction of fluids on the motion of pendulums. *Trans. Camb. Philos. Soc. Part II* **1851**, *9*, 8–106.
2. Schiller, V.L. Uber die grundlegenden berechnungen bei der schwerkraftaufbereitung. *Z. Ver. Deutsch. Ing.* **1933**, *77*, 318–321.
3. Yourcenar, M.; Brami, J.; Delcroix, M.; Gaudin, C.; Poignault, R.; Sarde, M. Une volonté sans fléchissement: Correspondance 1957–1960. 2007. Available online: https://cir.nii.ac.jp/crid/1130000795912835072 (accessed on 24 December 2024).
4. Thomas, D.; Lapple, C. Deposition of aerosol particles in fibrous filters. *AIChE J.* **1961**, *7*, 203–210. [CrossRef]
5. Kim, B.H. *Modeling of Hindered-Settling Column Separations*; The Pennsylvania State University: University Park, PA, USA , 2003 .
6. Faxen, H. Die bewegung einer starren kugel langs der achse eines mit zaher flussigkeit gefullten rohres. *Arkiv for Matemetik Astronomi och Fysik* **1923**, *17*, 1–28.
7. Kern, L.; Perkins, T.; Wyant, R. The mechanics of sand movement in fracturing. *J. Pet. Technol.* **1959**, *11*, 55–57. [CrossRef]
8. Clifton, R.; Wang, J. Multiple fluids, proppant transport, and thermal effects in three-dimensional simulation of hydraulic fracturing. In Proceedings of the SPE Annual Technical Conference and Exhibition, Houston, TX, USA, 2–5 October 1988; SPE: Richardson, TX, USA, 1988; p. SPE-18198.
9. Guo, D.; Ji, L.; Zhao, J. Calculation of migration and distribution of proppant in three-dimensional fractures. *Henan Pet.* **2001**, *15*, 32–34.
10. Wang, S.; Yang, Z.; Lu, H.; Li, X. Numerical simulation research of proppant transportation in hydraulic fracturing. *J. Oil Gas Technol. (J. Jianghan Pet. Inst.)* **2009**, *31*, 380–383.
11. Barree, R.; Conway, M. Experimental and numerical modeling of convective proppant transport. In Proceedings of the SPE Annual Technical Conference and Exhibition, New Orleans, LA, USA, 25–28 September 1994; SPE: Richardson, TX, USA, 1994; p. SPE-28564.
12. Barree, R. *Evaluation of Frac and Pack Completions and Future Outlook*; SPE: Richardson, TX, USA, 1997; p. SPE-13743.
13. Al-quraishi, A.A.; Christiansen, R.L. Dimensionless groups for interpreting proppant transport in hydraulic fractures. In Proceedings of the SPE Middle East Oil and Gas Show and Conference, Manama, Bahrain, 20–23 February 1999; SPE: Richardson, TX, USA, 1999; p. SPE-53262.
14. Malhotra, S.; Sharma, M.M. Settling of spherical particles in unbounded and confined surfactant-based shear thinning viscoelastic fluids: An experimental study. *Chem. Eng. Sci.* **2012**, *84*, 646–655. [CrossRef]
15. Wen, Q.; Zhuai, H.; Luo, M.; Hu, L.; Huang, Y.; Huang, B.; Li, J. Study on proppant settlement and transport rule in shale gas fracturing. *Pet. Geol. Recovery Effic.* **2015**, *19*, 104–107.
16. Lin, X.; Yang, Z.; Hu, Y.; Xu, X.; Liu, Y.; Luo, P. Simulation of proppant settlement and migration in shale volume fracturing and its application. *Pet. Geol. Oilfield Dev. Daqing* **2021**, *40*, 151–157.
17. Xiao, F.; Zhang, S.; Li, X.; Wang, F.; Liu, X. Settlement law and flow conductivity of sand-carrying combined particle sizes + slickwater: Taking jimsar shale oil reservoir as an example. *Pet. Geol. Oilfield Dev. Daqing* **2023**, *42*, 167–174.
18. Tehrani, M.; Hammond, P.; Unwin, A. Experimental study of particle migration in suspensions undergoing poiseuille flow. In Proceedings of the 66th Annual Meeting Society of Rheology, Philadelphia, PA, USA, 2–6 October 1994.
19. Dayan, A.; Stracener, S.M.; Clark, P.E. Proppant transport in slick-water fracturing of shale-gas formations. In Proceedings of the SPE Annual Technical Conference and Exhibition, New Orleans, LA, USA, 25–28 September 1994; SPE: Richardson, TX, USA, 2009; p. SPE-125068.
20. Tsai, K.; Fonseca, E.; Degaleesan, S.; Lake, E. Advanced computational modeling of proppant settling in water fractures for shale gas production. *SPE J.* **2013**, *18*, 50–56. [CrossRef]
21. Dontsov, E.; Peirce, A. Proppant transport in hydraulic fracturing: Crack tip screen-out in kgd and p3d models. *Int. J. Solids Struct.* **2015**, *63*, 206–218. [CrossRef]
22. Wen, Q.; Duan, X.; Zhang, Y.; Liu, H.; Li, Y.; Wang, S. Study on settlement and migration law of proppant in complex fracture network. *J. Xi'An Shiyou Univ. (Nat. Sci. Ed.)* **2016**, *31*, 79–84.
23. Shiozawa, S.; McClure, M. Simulation of proppant transport with gravitational settling and fracture closure in a three-dimensional hydraulic fracturing simulator. *J. Pet. Sci. Eng.* **2016**, *138*, 298–314. [CrossRef]
24. Zhang, G.; Gutierrez, M.; Li, M. A coupled cfd-dem approach to model particle-fluid mixture transport between two parallel plates to improve understanding of proppant micromechanics in hydraulic fractures. *Powder Technol.* **2017**, *308*, 235–248. [CrossRef]
25. Guo, T.; Gong, Y.; Liu, X.; Wang, Z.; Xu, J.; Sheng, M.; Chen, M.; Luo, Z. Numerical simulation of proppant migration and distribution in complex fractures. *J. China Univ. Pet. (Ed. Nat. Sci.)* **2022**, *46*, 89–95.

26. Lu, C.; Ma, L.; Li, Z.; Huang, F.; Huang, C.; Yuan, H.; Tang, Z.; Guo, J. A novel hydraulic fracturing method based on the coupled cfd-dem numerical simulation study. *Appl. Sci.* **2020**, *10*, 3027. [CrossRef]
27. Snider, D.M. An incompressible three-dimensional multiphase particle-in-cell model for dense particle flows. *J. Comput. Phys.* **2000**, *170*, 523–549. [CrossRef]
28. Zeng, J.; Li, H.; Zhang, D. Numerical simulation of proppant transport in propagating fractures with the multi-phase particle-in-cell method. *Fuel* **2019**, *245*, 316–335. [CrossRef]
29. Mao, S.; Zeng, J.; Wu, K.; Zhang, D. *Lagrangian Numerical Simulation of Proppant Transport in Channel Fracturing*; SPE: Richardson, TX, USA, 2023; p. SPE-136986.
30. Suri, Y.; Islam, S.Z.; Hossain, M. Proppant transport in dynamically propagating hydraulic fractures using CFD-XFEM approach. *Int. J. Rock Mech. Min. Sci.* **2020**, *131*, 104356. [CrossRef]
31. Wen, Z.; Tang, H.; Zhang, L.; Chen, S.; Zeng, J.; Qin, J.; Wang, L.; Zhao, Y. *Coupled Simulation of Fracture Propagation and Lagrangian Proppant Transport*; SPE: Richardson, TX, USA, 2024; p. SPE-466883.
32. Wang, X.; Yao, J.; Gong, L.; Sun, H.; Yang, Y.; Zhang, L.; Li, Y.; Liu, W. Numerical simulations of proppant deposition and transport characteristics in hydraulic fractures and fracture networks. *J. Pet. Sci. Eng.* **2019**, *183*, 106401. [CrossRef]
33. O'Rourke, P.J.; Snider, D.M. A new blended acceleration model for the particle contact forces induced by an interstitial fluid in dense particle/fluid flows. *Powder Technol.* **2014**, *256*, 39–51. [CrossRef]
34. O'Rourke, P.J.; Snider, D.M. An improved collision damping time for MP-PIC calculations of dense particle flows with applications to polydisperse sedimenting beds and colliding particle jets. *Chem. Eng. Sci.* **2010**, *65*, 6014–6028. [CrossRef]
35. Andrews, M.J.; O'Rourke, P.J. The multiphase particle-in-cell (MP-PIC) method for dense particulate flows. *Int. J. Multiph. Flow* **1996**, *22*, 379–402. [CrossRef]
36. Harris, S.E.; Crighton, D.G. Solitons, solitary waves, and voidage disturbances in gas-fluidized beds. *J. Fluid Mech.* **1994**, *266*, 243–276. [CrossRef]
37. Ergun, S.; Orning, A.A. Fluid flow through randomly packed columns and fluidized beds. *Ind. Eng. Chem.* **1949**, *41*, 1179–1184. [CrossRef]
38. Wen, C.Y.; Yu, Y.H. Mechanics of fluidization. *Fluid Part. Technol. Chem. Eng. Prog. Symp. Ser.* **1966**, *62*, 100–111.
39. Clark, P.E.; Zhu, Q. Convective transport of propping agents during hydraulic fracturing. In Proceedings of the SPE Eastern Regional Meeting, Columbus, OH, USA, 23–25 October 1996; SPE: Richardson, TX, USA, 1996; p. SPE-37358.
40. Ma, J. Fiber-Assisted Proppant-Carrying Mechansim in Channel Fracturing. Ph.D. Thesis, Southwest Petroleum University, Chengdu, China, 2017.

processes

MDPI

Article

An Optimal Model for Determination Shut-in Time Post-Hydraulic Fracturing of Shale Gas Wells: Model, Validation, and Application

Jianmin Li [1], Gang Tian [1], Xi Chen [1], Bobo Xie [1], Xin Zhang [1], Jinchi Teng [1], Zhihong Zhao [2,*] and Haozeng Jin [2]

[1] PetroChina Xinjiang Oilfield Company, Karamay 834000, China; zy-ljm@petrochina.com.cn (J.L.); tgll2014@petrochina.com.cn (G.T.); chenxi@petrochina.com.cn (X.C.); xiebb@petrochina.com.cn (B.X.); gcyzhangxin@petrochina.com.cn (X.Z.); tengjinchi@petrochina.com.cn (J.T.)

[2] State Key Laboratory of Southwest Petroleum University, Chengdu 610500, China; jinhaozeng@gmail.com

* Correspondence: zhaozhihong@swpu.edu.cn

Abstract: The global shale gas resources are huge and have good development prospects, but shale is mainly composed of nanoscale pores, which have the characteristics of low porosity and low permeability. Horizontal drilling and volume fracturing techniques have become the effective means for developing the shale reservoirs. However, a large amount of mining data indicate that the fracturing fluid trapped in the reservoir will inevitably cause hydration interaction between water and rock. On the one hand, the intrusion of fracturing fluid into the formation causes cracks to expand, which is conducive to the formation of complex fracture networks; on the other hand, the intrusion of fracturing fluid into the formation causes the volume expansion of clay minerals, resulting in liquid-phase trap damage. At present, the determination of well closure time is mainly based on experience without theoretical guidance. Therefore, how to effectively play the positive role of shale hydration while minimizing its negative effects is the key to optimizing the well closure time after fracturing. This paper first analyzes the shale pore characteristics of organic pores, clay pores, and brittle mineral pores, and the multi-pore self-absorption model of shale is established. Then, combined with the distribution characteristics of shale hydraulic fracturing fluid in the reservoir, the calculation model of backflow rate and shut-in time is established. Finally, the model is validated and applied with an experiment and example well. The research results show that the self-imbibition rate increases with the increase in self-imbibition time, and the flowback rate decreases with the increase in self-imbibition time. The self-imbibition of slick water is the maximum, the self-imbibition of breaking fluid is the minimum, and the self-imbibition of mixed fluid is the middle, and the backflow rates of these three liquids are in reverse order. It is recommended the shut-in time of Longmaxi Formation shale is 17 days according to the hydration and infiltration model.

Keywords: shale; hydraulic fracturing; self-imbibition; flowback; shut-in time

1. Introduction

The characteristics of the pore space in shale rocks can be summarized as having various features in terms of the extremely narrow pore throat, low porosity, high compactness, large specific surface area, complex minerals and organic matter, multiple pores coexisting, and natural cracks and weakly cemented surface development [1–3]. Consequently, conventional depressurization methods cannot achieve commercial exploitation of shale gas. Nevertheless, massive hydraulic fracturing has become a key technology for commercial exploitation of shale gas [4–6], as the horizontal wells associated with massive hydraulic fracturing can connect natural cracks to allow the form of flow channels for shale gas [7–9] through the bulk volume of reservoirs near to the wells. Ten thousand cubic meters of water-based fracturing fluid can be injected into shale formations during hydraulic fracturing [10]. Because of the differences in geological characteristics, in crushed fracture areas, and in fracturing fluid performance, the fracturing fluid in shale reservoirs typically has high flowback rate, as shown in Table 1 [11,12]. Currently, it is widely recognized that there is a negative correlation between gas production and flowback rate after hydraulic fracturing in shale [13–16].

Table 1. Flowback rate after fracturing in North America and China shale.

North America	Flowback Rate (%)	China	Flowback Rate (%)
Barrnet/Marcelus [17]	50	Weiyuan	40.3
Horn River [18]	25~55	Yongchuan	34.3
Haynesville [17]	5	Changning	12.7
Eagle Ford [19]	<20	Fuling	3.9
Average in North America [20]	6~10	\	\

The control factors of low flowback rate of fracturing fluid primarily include spontaneous imbibition effect of shale [18,21,22], unstable displacement of gas–water two-phase and gravity differentiation in fracturing fractures [23,24], and fracturing fluid retention in secondary fractures [25]. The spontaneous imbibition effect is the major factor, which is different from the conventional reservoir post-hydraulic fracturing [26]. Imbibition is a process in which the porous medium spontaneously draws in wetting liquid under a capillary pressure drop [27]. In addition to the capillary force, the imbibition in the shale reservoir is also powered by the osmotic pressure [20,28].

The shale microstructure is complex and unique, especially when the clay content is high, which is when pores and bedding in clay are developed. The traditional perspective is that shale water absorption causes an increase in the water saturation, the clay expands to fill the pores, and the permeability reduces in both matrix and fractures [29–31]. However, recent studies have suggested that the swelling of the shale, the increase in the pore pressure, and the weakening in mechanical strength of the rock can induce the micro-crack initiation, consequently improving the permeability of shale [1,13,29,32,33].

As shale imbibition can increase reservoir permeability, many researchers have studied the effect of well shut-in on yield. They found that well shut-in can enhance shale imbibition, which is favorable for the reduction in water saturation in the fractures, meaning it can increase the initial production, but has no significant impact on long-term production [13,14]. Simulating the effects of phase infiltration, capillary force, and effective stress on well shut-in reveals that wells should be shut-in right after fracturing and kept for a long period of time [34]. Well shut-in can increase gas and reduce water production; unfortunately this effect does not last quite long [35]. According to the previous studies, well shut-in can increase the production at early stage and for a short period. In practice, however, the shut-in time after a hydraulic fracturing has a lack of theoretical interpretation but just experimental understanding.

Due to the significant differences in pore shape, pore size, pore genesis, and wettability among organic pores, brittle mineral pores, and clay pores, the imbibition depth and imbibition amount of the three types of pores are significantly different. This article focuses on the mechanism of imbibing water for shale, and conducts research from the aspects of shale microstructure characteristics and imbibition force, establishing three imbibition models for three types of pores, i.e., organic pores, brittle mineral pores, and clay pores. A model is then introduced for calculating the shale shut-in time by combining with the direction of fracturing fluid flow during shale fracturing, which is validated by comparing with experimental data.

2. Shale Multiple-Pore Imbibition Model

Shale rock is a multi-porous medium; the pore type includes the organic pores, clay pores, and brittle mineral pores, and because the origins of the different types of the pores are different, the pore size, shape (Figure 1), and the distribution of the pore size have significant differences (Figure 2).

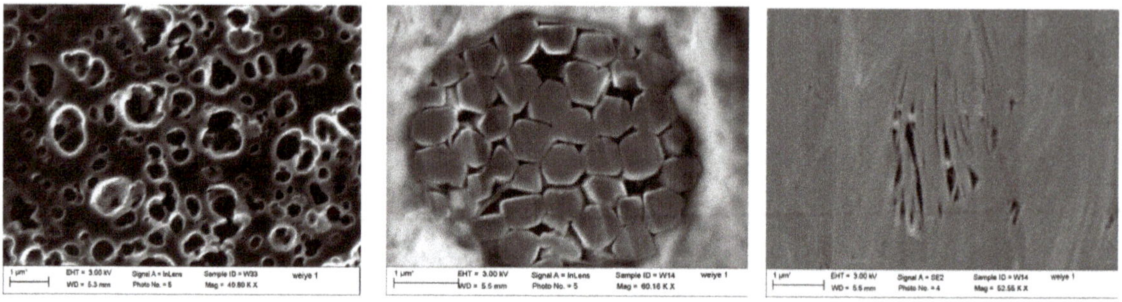

(a) (b) (c)

Figure 1. Types of pores in shale. (**a**) Organic pores, (**b**) brittle mineral pores, (**c**) clay pores.

Yuman et al. (2015) stated that the pores of shale can be subdivided into the clay pores, brittle mineral pores, and organic pore, as expressed in Equation (1) [36]. Based on the assumption, shale imbibition can then be decomposed into three types of pore imbibition for organic pores, brittle mineral pores, and clay pores. However, the forces of imbibition for three types of pores are different during the process of the imbibition. Thus, we establish the shale multi-pore imbibition model based on the force of the imbibition for three types of pores.

$$\phi = \phi_{tb} + \phi_{tc} + \phi_{to} \quad (1)$$

where ϕ is the total porosity (%) of shale, and ϕ_{tb}, ϕ_{tc}, and ϕ_{to} are the porosities accounting for brittle mineral pores, clay pores, and organic pores, respectively.

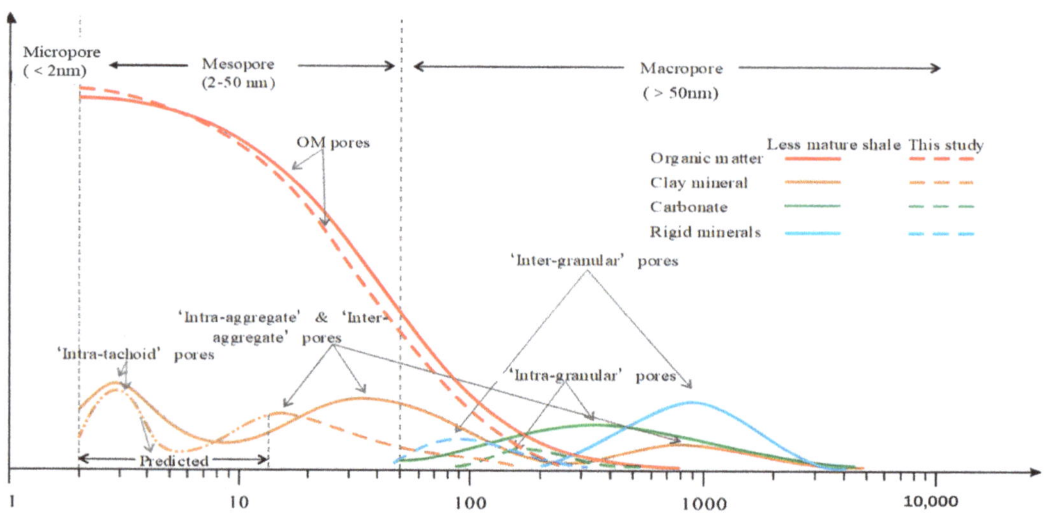

Figure 2. The relationship between pore size and type [37,38].

2.1. Pore Imbibition Models

(1) Organic pores and brittle pores

It can be seen from Figure 2 that the pore shape of organic pores and brittle minerals is similar to an ellipse, so the elliptical capillary fractal imbibition model can be used to describe the imbibition in these pores. There are no significant differences in the microstructure and physical property parameters for the organic pores and brittle pores. Under the condition of ignoring the influence of gravity on the imbibition of the water phase, the imbibition effect of the organic pores and brittle mineral pores is only controlled by the force of the capillary. Therefore, we have the same imbibition models for ease of calculation, as shown in Equation (2) [39].

$$L_e = \langle \frac{\sigma \cos\theta b_{max}^{D_f-2}(m+1)\left(2-D_f\right)(b_{max}-b_{min})}{8\mu D_f b_{min}^{1-D_T}(1-\varnothing)} \sum_{k=0}^{n} \omega_k \left\{ \frac{b_k^{D_T-D_f}\left[mk_b^2+6(m+1)L_s b_k+32L_s\right]}{(m^2+1)k_b+4(m+1)L_s} \right\} \rangle^{\frac{1}{2D_T}} \cdot t^{\frac{1}{2D_T}} \tag{2}$$

$$b_k = \frac{b_{max}-b_{min}}{2} t_k + \frac{b_{max}+b_{min}}{2} \tag{3}$$

where σ is interfacial tension between two phases, mN/m; ϕ porosity, %; θ is wetting contact angle, °; D_f is fractal dimension; D_T is flexibility fractal dimension; b_{min} is minimum pore diameter, nm; b_{max} is maximum pore diameter, nm; μ is aqueous phase viscosity, mPa·s; L_s boundary slip length, nm; t is imbibition time, min; m is aspect ratio of elliptical pores; n is the number of interval segments; ω_k is weight coefficient; and t_k is Gaussian point determined by n.

If a four-point formula is chosen, $t_k = \pm 0.8611363, \pm 0.3399810$, $\omega_k = 0.3478548, 0.6521452$.

(2) Clay minerals pores

It can be seen from Figure 2 that the shape of the clay pores is significantly different from the shapes of the organic and brittle pores. The clay pores exhibit a slender shape. Therefore, the model of the parallel fracture can be used to describe the imbibition phenomenon in these pores. Moreover, the force of imbibition for the clay pores is significantly different from the forces of imbibition of the organic pores and brittle mineral pores. It is not only affected by the capillary force, but also by the osmotic pressure. In the effective imbibition flow range of the reservoir, the influence of the gravity on the imbibition is very small compared with the capillary force and the osmotic pressure, because the 10-m high water column produces a gravity of 0.1 MPa, meaning the influence of the gravity can be ignored. The fractal model of clay pore imbibition length (L_c) is expressed as follows:

$$L_c = \left\{ \frac{(D_{Tc}+D_{fc}-1)(2-D_{fc})}{D_{fc} w_{\min}^{1-D_{Tc}}(1-\phi)} \left[\frac{\sigma \cos\theta_c}{3\mu}(A_2+6L_{sc}B_2) + \frac{p_\pi}{6\mu}(C_2+6L_{sc}A_2) \right] \right\}^{\frac{1}{2D_{Tc}}} \cdot t^{\frac{1}{2D_{Tc}}}, \tag{4}$$

$$A_2 = \frac{w_{\max}^{D_{Tc}}\left(1-\beta_c^{2+D_{Tc}-D_{fc}}\right)}{2+D_{Tc}-D_{fc}}; B_2 = \frac{w_{\max}^{D_{Tc}-1}\left(1-\beta_c^{1+D_{Tc}-D_{fc}}\right)}{1+D_{Tc}-D_{fc}}; C_2 = \frac{w_{\max}^{1+D_{Tc}}\left(1-\beta_c^{1+D_{Tc}-D_{fc}}\right)}{3+D_{Tc}-D_{fc}}; \beta_c = \frac{w_{\min}}{w_{\max}} \tag{5}$$

where $w_{\max}$ is the maximum clay pore width, nm; $w_{\min}$ is the minimum clay pore width, nm; subscripts c represents clay pore; and p_π is osmotic pressure, and can be expressed by [40]:

$$p_\pi = \varepsilon E_\pi RT(C_{sh}-C_f) \tag{6}$$

where R is a gas constant of 0.08206 (L·KPa)/(mol·K), usually; T is temperature of formation, K; C_{sh} is molar concentration of solute in the original formation water, mol/L; C_f is molar concentration of solute in fracturing fluid, mol/L; ε is number of ions after solute ionization (example.g., NaCl value is 2) ; and E_π is membrane efficiency, and can be expressed as follows [40]:

$$E_\pi = 1 - \frac{(R_{\text{ca-w}}+1)C_{\text{a}}/C_{\text{s}}}{\{R_{\text{ca-w}}(C_{\text{a}}/C_{\text{c}})+1+R_{a\text{-mw}}[R_{\text{ca-m}}(C_{\text{a}}/C_{\text{c}})+1]\}\phi_{\text{c}}} \tag{7}$$

$$\begin{cases} C_s = \frac{1}{2}(C_f + C_{sh}) \approx \frac{1}{2}C_{sh} \\ C_a = -\frac{1}{2}E_{CEC}\rho_{clay}(1-\phi_c) + \frac{1}{2}\left[E_{CEC}^2\rho_{clay}^2(1-\phi_c)^2 + 4C_s^2\phi_c^2\right]^{\frac{1}{2}} \\ C_c = C_a + E_{CEC}\rho_{clay}(1-\phi_c) \end{cases} \tag{8}$$

where C_s is arithmetic mean of solute in solution on both sides of semi-permeable membrane; C_a is the concentration of anions in the pores of the semi-permeable membrane; E_{CEC} is the cation exchange capacity of shale; ρ_{clay} is the density of clay minerals; $R_{ca\text{-}m}$ is ratio of friction coefficient of anions to water; $R_{ca\text{-}w}$ is the ratio of the friction coefficient of the anions to the membrane; and $R_{a\text{-}mw}$ is the ratio of anion–semi-permeable membrane to cation–water friction coefficient.

(3) Multiple pore self-imbibition model

According to the shale multi-porosity splitting model [36], the total porosity of shale can be decomposed into organic porosity (ϕ_{tb}), brittle mineral porosity (ϕ_{tc}), and clay porosity (ϕ_{to}). The cumulative imbibition volume V_{im} is, thus, the sum of the volumes imbibed from organic pores, brittle mineral pores, and clay pores. The shale imbibition model of multiple pores can be expressed as:

$$V_{im} = V_{imo} + V_{imb} + V_{imc} = A_f\phi_{to}L_o + A_f\phi_{tb}L_b + A_f\phi_{tc}L_c \tag{9}$$

where V_{im} is total imbibition volume, m^3; V_{imo}, V_{imb}, V_{imc} are imbibition volume of organic pores, clay pores, and brittle mineral pores, m^3, respectively; and A_f is contact area of fracturing fluid and shale, m^2, which can be obtained using the pressure drop analysis method in Equation (10).

$$A_f = 2\frac{V_{inj} - V_{leak}}{w_f} \tag{10}$$

where V_{leak} is the fluid loss volume, m^3, which can be obtained from pressure drop data [41]; V_{inj} is the volume injection fracturing fluid, shale reservoir filtration typically accounts for 10–40%, the Horn River shale is estimated at 30% [42]; and w_f is the average fracture width, which can be determined with pressure drop analysis method of Equation (11) [42].

$$f(w_f) = \frac{C_f^2 L_m}{2DC_{sh}\Delta t}\frac{dN_w}{dC_f}, \tag{11}$$

where Δt is the backflow curve discrete element time; N_w is non-dimensional backflow of fracturing fluid after fracturing (the ratio of the amount of phased reflux to the amount of cumulative reflux).

According to Equation (11), the average fracture width can be obtained, and the fracture area of each section can be calculated according to Equation (10).

2.2. Validation and Analysis

Experimental materials:

A shale core of the Longmaxi Formation in Sichuan Basin was chosen and the imbibed fluid is distilled water. Other physical and chemical properties of the experimental materials are shown in Table 2. In order to improve the accuracy of the data, the analytical scale automatically weighs and measures. The range of the analytical scale is 220 g, and the precision is 0.0001 g. The experimental setup is shown in Figure 3.

Table 2. Shale sample and fluid essential parameters.

Parameters	Value		
	Organic Pore	**Brittle Pore**	**Clay Pore**
Wetting contact angle θ (°) Mineral porosity (%)	80.40 20.08	16.20 0.88	11.50 7.84
Aspect ratio of elliptical pores m (dimensionless)	2.336	3.286	\
Boundary slip length L_s (nm)	4.41	1.56	1.53
Maximum pore diameter b_{max} (nm)	627	1020	461
Minimum pore diameter b_{min} (nm)	1.47	7.5	1.47
Fractal dimension D_f	2.681	2.549	2.619
Flexibility fractal dimension D_T	1.212	1.297	1.255
The porosity of the core $\phi_{to}/\phi_{tb}/\phi_{tc}$ (%)	1.534	0.676	1.600
Water–core contact area A_f (cm^2)		6.45	
Viscosity of water μ (mPa·s)		1.00	
Gas–water interfacial tension σ (mN/m)		74.10	
Molar concentration of solute in the original formation water C_{sh} (mol/L)		0.525	
Molar concentration of solute in fracturing fluid C_f (mol/L)		0.017	
Test temperature T (K)		293.15	
The cation exchange capacity of shale E_{CEC}		3.7	
Ratio of friction coefficient between anions and water $R_{ca\text{-}m}$		1.8	
The ratio of the friction coefficient between the anions and the membrane $R_{ca\text{-}w}$		1.63	
The ratio of anion–semi-permeable membrane to cation–water friction coefficient $R_{a\text{-}mw}$		1.37	

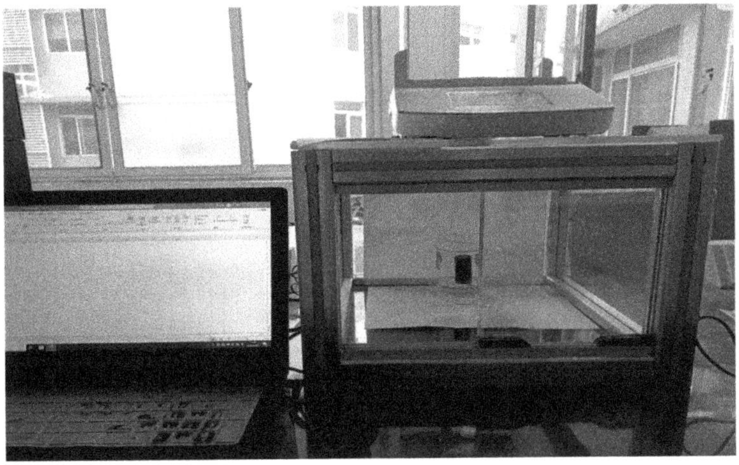

Figure 3. Schematic diagram of the forward imbibition experimental setup.

Experimental procedures:

(1) Prepare a standard shale core of 5.0 cm in length and measures the basic parameters of the core and fluid. The test results are shown in Table 2;
(2) Dry the core by place the core in an oven for more than 10 h, to weigh, and to wrap the core surface with a heat shrinkable tube, as shown in Figure 3;
(3) Monitor the change in imbibition weight over time using a scale, and carry out the same drying treatment in the end of the experiment.

It can be seen from Figure 4 that the calculation results are close to the experimental results, which proves that the imbibition model is accurate and reliable. When the shale imbibes for 70 min, the imbibition volume increases first and then decreases with the imbibition time passing. We can find that test results are 10~20% higher than calculation results in the second stage because the induced fractures promote shale imbibition (Figure 4). In the third stage, test results are lower than the calculated results due to core size effect and shale tending to be saturated resulting in reduced imbibition speed. Shale imbibition is a coupling process between liquid-phase flow and the dynamic change in shale microstructure; by contrast, the microstructure of traditional tight sandstone does not change during the imbibition process. In this work, we mainly study the imbibition in shale multi-porosity microstructures, and the dynamic changes in microscopic structure are not considered in the multiple-pore imbibition model.

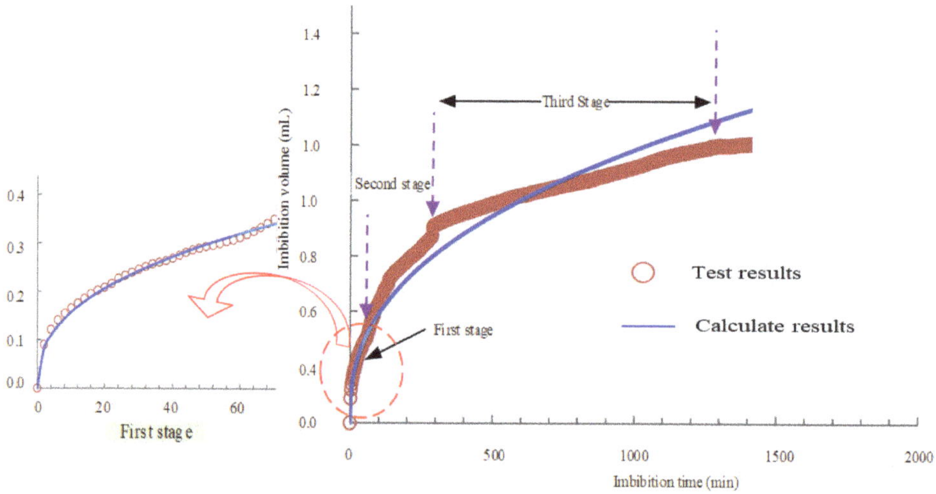

Figure 4. Comparison of calculation with experimental results.

3. Determination of Shut-In Time Model

Shale gas horizontal well staged fracturing has the characteristics of more fracturing segments and longer operation time. This results in a larger time interval between the end and toe fracturing segments, and the imbibition volume calculation needs to take into account the time difference. The imbibition process during fracturing can be divided into two steps/periods. The first step is operation, when the imbibition time of various segments is distinct. The second step is well shut-in, during which all segments have the same imbibition time. During the operation stage, each segment is separated by the bridge plug. Therefore, the imbibition of each segment is independent, and the total imbibition volume during operation stage (V_{im1}) can be expressed as follows:

$$V_{im1} = \sum_{i=1}^{n} V_{imf,i} = \sum_{i=1}^{n} A_f \left(\phi_{to} C_{imo} t_{f,i}^{m_o} + \phi_{tb} C_{imb} t_{f,i}^{m_b} + \phi_{tc} C_{imc} t_{f,i}^{m_c} \right) \quad (12)$$

During the shut-in well stage, the bridge plug between the segments has been drilled through, thus, the imbibition volume can be given by:

$$V_{im2} = nA_f(\phi_{to}C_{imo} + \phi_{tb}C_{imb} + \phi_{tc}C_{imc})t_{shut} \tag{13}$$

where $V_{\mathrm{imf},i}$, V_{im1}, V_{im2} are the imbibition volume contributed by the I segment during the operation stage, the total imbibition volume during the operation, and the total imbibition volume of the shut-in stage, m^3; $t_{f,i}$ is imbibition time of operation the i, s; t_{shut} is shut-in time, s; n is the total fracturing segments of whole well. C_{imo}, C_{imb}, C_{imc} are the coefficients of Equations (2), (3), and (5), and they are expressed as follows:

$$C_{imo} = \left\langle \frac{\sigma\cos\theta_o b_{\mathrm{maxo}}^{D_{fo}-2}(m_o+1)(2-D_{fo})(D_{To}+D_{fo}-1)(b_{\mathrm{maxo}}-b_{\mathrm{mino}})}{8\mu D_{fo} b_{\mathrm{mino}}^{1-D_T}(1-\phi)} \sum_{k=0}^{n} \omega_k \left\{ \frac{b_{ko}^{D_{To}-D_{fo}}\left[m_o b_{ko}^2 + 6(m_o+1)L_{so}b_{ko} + 32{L_{so}}^2\right]}{(m_o^2+1)b_{ko} + 4(m_o+1)L_s} \right\} \right\rangle^{\frac{1}{2D_{To}}} \tag{14}$$

$$C_{imb} = \left\langle \frac{\sigma\cos\theta_b b_{\mathrm{maxb}}^{D_{fb}-2}(m_b+1)(2-D_{fb})(D_{Tb}+D_{fb}-1)(b_{\mathrm{maxb}}-b_{\mathrm{minb}})}{8\mu D_{fb} b_{\mathrm{minb}}^{1-D_T}(1-\phi)} \sum_{k=0}^{n} \omega_k \left\{ \frac{b_{kb}^{D_{Tb}-D_{fb}}\left[m_b b_{kb}^2 + 6(m_b+1)L_{sb}b_{kb} + 32{L_{sb}}^2\right]}{(m_b^2+1)b_{kb} + 4(m_b+1)L_s} \right\} \right\rangle^{\frac{1}{2D_{Tb}}} \tag{15}$$

$$C_{imc} = \left\{ \frac{(D_{Tc}+D_{fc}-1)(2-D_{fc})}{D_{fc} w_{\min}^{1-D_{Tc}}(1-\phi)} \left[\frac{\sigma\cos\theta_c}{3\mu}(A_2 + 6L_{sc}B_2) + \frac{p_\pi}{6\mu}(C_2 + 6L_{sc}A_2) \right] \right\}^{\frac{1}{2D_{Tc}}} \tag{16}$$

The imbibition rate is defined as the ratio of the imbibition volume to the volume of liquid that effectively creates the fractures. The expression is as follows:

$$N_{\mathrm{loss}} = \frac{V_{\mathrm{im}}}{V_{\mathrm{inj}} - V_{\mathrm{leak}}} \tag{17}$$

Suppose that the water phase is in a bound state after self-imbibition and there is no discharge, the expression of fracturing fluid backflow rate is as follows:

$$N_{\mathrm{fb}} = \frac{V_{\mathrm{inj}} - V_{\mathrm{im}}}{V_{\mathrm{inj}}} \tag{18}$$

When then imbibition rate reaches 100%, it indicates that the fracturing fluid in the fractures is completely imbibed into the shale matrix. At this time, there is no fracturing fluid remaining in the fractures, and there is no two-phase flow in the crack during production to reduce the permeability of crack [13,14]. The imbibition time of water phase minus the fracturing operation time is the shut-in time.

$$t_{shut} = \frac{V_{inj} - V_{leak} - V_{im1}}{nA_f(\phi_{to}C_{imo} + \phi_{tb}C_{imb} + \phi_{tc}C_{imc})} \tag{19}$$

4. Model Application Analysis

(1) Basic parameters of reservoir and fracturing

Take a well, named as W1, of the shale reservoir in Sichuan Basin as an example. Table 2 lists basic formation parameters while Table 3 reveals fracturing parameters. The main ionic composition of the backflow solution is NaCl, so the value of the ionic diffusion coefficient of the backflow solution approximate is 1.484×10^{-9} m^2/s [15]. The fracturing liquid properties of sliding water, breaking fluid, and linear glue are shown in Table 4.

Table 3. Fracturing parameters.

Parameters	Value	Parameters	Value
Fracturing segments	16	Fracturing time t (d)	16
Injection fluid volume V_{inj} (m^3)	27,740	Average fluid volume per segment V_{injfr} (m^3)	1733.75
Slippage and linear gel ratio	7: 3	Matrix Cl- salinity	15,291.8
Discrete element time Δt (d)	1	Diffusion coefficient (m^2/s)	1.484×10^{-9}

Table 4. Performance parameters of different types of liquids.

Liquid Type	Contact Angle (°)	Viscosity (mPa·s)	Surface Tension (mN/m)	Salinity (mol/L)	Density (g/cm³)
Slick water	42.4	7	35.34	0	1
Breaking fluid	42.8	14	25.32	0	1.004

(2) Shut-in time calculation

Figure 5 shows the backflow rate under different self-imbibition rate and time with different fracturing fluid; it reveals that increasing the shut-in time, namely, increasing the self-imbibition time, causes a rise in the self-imbibition rate at the beginning and when decreasing the flowback rate thereafter, the self-imbibition of slick water reaches to its maximum, meanwhile the self-imbibition of breaking fluid achieves its minimum, and, furthermore, the self-imbibition of mixed fluid (the ratio of slick water to linear glue is 7:3) arrive at the middle. The backflow rates of these three liquids are in reverse order.

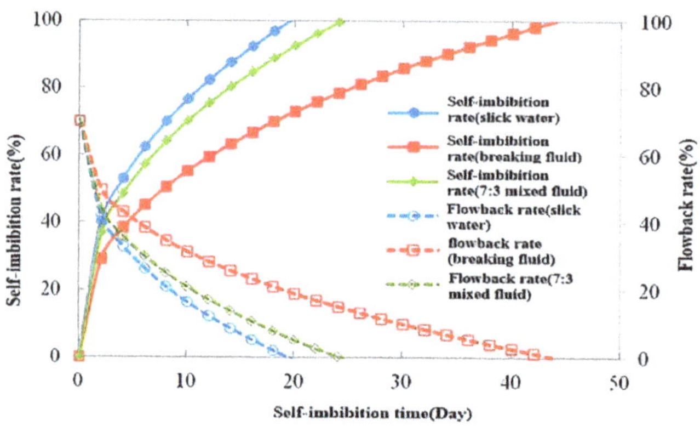

Figure 5. Backflow rate under different self-imbibition rate and time with different fracturing fluid.

Figure 6 is the self-absorption of each fracturing section during fracturing operation. The total self-imbibition is 7811.7 m^3 with a total inject fluid of 27,740 m^3, so the average self-imbibition fluid is 488.23 m^3 per single stage during the shut-in time; in other words, the needed self-imbibition rate is 85.4%. According to Figure 4 or Equation (19), the shut-in time can be determined to be 17 days.

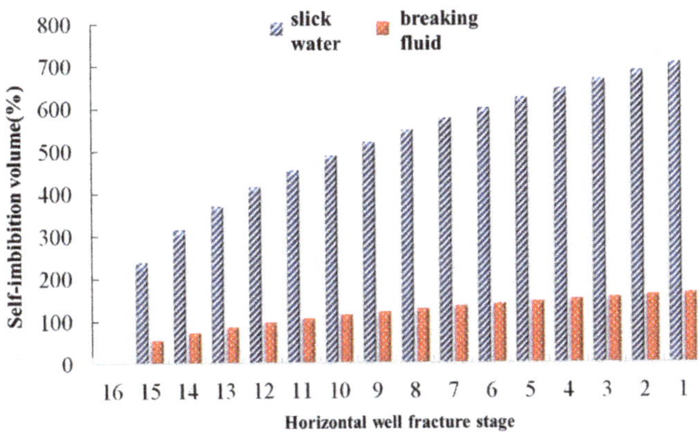

Figure 6. The self-absorption of each fracturing section during fracturing operation.

(3) Analysis of the flowback fluid data

Well W1 was actually closed for 7 days after fracturing operation, and then flowback was performed. According to the time of chlorine balance in the flowback fluid, the initial equilibrium of W1 was reached in 17–18 days, which was consistent with the calculated results by the proposed model. However, the shut-in of W1 lasted for 70 days, and after the long-time shut-in, the concentration of chlorine increased significantly (Figure 7). This indicates that ion exchange between fracturing fluid and formation fluid and the dissolution of shale matrix are still taking place during the shut-in after the initial equilibrium.

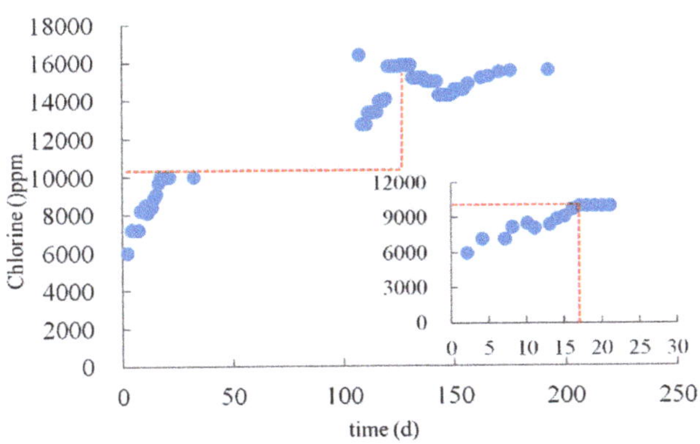

Figure 7. Backflow fluid curve in Well W1. The blue dots represent the recorded experimental data, while the red dashed line has no practical significance.

(4) Production dynamic analysis

From the production dynamic curve during the flowback process (Figure 8), it can be seen that the production of the water decreases rapidly after well W1 is shut-in, the trend of decrease is almost vertical, the production of the water almost decreases to 0 in a short period of time, indicating that the shut-in process increases the time of the imbibition. This results in a significant increase in cumulative imbibition. After 70 days, the W1 still produced a certain degree of water, and it indicates that the ability of displacing oil is limited and only relies on the imbibition, and other methods of increasing the production need to be combined to improve the production of the well.

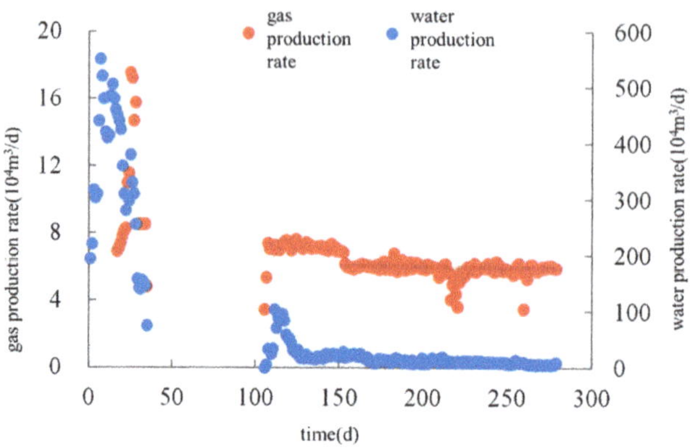

Figure 8. Production dynamic curve during flowback.

5. Conclusions

(1) This article establishes a shale multi-pore imbibition model based on the assumption that the shale pores can be divided into the organic pores, brittle mineral pores, and the clay pores. The model can effectively simulate the early spontaneous infiltration and absorption process of shale;

(2) The volume of the imbibition gradually increases with the hydration time: slick water > mixed liquid > gel-breaking liquid, while the infiltration rate gradually decreases. During the backflow process, the backflow rate and water imbibition of the three liquids show an opposite trend;

(3) According to the self-imbibition of each fracturing period during fracturing operation, the total self-imbibition amount is 7811.7 m^3 with a total inject fluid volume of 27,740 m^3, so the average self-imbibition fluid volume is 488.23 m^3 per single stage during the shut-in time. According to the shale multi-pore self-imbibition model established in this article, the shut-in time can be determined to be 17 days;

(4) According to the flowback data, the concentration of the chloride ions in the flowback fluid reaches initial equilibrium within 17–18 days, which is consistent with the results of the model calculation. However, the closure time of W1 lasted for 70 days, and the prolonged closure significantly increases the concentration of the chloride ions in the flowback fluid, indicating that there is still exchange of the ions between the fracturing fluid and the formation fluid during the initial ion equilibrium process of the shut-in;

(5) It can be seen from the field production data that after the water productivity of W1 decreases rapidly, the trend of the declining is almost vertical. The production of the water almost decreases to 0 in a short period of the time, indicating that the process of the shut-in increases the time for imbibition. This leads to a significant increase in cumulative imbibition. After 70 days, well W1 still produced a certain amount of the water, indicating that the ability of displacing oil is limited and relies only on the imbibition. Other methods for increasing the production need to be combined to achieve this.

Author Contributions: Conceptualization, J.L., G.T., X.C., B.X., X.Z. and J.T.; Methodology, validation, investigation and writing, Z.Z. and H.J. All authors have read and agreed to the published version of the manuscript.

Funding: National Natural Science Foundation of China (project No. 51525404).

Data Availability Statement: Data are contained within the article.

Acknowledgments: The authors would like to acknowledge that this work was successfully completed through the financial support by the National Natural Science Foundation of China (project No. 51525404).

Conflicts of Interest: Authors J.L., G.T., X.C., B.X., X.Z. and J.T. were employed by the company PetroChina Xinjiang Oilfield Company. The remaining authors declare that the research was conducted in the absence of any commercial or financial relationships that could be construed as potential conflicts of interest.

References

1. Yuan, B.; Wang, Y.; Wei, N. The effects of fracturing fluid retention on permeability of shale reservoirs. *Energy Procedia* **2019**, *158*, 5934–5939. [CrossRef]
2. Yuan, B.; Wang, Y.; Zeng, S. Effect of Slick Water on Permeability of Shale Gas Reservoirs. *J. Energy Resour. Technol.* **2018**, *140*, 112901–112907. [CrossRef]
3. Zhang, S.; Xian, X.; Zhou, J.; Liu, G.; Guo, Y.; Zhao, Y.; Lu, Z. Experimental study of the pore structure characterization in shale with different particle size. *J. Energy Resour. Technol.* **2018**, *140*, 54502–54511. [CrossRef]
4. Ahn, C.H.; Dilmore, R.; Wang, J.Y. Modeling of hydraulic fracture propagation in shale gas reservoirs: A three-dimensional, two-phase model. *J. Energy Resour. Technol.* **2017**, *139*, 2903–2912. [CrossRef]
5. Guo, T.; Li, Y.; Ding, Y.; Qu, Z.; Gai, N.; Rui, Z. Evaluation of Acid Fracturing Treatments in Shale Formation. *Energy Fuels* **2017**, *31*, 10479–10489. [CrossRef]
6. Vincent, M.C. The next opportunity to improve hydraulic-fracture stimulation. *J. Petrol. Technol.* **2012**, *64*, 118–127. [CrossRef]
7. He, Y.; Cheng, S.; Li, S.; Huang, Y.; Qin, J.; Hu, L.; Yu, H. A semianalytical methodology to diagnose the locations of underperforming hydraulic fractures through pressure-transient analysis in tight gas reservoir. *SPE J.* **2017**, *22*, 924–939. [CrossRef]
8. Nasriani, H.R.; Jamiolahmady, M. Flowback cleanup mechanisms of post-hydraulic fracturing in unconventional natural gas reservoirs. *J. Nat. Gas Sci. Eng.* **2019**, *66*, 316–342. [CrossRef]
9. Sun, J.; David, S. Investigating the effect of improved fracture conductivity on production performance of hydraulic fractured wells through field case studies and numerical simulations. *J. Can. Pet. Technol.* **2015**, *54*, 442–449. [CrossRef]
10. Hull, N.M.; Rosenblum, J.S.; Robertson, C.E.; Harris, J.K.; Linden, K.G. Succession of toxicity and microbiota in hydraulic fracturing flowback and produced water in the Denver–Julesburg Basin. *Sci. Total Environ.* **2018**, *644*, 183–192. [CrossRef]
11. Clarkson, C.R.; Qanbari, F. An approximate semianalytical multiphase forecasting method for multifractured tight Light-Oil wells with complex fracture geometry. *J. Can. Pet. Technol.* **2015**, *54*, 489–508. [CrossRef]
12. Estrada, J.M.; Bhamidimarri, R. A review of the issues and treatment options for wastewater from shale gas extraction by hydraulic fracturing. *Fuel* **2016**, *182*, 292–303. [CrossRef]
13. Ghanbari, E.; Dehghanpour, H. The fate of fracturing water: A field and simulation study. *Fuel* **2016**, *163*, 282–294. [CrossRef]
14. Cheng, Y. Impact of water dynamics in fractures on the performance of hydraulically fractures wells in gas-shale reservoirs. *J. Can. Pet. Technol.* **2012**, *51*, 143–151. [CrossRef]
15. Zolfaghari, A.; Dehghanpour, H.; Ghanbari, E.; Bearinger, D. Fracture characterization using flowback salt-concentration transient. *SPE J.* **2016**, *21*, 233–244. [CrossRef]
16. Abbasi, M.; Dehghanpour, H.; Hawkes, R.V. Flowback analysis for fracture characterization. In Proceedings of the SPE Canadian Unconventional Resources Conference, Calgary, AL, Canada, 30 October–2 November 2012. [CrossRef]
17. King, G.E. Hydraulic Fracturing 101: What every representative, environmentalist, regulator, reporter, investor, university researcher, neighbor and engineer should know about estimating frac risk and improving frac performance in unconventional gas and oil wells. In Proceedings of the SPE Hydraulic Fracturing Technology Conference, The Woodlands, TX, USA, 6–8 February 2012. [CrossRef]
18. Perapon, F.; Torcuk, M.A.; Wallace, J.; Bertoncello, A.; Kazemi, H.; Wu, Y.-S.; Honarpour, M. Managing Shut-in Time to Enhance Gas Flow Rate in Hydraulic Fractured Shale Reservoirs: A Simulation Study. In Proceedings of the SPE Annual Technical Conference and Exhibition, New Orleans, LA, USA, 30 September–2 October 2013. [CrossRef]
19. Nicot, J.-P.; Scanlon, B.R. Water Use for Shale-Gas Production in Texas, U.S. *Environ. Sci. Technol.* **2012**, *46*, 3580–3586. [CrossRef] [PubMed]
20. Singh, H. A critical review of water uptake by shales. *J. Nat. Gas Sci. Eng.* **2016**, *34*, 751–766. [CrossRef]
21. Dehghanpour, H.; Lan, Q.; Saeed, Y.; Fei, H.; Qi, Z. Spontaneous imbibition of brine and oil in gas shales: Effect of water adsorption and resulting microfractures. *Energy Fuels* **2013**, *27*, 3039–3049. [CrossRef]
22. Roshan, H.; Ehsani, S.; Marjo, C.E.; Andersen, M.S.; Acworth, R.I. Mechanisms of water adsorption into partially saturated fractured shales: An experimental study. *Fuel* **2015**, *159*, 628–637. [CrossRef]
23. Kuru, E.; Parmar, J.S.; Dehghanpour, H. Drainage against gravity: Factors impacting the load recovery in fractures. In Proceedings of the SPE Unconventional Resources Conference—USA, The Woodlands, TX, USA, 10–12 April 2013. [CrossRef]
24. Parmar, J.; Dehghanpour, H.; Kuru, E. Displacement of water by gas in propped fractures: Combined effects of gravity, surface tension, and wettability. *J. Unconv. Oil Gas Resour.* **2014**, *5*, 10–21. [CrossRef]

25. Fan, L.; Thompson, J.W.; Robinson, J.R. Understanding gas production mechanism and effectiveness of well stimulation in the haynesville shale through reservoir simulation. In Proceedings of the Canadian Unconventional Resources and International Petroleum Conference, Calgary, AL, Canada, 19–21 October 2010. [CrossRef]
26. Xu, M.; Gupta, A.; Dehghanpour, H. How significant are strain and stress induced by water imbibition in dry gas shales? *J. Pet. Sci. Eng.* **2019**, *176*, 428–443. [CrossRef]
27. Cai, J.; Yu, B. Advances in studies of spontaneous imbibition in porous media. *Adv. Mech.* **2012**, *42*, 735–754. [CrossRef]
28. Ghanbari, E.; Dehghanpour, H. Impact of rock fabric on water imbibition and salt diffusion in gas shales. *Int. J. Coal Geol.* **2015**, *138*, 55–67. [CrossRef]
29. Muntasheri, G. A critical review of hydraulic-fracturing fluids for moderate-to ultralow-permeability formations over the last decade. *Soc. Pet. Eng. J.* **2014**, *29*, 243–260. [CrossRef]
30. Gdanski, R.; Weaver, J.; Slabaugh, B.; Walters, H.; Parker, M. Fracture face damage—It matters. In Proceedings of the European Formation Damage Conference, Sheveningen, The Netherlands, 25–27 May 2005.
31. Li, Z.; Duan, Y.; Wei, M.; Peng, Y.; Chen, Q. Pressure performance of interlaced fracture networks in shale gas reservoirs with consideration of induced fractures. *J. Pet. Sci. Eng.* **2019**, *178*, 294–310. [CrossRef]
32. Zhang, S.; Sheng, J.J. Effect of water imbibition on hydration induced fracture and permeability of shale cores. *J. Nat. Gas Sci. Eng.* **2017**, *45*, 726–737. [CrossRef]
33. Bertoncello, A.; Wallace, J.; Blyton, C.; Honarpour, M.M.; Kabir, S. Imbibition and water blockage in unconventional reservoirs: Well-management implications during flowback and early production. *SPE Reserv. Eval. Eng.* **2014**, *17*, 497–506. [CrossRef]
34. Fakcharoenphol, P.; Torcuk, M.; Kazemi, H.; Wu, Y.-S. Effect of shut-in time on gas flow rate in hydraulic fractured shale reservoirs. *J. Nat. Gas Sci. Eng.* **2016**, *32*, 109–121. [CrossRef]
35. Yuman, W.; Jinliang, H.; Xinjing, L.; Dazhong, D.; Shufang, W.; Quanzhong, G. Quantitative characterization of fractures and pores in shale beds of the Lower Silurian, Longmaxi Formation, Sichuan Basin. *Nat. Gas Ind.* **2015**, *35*, 8–15. [CrossRef]
36. Chen, Q.; Zhang, J.; Tang, X.; Li, W.; Li, Z. Relationship between pore type and pore size of marine shale: An example from the Sinian–Cambrian formation, upper Yangtze region, South China. *Int. J. Coal Geol.* **2016**, *158*, 13–28. [CrossRef]
37. Kuila, U.; Prasad, M. Specific surface area and pore-size distribution in clays and shales. *Geophys. Prospect.* **2013**, *61*, 341–362. [CrossRef]
38. He, S.G. Research of Water Imbibition Model in Shale Multiple Pores. Ph.D. Thesis, Southwest Petroleum University, Chengdu, China, 2017. Available online: http://ss.zhizhen.com/detail_38502727e7500f26321d19d4b4d28a60bfe030dee0234f581921b0a3ea255101928fa69a765a3d2d94e8dd3172203ac61d7c9cc53f22c22a2e6be98ac67925dd13f3e2a4824b8a52b564a895a5c9df8a (accessed on 1 February 2024).
39. Wang, P.Q.; Zhou, S.L. (Eds.) *Drilling Fluid and its Interaction Theory*; Petroleum Industry Press: Beijing, China, 2003.
40. Bai, M.; Green, S.; Suarez-Rivera, R. Effect of Leakoff Variation on Fracturing Efficiency for Tight Shale Gas Reservoirs. In Proceedings of the Alaska Rocks 2005, The 40th U.S. Symposium on Rock Mechanics (USRMS), Anchorage, AL, USA, 25–29 June 2005.
41. Rogers, S.; Elmo, D.; Dunphy, R.; Bearinger, D. Understanding Hydraulic Fracture Geometry and Interactions in the Horn River Basin Through DFN and Numerical Modeling. *Int. J. Sports Med.* **2010**, *31*, 458–462.
42. Rahman, M.M.; Chen, Z.; Rahman, S.S. Experimental Investigation of Shale Membrane Behavior Under Tri-Axial Condition. *Pet. Sci. Technol.* **2005**, *23*, 1265–1282. [CrossRef]

processes MDPI

Article

Experimental Study on Factors Affecting Fracture Conductivity

Fuchun Tian [1], Yunpeng Jia [1], Liyong Yang [1], Xuewei Liu [1], Xinhui Guo [2,*] and Dmitriy A. Martyushev [3,*]

[1] Petroleum Engineering Research Institute, Petrochina Dagang Oilfield Company, Tianjin 300280, China; tianfchun@petrochina.com.cn (F.T.); jiaypeng@petrochina.com.cn (Y.J.); yanglyong@petrochina.com.cn (L.Y.); dg_liuxwei@petrochina.com.cn (X.L.)
[2] College of Environment, Liaoning University, Shenyang 110036, China
[3] Department of Oil and Gas Technologies, Perm National Research Polytechnic University, Perm 614990, Russia
* Correspondence: 15537920191@163.com (X.G.); martyushevd@inbox.ru (D.A.M.)

Abstract: The conductivity of propped fractures following hydraulic fracturing is crucial in determining the success of the fracturing process. Understanding the primary factors affecting fracture conductivity and uncovering their impact patterns are essential for guiding the selection of fracturing engineering parameters. We conducted experiments to test fracture conductivity and analyzed the effects of proppant particle size, closure pressure, and fracture surface properties on conductivity. Using the orthogonal experimental method, we clarified the primary and secondary relationships of the influencing factors on conductivity. The results indicate that proppant particle size, formation closure pressure, and fracture surface properties significantly affect fracture conductivity, with the order of influence being closure pressure > fracture surface properties > proppant particle size. Using large-particle-size proppants effectively increases interparticle porosity and enhances fracture conductivity. However, large-particle-size proppants reduce the number of contact points between particles, increasing the pressure on individual particles and making them more prone to crushing, which decreases fracture conductivity. Proppants become compacted under closure pressure, leading to a reduction in fracture conductivity. Proppant particles can embed into the fracture surface under closure pressure, further impacting fracture conductivity. Compared to non-laminated fracture surfaces, proppant particles are more likely to embed into laminated fracture surfaces under closure pressure, resulting in a greater embedding depth and reduced conductivity.

Keywords: hydraulic fracturing; conductivity; proppant; closure pressure; laminated rock

1. Introduction

Oil and gas resources are the primary energy sources globally, with increasing demand for these resources. With the increasing depths of oil and gas exploration, deep-seated reservoirs are exhibiting characteristics like tightness and low permeability [1–3]. These traits significantly constrain the augmentation of oil and gas production [4–7]. Hydraulic fracturing stands as a crucial technique for augmenting production and recovery rates in single-well operations within unconventional oil and gas reservoirs [4,8,9]. The aim of hydraulic fracturing is to establish a high-conductivity conduit linking the reservoir to the wellbore [10–13]. Proppants, along with high-pressure fracturing fluids, are injected into the fractures. They play a vital role in providing support, preventing the closure of hydraulic fractures, maintaining high conductivity, and facilitating the flow of oil and gas [14–16].

The conductivity of propped fractures directly determines the construction quality and production enhancement effects of hydraulic fracturing [17–19]. However, under high closure pressure conditions, the conductivity of propped fractures can rapidly decrease over time, with reductions of up to 95% [20,21]. Clarifying the key factors that affect fracture proppant performance and revealing their impact patterns are prerequisites for guiding engineering parameter selection. Zhang [22] conducted conductivity tests using an improved modified American Petroleum Institute (API) conductivity cell and found that larger proppant particle sizes and higher proppant concentrations resulted in stronger proppant conductivity. Long-term conductivity tests showed that fracture conductivity could can decrease by 20% within 20 h. In addition to proppant parameters, the embedding of proppants due to high stress is also a critical factor affecting fracture conductivity. Bandara et al. [23] utilized the discrete element numerical simulation method to analyze the impact of proppant embedding. The results indicated that conventional proppants could embed to a depth of up to 88% and were prone to crushing. Wen [21] studied the influence of different proppant concentrations and types on embedding and fracture conductivity using a long-term conductivity evaluation system. The results showed that proppant embedding had a significant detrimental effect on fracture conductivity, reducing it by 87.5%. Higher proppant concentrations were found to effectively prevent the reduction in fracture conductivity. Hou [24] discovered through conductivity testing experiments that fracture conductivity was directly proportional to proppant concentration but had a more complex relationship with proppant particle size. Within the closure pressure range of 10–40 MPa, fracture conductivity was directly proportional to proppant particle size. However, at higher closure pressures, proppant particles were more prone to breakage, resulting in debris that reduced proppant porosity and blocked the flow pathways for oil and gas. Consequently, fracture conductivity was inversely proportional to proppant particle size. It can be observed that factors such as proppant particle size, closure pressure, and embedding level can all influence the conductivity of propped fractures. Furthermore, the positional relationship between fracture surfaces and joint surfaces should not be overlooked as it affects fracture conductivity. During the hydraulic fracturing process, fractures can be generated parallel or perpendicular to bedding planes. Differences in fracture surface properties can lead to variations in proppant embedding levels, thereby affecting the conductivity of propped fractures [25]. Lastly, understanding the sensitivity of conductivity to various influencing factors is of great significance in guiding the selection of engineering parameters for hydraulic fracturing [26].

Therefore, we focus on fracture conductivity and analyze the impact of proppant particle size, closure pressure, and fracture surface properties through conductivity testing experiments. The objective is to elucidate the primary and secondary relationships among factors influencing conductivity, aiming to establish a theoretical foundation for the selection of engineering parameters in hydraulic fracturing.

2. Experiment

2.1. Material

To assess the impact of proppant particle size on fracture conductivity, we conducted conductivity testing experiments utilizing quartz sand with particle sizes ranging from 106 to 212 μm (70–140 mesh), 212 to 425 μm (40–70 mesh), and 300 to 600 μm (30–50 mesh). The metal samples used exhibited significantly higher strength compared to quartz sand, rendering it challenging for the sand to embed into the metal samples under closure pressure. Consequently, we conducted experiments using both metal samples and rock samples to analyze the influence of embedding depth on fracture conductivity. Moreover, fractures may exhibit surfaces that are either vertical or parallel to bedding structures, leading to variations in fracture surface properties that influence proppant embedding and consequently alterations in fracture conductivity. We conducted conductivity testing experiments using rock samples with both vertical and horizontal bedding orientations to investigate the effects of embedding depth and bedding structures on fracture conductivity. The experimental samples were collected from the Kongdian Formation in the Cangdong Sag of the Huanghua Depression, Bohai Bay Basin. The processing and treatment of the test rock samples strictly adhered to the standard NB/T 14023-2017 [27], titled "Recommended Method for Long-term Conductivity Testing of Shale Proppant Filling Beds". The dimensions of the rock samples were as follows: length 17.75 ± 0.04 cm, width 3.75 ± 0.05 cm, and height varying between 0.9 and 1.5 cm, as depicted in Figure 1.

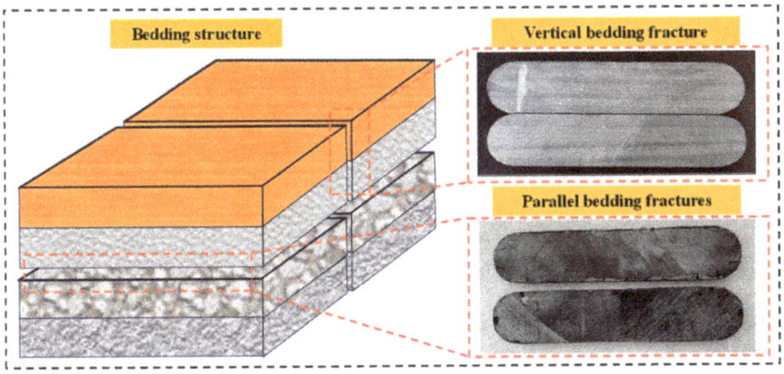

Figure 1. Vertical bedding fracture and parallel joint fracture.

2.2. Experimental Apparatus

In this experiment, the FCMS-V flow conductivity measurement system was employed to assess the flow conductivity of the propped fractures. The experimental apparatus is depicted in Figure 2a–c. The instrument can test fluid flow rates from 0 to 100 mL/min, with a maximum test pressure of 10 MPa, an operational temperature range up to 200 °C, and a maximum closure pressure of 150 MPa. The thickness of the proppant ranges from 0.25 to 1.27 cm. The measurement apparatus primarily comprises a pressure loading system and a flow conductivity testing system. The pressure loading system includes a pressure loading machine and a pressure compensation system, as depicted in the schematic diagram in Figure 2b. The hydraulic pressure sensor on the pressure machine is utilized to measure the hydraulic pressure applied, which can be converted into the load borne by the platens and the closure pressure of the proppant. The flow conductivity testing system primarily includes an advection pump, a conductivity cell, and a data acquisition control system. The conductivity cell (Figure 2c) is designed for linear flow, and the proppant placement area measures 64.5 cm^2. The design and production strictly conform to the standards of the American Petroleum Institute (API). The data acquisition control system includes differential pressure sensors, temperature sensors, displacement sensors, and more. The differential pressure sensor is employed to measure the fluid pressure difference across the proppant layer within the platens. The temperature sensor is utilized to measure the heating temperature of the platens and the intermediate container, allowing the simulation of reservoir temperature by setting the temperature manually. The displacement sensor is employed for measuring the width of the fracture.

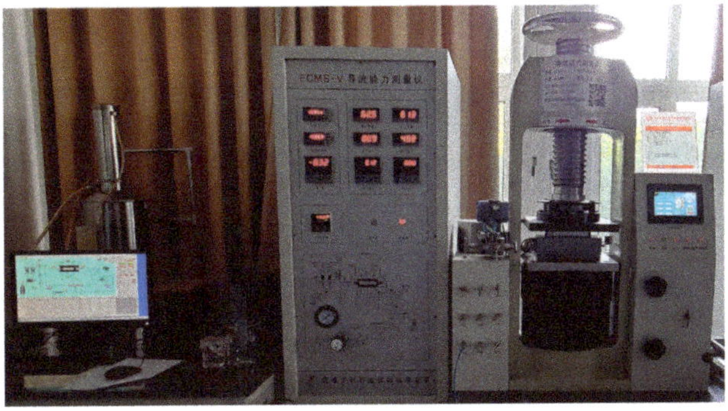

(**a**)

Figure 2. *Cont.*

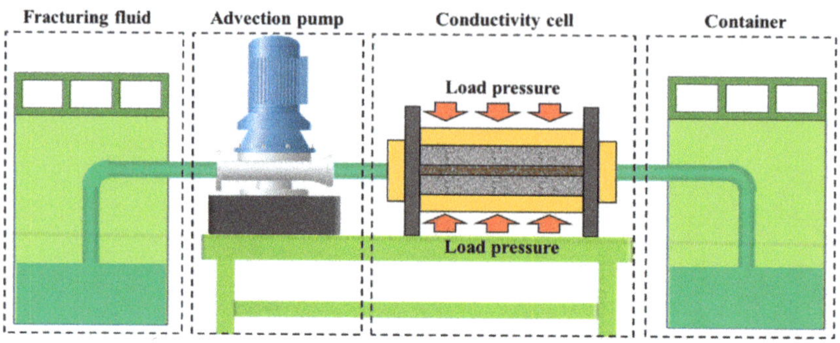

(**b**)

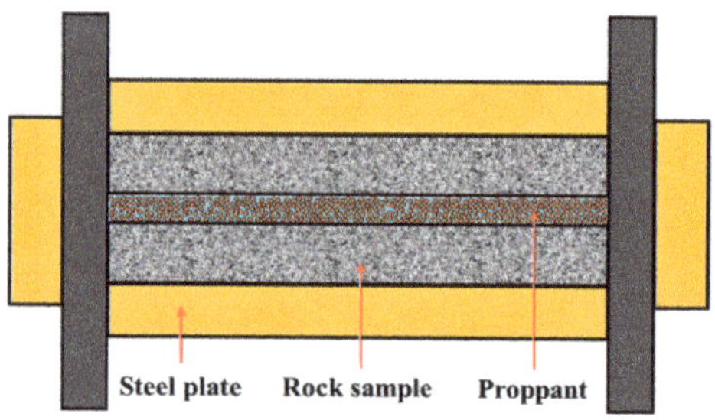

(**c**)

Figure 2. FCMS-V flow conductivity measurement system: (**a**) FCMS-V flow conductivity measurement experimental system; (**b**) Schematic diagram of FCMS-V flow conductivity measurement experimental system; (**c**) Conductivity cell.

After placing the rock sample into the conductivity cell, the gaps between the rock sample and the sidewalls of the conductivity cell were filled with room-temperature vulcanizing organic silicon adhesive. This step was taken to prevent the proppant from infiltrating the gaps between the rock sample and the inner walls of the conductivity cell, thus minimizing potential experimental errors. Following the placement of both upper and lower rock samples, the depths at both ends of the rock sample were meticulously adjusted to maintain a difference in depth of less than 0.1 mm.

2.3. Experimental Setup

According to the requirements of standard NB/T 14023-2017 [27], the testing duration for long-term conductivity should be within the range of 0 h to 50 h $\pm$ 2 h. To analyze the impact of proppant particle size, closure pressure, and fracture surface material on propped fracture conductivity, a series of 52 h long-term conductivity tests were conducted for each variable. The experimental schedule is elaborated in Table 1. To mitigate the impact of other variables on the experimental outcomes, proppant concentration was held constant at 10 kg/m^2, temperature at 100 °C, and flow rate at 2 mL/min.

The properties of the fracture surface can affect the depth of proppant embedding, thereby altering fracture conductivity. Therefore, following experiments 1-7 to 1-9, calculations were performed to assess the depth of proppant embedding and analyze the influence of fracture surface properties on conductivity. The determination of proppant embedment depth involves the following calculations:

(1) Assuming the rigid steel sample remains non-deformable, calculate the deformation of the crack width under various closure pressures by subtracting the initial crack width (W_{01}) from the crack width after the conductivity test (W_{02}). This calculation yields the deformation $D_0 = W_{01} - W_{02}$.

(2) In situations where the rock sample is without stratification, compute the crack deformation as $D_1 = W_{11} - W_{12}$, with W_{11} representing the initial crack width and W_{12} denoting the crack width after the conductivity test. For scenarios involving a laminated rock sample, determine the deformation of the crack width as $D_2 = W_{21} - W_{22}$, where W_{21} signifies the initial crack width, and W_{22} represents the crack width after the conductivity test.

(3) Calculate the embedment depth of the proppant as follows: for the case of non-laminated rock samples, the embedment depth is $D_{0\text{-}1} = D_1 - D_0$, and for the case of laminated rock samples, the embedment depth is $D_{0\text{-}2} = D_2 - D_0$.

Fracture conductivity is collectively influenced by a combination of various factors. To elucidate the influence of individual factors on conductivity performance, an orthogonal experimental design was employed using fracture equilibrium conductivity as the evaluation criterion. This design considered particle size, closure pressure, and fracture surface properties as variables. The levels of these three factors are presented in Table 2, and a comprehensive orthogonal experimental design is outlined in Table 3.

Table 1. Multi-factor sequential conductivity testing experiments.

Experiment Number	Particle Size (mesh)	Closure Pressure (MPa)	Fracture Surface Material
1-1 1-2 1-3	30–50 40–70 70–140	30	Steel plate
1-4 1-5 1-6	30–50	30 40 50	Steel plate
1-7 1-8 1-9	30–50	30	Steel plate Non-laminated rock sample Laminated rock sample

Table 2. Levels of factors in orthogonal experiment.

Levels	Particle Size (mesh)	Closure Pressure (MPa)	Fracture Surface Material
1	30–50	30	Steel plate
2	40–70	40	Non-laminated rock sample
3	70–140	50	Laminated rock sample

Table 3. Orthogonal experimental design for conductivity testing.

Experiment Number	Particle Size (mesh)	Closure Pressure (MPa)	Fracture Surface Material
2-1	30–50	30	Steel plate
2-2	30–50	40	Laminated rock sample
2-3	30–50	50	Non-laminated rock sample
2-4	40–70	30	Laminated rock sample
2-5	40–70	40	Non-laminated rock sample
2-6	40–70	50	Steel plate
2-7	70–140	30	Non-laminated rock sample
2-8	70–140	40	Steel plate
2-9	70–140	50	Laminated rock sample

2.4. Experimental Procedure

The experimental procedure consists of five main steps: the calibration of the conductivity cell, preparation of the conductivity cell, evacuation, conductivity testing, and heating. Each step is carefully controlled to ensure the accuracy and reliability of the results. Automated systems are used to regulate pressure and temperature, and continuous monitoring is employed to detect any deviations and make real-time adjustments.

Step 1: Calibration of the conductivity cell

In conditions where no proppant is present in the conductivity cell, raise the closure pressure by 30 MPa and record the initial reading of the displacement sensor. Then, progressively elevate the closure pressure to 40 and 50 MPa while recording displacement measurements at each pressure level. These measurements form the baseline values for determining proppant fill thickness.

Step 2: Preparation of the conductivity cell

Add the specified proppant mass into the conductivity cell, followed by elevating the closure pressure to the prescribed level.

Step 3: Evacuation

Utilize a vacuum pump to remove air from the conductivity cell and connected pipes. Once the evacuation process is finished, close the valves.

Step 4: Conductivity testing

Initiate the advection pump to introduce the test medium into the conductivity cell, configuring parameters such as closure pressure, test flow rate, and pressurization duration. Automated data recording for conductivity and displacement ensues.

Step 5: Heating

Power on the heating switch and adjust to the desired temperature. The temperature controller will autonomously maintain the conductivity cell's temperature at the specified level.

3. Experimental Results and Analysis

3.1. Proppant Particle Size

Figure 3 illustrates the influence of varying proppant particle sizes on fracture conductivity at a closure pressure of 30 MPa. It is evident that with an increase in the test duration, overall fracture conductivity demonstrates a declining pattern. Initially, fracture conductivity decreases rapidly, but it gradually stabilizes with time. Furthermore, the particle size of the proppant exerts a notable influence on fracture conductivity. For particle sizes within the 30–50 mesh range, the initial fracture conductivity measures 16.3 μm^2·cm, and the stabilized conductivity is 12.25 μm^2·cm, representing a decrease of 4.05 μm^2·cm. In the case of particle sizes ranging from 40 to 70 mesh, the conductivity diminishes from 13.3 μm^2·cm to 11.11 μm^2·cm, reflecting a decrease of 2.19 μm^2·cm. In cases where the particle size falls within the 70–140 mesh range, the initial fracture conductivity reduces from 8.64 μm^2·cm to 6.42 μm^2·cm, resulting in a decrease of 2.22 μm^2·cm. It is evident that as the proppant particle size increases, long-term fracture conductivity improves. Nevertheless, it is important to note that the increase in proppant particle size also leads to a more pronounced reduction in the conductivity of the supported fracture under the influence of closure pressure.

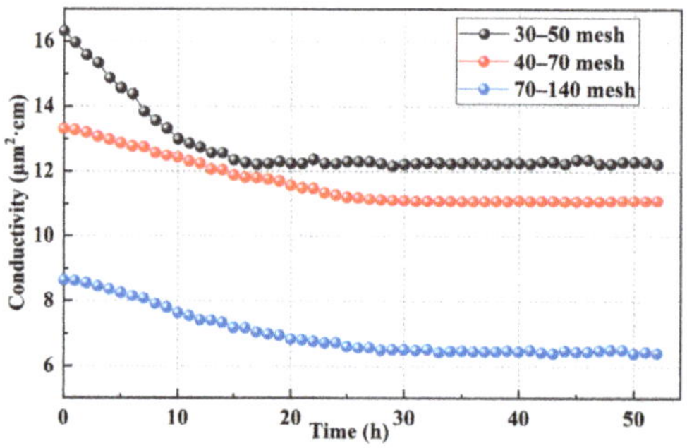

Figure 3. Effect of particle size on fracture conductivity.

During the initial phases of the experiment, the proppant experiences compaction and crushing due to the applied closure pressure [28]. Fine particles generated from the fragmentation of proppant grains are pressed together, resulting in the obstruction of pores and channels that facilitate fluid and gas transport. Consequently, this leads to a general reduction in the permeability and conductivity of the fracture. With the ongoing application of closure pressure, proppant grains undergo additional compaction, resulting in a minor decline in the overall conductivity of the fracture. Once the proppant grains attain a stable condition, their conductivity remains constant.

The utilization of larger-sized proppants effectively enhances intergranular porosity, resulting in the creation of more efficient flow pathways and consequently an improvement in fracture conductivity. Nonetheless, larger-sized support agents result in fewer contact points between particles, leading to heightened pressure on individual particles. This increased pressure facilitates support agent particles reaching their compressive strength and consequently undergoing fracture. These resulting fragments become compacted due to the impact of closure pressure, leading to a more tightly packed configuration of proppants. This, in turn, results in a swift decline in fracture conductivity. As the particle size increases, the consequences of fragmentation and compaction become increasingly prominent, resulting in a more substantial reduction in fracture conductivity. Nonetheless, enlarging the proppant particle size can still effectively enhance fracture conductivity.

3.2. Closure Pressure

Figure 4 depicts the impact of closure pressure on fracture conductivity under the condition of a particle size range of 30–50 mesh. It is evident that as the closure pressure increases, fracture conductivity consistently experiences a declining trend. At a closure pressure of 30 MPa, the initial fracture conductivity measures 16.31 $\mu m^2 \cdot cm$, and the stabilized conductivity is 12.25 $\mu m^2 \cdot cm$, representing a reduction of 24.8%. With the closure pressure increased to 40 MPa, the initial fracture conductivity decreases to 10.1 $\mu m^2 \cdot cm$, while stabilized conductivity reaches 5.56 $\mu m^2 \cdot cm$, indicating a reduction of 44.9%. Moreover, raising the closure pressure to 50 MPa leads to a decrease of 62.7% in fracture conductivity. With increasing closure pressure, the decrease in fracture conductivity becomes more significant. This phenomenon can be attributed to the rapid compaction of the porous medium created by proppant particles due to the effects of closure pressure. This leads to a reduction in the dominant flow channels and causes certain proppant particles to experience stress levels exceeding their compressive strength, leading to fragmentation. These smaller proppant fragments further obstruct transport pathways within the porous medium, resulting in a swift initial decline in fracture conductivity [29]. Higher closure pressure elevates the stress encountered by proppant particles, leading to a greater proportion of fractured proppant particles, decreased proppant porosity, and a more substantial reduction in fracture conductivity.

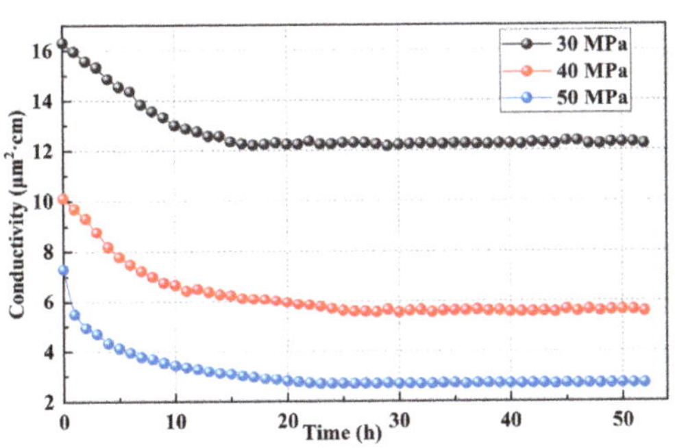

Figure 4. Effect of closure pressure on fracture conductivity.

Equilibrium conductivity is a critical parameter that determines the effectiveness of fracturing treatments. With an increase in closure pressure from 30 MPa to 40 MPa, the equilibrium conductivity decreases from 12.25 $\mu m^2 \cdot cm$ to 5.56 $\mu m^2 \cdot cm$, leading to a reduction of 6.69 $\mu m^2 \cdot cm$. Furthermore, when the closure pressure is raised from 40 MPa to 50 MPa, the equilibrium conductivity decreases from 5.56 $\mu m^2 \cdot cm$ to 2.72 $\mu m^2 \cdot cm$, indicating a reduction of 2.84 $\mu m^2 \cdot cm$. It is evident that augmenting the closure pressure can result in a decrease in long-term fracture conductivity. Nonetheless, as the closure pressure continues to rise, its restraining impact on fracture conductivity gradually lessens. This is because, as the closure pressure increases, the number of fractured particles gradually decreases. Furthermore, proppant particles become more densely packed, reducing the space available for the compression impact of the proppant. As a result, the inhibitory effect of closure stress on fracture conductivity gradually diminishes.

3.3. Fracture Surface Properties

Figure 5 illustrates the influence of fracture surface properties on fracture conductivity. At a closure pressure of 30 MPa and using a proppant in the 30–50 mesh size range, the utilization of steel as the fracture surface results in a long-term fracture conductivity of 12.2 $\mu m^2 \cdot cm$. Conversely, when the fracture surface consists of non-laminated rock samples, the long-term fracture conductivity measures 6.96 $\mu m^2 \cdot cm$. In scenarios where the fracture surface is composed of laminated rock samples, the long-term fracture conductivity amounts to 4.38 $\mu m^2 \cdot cm$. It is clear that variations in fracture surface properties exert a substantial influence on fracture conductivity.

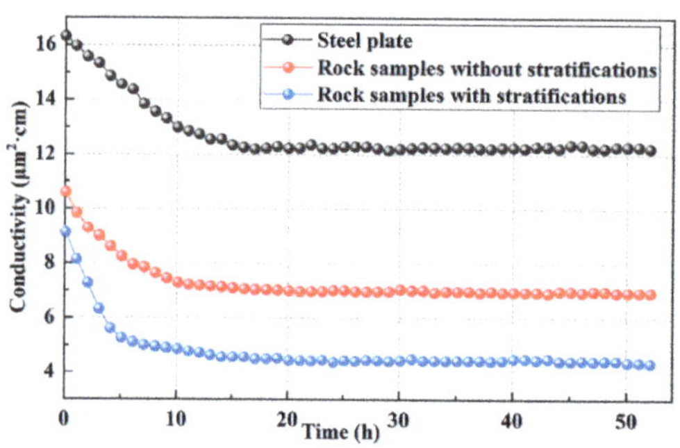

Figure 5. Effect of fracture surface properties on fracture conductivity.

To further investigate the factors contributing to the changes in fracture conductivity, electron microscopy scanning experiments were performed on the surfaces of rock samples both with and without layered structures following the experiments. The experimental results are shown in Figure 6a,b. It is evident that the proppant can become embedded within the fracture surfaces due to the effects of closure pressure. Moreover, disparities in fracture surface properties can result in differences in the extent of proppant embedment, consequently impacting fracture conductivity. Owing to the considerably higher strength of the steel sample in comparison to that of the proppant, there is minimal embedding of the proppant on the steel sample surface. In cases where non-laminated rock samples serve as the fracture surface, a portion of the proppant particles become embedded due to the influence of the closure pressure (Figure 6a). However, when experiments are carried out using laminated rock samples, the existence of layered surfaces results in a higher quantity and greater depth of proppant embedment (Figure 6b). This heightened level of proppant embedment inevitably leads to a decrease in fracture width, consequently reducing fracture conductivity [30–36].

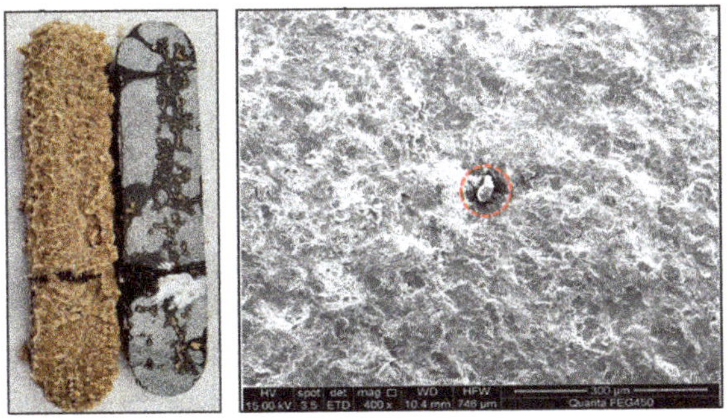

(a)

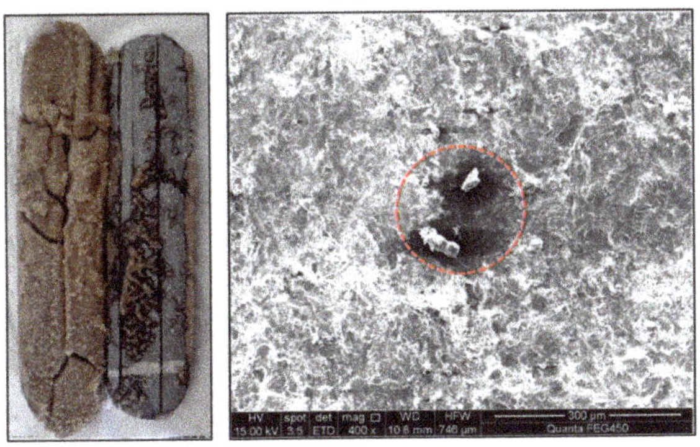

(b)

Figure 6. Electron microscopy scanning experiment results on rock samples. (The red circle in the figure indicates the proppant embedded in the rock sample): (a) Electron microscope scanning results of non-laminated rock sample; (b) Electron microscope scanning results of laminated rock sample.

Hence, in order to elucidate the impact of proppant embedment on crack width and conductivity in diverse crack surface conditions, we compared the variations in support thickness before and after conducting experiments with steel samples and rock samples. Subsequently, we calculated the depth of proppant embedment. Following this, we subtracted the equilibrium conductivity of the steel sample from that of the rock sample, both with and without bedding planes, to quantify the effect of proppant embedment on conductivity.

The impact of fracture surface properties on proppant embedment depth and conductivity is presented in Table 4 and Figure 7. The proppant embedment effect is negligible when using steel samples in these experiments. Hence, under a closure pressure of 30 MPa, the equilibrium conductivity is 12.25 $\mu m^2 \cdot cm$ in the absence of any embedment effect. For non-laminated rock samples, the embedment depth measures 0.048 mm, resulting in a decreased equilibrium conductivity of 6.96 $\mu m^2 \cdot cm$. Conversely, when employing laminated rock samples, it was observed that the embedding depth increased to 0.05 mm and the equilibrium conductivity was reduced to 6.28 $\mu m^2 \cdot cm$. It is evident that variations in fracture surface properties result in differences in proppant embedment depth, significantly impacting fracture conductivity. In contrast to non-laminated rock samples, the proppant exhibits a heightened propensity to embed within laminated rock samples under the influence of the closure pressure, ultimately resulting in a greater embedment depth. This heightened embedment depth in turn narrows the fracture width, causing a corresponding decrease in fracture conductivity.

Table 4. Embedment depths of the proppant under different fracture surface conditions.

Fracture Surface Material	Fracture Width Parameters (mm)	Value (mm)
Steel plate	W_{01} W_{02} D_0	5.724 5.607 0.117
Non-laminated rock sample	W_{11} W_{12} D_1 $D_{0\text{-}1}$	5.74 5.575 0.165 0.048
Laminated rock sample	W_{21} W_{22} D_2 $D_{0\text{-}2}$	5.758 5.591 0.167 0.05

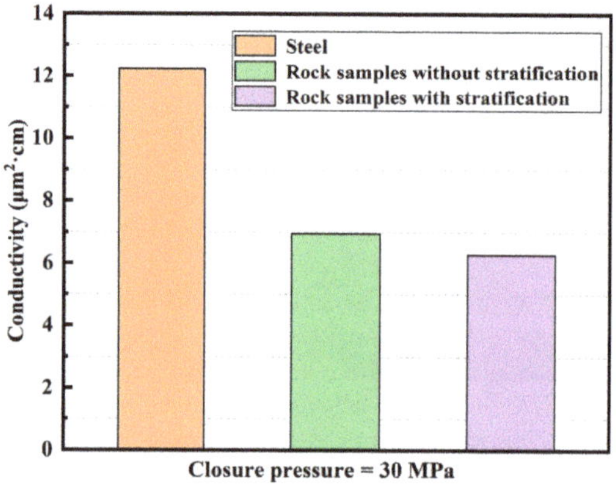

Figure 7. Fracture conductivity under different fracture surface conditions.

3.4. Analysis of Orthogonal Experimental Results

The results of the orthogonal experiments are presented in Table 5, where "R" signifies the range of each factor. The value of "R" is used to assess the impact of proppant sizes, closure pressure, and fracture surface properties on conductivity. It can be observed that the range of closure pressure is the largest, measuring 4.76 MPa. This indicates closure pressure as the primary factor impacting conductivity. The ranges for the fracture surface properties and particle sizes are relatively similar, measuring 3.639 MPa and 3.604 MPa, respectively. Hence, the order of influence on conductivity can be summarized as follows: closure pressure > fracture surface properties > particle sizes. Notably, closure pressure assumes a central role in compacting proppant particles within the fracture, thereby diminishing dominant flow pathways. This renders closure pressure the primary determinant of fracture conductivity. The fracture surface properties wield the capacity to influence the depth of proppant embedment, subsequently leading to variations in fracture width. Additionally, particle sizes can impact the porosity of the proppant layer, both of which have an impact on fracture conductivity to some extent.

Table 5. Orthogonal experimental scheme of conductivity test.

Experiment Number	Particle Size (Mesh)	Closure Pressure (MPa)	Fracture Surface Material	Conductivity ($\mu m^2 \cdot cm$)
2-1	30–50	30	Steel plate	12.25
2-2	30–50	40	Laminated rock sample	3.601
2-3	30–50	50	Non-laminated rock sample	2.154

Table 5. *Cont.*

Experiment Number	Particle Size (Mesh)	Closure Pressure (MPa)	Fracture Surface Material	Conductivity ($\mu m^2 \cdot cm$)
2-4	40–70	30	Laminated rock sample	4.211
2-5	40–70	40	Non-laminated rock sample	2.273
2-6	40–70	50	Steel plate	2.5935
2-7	70–140	30	Non-laminated rock sample	3.0865
2-8	70–140	40	Steel plate	3.587
2-9	70–140	50	Laminated rock sample	0.519
K_1	18.005	19.548	18.431	
K_2	9.078	9.461	7.514	
K_3	7.193	5.267	8.331	
k_1	6.001	6.516	6.143	
k_2	3.026	3.154	2.504	
k_3	2.398	1.756	2.777	
R	3.604	4.760	3.639	

4. Conclusions

(1) The use of larger-particle-size proppants can effectively increase interparticle porosity, thereby enhancing fracture conductivity. However, larger-particle-size proppants reduce the contact points between particles, resulting in an increased individual particle load-bearing pressure and a higher likelihood of particle breakage. This leads to a denser arrangement of proppants, causing a rapid decline in fracture conductivity. As the particle size increases, the impacts of particle breakage and compaction become more pronounced, resulting in a more significant decrease in fracture conductivity. Nevertheless, increasing the particle size of proppants can still effectively enhance fracture conductivity.

(2) Higher closure pressures can initially decrease long-term fracture conductivity. Nevertheless, as the closure pressure continues to rise, the available space for proppant compaction diminishes, gradually reducing its ability to inhibit fracture conductivity.

(3) Proppant particles can embed into the fracture surface due to closure pressure, and variations in fracture surface properties can lead to differences in embedment depth, consequently impacting fracture conductivity. Compared to non-laminated rock samples, proppant particles are more likely to embed into laminated rock samples under closure pressure, resulting in a greater embedment depth.

(4) Closure pressure, fracture surface properties, and particle size can all influence fracture conductivity, with respective ranges of 4.76, 3.639, and 3.604. The order of influence of these factors on fracture conductivity is as follows: closure pressure > fracture surface properties > particle size.

(5) Due to constraints, we conducted experiments only at the laboratory scale. Future plans include conducting large-scale field experiments involving hydraulic fracturing with proppant injections to validate and analyze the experimental results.

Author Contributions: Methodology, F.T.; validation, D.A.M. and X.G.; investigation, D.A.M., X.G., and F.T.; data curation, D.A.M. and X.G.; writing—original draft preparation, F.T., L.Y., and Y.J.; writing—review and editing, F.T.; project administration, X.L.; funding acquisition, F.T. All authors have read and agreed to the published version of the manuscript.

Funding: This research was funded by the National Key Research and Development Program of China (No. 2022YFF0801204) and the Ministry of Science and Higher Education of the Russian Federation (Project No. FSNM-2024-0005).

Data Availability Statement: The raw/processed data required to reproduce the above findings cannot be shared at this time as the data also form part of an ongoing study.

Conflicts of Interest: Authors Fuchun Tian, Yunpeng Jia, Liyong Yang and Xuewei Liu were employed by the company Petrochina Dagang Oilfield Company. The remaining authors declare that the research was conducted in the absence of any commercial or financial relationships that could be construed as a potential conflict of interest.

References

1. Zou, C.; Zhu, R.; Liu, K.; Su, L.; Bai, B.; Zhang, X.; Yuan, X.; Wang, J. Tight Gas Sandstone Reservoirs in China: Characteristics and Recognition Criteria. *J. Pet. Sci. Eng.* **2012**, *88*, 82–91. [CrossRef]
2. Zhang, J. Theoretical Conductivity Analysis of Surface Modification Agent Treated Proppant. *Fuel* **2014**, *134*, 166–170. [CrossRef]
3. Gu, M.; Kulkarni, P.; Rafiee, M.; Ivarrud, E.; Mohanty, K. Optimum Fracture Conductivity for Naturally Fractured Shale and Tight Reservoirs. *SPE Prod. Oper.* **2016**, *31*, 289–299. [CrossRef]
4. Liu, Y.; Zheng, X.; Peng, X.; Zhang, Y.; Chen, H.; He, J. Influence of Natural Fractures on Propagation of Hydraulic Fractures in Tight Reservoirs during Hydraulic Fracturing. *Mar. Pet. Geol.* **2022**, *138*, 105505. [CrossRef]
5. Bolintineanu, D.S.; Rao, R.R.; Lechman, J.B.; Romero, J.A.; Jove-Colon, C.F.; Quintana, E.C.; Bauer, S.J.; Ingraham, M.D. Simulations of the Effects of Proppant Placement on the Conductivity and Mechanical Stability of Hydraulic Fractures. *Int. J. Rock Mech. Min. Sci.* **2017**, *100*, 188–198. [CrossRef]
6. Wang, X.; Zhang, X.; Zhang, M.; Zhang, Q.; Dong, P.; Ding, H.; Liu, X. Proppant Settlement and Long-Term Conductivity Calculation in Complex Fractures. *ACS Omega* **2024**, *9*, 12789–12800. [CrossRef] [PubMed]
7. Li, Y.; Peng, G.; Tang, J.; Zhang, J.; Zhao, W.; Liu, B.; Pan, Y. Thermo-Hydro-Mechanical Coupling Simulation for Fracture Propagation in CO_2 Fracturing Based on Phase-Field Model. *Energy* **2023**, *284*, 128629. [CrossRef]
8. Han, X.; Feng, F.-P.; Zhang, X.-C.; Cao, J.; Zhang, J.; Suo, Y.; Yan, Y.; Yan, M.-S. An Unequal Fracturing Stage Spacing Optimization Model for Hydraulic Fracturing That Considers Cementing Interface Integrity. *Pet. Sci.* **2023**, *20*, 2165–2186. [CrossRef]
9. Zou, J.; Jiao, Y.-Y.; Tan, F.; Lv, J.; Zhang, Q. Complex Hydraulic-Fracture-Network Propagation in a Naturally Fractured Reservoir. *Comput. Geotech.* **2021**, *135*, 104165. [CrossRef]
10. Fan, M.; Han, Y.; Gu, M.; McClure, J.; Ripepi, N.; Westman, E.; Chen, C. Investigation of the Conductivity of a Proppant Mixture Using an Experiment/Simulation-Integrated Approach. *J. Nat. Gas Sci. Eng.* **2020**, *78*, 103234. [CrossRef]
11. Cong, Z.; Li, Y.; Liu, Y.; Xiao, Y. A New Method for Calculating the Direction of Fracture Propagation by Stress Numerical Search Based on the Displacement Discontinuity Method. *Comput. Geotech.* **2021**, *140*, 104482. [CrossRef]
12. Li, Y.; Hubuqin; Zhang, J.; Yang, H.; Zeng, B.; Xiao, Y.; Liu, J. Optimization Method of Oriented Perforation Parameters Improving Uneven Fractures Initiation for Horizontal Well Fracturing. *Fuel* **2023**, *349*, 128754. [CrossRef]
13. Li, Y.; Long, M.; Zuo, L.; Li, W.; Zhao, W. Brittleness Evaluation of Coal Based on Statistical Damage and Energy Evolution Theory. *J. Pet. Sci. Eng.* **2019**, *172*, 753–763. [CrossRef]
14. Sun, H.; He, B.; Xu, H.; Zhou, F.; Zhang, M.; Li, H.; Yin, G.; Chen, S.; Xu, X.; Li, B. Experimental Investigation on the Fracture Conductivity Behavior of Quartz Sand and Ceramic Mixed Proppants. *ACS Omega* **2022**, *7*, 10243–10254. [CrossRef] [PubMed]
15. Chen, Q.; Huang, Z.; Huang, H.; Chen, Q.; Ling, X.; Xin, F.; Kong, X. An Experimental Study on the Impact of the Particle Size and Proportion of Composite Proppant on the Conductivity of Propped Fractures in Coalbed Methane Reservoirs Following Pulverized Coal Fines Infiltration. *Processes* **2023**, *11*, 2205. [CrossRef]
16. Miedzinska, D.; Poplawski, A.; Kwietniewski, M.; Lutynski, M. The Impact of CO2 Injection on Steel Balls Embedment in Shale Rock—Experimental Research. *J. Nat. Gas Sci. Eng.* **2018**, *56*, 619–628. [CrossRef]
17. Fan, M.; Li, Z.; Han, Y.; Teng, Y.; Chen, C. Experimental and Numerical Investigations of the Role of Proppant Embedment on Fracture Conductivity in Narrow Fractures. *SPE J.* **2021**, *26*, 324–341. [CrossRef]
18. Liu, Y.; Guo, J.; Jia, X.; Duan, Y.; Ma, J.; Hu, J. Long Term Conductivity of Narrow Fractures Filled with a Proppant Monolayer in Shale Gas Reservoirs. *J. Eng. Res.* **2017**, *5*, 236–249.
19. Fan, M.; McClure, J.; Han, Y.; Li, Z.; Chen, C. Interaction Between Proppant Compaction and Single-/Multiphase Flows in a Hydraulic Fracture. *SPE J.* **2018**, *23*, 1290–1303. [CrossRef]
20. Yu, J.; Wang, J.; Wang, S.; Li, Y.; Singh, A.; Rijken, P.; Elsworth, D. Conductivity Evolution in Propped Fractures During Reservoir Drawdown. *Rock Mech. Rock Eng.* **2022**, *55*, 3583–3597. [CrossRef]
21. Wen, Q.; Zhang, S.; Wang, L.; Liu, Y.; Li, X. The Effect of Proppant Embedment upon the Long-Term Conductivity of Fractures. *J. Pet. Sci. Eng.* **2007**, *55*, 221–227. [CrossRef]
22. Zhang, J.; Kamenov, A.; Zhu, D.; Hill, A.D. Laboratory Measurement of Hydraulic-Fracture Conductivities in the Barnett Shale. *SPE Prod. Oper.* **2014**, *29*, 216–227. [CrossRef]
23. Bandara, K.M.A.S.; Ranjith, P.G.; Zheng, W.; Tannant, D.D.; De Silva, V.R.S.; Rathnaweera, T.D. Grain-Scale Analysis of Proppant Crushing and Embedment Using Calibrated Discrete Element Models. *Acta Geotech.* **2022**, *17*, 4837–4864. [CrossRef]
24. Hou, T.; Zhang, S.; Ma, X.; Shao, J.; He, Y.; Lv, X.; Han, J. Experimental and Theoretical Study of Fracture Conductivity with Heterogeneous Proppant Placement. *J. Nat. Gas Sci. Eng.* **2017**, *37*, 449–461. [CrossRef]
25. Bandara, K.M.A.S.; Ranjith, P.G.; Rathnaweera, T.D. Laboratory-Scale Study on Proppant Behaviour in Unconventional Oil and Gas Reservoir Formations. *J. Nat. Gas Sci. Eng.* **2020**, *78*, 103329. [CrossRef]
26. Wang, J.; Xiong, Y.; Jiao, G.; Lei, D. Assessment of Long-Term Proppant Conductivity. *Front. Energy Res.* **2022**, *10*, 813571. [CrossRef]

27. *NB/T 14023-2017*; National Energy Administration of the People's Republic of China. Recommended Method for Long-term Conductivity Testing of Shale Proppant Filling Beds. National Energy Administration: Beijing, China, 2017.
28. Xiang, Z.; Ding, Y.; Ao, X.; Zhong, Z.; Li, Z.; Xiao, H.; Chen, Z.; Xiao, Q.; Ma, X. Experimental Study on Fracturing Fracture Deformation Mechanism of Shale Reservoir. *Front. Energy Res.* **2022**, *9*, 813978. [CrossRef]
29. Liu, Y.; Mu, S.; Guo, J.; Chen, C.; Wang, S. Model for Fracture Conductivity Considering Particle Size Distribution in a Proppant Monolayer. *J. Nat. Gas Sci. Eng.* **2021**, *95*, 104188. [CrossRef]
30. Yan, C.; Zhang, P.; Dong, L.; Zhang, C.; Cheng, Y. Long-Term Fracture Conductivity in Tight Reservoirs. *Energy Sci. Eng.* **2024**, *12*, 1187–1199. [CrossRef]
31. Martyushev, D.A.; Yang, Y.; Kazemzadeh, Y.; Wang, D.; Li, Y. Understanding the Mechanism of Hydraulic Fracturing in Naturally Fractured Carbonate Reservoirs: Microseismic Monitoring and Well Testing. *Arab. J. Sci. Eng.* **2024**, *49*, 8573–8586. [CrossRef]
32. Filippov, E.V.; Zakharov, L.A.; Martyushev, D.A.; Ponomareva, I.N. Reproduction of reservoir pressure by machine learning methods and study of its influence on the cracks formation process in hydraulic fracturing. *J. Min. Inst.* **2022**, *258*, 924–932. [CrossRef]
33. Ponomareva, I.N.; Martyushev, D.A. Evaluation of hydraulic fracturing results based on the analysis of geological field data. *Georesursy* **2020**, *22*, 8–14. [CrossRef]
34. Wang, D.-B.; Zhou, F.-J.; Li, Y.-P.; Yu, B.; Martyushev, D.; Liu, X.-F.; Wang, M.; He, C.-M.; Han, D.-X.; Sun, D.-L. Numerical simulation of fracture propagation in Russia carbonate reservoirs during refracturing. *Pet. Sci.* **2022**, *19*, 2781–2795. [CrossRef]
35. Savelev, V.V.; Ognev, I.N. Sensitivity analysis of the fracturing fluid rheology effect on the hydraulic fracture geometry in the terrigenous reservoirs. *Georesursy* **2023**, *25*, 138–148. [CrossRef]
36. SYamkin, M.A.; Safiullina, E.U.; Yamkin, A.V. Analysis of the results of modeling fluid inflow to a hydraulic fractur-ing. *Bull. Tomsk Polytech. Univ. Geo Assets Eng.* **2024**, *335*, 14–21. [CrossRef]

processes

MDPI

Article

A Numerical Investigation on Kick Control with the Displacement Kill Method during a Well Test in a Deep-Water Gas Reservoir: A Case Study

Qingchao Li [1,2], Qiang Li [2,3,*] and Ying Han [1,*]

[1] School of Energy Science and Engineering, Henan Polytechnic University, Jiaozuo 454000, China; liqingchao2020@hpu.edu.cn
[2] School of Petroleum Engineering, China University of Petroleum (East China), Qingdao 266580, China
[3] College of Engineering, China University of Petroleum-Beijing at Karamay, Karamay 834000, China
[*] Correspondence: liqiang202403@cupk.edu.cn (Q.L.); hyhpu@hpu.edu.cn (Y.H.)

Abstract: The efficient exploitation of marine oil and gas resources holds significant potential to mitigate the current severe energy crisis. Regrettably, incidents, such as gas kick and even blowouts, can significantly impact normal development activities. The displacement kill method is one effective strategy for well control in deep-water areas. In this study, the detailed mathematical method for determining kill parameters involved in the kill operation by using the displacement kill method was proposed. Of course, this includes both cases: one where the kill fluid leaks during the kill process and another where no leakage occurs. Meanwhile, its applicability was verified through comparison with experimental results. Then, evolution characteristics of kill parameters, when killing fluid leakage occurs and when it does not occur, were analyzed. Finally, factors, such as pit gain and shut-in casing pressure, affecting the kill parameters of kill operation, were explored. It was found that the experimental and calculated results show great similarity, although there are slight differences between them. The total kill time in the simulation is 44 s shorter than that in the verification experiment. This indicates that the model established in this study is suitable for simulating the process of kill operation using the displacement kill method. In addition, the investigation results show that leakage of kill fluid increases the difficulty of the kill operation and prolongs the operation time. The number of kill cycles in the presence of kill fluid leakage is one more than that when there is no fluid leakage, resulting in an additional 70 min of total duration. Furthermore, the increase in pit gain and the rise in shut-in casing pressure can also pose challenges to the kill operations. The total kill time will be extended by 164 min when the mud pit gain increases from 20 m^3 to 50 m^3. The number of kill cycles rises by two when the shut-in casing pressure is increased from 5 MPa to 20 MPa. To ensure the safety of the drilling operation in abnormally high-pressure reservoirs, it is crucial to monitor parameters such as casing pressure during the drilling process and timely well control measures.

Keywords: displacement kill method; gas kick; well control; oil and gas production; deep-water area

1. Introduction

It is widely acknowledged that fossil fuels, such as oil and gas, will remain the primary energy sources constraining the development of human society for a long time in the future [1–3]. However, numerous mature onshore oil fields in the central and eastern regions of China have experienced a significant decline in oil production in recent years [4–7]. Taking Daqing Oil Field as an example, its oil production has dropped year by year from 56 million tons in the peak period (in 1996) to 30 million tons in 2022 [8]. It is predicted that its oil production will drop sharply to 8 million tons in 2060 [9–12]. Meanwhile, the increase in production of the newly discovered oil fields in Western China failed to make up for the widespread production decline of these oil fields [13–15]. In recent years, China has had to rely on imports for more than 70% of its crude oil demands and more than 40% of its natural gas demands [16,17]. In 2023, the overseas dependence of crude oil was expected to exceed 80%. It is no exaggeration to say that China's domestic energy security has clearly been threatened [18]. Fortunately, the efficient exploration and development of unconventional oil and gas resources, such as shale gas and offshore oil and gas resources, has become an effective way to alleviate this situation [19–21].

Nevertheless, a series of issues or accidents that endanger the safety of the development operation are extremely likely to occur during oil and gas production in the ocean (especially in deep-water areas) [22–25]. Among them, gas kick (or blowout), which is likely to occur during the drilling operation in high-pressure gas reservoirs, is a typical one [26–28]. This issue is caused by the invasion of high-pressure gas from the reservoir into the wellbore annulus under the negative pressure difference between bottom-hole pressure and reservoir pressure [29,30]. For gas wells in deep-water areas, the circulation pipeline of drilling fluids is extremely long, and high-pressure gas in the annulus continuously expands during its migration to the platform [31,32]. The excessive annular pressure caused by gas expansion not only affects the integrity of the wellbore in weak positions, such as casing shoes, but also leads to blowout. Undoubtedly, the severe gas kick naturally poses a significant threat to the drilling operation. For most ordinary wells, gas kick can be effectively controlled by the conventional well killing methods (such as the driller's method) as long as it was detected in time [33,34]. However, conventional well killing methods are occasionally inadequate for controlling gas kick and blowout occurs in gas wells in deep-water areas. The common conventional methods, such as the driller method, are suitable for drilling operations in areas with simple geological conditions and a well-defined pressure system. However, the geological conditions and pressure systems of oil and gas reservoirs, especially gas reservoirs in deep-water areas, are complex [35,36]. This is especially true in scenarios where the drill string is absent from the wellbore or the drill bit is positioned near the wellhead. In other words, conventional well killing methods are not suitable for well control of oil and gas wells in deep-water areas [37].

Up till now, numerous investigations on the use of unconventional well killing methods for kill operations in deep-water areas have been conducted, and some progress has been made. To name a few, Noynaert and Schubert (2005) simulated blowout control of gas wells in an ultra-deep-water area with the self-developed simulator when dynamic well killing technology was used [38]. This investigation determined the dynamic kill requirements for controlling blowout based on a series of relief well parameters. Yuan et al. (2023) optimized the operational parameters of the kill operation for well kill in a dee-water area based on the dynamic well killing method [11]. It reveals that this method was effective at controlling blowout in worst-case scenarios. Ren et al. (2015) calculated the displacement amount during the dynamic kill process in a deep-water drilling operation after the gas kick occurred [39]. It was found that the investigation can provide theoretical support for solving the complex issue of deep-water gas kick and blowout. The displacement kill method is one of the unconventional well killing methods commonly used in high-pressure gas wells [40]. Figure 1 presents the schematic diagram of the displacement kill method. As can be seen in Figure 1 left, the kill fluid is first injected into the annular space, and its droplets gradually fall to the bottom. Meanwhile, the liquid level in the annulus gradually rises, and gas discharge in the annulus leads to a decrease in annulus pressure (see Figure 1 right). The above-mentioned process is continuously circulated to gradually reduce the annular pressure. This process needs to continue until the wellbore is filled with killing fluid. Unfortunately, few studies on kill operation of high-pressure gas wells in deep-water area based on the displacement killing method have been intensively conducted. Even if there are, most of the relevant investigations are analyses of kill engineering cases that have occurred. It is impossible to provide effective and substantive support for subsequent kill operations at all. Moreover, there are no investigations on a kill strategy for a well test of a high-pressure gas well in a deep-water area.

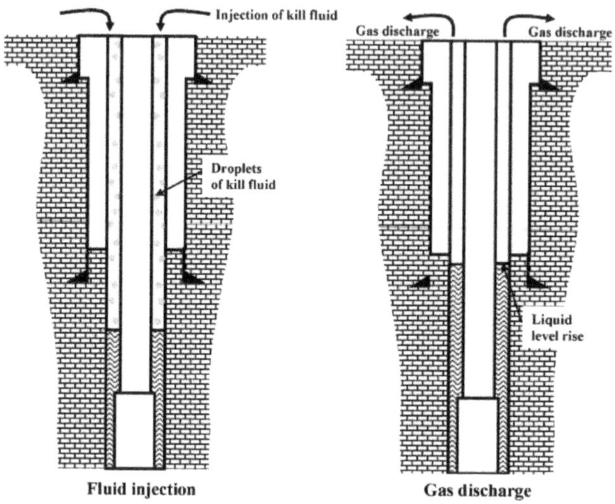

Figure 1. Mechanism diagram of the displacement well killing method.

There are many challenges (such as the narrow safe mud weight window) in the test process of offshore high-pressure gas wells. These challenges can easily lead to gas kick or even blowout, threatening the smooth implementation of the well test operation. At present, there is a scarcity in the number of kill techniques that can be used for reference. Therefore, it is urgent to propose a calculation method for kill parameters during the well test of gas wells in deep-water areas. In the present work, the method for determining kill parameters, such as injection volume, injection pressure, and replacement period during the kill operation with the displacement well kill method, was proposed. Then, the variation patterns of kill parameters mentioned above under different working conditions (such as presence or absence of kill fluid leakage) were simulated. At the same time, its mechanism is analyzed. In fact, the method for determining the well killing parameters used in the well testing process is also applicable to the normal gas reservoir development process, with the only difference being the research objects. Therefore, this study also intends to provide references for well control and the kill operation in normal deep-water oil and gas development.

To highlight the innovation of this study, the highlights are summarized as follows:

(1) The determination method of the parameters related to the kill operation using the displacement method with and without considering leakage has been proposed.
(2) A sensitivity analysis was conducted on the kill case during the well test process of the deep-water gas reservoir.

2. Determination of Kill Parameters in Kill Operation

Three kill parameters need to be determined for the kill operation with the displacement kill method: (1) injection volume and injection pressure, (2) single gas release, and (3) displacement period. In this section, the determination method of kill parameters for two situations, kill fluid leakage and no fluid leakage, were discussed or illustrated separately. In this study, the dissolution of gas in liquids needs to be ignored. In addition, the compressibility of liquids in the annulus is not taken into account.

It is necessary to make some assumptions before numerical modeling. In this study, the following assumptions were made

(1) Throughout the kill operation, both the gas and liquid phases in the wellbore remain macroscopically stationary, and do not flow.

(2) The compressibility of the gas and liquid phases in the wellbore should be ignored. Furthermore, their properties (such as the rheology) are not affected by changes in temperature and pressure.
(3) The rate is constant for both the fluid injection and the gas discharge operations.

2.1. Kill Parameters with Kill Fluid Leakage

(1) Calculation of kill parameters in the first round

If the reservoir is unconsolidated, kill fluid leakage accidents are easy to occur in the kill operation [39,41]. In this case, the injection volume should be properly controlled, and the maximum annulus pressure should be lower than the leakage/fracture pressure [42,43]. The first injection volume of the kill fluid can be determined by Equation (1).

$$\begin{cases} V_{\text{Inj1}} = V_1 - \frac{-B_1+\sqrt{B_1^2+4A_1C_1}}{2A_1} \\ V_1 = \frac{Z_1 nRT_1}{P_1} \\ A_1 = \frac{0.001\rho_{\text{m}}g}{q} \\ B_1 = P_L - \frac{0.001\rho_{\text{m}}gV_1}{q} \\ C_1 = \frac{Z_2P_1V_1}{Z_1} \end{cases} \quad (1)$$

The height of liquid column formed in the annulus by the kill fluid injected for the first time can be calculated.

$$H_1 = \frac{V_{\text{Inj1}}}{q} \quad (2)$$

The hydrostatic pressure generated by the injected kill fluid can be represented as follows:

$$\Delta P_1 = 0.001\rho_{\text{m}}gH_1 \quad (3)$$

During the first round of fluid injection, gas in annulus is discharged from wellhead when kill fluid reaches the bottom of wellbore. The volume of gas (standard state) discharged from annulus in the first round of fluid injection can be determined.

$$V_{\text{dis1}} = \frac{Z_0T_0V_{\text{total}}}{P_0T_1}\left(\frac{P_2}{Z_2} - \frac{(P_{\text{p}} - \Delta P_1)}{Z_{21}}\right) \quad (4)$$

(2) Calculation of kill parameters in the second round

When the bottom-hole pressure is equal to the pore pressure, the first round of gas discharge should be stopped. Then, the second round of kill fluid injection can be started. The amount of kill fluid required to be injected into the annulus in the second round is calculated as follows:

$$\begin{cases} V_{\text{Inj2}} = -\frac{-B_2+\sqrt{B_2^2+4A_2C_2}}{2A_2} \\ A_2 = \frac{0.001\rho_{\text{m}}g}{q} \\ B_2 = P_{\text{L}} - \Delta P_1 - \frac{0.001\rho_{\text{m}}gV_2}{q} \\ C_1 = \frac{Z_3P_{21}V_2}{Z_2} \end{cases} \quad (5)$$

Correspondingly, the height of the liquid column and the hydrostatic pressure generated by injected the kill fluid in the annulus in the second round are expressed as Equations (6) and (7), respectively.

$$H_2 = \frac{V_{\text{Inj2}}}{q} \quad (6)$$

$$\Delta P_2 = 0.001\rho_{\text{m}}gH_2 \quad (7)$$

The volume of gas (standard state) discharged from the annulus in the second round of fluid injection can be determined by Equation (8).

$$V_{\text{dis2}} = \frac{Z_0 T_0 V_3}{P_0 T_1}\left(\frac{P_3}{Z_3} - \frac{(P_p - \Delta P_1 - \Delta P_2)}{Z_{31}}\right) \tag{8}$$

Herein, only the determination method of kill parameters in two displacement cycle is given. Notably, the kill operation needs to continue until all the gases in the annulus are discharged.

(3) Determination of kill times and the kill cycle

The minimum number of well killing operations (c) and the longest kill cycle (t) can be estimated with Equations (9) and (10).

$$c = \frac{H}{H_1} \tag{9}$$

$$t = c(t_1 + t_2) \tag{10}$$

2.2. Kill Parameters without Kill Fluid Leakage

If the reservoir sediments are well cemented, leakage accidents of the kill fluid are less likely to occur during the kill operation [44,45]. In this case, there is no need to consider the risk of kill fluid leakage. However, the situation where maximum annulus pressure (P_{amax}, in MPa) exceeds the pressure-bearing capacity of the wellhead equipment should be avoided. For calculation of the kill parameters in both cases of kill fluid leakage and no leakage, the only difference lies in the calculation method of the fluid injection volume.

(1) Calculation of kill parameters in the first round

The injection volume of kill fluid in the first injection round can be determined by the following:

$$V_{\text{Inj1}} = V_{\text{amax}}\left(1 - \frac{Z_2 P_1}{Z_1 P_{\text{amax}}}\right) \tag{11}$$

Calculation of liquid column height in the annulus (H_1) generated by the injected kill fluid and the hydrostatic pressure can be implemented by Equations (2) and (3), respectively. When the first round of injection of the kill fluid reaches the wellbore bottom, some gas in the annulus needs to be discharged. Similarly, gas volume required to be discharged after the first round of fluid injection can also be determined by Equation (4).

(2) Calculation of kill parameters in the second round

The volume of kill fluid that needs to be injected into annulus in the second round can be expressed as follows:

$$V_{Inj2} = V_{\text{amax}} - \frac{Z_3 P_{21} V_2}{Z_2 P_{amax}} \tag{12}$$

where, all parameters involved in Equation (12) have already been illustrated in Section 2.1.

Then, both height of the liquid column and the hydrostatic pressure generated by the second round of kill fluid injection can be calculated using Equations (6) and (7), respectively. Similarly, the volume of gas discharged from the annulus after the completion of the second round of kill fluid injection can also be obtained through Equation (8).

(3) Determination of kill times and the kill cycle

The minimum number of well killing operations (c) and the longest kill cycle (t) can be estimated with the equations (Equations (9) and (10)) presented in Section 2.1.

Based on the above model, kill parameters can be determined, and the workflow for it is displayed in Figure 2.

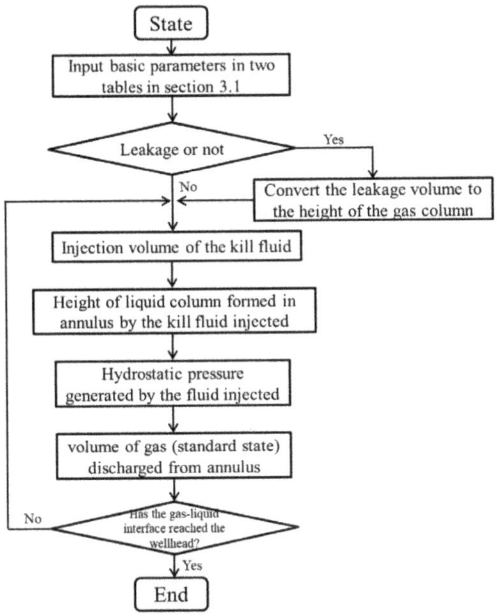

Figure 2. The flow chart for calculation of the kill parameters.

3. Applicability Verification of Investigation Methodology

3.1. Basic Data for Kill Simulation

The investigation well involved in this study is located in the northern slope of Baiyun Sag, Pearl River Mouth Basin of the Northern South China Sea. In this area, abnormal high-pressure gas reservoirs are widely spread, and well control during drilling operation is an important issue [46–48]. For the investigation well, the wellbore structure is shown in Figure 3, and the basic information of both the wellbore and reservoir are displayed in Table 1. During the well test process of this gas well, the TCP (Tubing Conveyed Perforating) + APR (Annulus Pressure Responsive) combined test technology was used. Therefore, the basic data of the test string in the investigation well is shown in Table 2.

Table 1. Basic data for simulating the kill operation.

Parameters		Value
Well depth, m		3550
Water depth, m		186.5
30″ conductor	Size, mm	762
	Setting depth, m	253.4
13-5/8″ casing	Size, mm	340
	Setting depth, m	805
9-7/8″ casing	Size, mm	248.48
	Setting depth, m	3495.5
Reservoir depth, m		3143
Internal pressure strength, MPa		32.98
Casing collapse resistance, MPa		47.38
Permissible casing pressure, MPa		21.0
Surface temperature of seawater, °C		26.6
Geothermal gradient, °C/100 m		3.0
Pore pressure, MPa		32.98
Leakage/fracture pressure, MPa		45.0
Mud weight, g/cm^3		1.18
Initial drilling fluid height, m		2050

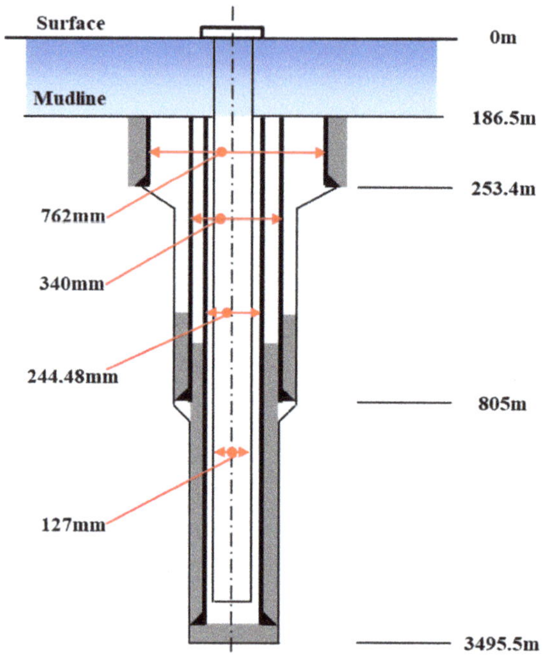

Figure 3. Wellbore structure of a well drilled in the high-pressure gas reservoir in the South China Sea.

Table 2. The string parameters for well test of a gas well in the South China Sea.

No.	Test String	Outer Diameter, mm	Inner Diameter, mm	Length, m	Depth, m
1	Flowhead	-	-	-	−1.11
2	Lubricator valve	311.15	76.20	1.78	10.08
3	5″ drill pipe	127.00	108.60	191.85	201.93
4	Centraliser	311.15	76.20	1.82	203.75
5	Shear joint	127.00	76.20	0.71	204.46
6	Subsea tree	311.15	76.20	1.26	205.72
7	Expansion joint	127.00	57.15	18.08	3227.10
8	RD circulation valve	127.00	56.00	1.73	3286.82
9	OMNI valve	127.00	56.00	7.46	3294.28
10	LPR-N valve	127.00	56.00	4.83	3310.14
11	Safety joint	154.00	60.00	1.23	3318.89
12	RTTS packer	200.00	60.00	2.18	3321.07
13	Perforating gun	177.80	-	18.00	3373.00

In this study, a preliminary program for design and calculation of kill parameters was compiled with VB.NET when the displacement kill method was used, and the kill operation was simulated. With this program, the variation pattern of kill parameters (such as annulus pressure and total injection volume) in the kill operation when the displacement kill method was used were explored. The program interface is shown in Figure 4. From Figure 4, it can be identified that although the interface of this program is simple, it is fully functional. Based on this program, the influence of various factors affecting kick control with the displacement kill method can be explored. Moreover, the simulation results obtained using this program can provide reference for the design of the well killing operation.

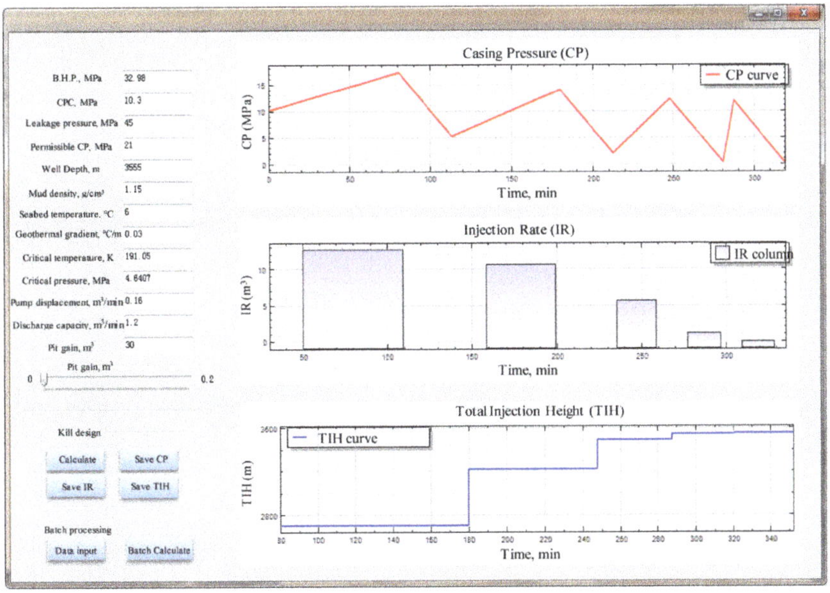

Figure 4. Program interface for calculation of kill parameters in the displacement kill method.

3.2. Applicability of the Investigation Methodology

It is necessary to verify the applicability of both the self-designed computer program and the investigation methodology before using it for a case study. In this section, this purpose is achieved by comparing the results obtained using the program with the experimental results obtained using the self-made well control system. The self-made well control system was illustrated as Figure 5. It can be clearly seen from Figure 5 that the experimental system actually occupies an extremely large area, and it is placed in a laboratory covering an area of 200 m^2. Furthermore, the composition of this experimental system is also extremely complex. As was observed in Figure 5, the well control experimental system mainly includes six subsystems. They are the fluid (gas, liquid) storage system, power system, fluid metering and control system, high-pressure annulus experiment module, gas–liquid separation system and back pressure control system. Among them, the fluid storage system includes a water tank (effective volume: 2 m^3) and a gas tank (pressure: 10 MPa, volume: 100 L). The main component of the power system is a high-pressure air compressor with an outlet pressure of up to 10 MPa. In addition, the flow rate of gas and liquid in the system is realized by an electromagnetic flowmeter whose measurement accuracy is 0.5 L/min. The high-pressure annulus experimental module is a transparent plexiglass tube with a length of 15.0 m placed vertically. The inner diameter of the outer tube within the annular space is 120.0 mm, and the outer diameter of the inner tube is 63.0 mm. Based on this experimental system, the applicability of the investigation methodology in this work was verified.

For ensuring experimental safety, nitrogen (N_2) was injected into the wellbore annulus to approximate the gas kick experienced during the drilling operations. In the experiment, the injection rate of the kill fluid and the discharge rate of gas were 10 L/s and 0.32 m^3/min, respectively. Additionally, the height of the gas phase in the annulus is 12.5 m. In the experiment, the position of the gas–liquid interface in the annulus was measured every 2.5 min. The comparative results between the simulation and experimental data are illustrated in Figure 6. As can be observed in Figure 6, almost all the experimental results are close to the variation curve of the gas–liquid interface obtained by simulation. This demonstrates that the model and investigation methodology proposed in this study are capable of simulating well kill operations using the displacement method. However, the end time of the well kill operation in the experiment was slightly later than that in the simulation. In the simulation, it takes approximately 856 s for the invading gas within the annulus to be completely discharged. However, in the experiment, this time is extended to at least 900 s. This phenomenon can be attributed to the assumption that the injection rate, gas discharge rate, and other parameters were assumed to remain constant throughout the simulation. However, these parameters fluctuated within a certain range in experiment.

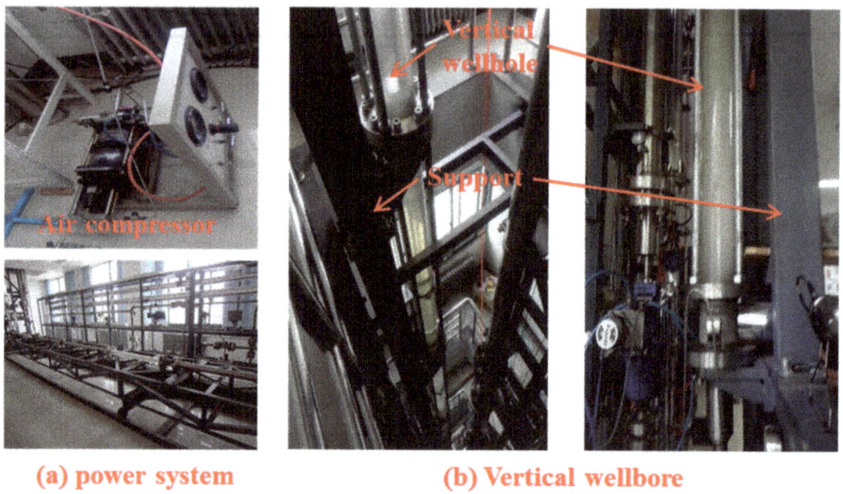

Figure 5. The experimental system for verifying the applicability of investigation methodology.

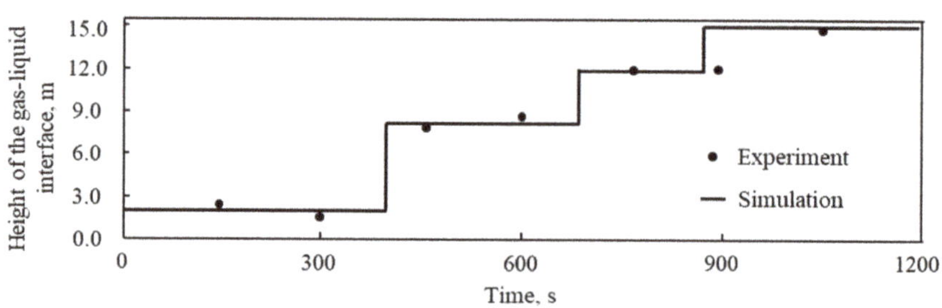

Figure 6. The comparative results between the simulation and experimental data.

4. Evolution Characteristics of Kill Parameters with and without Kill Fluid Leakage

Only through an in-depth analysis of the variation patterns/characteristics of kill parameters during kill operations can the related design of kill operations be executed more effectively. The evolution of casing pressure, injection volume, and liquid column height during kill operations was investigated for the case of considering kill fluid leakage, and the results was presented in Figure 7. As observed in Figure 7, three stages are involved in any kill cycle: the injection of kill fluid, fluid fall, and the discharge of invading gas in the annulus. The function of the first two stages is to allow the kill fluid to be injected into the annulus and occupy part of its space. However, it should be noted that the sum of the hydrostatic pressure and the casing pressure must not exceed the leakage pressure. The final stage of each cycle is to release the compressed gas in the annular space. Notably, when the gas is released in each cycle, the sum of the casing pressure and the hydrostatic pressure cannot be lower than the formation pressure. The duration of the three stages involved in one cycle varies. In the first cycle in Figure 7, the duration of the injection stage is 94.9 min, and the duration of the discharge is only 32.5 min. The latter is obviously faster than the former. The fall of the kill fluid from the wellhead to the gas–liquid interface in the annulus occurs along with the first stage, the duration of this stage is 19.5 min in the first cycle. For the default investigation case, almost all the gas invading the annulus can be completely displaced through four rounds of the above-mentioned kill cycle using the displacement kill method. The entire kill operation takes about 420 min.

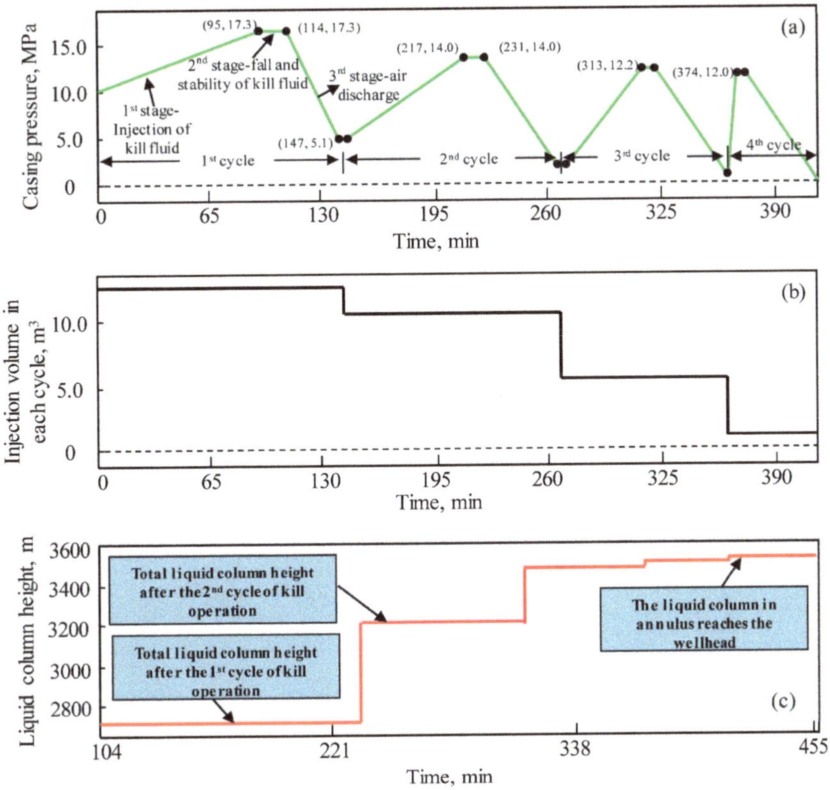

Figure 7. Evolution of (**a**) casing pressure, (**b**) injection volume, and (**c**) liquid column height during the kill operation when kill fluid leakage occurs.

In Figure 7, we can also see that the casing pressure increases continuously during the injection of kill fluid into the annulus in each cycle. This occurs because, as the kill fluid is injected, the space occupied by the gas in the annulus is continuously compressed. Although during this process, a portion of the mixture of the kill fluid and drilling fluid leaks into the reservoir. Taking the first cycle as an example, 13.4 m^3 of kill fluid was injected into the annulus, filling a length of 663 m in it. The gas above the liquid level in the annulus is compressed, and the casing pressure increases from 10 MPa to 17.3 MPa during this process. After the throttle valve at the wellhead is opened, the compressed gas inside the annulus is released, causing the casing pressure to decrease accordingly. At the end of the first cycle, the casing pressure decreased to 5.1 MPa, and the distance between the liquid level and the down-hole increased to 2713 m. With the continuous injection of kill fluid during the subsequent kill operation, the position of the gas–liquid interface in the annulus gradually rises. Following four cycles of kill operations, the gas–liquid interface advanced to a position 12 m from the wellhead. Finally, the 12 m long gas column near the wellhead was discharged by injecting the kill fluid while keeping the valve open.

The highest and lowest values of casing pressure gradually decrease during each injection exhaust cycle of the entire kill operation.

Figure 8 presents the evolution of casing pressure, injection volume and liquid column height during kill operation when no kill fluid leakage occurs. By comparing Figures 7 and 8, it can be clearly observed that the duration of the kill operation is significantly shortened in the absence of kill fluid leakage. Without kill fluid leakage, the gas in the annulus can be completely discharged in just 350 min, which is 70 min shorter than the result displayed in Figure 7. This should be attributed to the lower amount of injected kill fluid when there is no leakage. Consequently, the number of required kill cycles will be reduced. The number of required kill cycles in the absence of kill fluid leakage is four, one fewer than in the presence of leakage. Moreover, another difference between the results in Figures 7 and 8 is the casing pressure after fluid injection in each kill cycle. Without kill fluid leakage, this casing pressure is a constant value (i.e., permissible casing pressure, 21.0 MPa). This is because, in formations where kill fluid leakage is less likely to occur, the limiting factor for the injection volume of kill fluid is the permissible casing pressure. On the premise of ensuring that the bottom-hole pressure is lower than the leakage pressure, the maximum casing pressure should reach its permissible value at the end of each injection of the kill fluid. However, in formations prone to kill fluid leakage, the limiting factor for the injection volume of kill fluid is the leakage pressure. Under the premise that the maximum bottom-hole pressure does not exceed the leakage pressure, the casing pressure gradually decreases with the increase in hydrostatic pressure during the kill operation (see Figure 7).

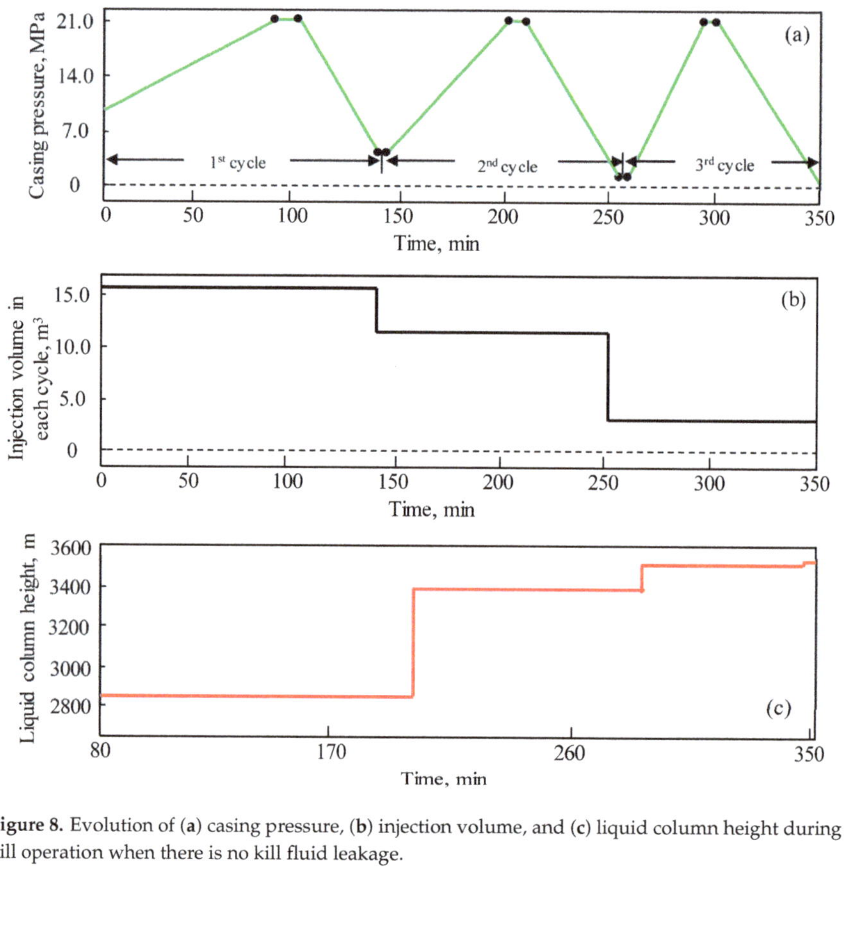

Figure 8. Evolution of (**a**) casing pressure, (**b**) injection volume, and (**c**) liquid column height during kill operation when there is no kill fluid leakage.

5. Sensitivity Analysis and Discussion

5.1. Effect of Pit Gain

Factors such as gas invaded volume and initial casing pressure will seriously affect the construction parameters of kill operations [49,50]. Conducting a comprehensive analysis of these factors and elucidating their mechanisms of influence can offer valuable insights for the engineering design of kill operations. Pit gain refers to the volume of drilling fluid in the annulus that is replaced by the gas kick before shut-in. Generally speaking, this parameter (i.e., pit gain) can indirectly reflect the volume of gas kick in the annulus. A higher pit gain means more gas invaded into the annulus during the drilling operation. Figure 9 illustrates the schematic diagram that reveals the mechanism underlying the discrepancy in pit gain when the amount of invading gas is varied. From what we can see in Figure 9, the amount of drilling fluid in the mud pit is constant, and the pit gain is almost zero throughout the normal drilling operation. However, if a lot of reservoir gas invades the annulus during the drilling operation, there will be a long gas column in the annulus. In this way, the space previously occupied by the drilling mud in the annulus will be occupied by these gas columns. As a result, the drilling mud will be displaced into the mud pit, which will lead to a notable increase in the mud pit gain. Accordingly, more kill fluid is needed to displace and discharge these gases.

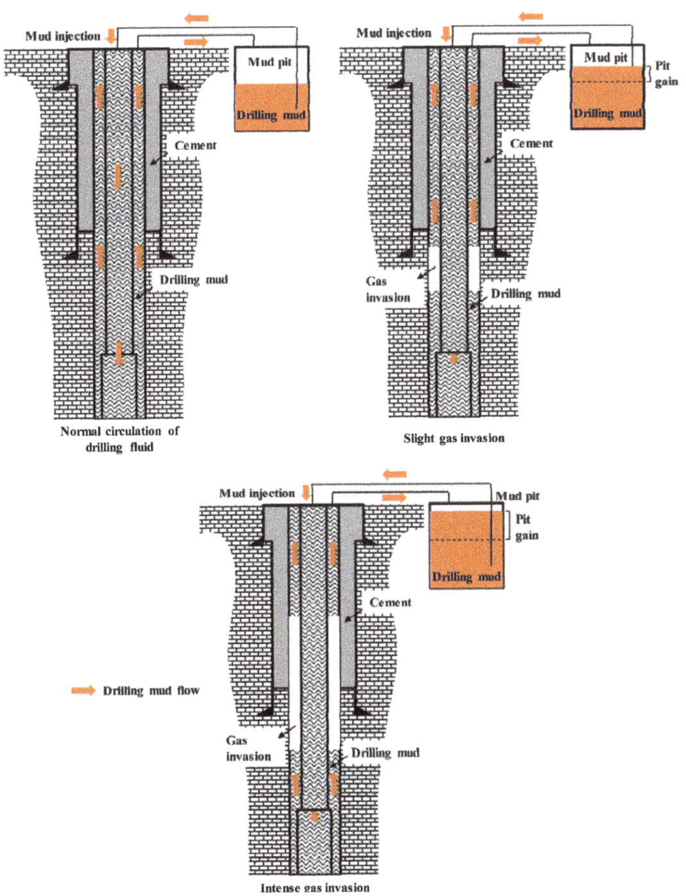

Figure 9. Schematic diagram of drilling fluid circulation and mud pool increment changes during normal drilling operation and when gas invades the annulus.

The kill parameters, when the quantity of gas invading into the annulus (i.e., pit gain) is different, were determined, and the results are shown in Figure 10. As illustrated in Figure 10, when the pit gain is only 20 m^3, the length of the gas column in the annulus is relatively short. Only two kill cycles are needed to completely discharge the reservoir gases that have invaded the annulus, and the total kill time is 296 min. In this instance, the injection of 10.5 m^3 of kill fluid raised the liquid level in the annulus to 3080 m from the bottom of the well in the first kill cycle. In the second cycle, only about 9.5 m^3 of kill fluid needs to be injected to discharge the remaining gas in the annulus. As the pit gain increases, the difficulty of the kill operation will correspondingly rise, and the requisite total time will also be significantly extended. When the increment of the mud pit increases to 30 m^3 and 50 m^3, three and four kill cycles are required to do that, respectively. The total time of the corresponding kill operation is extended to 350 min and 460 min, respectively, which is 54 min and 164 min longer than that when pit gain is 20 m^3. Furthermore, as the gas column rises within the circulating drilling fluid, it will undergo an expansion in volume and experience a surge in pressure, further posing a threat to the safety of drilling operations. The more gas that invades the annulus, the greater the threat it poses to drilling safety [51]. Therefore, it is crucial to monitor the drilling parameters, such as casing pressure during the drilling process, and promptly address the drilling accidents, such as gas kick [52].

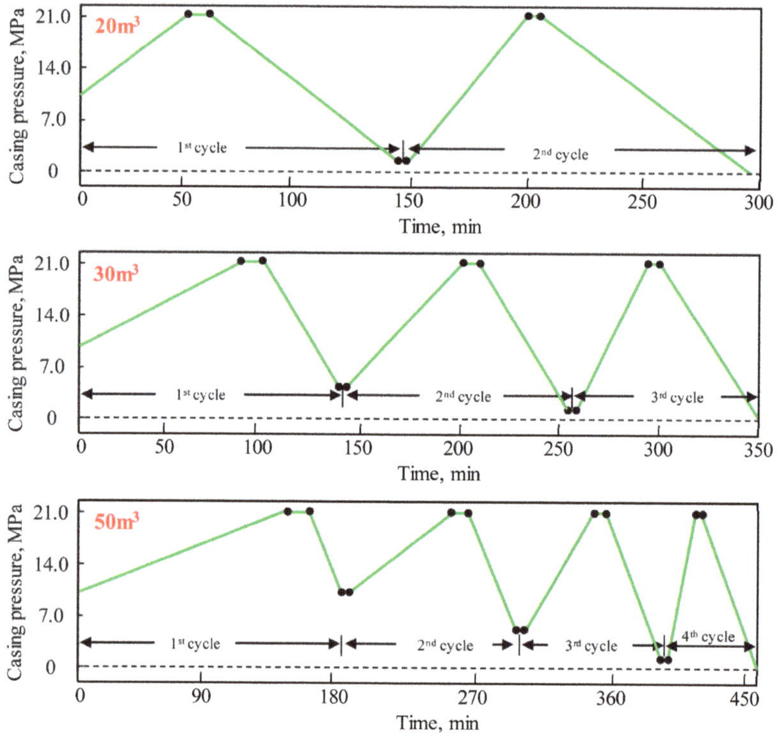

(a)

Figure 10. *Cont.*

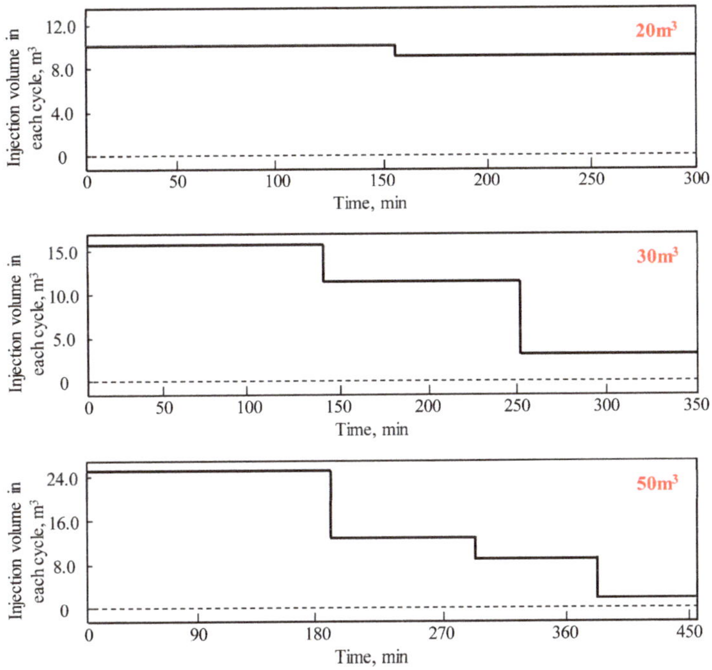

(b)

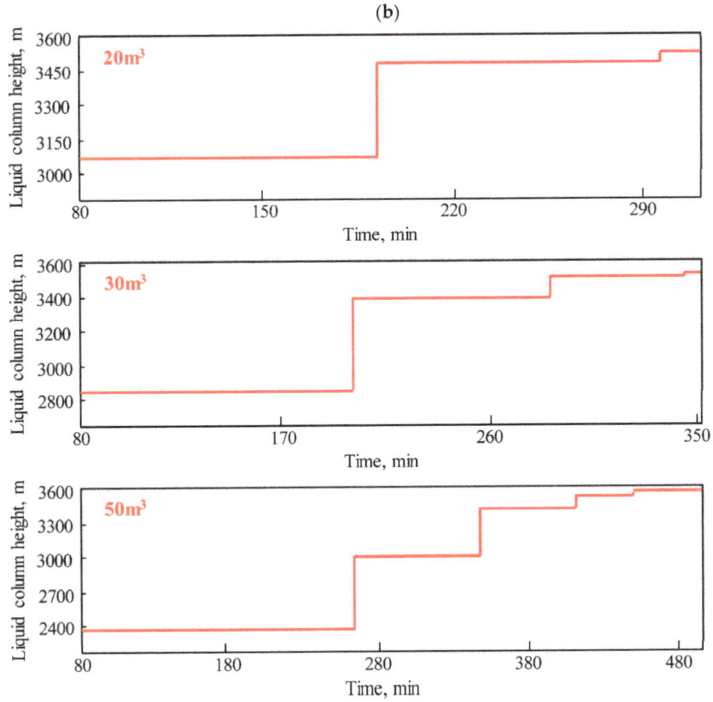

(c)

Figure 10. Evolution of kill parameters during the kill operation when the pit gain is 20 m^3, 30 m^3, and 50 m^3, respectively. (**a**) Casing pressure; (**b**) injection volume; (**c**) liquid column height.

It can also be seen from Figure 10 that when the pit gain is low, the minimum casing pressure of each cycle during the entire kill operation will decrease rapidly. When the pit gain is only 20 m^3, the casing pressure decreased to 1.8 MPa at the end of the first kill cycle (i.e., when the first gas discharge is completed). By the end of the second gas discharge, the casing pressure was close to zero. However, when the pit gain is high, this casing pressure drop is not so rapid, but gradual. As observed in Figure 10a, if the mud pit gain increases to 30 m^3, the minimum casing pressure at the end of the first kill operation is 5.1 MPa. By the end of the second round of the kill cycle, it dropped to the level at the first round of kill operation in the previous case (i.e., the case where the pit gain is 20 m^3). For the case where the pit gain is 50 m^3, this phenomenon is more significant. Attributed to the low casing pressure at the end of the initial kill cycle under low pit gain, the gas discharge stage in each kill cycle requires an extremely long duration. However, due to the small volume of kill fluid required in each cycle, the time required for injection of kill fluid is significantly short. On the contrary, when the pit gain is high, the injection time and gas discharge time in kill cycle are opposite to it.

5.2. Effect of the Shut-In Casing Pressure

Shut-in casing pressure is another factor affecting the kill parameters of the kill operation in the displacement kill method [53,54]. The effect of shut-in casing pressure on the kill operation in the displacement kill method was investigated, and the results were presented in Figure 11. As can be seen in Figure 11, if the shut-in casing pressure is high, more kill cycles are required. Correspondingly, the duration of the entire kill operation will be long. When the shut-in casing pressure is only 5.0 MPa, only two kill cycles are needed to discharge the gas in the annulus, and the duration of the entire kill operation is 313 min. If the shut-in casing pressure increases to 10 MPa, the number of complete well killing cycles required becomes three, and the duration of the well killing operation is extended to 350 min. When the shut-in casing pressure continues to increase to an astonishing 20 MPa, the required number of complete kill cycles and the duration of the kill operation increases to 4 and 414 min, respectively. This is because a higher shut-in casing pressure means more reservoir gas is locked in the gas column within the annulus. The discharge of more gas from the wellbore will inevitably take longer. Meanwhile, the casing pressure does not decrease instantaneously, but gradually decreases with the reciprocating cycle of fluid injection and gas discharge. Therefore, for the case of higher shut-in casing pressure, more instances of step-by-step pressure reduction (i.e., kill cycles) are required.

This also illustrates a fact from another perspective, namely that the drilling process of an abnormal high-pressure reservoir should be well monitored to avoid uncontrollable gas kick or blowout. In addition to the above-mentioned difference in the time required for kill operation, there is another difference that needs to be noted: the first kill cycle. As observed in Figure 11, when the shut-in casing pressure is low, it takes a long time to increase the casing pressure to the permissible value by injecting the kill fluid. If the shut-in casing pressure is only 5.0 MPa, it takes 138 min for the casing pressure to increase from the shut-in casing pressure to its permissible value. Even when the shut-in casing pressure is 10 MPa, this time still lasts for 91 min. On the contrary, if the shut-in casing pressure is high, the process of increasing the casing pressure to the allowable safe casing pressure by injecting the well control fluid is very simple. Accordingly, this process will not take a long time. When the shut-in casing pressure is 20.0 MPa (which is close to the permissible value), the injection of 1.7 m^3 of kill fluid within 10 min allows the casing pressure to reach the permissible value.

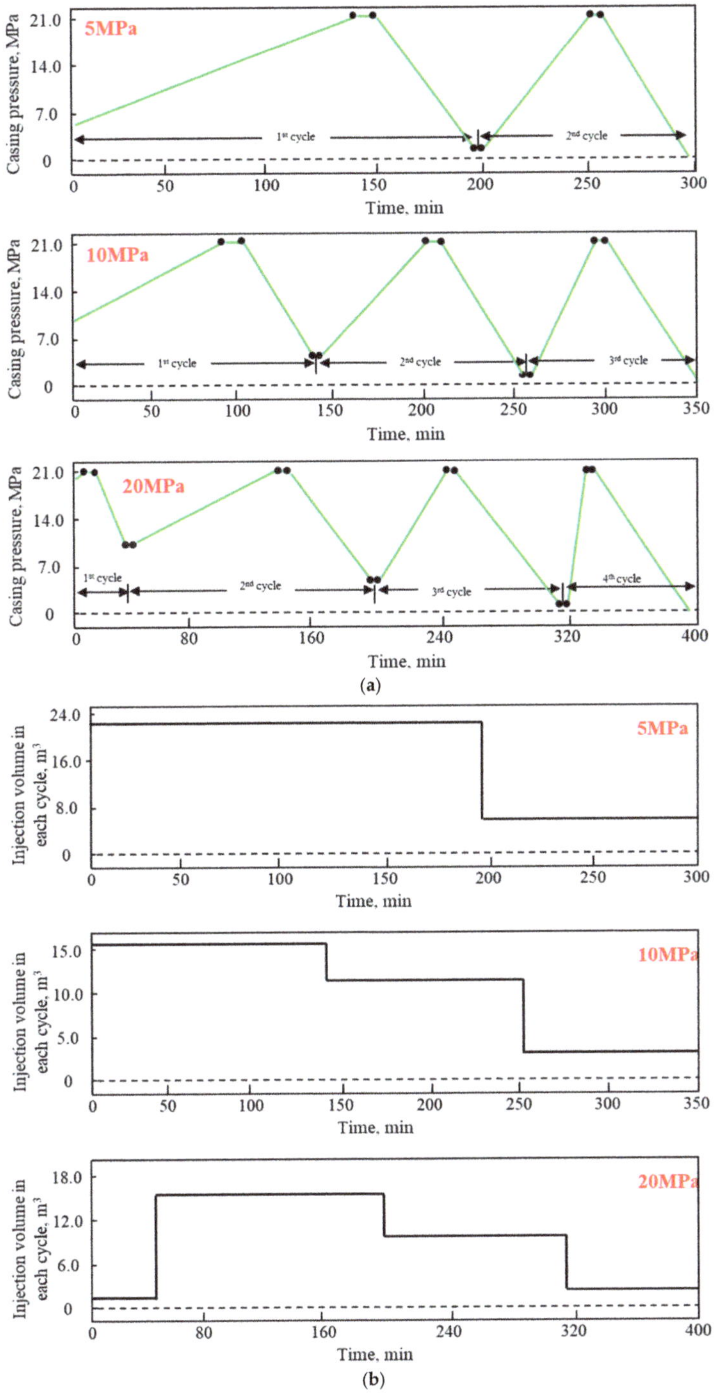

Figure 11. *Cont.*

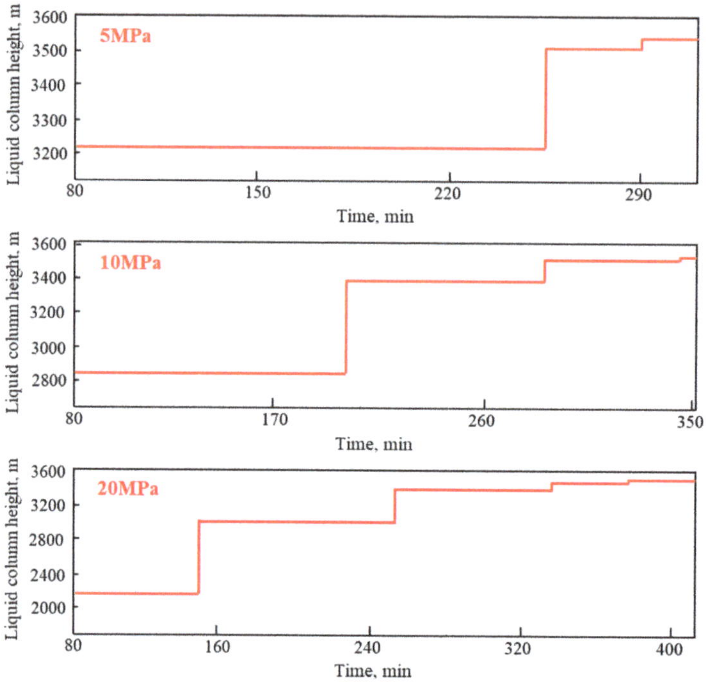

(c)

Figure 11. Evolution of kill parameters during the kill operation when the shut-in casing pressure is 5 MPa, 10 MPa, and 20 MPa, respectively. (**a**) Casing pressure; (**b**) injection volume; (**c**) liquid column height.

6. Conclusions

In the present work, the method for determining the kill parameters in the kill operation with the displacement method was developed, and its applicability was then verified. Moreover, evolution characteristics of kill parameters when killing fluid leakage occurs and when it does not occur were analyzed, and factors affecting kill operation were also investigated. The main conclusions obtained are as follows:

(1) The method for determining kill parameters mentioned in the study is applicable for the corresponding simulation analysis. In simulation and experiment, it takes 856 s and 900 s, respectively for the gas inside the annulus to be completely discharged, with a time difference of only 44 s. This is due to fluctuations in the control of experimental conditions during the experimental process. So, this time difference is acceptable.

(2) The fluid leakage during the kill process will not only increase the cost but also cause an increase in the number of well killing cycles and an extension of the operation time. When there is no leakage of kill fluid, injecting 30 m^3 of kill fluid three times within 350 min can completely discharge the gas and relieve the danger. However, the total operation time is extended to 420 min, and the total volume of injected fluid also increases to 32.4 m^3 in the presence of fluid leakage.

(3) A high pit gain indicates a significant amount of gas and a long gas column in the annulus, thereby complicating the kill operations. If the pit gain increases from 20 m^3 to 50 m^3, the total time used for the kill operation extends by 164 min, and the decrease in casing pressure significantly slows down. This indicates that accurate and real-time monitoring of drilling parameters, such as mud pit increment, during the drilling process is of great significance for ensuring drilling safety.

(4) For high shut-in casing pressure, although the casing pressure can be raised to its permissible value in a very short time in the first kill cycle, the total operation time was long. This occurs because the casing pressure gradually decreases with each successive kill cycle during the kill process by using the displacement method.

Nevertheless, there are still many deficiencies that need to be supplemented in subsequent studies within this study. Among them, it is necessary to compare the results of this study with those of other kill methods (such as operation time, etc.) in order to highlight the superiority of this research method. In addition, the authors also want to take into account the rheological properties of the gas and liquid phases involved in study for further analysis. In sensitivity analysis, some factors can have an impact on the cost (direct and indirect) of the well killing process. Finally, in future research, the authors will also take economic analysis into consideration.

Author Contributions: Conceptualization, Q.L. (Qingchao Li); methodology, Q.L. (Qingchao Li) and Q.L. (Qingchao Li); formal analysis, Y.H.; investigation, Q.L. (Qingchao Li), Q.L. (Qiang Li) and Y.H.; writing—original draft preparation, Q.L. (Qingchao Li) and Q.L. (Qingchao Li); writing—review and editing, Q.L. (Qiang Li) and Y.H.; visualization, Q.L. (Qingchao Li); supervision, Q.L. (Qiang Li) and Y.H. All authors have read and agreed to the published version of the manuscript.

Funding: This research was funded by the Fundamental Research Funds for the Universities of Henan Province (NSFRF240616), Henan Provincial Science and Technology Research Project (232102321128, 242102320342), and the Postdoctoral Program of Henan Polytechnic University (Grant No. 712108/210).

Data Availability Statement: The original contributions presented in the study are included in the article, further inquiries can be directed to the corresponding author.

Acknowledgments: The conception and launch of this work are also supported by the Rock Mechanics Laboratory (RML) of China University of Petroleum (East China).

Conflicts of Interest: The authors declare no conflicts of interest.

Nomenclature

V_1	Initial volume of gas in annulus after shut-in, m^3
Z_1	Compressibility factor of gas in annulus after well shut-in, dimensionless
R	Gas constant, 8.314 J/(K·mol)
T_1	Temperature of gas in annulus after shut-in, K
P_1	Pressure of gas in annulus after shut-in, MPa
ρ_m	Density of kill fluid, kg/L
q	Annulus volume per unit length, m^3/m
Z_2	Compressibility factor of gas in annulus after the first injected kill fluid reaches the wellbore bottom, dimensionless
V_{inj1}	First injection volume of kill fluid, m^3
Z_0	Compressibility factor of gas in standard state, $Z_0 = 1$
T_0	Temperature in standard state, K
P_0	Pressure in standard state, MPa.
V_{total}	Initial total volume of gas in annulus before gas discharge, m^3
P_2	Gas pressure in annulus when injected fluid in the first round reaches the wellbore bottom, MPa
P_p	Pore pressure, MPa
Z_{21}	Compressibility factor of gas when the first round of gas discharge has completed, dimensionless
T_1	Reservoir temperature, K
Z_3	Compression factor of gas in annulus when the kill fluid injected in the first round reaches the wellbore bottom, dimensionless
P_{21}	Pressure of gas in annulus at the beginning of the second round of fluid injection, MPa
V_2	Volume of gas in annulus when kill fluid injected for the first time reaches the wellbore bottom, m^3

V_3	Volume of gas in annulus before gas discharge, m^3
P_3	Gas pressure in annulus when injected fluid in the second round has reached the wellbore bottom, MPa
Z_3	Compressibility factor of gas in annulus when injected fluid in the second round reaches the wellbore bottom, dimensionless
Z_{31}	Compressibility factor of gas in annulus when injected fluid when the second round of gas discharge has completed, dimensionless
H	Well depth, m
t_1	Time taken for the first injection of kill fluid, h
t_2	Time taken for gas discharge, h
V_{amax}	Initial volume of gas column in annulus after well shut-in at the beginning of the first round, m^3

References

1. Zhao, X.; Geng, Q.; Zhang, Z.; Qiu, Z.; Fang, Q.; Wang, Z.; Yan, C.; Ma, Y.; Li, Y. Phase change material microcapsules for smart temperature regulation of drilling fluids for gas hydrate reservoirs. *Energy* **2023**, *263*, 125715. [CrossRef]
2. Li, Q.; Wu, J. Factors affecting the lower limit of the safe mud weight window for drilling operation in hydrate-bearing sediments in the Northern South China Sea. *Geomech. Geophys. Geo-Energy Geo-Resour.* **2022**, *8*, 82. [CrossRef]
3. Wang, Z.; Fan, Z.; Chen, X.; Fan, Z.; Wei, Q.; Wang, X.; Yue, W.; Liu, B.; Wu, Y. Global oil and gas development in 2022: Situation, trends and enlightenment. *Pet. Explor. Dev.* **2023**, *50*, 1167–1186. [CrossRef]
4. Tang, X.; Zhang, B.; Höök, M.; Feng, L. Forecast of oil reserves and production in Daqing oilfield of China. *Energy* **2010**, *35*, 3097–3102. [CrossRef]
5. He, Y.; Qiao, Y.; Qin, J.; Tang, Y.; Wang, Y.; Chai, Z. A novel method to enhance oil recovery by inter-fracture injection and production through the same multi-fractured horizontal well. *J. Energy Resour. Technol.* **2022**, *144*, 043005. [CrossRef]
6. Lei, Q.; Wang, D.; Guan, B.; Shi, J.; Cai, B.; He, C.; Sun, Q.; Huang, R. Shale oil and gas exploitation in China: Technical comparison with US and development suggestions. *Pet. Explor. Dev.* **2023**, *50*, 944–954. [CrossRef]
7. Li, Q.; Wang, Y.; Wang, F.; Wu, J.; Usman, T.; Li, Q.; Yuan, L.; Liu, Z. Effect of thickener and reservoir parameters on the filtration property of CO_2 fracturing fluid. *Energy Sources Part A Recovery Util. Environ. Eff.* **2020**, *42*, 1705–1715. [CrossRef]
8. Er, C.; Zhao, J.; Li, Y.; Si, S.; Bai, Y.; Wu, W.; Han, Q. Relationship between tight reservoir diagenesis and hydrocarbon accumulation: An example from the early Cretaceous Fuyu reservoir in the Daqing oil field, Songliao Basin, China. *J. Pet. Sci. Eng.* **2022**, *208*, 109422. [CrossRef]
9. Zhou, Z.; Zhao, J.; Zhou, T.; Huang, Y. Study on in-depth profile control system of low-permeability reservoir in block H of Daqing oil field. *J. Pet. Sci. Eng.* **2017**, *157*, 1192–1196. [CrossRef]
10. Jin, Z.; Liang, X.; Bai, Z. Exploration breakthrough and its significance of Gulong lacustrine shale oil in the Songliao Basin, Northeastern China. *Energy Geosci.* **2022**, *3*, 120–125. [CrossRef]
11. Yuan, S.; Lei, Z.; Li, J.; Yao, Z.; Li, B.; Wang, R.; Liu, Y.; Wang, Q. Key theoretical and technical issues and countermeasures for effective development of Gulong shale oil, Daqing Oilfield, NE China. *Pet. Explor. Dev.* **2023**, *50*, 638–650. [CrossRef]
12. Li, Q.; Liu, L.; Yu, B.; Guo, L.; Shi, S.; Miao, L. Borehole enlargement rate as a measure of borehole instability in hydrate reservoir and its relationship with drilling mud density. *J. Pet. Explor. Prod.* **2021**, *11*, 1185–1198. [CrossRef]
13. Zhou, L.; Zhang, D.; Sun, J.; Gu, Y.; Zhang, C.; Zhang, C. Status-quo of proved oil/gas initially-in-place in China as of 2020 and its variation trend in the past decade. *Energy Geosci.* **2022**, *3*, 343–348. [CrossRef]
14. He, D.; Jia, C.; Zhao, W.; Xu, F.; Luo, X.; Liu, W.; Tang, Y.; Gao, S.; Zheng, X.; Li, D.; et al. Research progress and key issues of ultra-deep oil and gas exploration in China. *Pet. Explor. Dev.* **2023**, *50*, 1333–1344. [CrossRef]
15. Zhao, S.; Zhao, Q.; Zhang, X.; Jiang, X.; Wu, H.; Yan, J. Key policy needs for the success of China's shale gas revolution. *Energy Rep.* **2024**, *11*, 4015–4020. [CrossRef]
16. Wang, Q.; Li, S.; Li, R. China's dependency on foreign oil will exceed 80% by 2030: Developing a novel NMGM-ARIMA to forecast China's foreign oil dependence from two dimensions. *Energy* **2019**, *163*, 151–167. [CrossRef]
17. Guo, J.; Gao, J.; Gao, S.; Yan, K.; Zhang, B.; Guan, C. Increasing impacts of China's oil and gas demands on global CH4 emissions. *Sci. Total Environ.* **2024**, *912*, 169624. [CrossRef]
18. Chen, S.; Ding, Y.; Song, Y.; Zhang, M.; Nie, R. Study on China's energy system resilience under the scenarios of long-term shortage of imported oil. *Energy* **2023**, *270*, 126831. [CrossRef]
19. Yi, J.; Qi, Z.; Li, X.; Liu, H.; Zhou, W. Spatial correlation-based machine learning framework for evaluating shale gas production potential: A case study in southern Sichuan Basin, China. *Appl. Energy* **2024**, *357*, 122483. [CrossRef]
20. Zeng, B.; Chen, G.; Meng, W.; Wang, J. Prediction, analysis and suggestions of shale gas production in China based on a new grey model with four parameters. *Alex. Eng. J.* **2024**, *86*, 258–276. [CrossRef]
21. Li, Z.; Xie, R.; Wu, Y.; Yuan, J. Progress and prospect of CNOOC's oil and gas well drilling and completion technologies. *Nat. Gas Ind. B* **2022**, *9*, 209–217. [CrossRef]
22. Zhu, J.; Chen, G.; Yin, Z.; Khan, F.; Meng, X. An integrated methodology for dynamic risk evaluation of deepwater blowouts. *J. Loss Prev. Process Ind.* **2022**, *74*, 104647. [CrossRef]

23. Li, Q.; Cheng, Y.; Ansari, U.; Han, Y.; Liu, X.; Yan, C. Experimental Investigation on Hydrate Dissociation in Near-Wellbore Region Caused by Invasion of Drilling Fluid: Ultrasonic Measurement and Analysis. *Environ. Sci. Pollut. Res.* **2022**, *29*, 36920–36937. [CrossRef] [PubMed]
24. Li, Q.; Zhao, D.; Yin, J.; Zhou, X.; Li, Y.; Chi, P.; Han, Y.; Ansari, U.; Cheng, Y. Investigation on Sediment Instability Caused by Gas Production from Hydrate-Bearing Sediment in Northern South China Sea by Horizontal Wellbore: Evolution and Mechanism. *Nat. Resour. Res.* **2023**, *32*, 1595–1620. [CrossRef]
25. Liu, H.; Liu, H.; Zhang, Q.; Fan, M.; Yin, B.; Wang, X.; Sun, X.; Wang, Z. Effect of liquid viscosity on the gas-liquid two phase countercurrent flow in the wellbore of bullheading killing. *Geoenergy Sci. Eng.* **2023**, *221*, 111274. [CrossRef]
26. Zhang, Z.; Sun, B.; Wang, Z.; Pan, S.; Lou, W.; Sun, D. Early monitoring method of downhole accident driven by physics based model and data driven methods coupling. *Geoenergy Sci. Eng.* **2023**, *221*, 111296. [CrossRef]
27. Li, Q.; Zhang, C.; Yang, Y.; Ansari, U.; Han, Y.; Li, X.; Cheng, Y. Preliminary experimental investigation on long-term fracture conductivity for evaluating the feasibility and efficiency of fracturing operation in offshore hydrate-bearing sediments. *Ocean Eng.* **2023**, *281*, 114949. [CrossRef]
28. Wang, Z.; Chen, G.; Zhang, R.; Zhou, W.; Hu, Y.; Zhao, X.; Wang, P. Early monitoring of gas kick in deepwater drilling based on ensemble learning method: A case study at South China Sea. *Process Saf. Environ. Prot.* **2023**, *169*, 504–514. [CrossRef]
29. Li, X.; Zhang, J.; Tang, X.; Li, C.; Li, B.; Wang, Y.; Zhao, Z. Propagation characteristics and application effects of measurement-while-drilling pressure wave for early gas-kick detection. *J. Loss Prev. Process Ind.* **2022**, *76*, 104741. [CrossRef]
30. Xu, Z.; Song, X.; Li, G.; Zhu, Z.; Zhu, B. Gas kick simulation in oil-based drilling fluids with the gas solubility effect during high-temperature and high-pressure well drilling. *Appl. Therm. Eng.* **2019**, *149*, 1080–1097. [CrossRef]
31. Yin, B.; Liu, G.; Li, X. Multiphase transient flow model in wellbore annuli during gas kick in deepwater drilling based on oil-based mud. *Appl. Math. Model.* **2017**, *51*, 159–198. [CrossRef]
32. Chen, X.; He, M.; Xu, M.; Zhou, C. Fully transient coupled prediction model of wellbore temperature and pressure for multi-phase flow during underbalanced drilling. *Geoenergy Sci. Eng.* **2023**, *223*, 211540. [CrossRef]
33. Yang, H.; Li, J.; Jiang, J.; Zhang, H.; Guo, B.; Zhang, G.; Chen, W. A dynamic managed pressure well-control method for rapid treatment of gas kick in deepwater managed pressure drilling. *Pet. Sci.* **2022**, *19*, 2297–2313. [CrossRef]
34. Chen, X.; He, M.; Xu, M.; Wang, S.; Dai, B. Early gas kick detection-inversion-control integrated system: The significance of applications of managed pressure drilling: A review. *Geoenergy Sci. Eng.* **2023**, *229*, 212134. [CrossRef]
35. Zhou, L.; Upchurch, E.R.; Liu, Y.; Anfinsen, B.T.; Hashemian, Y.; Yuan, Z. Evaluating Subsea Capping Stack Usage for CO_2 Blowouts. In Proceedings of the Offshore Technology Conference, Houston, TX, USA, 6–9 May 2024. OTC-35246-MS.
36. Yuan, Z.; Liu, Y.; Zhou, L.; Hashemian, Y. Dynamic Simulation of Riser Gas Handling with a Fixed-Choke Constant Outflow. In Proceedings of the SPE/IADC Drilling Conference and Exhibition, Galveston, TX, USA, 7–8 August 2024. SPE-217706-MS.
37. Gu, Q.; Fallah, A.; Feng, T.; Bakshi, S.; Chen, D.; Ashok, P.; Moore, D.; Oort, E. A novel dilution control strategy for gas kick handling and riser gas unloading mitigation in deepwater drilling. *J. Pet. Sci. Eng.* **2021**, *196*, 107973. [CrossRef]
38. Noynaert, S.; Jerome, J. Modeling ultra-deepwater blowouts and dynamic kills and the resulting blowout control best practices recommendations. In Proceedings of the SPE/IADC Drilling Conference and Exhibition, Amsterdam, The Netherlands, 23–25 February 2005; SPE: Amsterdam, The Netherlands, 2005. SPE-92626-MS.
39. Ren, W.; Fan, H.; Deng, S.; Cui, C.; Peng, Q.; Liu, X.; Dou, X. Displacement Calculation of Dynamic Killing Drilling. In Proceedings of the SPE Nigeria Annual International Conference and Exhibition, Lagos, Nigeria, 1–6 August 2015. SPE-178362-MS.
40. Yuan, Z.; Hashemian, Y.; Morrell, D. Ultra-deepwater blowout well control analysis under worst case blowout scenario. *J. Nat. Gas Sci. Eng.* **2015**, *27*, 122–129. [CrossRef]
41. Zhang, X.; Ding, T.; Zhou, B.; Yin, B. Mechanism of displacement gas kick in horizontal well drilling into deep fractured gas reservoir. *Front. Energy Res.* **2024**, *12*, 1412038. [CrossRef]
42. May, F.; Warsitzka, M.; Kukowski, N. Analogue modelling of leakage processes in unconsolidated sediments. *Int. J. Greenh. Gas Control* **2019**, *90*, 102805. [CrossRef]
43. Zhao, X.; Qiu, Z.; Wang, M.; Xu, J.; Huang, W. Experimental investigation of the effect of drilling fluid on wellbore stability in shallow unconsolidated formations in deep water. *J. Pet. Sci. Eng.* **2019**, *175*, 595–603. [CrossRef]
44. Hosseini, S.M. Hydraulic Fracture Mechanism in Unconsolidated Formations. Master's Thesis, University of Texas at Austin, Austin, TX, USA, 2012.
45. Yarushina, V.M.; Bercovici, D.; Oristaglio, M.L. Rock deformation models and fluid leak-off in hydraulic fracturing. *Geophys. J. Int.* **2013**, *194*, 1514–1526. [CrossRef]
46. Magzoub, M.I.; Salehi, S.; Hussein, I.A.; Nasser, M.S. Loss circulation in drilling and well construction: The significance of applications of crosslinked polymers in wellbore strengthening: A review. *J. Pet. Sci. Eng.* **2020**, *185*, 106653. [CrossRef]
47. Alsaba, M.; Nygaard, R.; Saasen, A.; Nes, O.M. Lost circulation materials capability of sealing wide fractures. In Proceedings of the SPE Deepwater Drilling and Completions Conference, Galveston, TX, USA, 10–11 September 2014. SPE-170285-MS.
48. Feng, J.; Wang, Q.; Li, M.; Li, X.; Zhou, K.; Tian, X.; Niu, J.; Yang, Z.; Zhang, Q.; Sun, M. Pore Pressure Prediction for High-Pressure Tight Sandstone in the Huizhou Sag, Pearl River Mouth Basin, China: A Machine Learning-Based Approach. *J. Mar. Sci. Eng.* **2024**, *12*, 703. [CrossRef]
49. Xiong, T.; Cheng, L.; Guo, D.; Guan, L.; Jiang, Q.; Hu, Z.; Yin, S.; Chen, G. Overpressure characteristics and geological origin of the paleogene in Baiyun Sag, Pearl River Mouth Basin. *Front. Earth Sci.* **2023**, *11*, 1118443. [CrossRef]

50. Li, Q.; Liu, J.; Wang, S.; Guo, Y.; Han, X.; Li, Q.; Cheng, Y.; Dong, Z.; Li, X.; Zhang, X. Numerical insights into factors affecting collapse behavior of horizontal wellbore in clayey silt hydrate-bearing sediments and the accompanying control strategy. *Ocean Eng.* **2024**, *297*, 117029. [CrossRef]
51. Zhang, J.; Zhao, Z.; Li, X.; Zheng, Y.; Li, C.; Li, Z.; Liu, X. Research on the mechanism of the influence of flooding on the killing of empty wells. *J. Pet. Explor. Prod. Technol.* **2021**, *11*, 3571–3598. [CrossRef]
52. Gao, Y.; Chen, Y.; Zhao, X.; Wang, Z.; Li, H.; Sun, B. Risk analysis on the blowout in deepwater drilling when encountering hydrate-bearing reservoir. *Ocean Eng.* **2018**, *170*, 1–5. [CrossRef]
53. Li, Q.; Li, Q.C.; Wang, F.L.; Wu, J.J.; Wang, Y.L. The carrying behavior of Water-based fracturing fluid in shale reservoir fractures and Molecular dynamics of sand carrying mechanism. *Processes* **2024**, *12*, 2051. [CrossRef]
54. Li, Q.; Wang, Y.L.; Wang, Y.J.; San, J.S.; Li, Q.C.; Foster, G. Synthetic process on hydroxyl-containing polydimethylsiloxane as a thickener in CO_2 fracturing and thickening performance test. *Energy Sources Part A Recovery Util. Environ. Eff.* **2018**, *40*, 1137–1143. [CrossRef]

processes

MDPI

Article

A Novel Slickwater System with Strong-Polarity Fibers for High-Efficiency Proppant Flowback Mitigation

Yang Xu [1,2,*], Ping Chen [1,2], Kun Wang [1,2], Suoliang Wang [1,2], Qingcong Meng [1,2], Mingqi Li [1,2], Yingxian Ma [3,4] and Jie Zeng [3,4,*]

[1] Research Institute of Drilling & Production Engineering, CNPC Chuanqing Drilling Engineering Company Limited, Xi'an 710018, China; chen_ping@cnpc.com.cn (P.C.); wk_gcy@cnpc.com.cn (K.W.); wangsl_gcy@cnpc.com.cn (S.W.); mqc_gcy@cnpc.com.cn (Q.M.); czzcylmq@cnpc.com.cn (M.L.)

[2] National Engineering Laboratory for Exploration and Development of Low-Permeability Oil & Gas Fields, Xi'an 710018, China

[3] Petroleum Engineering School, Southwest Petroleum University, Chengdu 610500, China; mayingxian@swpu.edu.cn

[4] National Key Laboratory of Oil and Gas Reservoir Geology and Exploitation, Southwest Petroleum University, Chengdu 610500, China

* Correspondence: xy_gcy@cnpc.com.cn (Y.X.); jie.zeng@swpu.edu.cn (J.Z.)

Abstract: To avoid or mitigate proppant flowback after a massive hydraulic fracturing of tight formations and to reduce its impairment to well productivity, this study developed a new type of fiber material with strong polarity based on polyester fiber. This fiber material is modified by introducing a strong-polar functional monomer into the molecular structure and adopting the means of surface grafting. On the basis of this fiber material, a fiber slip-water system with excellent dispersion performance can be established to prevent proppant backflow. Laboratory experiments were performed to analyze the specific function of the fibers with strong polarity and its working mechanisms. The results indicate that strong-polarity fibers have excellent dispersion performance. The fibers and resistance-reducing agents form an interwoven structure that can carry proppants, resulting in the enhancement of the sand-carrying capacity of the fracturing fluid system and the overall strength of the sand bank. In terms of the sand-carrying capacity and mitigation of proppant flowback, strong-polar fibers have significantly improved compared to unmodified fibers. In a 5 mm simulated crack, strong-polar fibers can increase the static settling time of 70/140 mesh quartz sand proppant by 26.5%. Meanwhile, the placement height of the sand embankment increased by 23.4% after the settlement of the proppant. In proppant transport experiments, strong-polar fibers with a mass fraction of 0.4% can increase the transport distance of proppants by more than 50%. Within the closed stress range of 2–10 MPa, the concentration of 0.5% strong-polar fibers increases the critical sand flow rate of the proppant by more than twice. The strong-polarity fiber system introduced in this study can be used to develop a fiber slickwater fracturing fluid system suitable for the massive hydraulic fracturing of tight reservoirs and has broad application prospects in the field of proppant flowback mitigation in tight reservoirs.

Keywords: strong-polarity fiber; fiber fracturing fluid; proppant carrying capacity; proppant flowback mitigation

1. Introduction

Horizontal-well massive hydraulic fracturing is currently the most popular technology for tight oil and gas extraction [1,2]. This technology enables the creation of complex fracture networks within the reservoir, enhancing the fracture–reservoir contact area and achieving long-term stable production of oil and gas wells [3]. In the process of reservoir fracturing, a large amount of fracturing fluid and proppant are injected into the reservoir, forming a complex artificial fracture network to improve the permeability of the reservoir.

After the completion of fracturing construction, a large amount of fluid injected into oil and gas wells needs to be flowed back before they can be put into production. In the process of drainage and production, proppant particles will destabilize and move in fractures and gradually migrate to the wellbore, resulting in sand production in oil and gas wells [4]. One field example in the Changqing Oilfield, the largest oil and gas field in China, indicates that the produced proppant volume for a single well varies from 9 m^3 to 70 m^3, while the average value is 23.3 m^3. Proppant back production not only causes damage in the proppant bank [5] but also reduces the proppant supporting efficiency, which generates certain impairment on the long-term conductivity and productivity [6,7]. Apart from that, the back-produced proppants within the wellbore are deposited at the well bottom and also plug the production layer, which increases the difficulty for dealing with the wellbore [8]. When the fracturing fluid flowback rate or the production rate is high, proppants may flow out from the wellhead and cause damage to the choke and pipelines [9], leading to potential safety issues.

The generation of proppant backflow is closely related to the fluid flow in the fracture. The fluid flow in the reservoir after fracturing mainly covers the matrix flow, natural micro-fracture flow, and artificial fracture flow [10]. The fracturing flowback fluid and oil and gas seepage in fractures and porous media are extremely complex [11]. The process of proppant movement in fractures is a fluid–solid coupling physical process. The movement state of proppant particles depends on the physical properties of particles and fluids, the flow field distribution in fractures, the pressure field distribution, and the distribution of solid particles [12]. Affected by factors such as fluid velocity and viscosity [13,14], the proppant moves in the fracture and flows back into the wellbore with the flow of oil, gas, and fracturing fluid [15]. This backflow may occur after hydraulic fracturing or during the pressure redistribution process during well switching [16].

Using fibers can effectively improve the proppant settling profile [17,18], and eventually mitigate or avoid proppant back production [19]. Hu et al. [20] added polypropylene fiber to the fracturing fluid to resist proppant backflow. The complex three-dimensional network structure formed by fibers and ceramsite was observed with microstructure. A single fiber can intercept and fix 6~8 20/40 mesh ceramsite along the length direction, which effectively improves the overall compressive strength of proppant particles. Chen et al. found that the fiber can significantly improve the strength of the proppant filling layer through experimental research, and the addition of the fiber makes the proppant filling layer better resist the erosion of the backflow fluid. The critical velocity with fiber is 2.25 times higher than that without fiber. In recent years, the fiber-based proppant flowback mitigation technology has been applied to many tight gas and oil horizontal wells. However, the surface of classical fibers lacks active functional groups, and these fibers have poor dispersibility in fracturing fluids [21] and poor suitability for proppants. This fiber–proppant mixture can hardly form a stable structure so that proppants can easily flow out. In addition, due to the relatively large density difference between proppants and fibers and the use of low-viscosity slickwater in tight formations [22], the gravity differentiation of proppants and fibers frequently occurs along the fracture height direction, impairing the application efficiency.

To address the above-mentioned challenges, this study developed a new type of fiber material with strong polarity. Through the introduction of highly polar functional monomers and surface modification, the coupling among fibers, drag reducers, and proppant particles is greatly enhanced, improving the stability of the fiber–proppant interwoven structure. The new strong-polarity fiber system can be used to develop a fiber slickwater fracturing fluid system that has good dispersibility and can mitigate or avoid proppant back production. This technology will strongly support the efficient development of tight oil and gas reservoirs and solve the problem of proppant flowback after massive hydraulic fracturing in tight reservoirs.

2. Basic Mechanisms

2.1. Proppant Flowback Mechanisms

After a fracturing treatment, the well starts to produce fracturing fluids and hydrocarbon. When the fluid flow rate approaches a critical value, the injected proppant particles are influenced by the drag force of the fluid. The proppant particles gradually transition from a steady state to an unstable state, resulting in the phenomenon of "fluidization" [23]. Thus, the fluid carries proppants and flows into the wellbore and finally flows out from the wellbore [24]. This process refers to proppant backflow or back production. With the continuous backflow of proppants, the stability of the proppant pack is further destroyed, leading to more serious proppant back production [25].

Figure 1 shows the specific force analysis of a single proppant particle during the later stage of oil and gas well backflow and the production process. During the processes of fracturing fluid flowback and oil and gas production, the force that acts on proppants in the horizontal direction can be divided into two types: backflow driving force F and backflow resistance f. The driving force F mainly includes the drag force acting on the proppant particles as the fluid flows P_{drag} and the equivalent capillary force of the residual fluid within the fracture acting on the proppant σ_c [26]. The drag force exerted by the fluid on the particles can fluidize the proppants and is the main force causing the proppant flowback. The magnitude of equivalent capillary force is equal to that of the capillary force but is in an opposite direction compared with the capillary force. As for the backflow resistance f, it includes the force from the fracture wall acting on the proppant pack f_c and the interaction forces between proppant particles f_p. The force from the fracture wall mainly refers to the closure pressure perpendicular to the wall. It can crush the proppants or induce an axial extrusion force to extrude the proppants. The interaction forces involve cohesion and friction between particles [27]. The interaction forces make proppants attract each other and limit their movement [28].

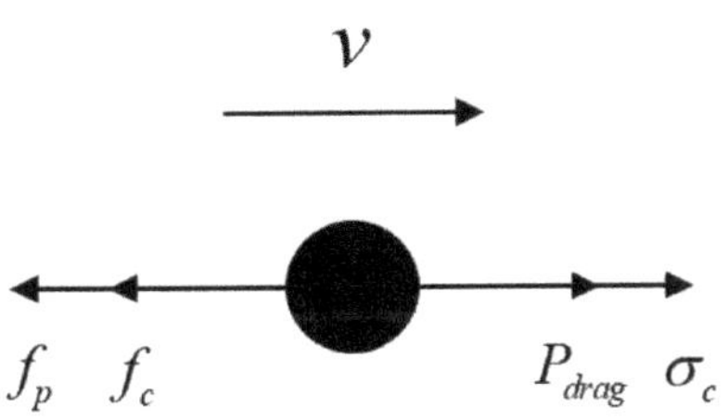

Figure 1. Force diagram of a single proppant.

Based on the force distribution of the proppant particles, a physical model for the force distribution of a single proppant particle in a crack is proposed, as shown in Equation (1). The critical condition under which the proppant will flow back is that the backflow driving force is larger or equal to the backflow resistance. When the driving force is smaller than the backflow resistance, the proppant pack remains structurally stable, and the proppant flowback will not happen. In contrast, when the driving force is larger than the backflow resistance, proppant particles are controlled by the fluid drag force, promoting proppant migration with the fluid flow. The structure of the proppant pack becomes unstable, which further causes more proppants to flow back into the wellbore and induces sand production from the wellbore.

$$\begin{cases} F = P_{drag} + \sigma_c \\ f = f_c + f_p \end{cases} \tag{1}$$

where P_{drag} is the fluid drag force (MPa), σ_c is the equivalent capillary force (MPa), and f_c is the force component of the closure stress acting on the proppants in the flow velocity direction (MPa). Figure 1 shows the corresponding force diagram.

2.2. Mechanism of Proppant Backflow Control Using Fibers

The fiber-containing fracturing technology is currently the most widely-used technology to prevent proppant flowback. This technology involves adding a fibrous material with certain flexibility to the sand-carrying fluid and pumping the mixture into the formation. The proppant sand bank serves as the base phase, while the fibers are the reinforcing phase, forming a composite proppant pack structure within artificial fractures [29,30]. The fiber materials make the proppant pack stable, and it is chemically stable. Thus, the wells can directly start the fracturing fluid flowback process without a shut-in. In this way, the critical fluid flow rate in the hydraulic fracture can be significantly increased. This technology has advantages, such as broad applicability, low cost, and simplicity of use. Consequently, in oil and gas wells with a serious proppant-backflow phenomenon, the fiber-containing fracturing technology has become a common choice to solve this issue.

The mechanism of the reinforcement of the proppant pack by adding the fibrous materials can be explained by the composite material theory [31]. The mixing of fibers into the proppant pack plays a role similar to the addition of a steel skeleton to concrete, as shown in Figure 2. The fibers are intertwined with each other and distributed in the sand body to form a 3D network structure [32]. Meanwhile, each fiber is in contact with several proppant particles [33] and interacts with each other through contact pressure and friction, thereby enhancing the cohesion of the proppants and improving the stability of the proppant pack. During the fracturing fluid backflow process, gel-breaking fracturing fluid and hydrocarbons flow freely through the pore structure of the fiber–proppant composite sand bank. The fluid flow has a scouring effect on proppants, which causes the proppant particles to undergo shear deformation and form a sand arch, and the shear deformation of the sand arch induces fiber deformation [34]. The stress transmission of the fiber makes the axial force of the fiber decompose into two parts: the tangential component and the normal component, which can resist the shear deformation of the sand arch and stabilize the proppant pack, thereby enhancing the anti-scouring ability of the proppants and the critical flow rate and achieving the goal of proppant backflow mitigation or prevention.

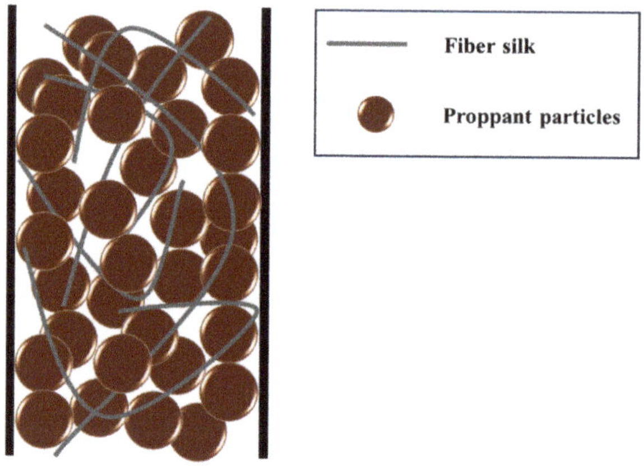

Figure 2. Schematic diagram of the mechanism of proppant flowback prevention using fibers.

3. Experimental Research

First, by introducing strong-polar functional monomers and surface modification, strong-polarity fiber materials were prepared in the laboratory. After the preparation of fibers, the functional characteristics and mechanism of action of the fibers were studied through laboratory experiments. The experimental process is shown in the Figure 3. The degree of performance improvement relative to unmodified fibers was evaluated by testing the microstructure characteristics, dispersion performance, dynamic and static sand carrying performance, and proppant flowback mitigation of strongly polar fiber systems.

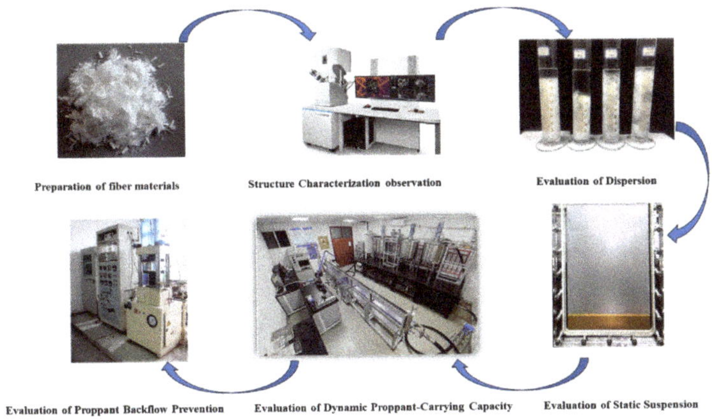

Figure 3. Experimental process for evaluating the performance of fiber systems.

3.1. Materials and Apparatus

The experimental materials mainly include phthalic acid (PTA), purchased from Shanghai Huichao New Materials Co., Ltd. (Shanghai, China); N-acryloyl glycinamide (NAGA), purchased from Guangdong Wengjiang Chemical Reagent Co., Ltd. (Shanghai, China); white oil, purchased from Shandong Xinsheng Chemical Co., Ltd. (Weifang, China); ethylene glycol (EG), dichloromethane (DCM), and palmitic acid monoglyceride (PAG) purchased from Chengdu Cologne Chemical Reagent Factory (Chengdu, China); and polyester fiber, purchased from Shandong Chenyu Engineering Materials Co., Ltd. (Jinan, China). The drag-reducing agent 0933-S5X was purchased from Chengdu Laoenpusi Technology Co., Ltd. (Chengdu, China). The 40/70 mesh quartz sand and 70/140 mesh quartz sand proppant with compressive strength of 70 MPa were purchased from Sichuan Chuanqing Underground Service Co., Ltd. (Chengdu, China). The purified ultrapure water was used to configure the solution.

The experimental instruments mainly included the following: high temperature esterification polycondensation esterification reactor produced by Weihai Huanyu Chemical Machinery Co., Ltd. (Weihai, China); zNN-D6 six-speed rotary viscometer produced by Ken Instrument (Shanghai) Co., Ltd. (Shanghai, China); Quanta 450 environmental scanning electron microscope produced by FEI, Hillsborough, Oregon, USA; rCT basic magnetic stirrer produced by IKA, Staufen, Germany; and electronic balance produced by Sartorius Scientific Instrument Co., Ltd. (Goettingen, Germany). Nanjing Lvding Machinery Equipment Co., Ltd. (Nanjing, China) produced hot air circulation oven. Other test instruments included fracture conductivity test system, proppant plate conveying device, and visual proppant settling device, all of which were laboratory self-assembled instruments.

3.2. Synthesis Process of Strong-Polarity Fibers

3.2.1. Principle of Synthesis Design

Polyester fibers are the raw materials of the strong-polarity fibers. In the polyester fiber polycondensation process, the method of controllable free radical polymerization is used to introduce the strong-polarity functional monomer (N-acryloyl glycinamide) into the molecular chain in a controllable manner, making the fiber-based polymer have uniform molecular weight and a regular structure. The fibers are made by melting and spinning the polymers. The surface grafting modification of the fiber was further carried out. Figure 4 shows the schematic diagram of the fiber surface grafting treatment. With the help of long-chain strong-polar surface modifier (Palmitic acid monoglyceride), specific strong-polar long-chain groups were grafted on the fiber surface to further enhance the hydrophilicity of the modified fiber while balancing the density of the material to synthesize surface-modified strong-polar fiber materials, thereby solving the problem of poor compatibility between conventional fibers and proppants and liquid systems.

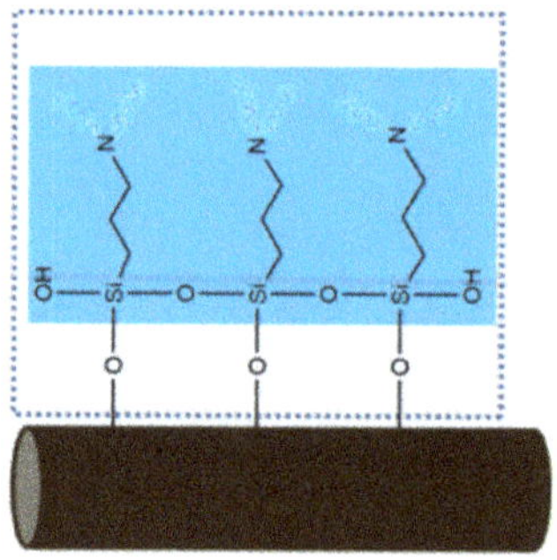

Figure 4. Schematic of the fiber surface grafting treatment. (The blue box contains strong polar long-chain functional groups grafted onto the fiber surface.).

The surface properties of the fiber materials were changed by the two-step treatment so that the fiber surface had better hydrophilicity, which could help improve fiber's wetting and dispersion properties in the fracturing fluid and also enhance the coupling with the hydrophilic proppant particles. Due to the change in surface properties, the fibers that are clumped together are more easily dispersed. Apart from that, the density of strong-polarity fibers can reach 1300 kg/m^3, which is denser than that of conventional fiber materials, reducing the density difference between fibers and proppants and mitigating the fiber escape phenomenon. As a result, in the fiber-containing fracturing field applications, these fibers can help to form a more stable fiber–proppant composite structure.

3.2.2. Strong-Polarity Fiber Preparation Process

The schematic diagram of the preparation process of the strong-polarity fiber is shown in Figure 5. The synthesis process included the following five steps.

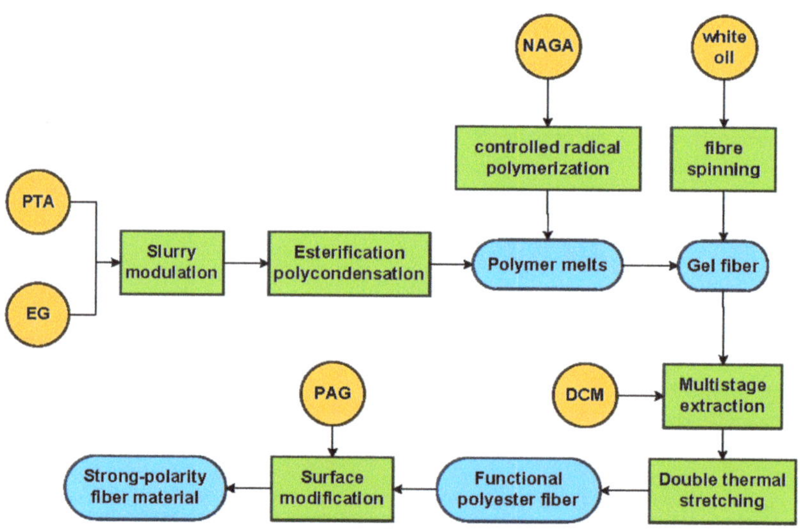

Figure 5. Flow chart of strong-polarity fiber preparation.

Step 1: polymer polymerization. The phthalic acid was sent into a U-shaped esterification kettle for esterification reaction under heating conditions of 240–290 °C. By controlling the amount of ethylene glycol and changing the reaction conditions, the esterification rate of the product was improved. By employing the controllable radical polymerization method, the N-acryloyl glycinamide was introduced into the molecular chain in a controllable manner, which made the molecular structure of the polymer more regular.

Step 2: spinning. The organic solvent wet gel spinning process was used to add the functional polyester fibers to the white oil. The polymer was swelled and dissolved in a twin-screw extruder and then extruded from the spinneret and quenched into a coagulation bath to form gel fibers. Methylene chloride was used as an extractant for multi-stage extraction. Dry fibers were prepared by extraction and drying. Then, the functional polyester fiber materials with diameters ranging from 7 μm to 20 μm and lengths ranging from 3 mm to 20 mm were obtained by superheating stretching. These materials have a large length-to-diameter ratio.

Step 3: surface modification. The palmitic acid monoglyceride was transferred into a three-neck flask. The flask was heated and the solution was stirred quickly to warm up to 40 °C to 60 °C. The functional polyester fibers were added to the three-neck flask. Then, the solution was fully stirred for 60 min to 80 min under a constant temperature condition. The above-mentioned constant temperature was set in the range of 40 °C to 60 °C.

Step 4: suction filtration. After surface modification, the functional polyester fibers were moved to the filter paper and cooled to room temperature. Then, suction filtration was performed.

Step 5: drying. The filtered fibers were placed in an oven at 60 °C to 80 °C for drying, and the drying time was between 16 h to 24 h. After that, the fibers preventing proppant backflow were obtained.

3.3. *Experimental Method*

3.3.1. Evaluation of Fiber Dispersion Properties

Unmodified fibers and strong-polarity fibers with different length-to-diameter ratios satisfying 0.5% mass fraction were added to 200 mL water and 200 mL slickwater with 0.1% mass fraction respectively. The fibers were stirred and dispersed evenly by a magnetic stirrer at a speed of 1000 r/min for 2 min. Then, the fiber suspension was quickly poured into a 250-mL graduated cylinder. After 2 h of settling, the dispersion state of the fibers in the solution was analyzed. The dispersity calculation equation was used to evaluate the dispersion properties of the fibers [35], see Equation (2).

$$y = \frac{h_0 - h_1}{h_0} \tag{2}$$

where h_0 is the height at which the fibers begin to precipitate in the fiber suspension (mm), h_1 is the height of the upper layer of water in the fiber suspension (mm), and y is the dispersity (%).

3.3.2. Evaluation of Proppant-Suspension Capacity of Fibers under Static Conditions

Static sand carrying performance test reference 'SY/T 5185-2016' [36] for testing. The unmodified fibers and 06-Q3-12-mm fibers with the best dispersion performance were added to 320 mL of ultrapure water. The mass fraction values of unmodifiedd fibers and 06-Q3-12mm fibers were 0.1%, 0.3%, and 0.5%. The fibers were stirred and dispersed by a magnetic stirrer for 2 min.

After the fibers were evenly dispersed, 70/140 mesh quartz-sand proppants with a proppant concentration of 10% were added and stirred evenly. Then, drag-reducing agents with a mass fraction of 0.1% were further added. The mixtures with certain mass concentration values were prepared and stirred for 5 min to achieve uniform dispersion.

After the solution preparation was completed, the fiber–sand mixture was quickly transferred to the proppant settling visualization device (300 mm × 200 mm × 5 mm) to simulate the static settling of proppant particles within a 5 mm wide fracture in a tight formation, analyze the settling state and placement efficiency, and calculate the static settling velocity and height of the proppant pack.

3.3.3. Evaluation of Proppant-Carrying Capacity

The strong-polar fiber with a mass fraction of 0.4%, the 40/70 mesh quartz sand proppant with a sand ratio of 10%, the resistance reducing agent with a mass fraction of 0.1%, and the rest of the water were added to the stirred tank to configure the fiber slickwater proppant-carrying fluid and configuration without fiber slickwater proppant-carrying fluid for comparison.

After preparation of the proppant carrying fluid, a certain amount of water was pumped into the 6 mm wide fracture to circulate the water and achieve a stable condition. Then, the uniformly mixed fluid containing fibers, drag reducer, proppants, and water was pumped into the flat fracture (300 cm × 60 cm × 6 mm) with an 18 L/min pumping rate.

When the proppant particles flowed through the observation area, the dynamic data in the experimental process were recorded through the high-definition video acquisition system. The transport and settlement states of the particles were observed through image analysis in a later stage.

3.3.4. Evaluation of Proppant Back Flow Prevention Performance

The method of measuring the critical flow rate was used to investigate the stability of the proppant pack [33]. The standard API conductivity tester was used, and the wall of the conductivity tester was made of CD-4MCu duplex stainless steel. In the proppant flowback control experiment, 40/70 mesh and 70/140 mesh proppants were used. Slickwater fracturing fluid with a mass fraction of 0.01% and 3 mPa·s viscosity was used to simulate the gel breaking and flowback stage of fracturing fluid. Specifically, the process of determining the critical flow rate of the proppant pack was as follows.

Step 1: After mixing the fiber materials and proppants uniformly with different concentrations, the mixture was placed between the two plates in the conductivity testing chamber with a placement concentration of 5 kg/m^2 to create a 2.5 mm wide artificial fracture.

Step 2: The loading on the plate was slowly added until the pressure reached the designed closure pressure. The slickwater was used to flush the artificial fracture. The displaced fluid amount gradually increased.

Step 3: Under different pressure conditions, slickwater flows in the proppant filling layer. When the flow rate reaches a certain speed, it causes the migration of proppant particles, and the proppant particles begin to flow out from the outlet of the diversion chamber. The sand production at the outlet of the conductivity testing chamber was observed, and the corresponding fluid flow rate at that time, which was the critical sand flow rate, was recorded.

3.3.5. Environmental Scanning Electron Microscope Test

Step 1: The drag reducer satisfying 0.1% mass fraction was added to 100 mL of ultrapure water. It was stirred and dispersed by the magnetic stirrer for 5 min, and the low-viscosity slickwater was configured. Then, 70/140 mesh quartz-sand proppants with a 20% concentration were further added to the slickwater. The fiber-free slickwater mixed sand liquid was obtained after being stirred and mixed uniformly.

Step 2: The strong-polarity fibers satisfying 0.5% mass fraction were added to 50 mL of proppant–slickwater mixture. Then, it was stirred for 5 min. Consequently, the low viscosity fiber–slickwater mixed sand liquid was obtained.

Step 3: A small amount of fiber-free slickwater mixed sand liquid and fiber-free slickwater mixed sand liquid were collected as samples for scan. The microscopic morphology of the mixed system of fibers, drag reducer, and proppants was observed and analyzed through environmental scanning electron microscope (ESEM).

4. Experimental Results and Discussion

4.1. Structure Characterization

The scanning electron microscope results of the fiber-free slickwater mixed sand liquid with a mass concentration of 0.1% are shown in Figure 6a. Without fibers, the molecular chain of the polymer of the drag reducer swells and stretches to form a network structure, which is wrapped and covered on the surface of the quartz sand particles, generating a certain fixing and connecting effect on the proppant. However, due to the weak network structure formed by the low-concentration drag reducer, the above effect is also weak. The scanning electron microscope results of the fiber–slickwater mixed sand liquid with the same concentration are shown in Figure 6b. When the strong-polarity fibers are added, the fibers are dispersed into filaments and interspersed in the proppant particles, and the drag reducer molecules encapsulate the proppant. The fiber is used as a rigid support body. Combined with the drag reducer, the mixture forms a network structure to wrap and carry proppants, and the coupling effect of fibers and proppants greatly enhances the overall structural strength of the proppant pack. The bonding of the fibers and the drag reducer forms a three-dimensional network structure with high strength, which greatly enhances the overall structural strength of the proppant bank, helps resist the external interference from the fluid, fixes the proppants network in their original position, and consequently maintains the stability of the sand body and prevents the backflow of the proppants.

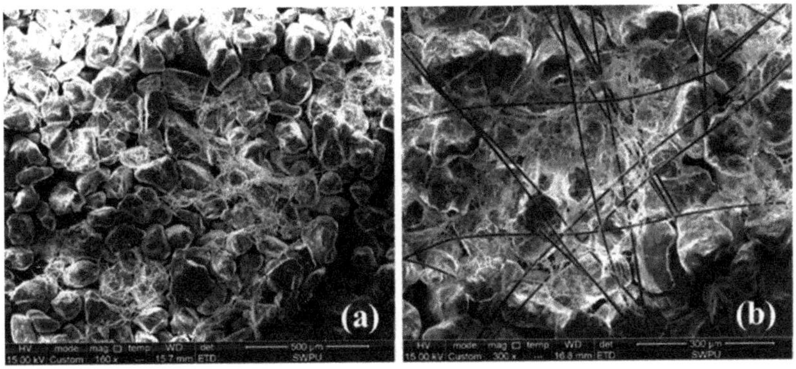

Figure 6. Scanning electron microscope photos of mixed sand slickwater system with and without strong-polarity fiber.(**a**) Fiber-free slickwater mixed sand liquid. (**b**) fiber–slickwater mixed sand liquid.

4.2. Evaluation of Fiber Dispersion Properties

The dispersion performance of the fiber materials affects the fracturing fluid preparation time in the field and also determines its effect on the proppant particles in the liquid system. Therefore, fiber dispersion experiments were carried out in clear water and slickwater fracturing fluids. The dispersion properties of different fibers in clear water and slickwater fracturing fluids were studied. Figure 7 shows the comparison of the dispersion behavior of unmodified fibers and strong-polarity fibers in clean water and slickwater with 0.1% mass fraction. Table 1 shows the dispersion performance of different fibers in clean water and slickwater. Taking the dispersion degree as the evaluation criterion, the more thorough the fiber dispersion is, the better the dispersion effect of the fiber in the solution would be.

Table 1. Evaluation results of dispersion properties of different fibers in clean water and slickwater.

Fibers	Solution	Dispersion Time of Fibers /s	Height of the Fibers Layer h_0/cm	Height of the Water Layer h_1/cm	Dispersity/%
Unmodified fiber	Ultra-pure water	30	18.1	10.4	42.54
Strong-polarity fiber	Ultra-pure water	<10	17.7	0	100.00
Unmodified fiber	Slickwater	30	18.1	6.2	65.75
Strong-polarity fiber	Slickwater	<10	17.7	0	100.00

Figure 7. Comparison of the dispersion of different fibers in ultra-pure water and slickwater. (**a**) Unmodified fiber. (**b**) strong polar fiber.

The dispersion of unmodified fibers in ultra-pure water and slickwater is shown in Figure 7a. The dispersion time of unmodified fiber is about 30 s. The unmodified fibers were distributed by filament in the solution. Due to the small fiber density, most of the fibers floated to the upper part of the clean water-based solution after 2 h, and a large section of clear water stratification appeared in the lower part of the system; the fiber dispersion degree was only 42.52%. After standing in slickwater with a mass concentration of 0.1% for 2 h, most of the fibers still floated to the upper part of the solution, and the lower part of the system showed a large layer of stratification. The fiber dispersion only increased to 65.75%. The dispersion stability of the unmodified fibers in water and slickwater is poor.

The dispersion of the developed strong-polar fiber in clear water and slick water is shown in Figure 7b. The dispersion time of the strong-polar fiber is 10 s, which can be quickly dispersed, and the fiber is flocculently distributed in the solution. The density of the strong-polarity fiber was slightly higher than that of water. The fibers were mainly distributed in the lower part of the solution after standing for 2 h, and a small section of clear water stratification appeared in the upper part of the system. The fibers were not stratified at all in water with a 100% dispersion degree. After standing in slickwater with a mass fraction of 0.1% for 2 h, the fibers still achieved rapid, uniform, and stable dispersion. The dispersion behavior was excellent.

This is due to the fact that after the fibers have been modified by two key improvement processes, the introduction of strong-polar groups and surface grafting, the fibers have stronger hydrophilicity. The density of the fiber materials is close to the density of the solution. The dispersion stability in the solution has been greatly improved. The two improvement processes are indispensable. The strong-polarity fibers obtained by this method can better satisfy the requirements of the rapid and uniform dispersion of fiber for fiber-accompanied injection fracturing.

4.3. Evaluation of Fiber's Static Suspension Performance

When fibers are added to the slickwater, the fibers disperse in the fracturing fluid to form a fiber-based network, which prevents the proppant from settling and thus improves the proppant carrying capacity of the system. The settling and placement of proppants in different concentrations of unmodified fiber–slickwater and strong-polarity fiber–slickwater are shown in Figure 8. The corresponding parameters are listed in Table 2.

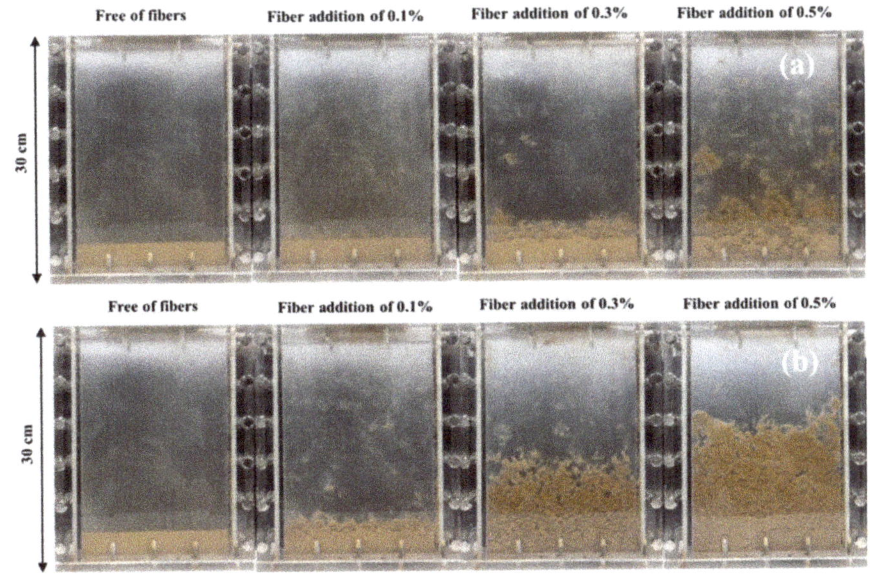

Figure 8. Static settling results of proppants in unmodified fiber–slickwater and strong-polarity fiber–slickwater with different concentrations. (**a**) unmodified-fiber-slickwater. (**b**) strong-polarity-fiber-slickwater.

Table 2. Static settling parameters of proppants in slickwater with different fibers.

Fibers	Concentration of Fibers	Settling Time of Proppant/s	The Laying Height of the Proppant/cm
/	0.00%	45	4.7
Unmodified fiber	0.10%	102	4.7
Unmodified fiber	0.30%	160	6.2
Unmodified fiber	0.50%	232	8.7
Strong-polarity fiber	0.10%	129	5.8
Strong-polarity fiber	0.30%	195	14.2
Strong-polarity fiber	0.50%	283	19.1

After adding the fiber materials to the slickwater, the fibers form a floc structure in the system. The addition of fibers can change the settlement behavior of proppants in the fracturing fluid, and the proppant settlement no longer obeys Stokes' law. The interaction between the fibers and proppants prevents the proppants from settling, thereby improving the proppant-carrying capacity of the system. With the gradual increase in suspended proppants, the flocculent fibers and proppant settle synchronously. This composite structure is gradually compacted during the settlement process and eventually becomes a part of the proppant pack. The rest of the flocculent fiber–proppant mixture is suspended in the system, which is beneficial for sufficient proppant placement in the whole artificial fracture.

As shown in Figure 8a and Table 2, in the slickwater without adding any fibers, the proppants settle rapidly in the system and can completely settle after 45 s. The height of the sand bank is low, only 4.7 cm. When 0.1% unmodified fibers were added to the slickwater, the settling velocity of the proppants was significantly reduced. The settling time was increased to 102 s, but the height of the sand bank was basically unchanged (4.7 cm). With a continuous increase of fiber amount, both the settling time and the height of proppant bank increase. When the mass fraction of the fibers reached 0.5%, the fiber-coated proppants formed a heterogeneous and loose cluster-like sand bank distribution. The proppant settlement time reached 232 s, which was 4.1 times higher than that of the non-fiber system. The proppant bank height was 8.7 cm, which was 85.1% higher than that of the non-fiber system.

As shown in Figure 8b and Table 2, the settling time of proppants was increased to 129 s in the slickwater system with 0.1% mass concentration of strong-polarity fibers. The settling time was 26.5% higher than that of the unmodified-fiber system with the same concentration. The proppant bank height was 5.8 cm, which was 23.4% higher than that of the unmodified-fiber system. With a continuous increase in the fiber amount, both the settling time and the height of proppant bank increase markedly. When the mass fraction of the fibers reached 0.5%, the fiber-coated proppants formed a uniform and dense cluster-like sand bank distribution. The settling time was increased to 283 s, which was 22.0% higher than that of the unmodified-fiber system with the same concentration. The proppant bank height was 19.1 cm, which was about 119.5% higher than that of the unmodified-fiber system. Compared with unmodified fibers, strong-polarity fiber materials can achieve higher proppant suspension capacity and better placement results at a lower dosage, showing the obvious advantages of strong-polarity fibers.

4.4. Evaluation of Dynamic Proppant-Carrying Capacity of Fibers

The experiment of proppant transport in the fiber–slickwater system was carried out using the plate fracture visualization device. The impact of adding strong-polarity fibers on the proppant transport and the proppant bank morphology was investigated. The results are shown in Figure 9. The placement of 40/70 mesh quartz sand in slickwater and fiber–slickwater is shown in Figure 9a and b, respectively.

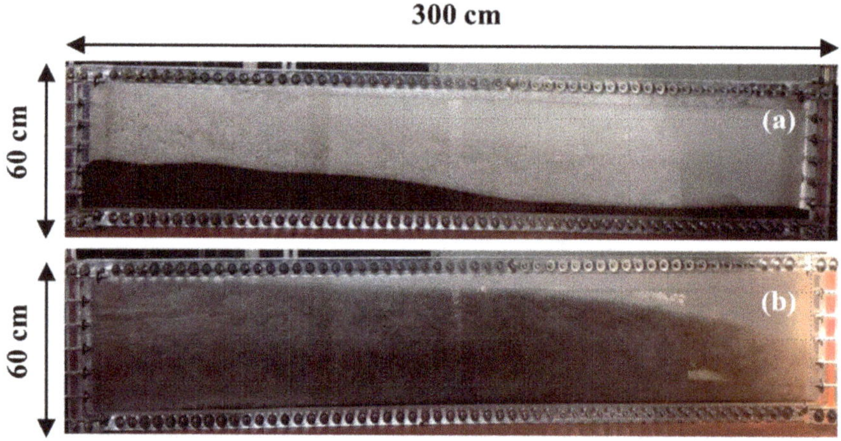

Figure 9. Placement of 40/70 mesh quartz sand. (**a**) Slickwater without fibers. (**b**) Slickwater with 0.4% strong- polarity fibers.

Figure 9a shows the resulting proppant bank with 40/70 mesh proppants in slickwater without fibers. Due to the weak proppant-carrying capacity of single slickwater, the proppant particles continue to settle from the fluid to the bottom of the fracture under the action of gravity. The proppants settled more at the fracture inlet and then were pushed to the tip of the fracture to form a sand bank. The further injection of proppants increased the length and height of the sand bank. The height growth process of the whole sand bank was slow. The equilibrium height was not reached after over 50 min.

Figure 9b shows the resulting proppant bank with 40/70 mesh proppants in slickwater containing 0.4% strong-polarity fibers. Other conditions are the same as those of Figure 8. In the slickwater fracturing fluid with strong-polarity fibers, the fibers were wound to form clusters, which can suspend and drag the proppants and carry the proppants into the fracture tip, improving the supporting efficiency at the inner fracture. By using the fluid with 0.4 mass fraction of strong-polarity fibers, the proppant transport distance can be increased by more than 50%. The growth of the sand bank is faster, and the height of the proppant bank is higher during the same accumulation time. The proppant bank consists of a large number of uncompacted fiber–proppant masses. At the same time, the proppant placement profile is greatly improved, making the sand bank more uniform.

4.5. Evaluation of Proppant Backflow Prevention Performance of Fibers

The variation of the critical sand flow-out rate of the proppant pack was tested at different closing pressure conditions. This parameter was used to characterize the proppant backflow prevention ability of the strong-polarity fibers. The impacts of 0.5% mass-fraction strong-polarity fibers and unmodified fibers on the critical sand flow flow-out of the 40/70 mesh proppant pack are shown in Figure 10. As can be seen from Figure 10, the closing pressure has a great influence on the critical sand flow-out rate of the proppant bank. Under the 2 MPa closure pressure condition, the critical sand flow-out rate of the proppant bank generated using 0.5% mass-fraction unmodified fibers was 117 mL/min. Under the same condition, the critical sand flow-out rate of that generated using strong-polarity fibers was 246 mL/min. With the increase in the closure pressure, the critical sand flow-out rate of the two proppant banks increased rapidly. Under the closure pressure of 8 MPa, the critical sand flow-out rate of the proppant bank generated using 0.5% mass-fraction unmodified fibers was 341 mL/min. Under the same condition, the critical sand flow-out rate of that generated using strong-polarity fibers was 723 mL/min. Under the tested closure pressure ranging from 2–10 MPa, the critical sand flow-out rate of the proppant bank can be increased by more than one time by adding strong-polarity fibers compared with using unmodified fibers.

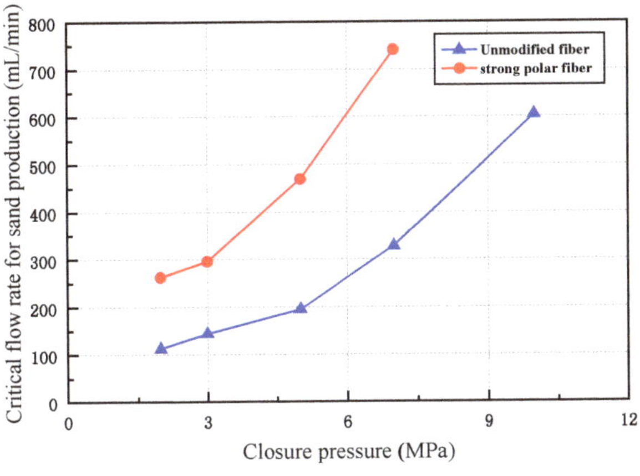

Figure 10. Impacts of different fiber materials on the critical sand flow-out rate of 40/70 mesh proppant pack.

The placement results of the two types of proppants with different fibers within the artificial fractures after flooding are shown in Figure 11. The proppant pack formed by adding 0.5% mass-fraction unmodified fibers shows that the proppant particles were eroded by the fluid, and the proppant profile at the inlet section was missing after flooding. At the same mass fraction, the proppant pack generated using the strong-polarity fibers exhibits better structural strength and is more stable. After the flooding test, the proppant profile can still be kept uniform and complete. There is no obvious missing part in the proppant pack. The post-flooding proppant profile of the sand pack intuitively indicates the outstanding advantages of the strong-polarity fiber material developed in this paper in preventing proppant backflow. It can better improve the stability of the proppant bank and effectively solve the sand production problem in the post-fracturing production process.

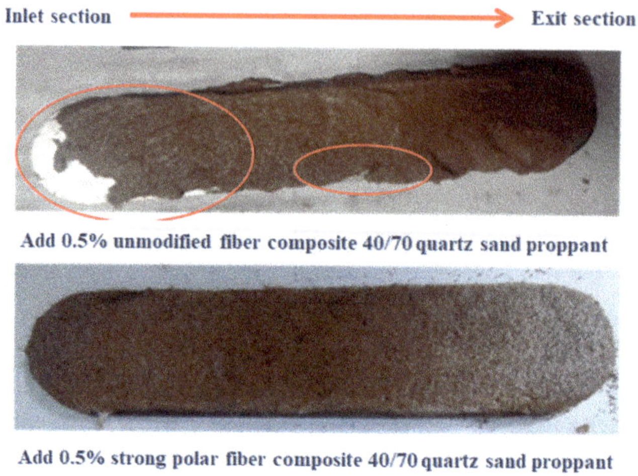

Figure 11. Proppant particle distribution after flooding of the proppant pack. (The red circle represents the missing part of the proppant pack after the experiment.).

5. Conclusions

In this paper, a new type of strong-polarity fiber was developed for the purpose of preventing proppant back production. Through the introduction of strong-polarity functional monomers and surface modification, the coupling effect between fibers and proppant particles was strengthened, improving the stability of the fiber and proppant composite structure and the ability of preventing proppant backflow. Experiments were performed to analyze the functional characteristics and backflow mitigation mechanisms. The following conclusions can be drawn:

(1) The proppant particles injected into the fracture are affected by the drag force of the fluid. When the backflow force is greater than the backflow resistance, the proppant particles transition from a stable state to an unstable one and were carried into the wellbore by the fluid, resulting in proppant backflow. Adding fibers can enhance the cohesion of the proppants, improve the stability of the proppant pack, greatly increase the critical sand flow-out rate, and achieve the purpose of preventing and controlling the backflow of proppant.

(2) By introducing strong-polarity functional monomers and surface grafting modifications, the hydrophilic properties and material density of the modified fibers were improved, and the coupling between fibers and proppant particles was enhanced. Meanwhile, the density of the material is also close to the density of the solution. Thus, the dispersion performance of the developed strong-polarity fibers is excellent. The fibers can be quickly dispersed in less than 10 s, and 100% dispersion can be achieved in clean water and slickwater, which is significantly better than that of unmodified fibers.

(3) Adding the strong-polarity fibers significantly improves the proppant suspension ability of slickwater-based fracturing fluid. The filament winding forms a flocculent structure that plays a key role in the suspension and dragging of the proppants, which can effectively reduce the settling velocity of proppants in the fracture. The strong-polarity fibers with a 0.4% mass fraction can increase the proppant transport distance by more than 50%, improve the supporting effect of proppants in the inner fracture, and greatly enhance the proppant accumulation volume. The proppant placement profile can be improved, and the placement becomes more uniform.

(4) Compared to unmodified fibers, strong-polarity fiber materials can effectively solve the sand production problem during the production process of fractured wells. By

using strong-polarity fiber materials, the coupling effect between fibers and proppant particles is improved, and the cohesive force inside the proppant sand is enhanced. Within the closed pressure range of 2–10 MPa, the critical sand flow-out rate of the proppant bank can be increased by more than twice. After flooding, the proppant can still be uniformly laid in artificial cracks, greatly enhancing the stability of the proppant pack.

Author Contributions: Conceptualization, Y.M.; Methodology, P.C., K.W. and S.W.; Investigation, Y.X., S.W., Q.M. and M.L.; Data curation, M.L.; Writing—original draft, Y.X., P.C., K.W., Y.M. and J.Z.; Writing—review & editing, Q.M. and J.Z. All authors have read and agreed to the published version of the manuscript.

Funding: This research received no external funding.

Data Availability Statement: The data presented in this study are available on request from the corresponding author.

Conflicts of Interest: Authors Yang Xu, Ping Chen, Kun Wang, Suoliang Wang, Qingcong Meng and Mingqi Li were employed by the Research Institute of Drilling & Production Engineering, CNPC Chuanqing Drilling Engineering Company Limited. The remaining authors declare that the research was conducted in the absence of any commercial or financial relationships that could be construed as a potential conflict of interest.

References

1. Jie, Y.; Yang, J.; Zhou, D.; Wang, H.; Zou, Y.; Liu, Y.; Zhang, Y. Study on the Optimal Volume Fracturing Design for Horizontal Wells in Tight Oil Reservoirs. *Sustainability* **2022**, *14*, 15531. [CrossRef]
2. Xu, J.; Liu, R.; Liu, H. Optimization of shut-in time based on saturation rebalancing in volume-fractured tight oil reservoirs. *Pet. Explor. Dev.* **2023**, *50*, 1445–1454. [CrossRef]
3. Zeng, Q.; Bo, L.; Liu, L.; Li, X.; Sun, J.; Huang, Z.; Yao, J. Analysis of fracture propagation and shale gas production by intensive volume fracturing. *Appl. Math. Mech.* **2023**, *44*, 1385–1408. [CrossRef]
4. Wen, Y.; Liu, L.; Huang, Y.; Liu, H.; Sui, M. Cause Analysis and Preventive Measures for Sand Production in Gas Wells of Sulige Gas Field. *ACS Omega* **2023**, *8*, 30590–30597. [CrossRef] [PubMed]
5. Zhang, T.; Tang, T.; Guo, J.; Zhou, H.; Liu, Y.; Zhang, Y.; Zeng, X. Experimental study on proppant-carrying migration and settlement of slickwater in narrow plate fractures based on PIV/PTV. *Petroleum* **2023**, *in press*. [CrossRef]
6. Nguyen, P.D.; Weaver, J.D.; Rickman, R.D.; Sanders, M.W. Remediation of Proppant Flowback—Laboratory and Field Studies. In Proceedings of the European Formation Damage Conference, Scheveningen, The Netherlands, 30 May–1 June 2007; p. 106108. [CrossRef]
7. Canon, J.M.; Romero, D.J.; Pham, T.T.; Valko, P.P. Avoiding Proppant Flowback in Tight-Gas Completions with Improved Fracture Design. In Proceedings of the SPE Annual Technical Conference and Exhibition, Denver, CO, USA, 5–8 October 2003; p. 84310. [CrossRef]
8. Zhang, H.; Yang, S.; Liu, D.; Li, Y.; Luo, W.; Li, J. Wellbore cleaning technologies for shale-gas horizontal wells: Difficulties and countermeasures. *Nat. Gas Ind. B* **2020**, *7*, 190–195. [CrossRef]
9. Wee, S.K.; Yap, Y.J. CFD study of sand erosion in pipeline. *J. Pet. Sci. Eng.* **2019**, *176*, 269–278. [CrossRef]
10. Li, B.L. *Study on Mutiple Temporal and Spatial Scale Flow Mechanism in Stimulated Tight Reservoir*; China University of Petroleum (East China): Qingdao, China, 2020.
11. Yan, G.; Li, Z.; Torres, S.A.G.; Scheuermann, A.; Li, L. Transient Two-Phase Flow in Porous Media: A Literature Review and Engineering Application in Geotechnics. *Geotechnics* **2022**, *2*, 32–90. [CrossRef]
12. Liu, X.J. *Research on the Law of Proppant Migration In hydraulic Fracturing Fractures*; China University of Petroleum (Beijing): Beijing, China, 2021. [CrossRef]
13. Chen, Y.; Sang, Y.; Guo, J.; Yang, J.; Chen, W.; Tang, B.; Feng, F.; Gou, X.; Zhang, Y. Experimental Study on the Backflow Mechanism of Proppants in Induced Fractures and Fiber Sand Control Under the Condition of Large-Scale and Fully Measurable Flow Field. *ACS Omega* **2023**, *8*, 42467–42478. [CrossRef]
14. Salahi, A.; Dehghan, A.N.; Sheikhzakariaee, S.J.; Davarpanah, A. Sand production control mechanisms during oil well production and construction. *Pet. Res.* **2021**, *6*, 361–367. [CrossRef]
15. Yang, X.; Wu, T.; Ren, L.; Huang, S.; Wang, S.; Li, J.; Liu, J.; Zhang, J.; Chen, F.; Chen, H. Mechanisms of Stress Sensitivity on Artificial Fracture Conductivity in the Flowback Stage of Shale Gas Wells. *Processes* **2023**, *11*, 2760. [CrossRef]
16. McLennan, J.; Walton, I.; Moore, J.; Brinton, D.; Lund, J. Proppant backflow: Mechanical and flow considerations. *Geothermics* **2015**, *57*, 224–237. [CrossRef]

17. Zhou, F.; Zong, Y.; Liu, Y.; Yang, X.; Xiong, C.; Zhang, S.; Sun, L.; Wang, J.; Li, J.; Guang, Z. Application and Study of Fine-Silty Sand Control Technique Using Fiber-Complex High-Pressure Pack in Sebei Gas Reservoir. In Proceedings of the SPE International Symposium and Exhibition on Formation Damage Control, Lafayette, LA, USA, 15–17 February 2006; p. 97832. [CrossRef]
18. Al-Ghurairi, F.; Solares, R.; Bartko, K.; Sierra, L. Results From a Field Trial Using New Additives for Fracture Conductivity Enhancement in a High-Gas Screenless Completion in the Jauf Reservoir, Saudi Arabia. In Proceedings of the SPE International Symposium and Exhibition on Formation Damage Control, Lafayette, LA, USA, 15–17 February 2006; p. 98088. [CrossRef]
19. Zhu, K.; Guo, D.; Zeng, X.; Li, S.; Liu, C. Proppant flowback control in coal bed methane wells: Experimental study and field application. *Int. J. Oil Gas Coal Technol.* **2014**, *7*, 189. [CrossRef]
20. Hu, Y.M.; Zhao, Z.F.; Li, X.W.; Chen, B.C.; Li, Z. Experimental study on proppant reflux resistance of fiber network under high gas injection rate in Yulin gas storage. *Fault-Block Oil Gas Fieid* **2021**, *28*, 786–791. [CrossRef]
21. Zhao, H.F.; Wang, W.; Jiang, C.Y.; Tian, J.C.; Mu, E.F. Study on fibrous fracturing fluid properties. *Drill. Prod. Technol.* **2017**, *40*, 90–92.
22. Jiang, Q.H.; Yang, X.T.; Yu, X.X.; Liu, J.Q.; Tang, L. Research progresses of slick-water drag reducer at home and abroad. *Chem. Ind. Eng.* **2022**, *39*, 76–83. [CrossRef]
23. Tang, B.T. Optimization of Backflow Parameters for Fractured Shale Gas Wells with Prevention of Proppant Reflow. Master' s Thesis, Southwest Petroleum University, Chengdu, China, 2019. [CrossRef]
24. Sun, Z.M.; Lu, X.C.; Jia, X.C.; Bai, Y.S.; Hu, W.X. Optimization of Mercury Intrusion Capillary Pressure Measurement for Characterizing the Pore Structure of Tight Rocks. *J. Nanosci. Nanotechnol.* **2017**, *17*, 6242–6251. [CrossRef]
25. Nguyen, P.D.; Weaver, J.D.; Dewprashad, B.T.; Parker, M.A.; Terracina, J.M. Enhancing Fracture Conductivity Through Surface Modification of Proppant. In Proceedings of the SPE Formation Damage Control Conference, Lafayette, LA, USA, 18–19 February 1998; p. 39428. [CrossRef]
26. Cheng, Y.S. Study on the Mechanism of Proppant Backflow after Fracturing and Its Influence. Master' s Thesis, China University of Geosciences (Beijing), Wuhan, China, 2021.
27. Ramones, M.; Rachid, R.; Milne, A.; Pelaez, K.; Gutierrez, L.; Madrid, O.; Ponce, G.; Ponce, O.; Cruz, S. Innovative Fiber-Based Proppant Flowback Control Technique Unlocks Reservoir Potential. In Proceedings of the SPE Latin America and Caribbean Petroleum Engineering Conference, Maracaibo, Venezuela, 21–23 May 2014; p. D011S003R002. [CrossRef]
28. Pitoni, E.; Devia, F.; James, S.G.; Heitmann, N. Screenless Completions: Cost-Effective Sand Control in the Adriatic Sea. *SPE Drill. Complet.* **2000**, *15*, 293–297. [CrossRef]
29. Bedford, A.; Stern, M. Toward A Diffusing Continuum Theory of Composite Materials. *J. Appl. Mech.* **1971**, *38*, 8–14. [CrossRef]
30. Lv, F.M.; Zhou, J.; Song, Q.W.; Zhong, Y. Research on fibrous composite fracturing sand control technique and its application. *Oil Drill. Prod. Technol.* **2008**, *30*, 97–100. [CrossRef]
31. Chen, X.L.; Ding, S.H.; Yuan, X.H. Application of fibre filler fracturing technology in the third region of xinjiang oilfield. *Drill. Prod. Technol.* **2013**, *36*, 61–63.
32. Wu, G.D. Laboratory research on control fracturing proppant backflow with fiber. *Drill. Prod. Technol.* **2012**, *35*, 88–90. [CrossRef]
33. Guo, G.; Xue, X.J.; Fan, H.B.; Liu, J.; Li, K.; Wu, J. Development and application of fiber slick water fracturing fluid for dense reservoirs. *Fault-Block Oil Gas Field* **2019**, *26*, 269–272. [CrossRef]
34. Yang, G.W.; Bai, J.W.; Chi, X.M.; Luo, M.L.; Yi, D.Q.; Xue, Y.F.; Zhang, J.F. Study on performance and surface modification of the degradable fiber plugging agent hydraulic fracturing. *Appl. Chem. Ind.* **2014**, *43*, 1431–1434. [CrossRef]
35. Song, B.; Bismarck, A.; Tahhan, R.; Springer, J. A Generalized Drop Length–Height Method for Determination of Contact Angle in Drop-on-Fiber Systems. *J. Colloid Interface Sci.* **1998**, *197*, 68–77. [CrossRef]
36. *SY/T 5185-2016*; Performance Evaluation Method of Gravel Packing Sand Control Water-Based Sand-Carrying Fluid. China National Energy Administration: Beijing, China, 2016. Available online: https://d.wanfangdata.com.cn/standard/SY/T%205185-2016 (accessed on 2 March 2024).

processes

Article

Effect of Acid-Injection Mode on Conductivity for Acid-Fracturing Stimulation in Ultra-Deep Tight Carbonate Reservoirs

Jiangyu Liu [1,2,3], Dengfeng Ren [1,2,3], Shaobo Feng [1,2,3], Ju Liu [1,2,3], Shiyong Qin [1,2,3], Xin Qiao [1,2,3] and Bo Gou [4,*]

1 Oil and Gas Engineering Research Institute, Tarim Oilfield Company, PetroChina, Korla 841000, China; ljyu-tlm@petrochina.com.cn (J.L.); rdf-tlm@petrochina.com.cn (D.R.); fengsb-tlm@petrochina.com.cn (S.F.); liuju-tlm@petrochina.com.cn (J.L.); qinsy-tlm@petrochina.com.cn (S.Q.); qiaoxin-tlm@petrochina.com.cn (X.Q.)
2 R&D Center for Ultra Deep Complex Reservoir Exploration and Development, CNPC, Korla 841000, China
3 Engineering Research Center for Ultra-Deep Complex Oil and Gas Reservoir Exploration and Development, Xinjiang Uygur Autonomous Region, Korla 841000, China
4 National Key Laboratory of Oil and Gas Reservoir Geology and Exploitation, Southwest Petroleum University, Chengdu 610500, China
* Correspondence: bogou@swpu.edu.cn

Abstract: The conductivity of acid-etched fractures and the subsequent production response are influenced by the injection mode of the fracturing fluid and acid fluid during acid fracturing in a carbonate reservoir. However, there has been a lack of comprehensive and systematic experimental research on the impact of commonly used injection modes in oilfields on conductivity, which directly affects the optimal selection of acid-fracturing injection modes. To address this gap, the present study focuses on underground rock samples, acid systems, and fracturing fluid obtained from ultra-deep carbonate reservoirs in the Fuman Oilfield. Experimental investigations were conducted to examine the conductivity of hydraulic fractures etched by various types of acid fluids under five different injection modes: fracturing fluid + self-generating acid or cross-linked acid; fracturing fluid + self-generating acid + cross-linked acid. The findings demonstrate that the implementation of multi-stage alternating acid injection results in the formation of communication channels, vugular pore space, and natural micro-cracks, as well as grooves and fish-scales due to enhanced etching effects. The elevation change, amount of dissolved rock, and conductivity exhibited by rock plates are significantly higher in comparison to those achieved through the single-acid injection mode while maintaining superior conductivity. It is recommended for optimal conductivity and retention rate in the Fuman Oilfield to adopt two stages of alternating acid-fracturing injection mode. Field application demonstrated that two-stages of alternating acid-fracturing generate more pronounced production response than the adjacent wells.

Keywords: ultra-deep carbonate reservoirs; acid-injection mode; multi-stages alternating acid fracturing; acid-fracture morphology; conductivity; field application

1. Introduction

The deep and ultra-deep carbonate reservoirs currently represent the focal areas of oil and gas exploration in China. These reservoirs are characterized by extensive lithologic or structure-lithologic complex gas or oil fields, such as those found in the Puguang, Yuanba Changxing-Feixianguan, and Anyue-Penglai Dengying formations in the Sichuan Basin, as well as the Shunbei and Fuman Ordovician formations in the Tarim Basin. Additionally, significant exploration is taking place around the Tongyuan fault within the Moxi Longwangmiao formation in the Sichuan Basin. Moreover, deep and ultra-deep carbonate reservoirs have emerged as the cornerstone for the expansion of oil and gas production in these basins [1]. The deep carbonate reservoir exhibits strong heterogeneity, with uneven distribution of natural oil- and gas-rich fracture-vug in space, which often poses challenges in achieving industrial production following drilling. Acid fracturing emerges as a pivotal technology for enhancing production in deep carbonate reservoirs. By creating acid-fractures, this technique effectively connects the discontinuous oil and gas reservoirs, enabling the commercial production of oil and gas wells [2].

The acid penetration range and fracture conductivity are crucial factors impacting the productivity of a fractured well. However, conductivity holds greater significance as only acid-etched fracture with a specific level of conductivity makes a significant contribution to productivity [3]. The factors influencing conductivity primarily encompass geological aspects, such as reservoir temperature, rock and mineral composition and distribution, as well as natural fracture development [4]. These inherent reservoir characteristics are difficult to adjust. On the other hand, engineering factors involve the acid system, acid-injection parameters (including injection rate and time), and acid-injection processes (such as injection mode selection). Enhancing the parameters of acid-fracturing engineering factors represents a crucial approach for regulating the acid-fracture conductivity.

The enhancement of acid-fracture conductivity has been extensively investigated by many scholars, and can generally be categorized into three main areas. First, hybrid fracturing with the combination of acid fracturing and proppant fracturing can enhance the acid-fracture conductivity [5,6]. However, proppant fracturing may pose challenges in terms of proppant injection and even introduce potential engineering risks associated with proppant plugging in ultra-deep tight carbonate reservoirs. Second, the utilization of additional agents enhances the mechanical strength of rock support points on the acid-fracture surface, thereby improving conductivity [7,8]. This method is relatively novel and is currently undergoing exploration and trial. Last, adjusting both acid-injection parameters and their corresponding mode helps to increase roughness on acid-fracture topography, thereby enhancing fracture conductivity [9–11]. This simple yet effective method can be easily implemented in the field.

Conventional gelled acid fracturing exhibits limited penetration distance and conductivity in deep carbonate reservoirs subjected to high formation temperatures (>150 °C) and closure pressure (>50 MPa), posing challenges in meeting long-term production requirements [12]. In order to enhance the acid penetration distance and conductivity, the pre-pad acid-fracturing mode is commonly employed. This pre-fluid typically consists of a fracturing fluid or weakly reactive autogenic acid, primarily utilized for creating and cooling a fracture. The frequently used acid systems include gelling acid and crosslinked acid, which are mainly applied for etching hydraulic fractures and generating conductivity. This injection mode is straightforward to implement on-site and extensively utilized in the initial stages of acid-fracturing development [13,14]. Multistage alternating acid fracturing is an enhanced mode of pre-pad acid fracturing, employing alternate stages of pad fluid and acid. It currently stands as the primary deep-penetrating acid-fracturing technology [9,15,16]. Compared to conventional gelled acid fracturing and pre-pad acid fracturing, multi-stage alternating acid fracturing has the following obvious advantages: first, the introduction of high viscosity fracturing fluid into the fracture reduces the acid leakoff in wormholes and natural fractures [9,15]; second, the non-reactive fracturing fluid or weakly reactive autogenic acid can effectively lower the acid-rock reaction temperature and increase the acid penetration distance [17,18]; third, the viscosity difference between the fracturing fluid and acid facilitates the occurrence of a viscous fingering phenomenon, creating non-uniform etching on the fracture surfaces and thereby improving the conductivity of acid-etched fractures [10].

The stimulation fluids and injection modes employed in different acid fracturing exhibit significant variations, and many scholars have demonstrated that these disparities directly impact the acid-fracture morphology as well as the conductivity. Lungwitz, B. et.al. found that differential etching resulted in higher conductivity with self-diverting-acid than that observed with straight HCl [19]. Navarrete et al. found emulsified acid produced higher conductivities than straight acid at high closure stresses, in spite of larger etched widths, providing a more efficient use of the acid [20]. Tariq et al. observed that rock hardness and roughness had a significant impact on fracture conductivity. Also, the type of treating fluid and rock determined the generated rock roughness, where higher reactivity resulted in higher roughness and hence conductivity [21]. Sui et al. found that as the experimental ambient temperature increased, the conductivity of the acid-etching fracture gradually increased [22].

The experimental investigation of the aforementioned single acid system demonstrates that the acid type and temperature exert an influence on both the etching morphology and conductivity. The multistage alternating acid-fracturing mode also has an impact on the acid-etching morphology and conductivity. Ren et al. found that the acid types can strongly affect both the fracture conductivity and etching patterns during alternating injection [23]. The acid-fracture conductivity generated by alternative stages with autogenic acid and gelled acid was influenced by the initial fracture morphology and temperature, as observed in the study conducted by Gou et al. [17], while the alternate injection of gelled acid and cross-linked acid combination, as discovered by Zhu et al., can create complex etching channels, resulting in the generation of high fracture conductivity in the porous carbonate reservoir [24].

The above experiments reveal that both the selection of the acid system and mode of acid injection significantly affect not only the acid-etched fracture morphology but also its conductivity. Nevertheless, current research predominantly concentrates on investigating a single mode of acid injection while seldom comparing variations in acid-fracture conductivity across different modes within rock samples from identical blocks. This lack of comprehensive and systematic analysis hinders our understanding of how different acid-injection modes influence conductivity and impede effective selection of appropriate acid-injection modes in oilfields. To address this knowledge gap, acid etching and conductivity tests were conducted on the underground cores of ultra-deep carbonate reservoirs in the Fuman Oilfield. These tests revealed valuable insights into the characteristics of acid-etching fracture morphology and conductivity under various conditions of acid-injection modes, thereby providing crucial support for the selection of acid-fracturing treatment in the field.

2. Experimental Methodology

2.1. Experimental Materials

The Fuman Oilfield in the Tarim Basin is situated at the subsidence center of both the Nanhua fault basin and the Sinian depression basin, with its target layer being Yijianfang Formation limestone buried at a depth ranging from 6500 to 8200 m. The formation temperature ranges between 146.9 °C and 163.9 °C, while the porosity of the downhole core varies from 0.54% to 7.46%, and the permeability ranges from 0.0079 to 91 mD [25].

The experimental rock samples were selected from downhole cores of the Yifangfang Formation in the Fuman Oilfield (Figure 1a). According to API standards, these samples were cut into rectangle shapes measuring 175.0 ± 0.2 mm in length, 35.0 ± 0.2 mm in width, and 50.0 ± 0.2 mm in thickness. In order to simulate underground rough hydraulic fractures, researchers often employ the Brazilian splitting method for rock sample preparation [17]. However, due to the small size and the susceptibility to damage of downhole cores, these cores have to be processed into rectangular polished rock slabs as in Figure 1b. This processing not only eliminates interference from initial hydraulic fracture morphology in experiments but also maximizes the impact of different acid-injection modes on etching morphology and flow conductivity.

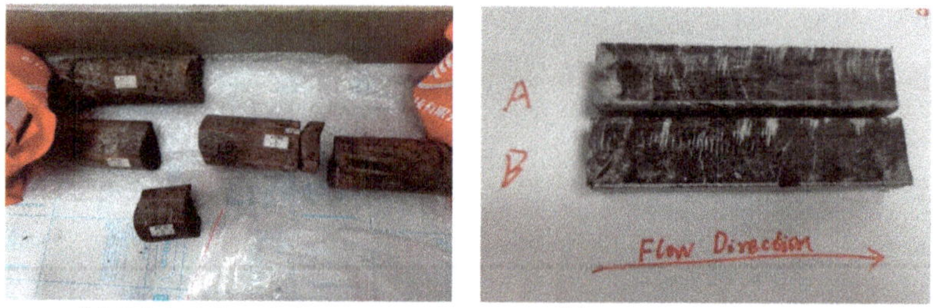

(**a**) Original downhole cores (**b**) Processed API rock slabs

Figure 1. Downhole cores and processed rock slabs. Note: A is upper slab, and B is lower slab.

The self-generating acid, cross-linked acid, gelled acid, and fracturing fluid utilized in the experiment were sourced from the Fuman Oilfield field, with their respective basic formulas presented in Table 1.

Table 1. Formulation of fracturing fluid and acid system using in Fuman Oilfield.

Acid System	Formulation
Fracturing fluid	0.45% super gum + 1% demulsifier +1% cleanup additive + 0.5% temperature stabilizer + 0.08% pH regulator + 0.1% bactericide + 0.8% cross-linking agent + 0.02% ammonium persulfate
Self-generating acid	A agent:B agent = 1:1
Cross-linked acid	20.0% HCl + 0.8% thickener + 0.5% demulsifier + 0.5% diverting agent + 2.0% corrosion inhibitors +1.0% iron ion stabilizer + 0.5% conditioner + 0.8% cross-linking agent
Gelled acid	20.0% HCl + 0.8% thickener + 0.5% demulsifier + 0.5% diverting agent + 2.0% corrosion inhibitors +1.0% iron ion stabilizer + 0.5% conditioner

2.2. Experimental Sets

Acid-fracture conductivity test equipment consists of three modules, namely, the acid-etching hydraulic fracture module, the fracture laser scanning 3D imaging module and the conductivity test module in Figure 2 [17]. The three equipment modules serve different simulation functions, with the acid-etching simulation module designed to replicate the process of acid etching during acid fracturing. The laser scanning 3D imaging module is used to capture the hydraulic fracture morphology before and after acid etching. Lastly, the conductivity test module assesses the acid-fracture conductivity under closure pressure.

(**a**) Acid-etching module (**b**) Fracture laser scanning 3D imaging module (**c**) Conductivity test module

Figure 2. Acid-fracture conductivity test equipment.

2.3. Experimental Schedule Design

In order to match the reservoir conditions with the experimental test conditions of acid-fracture conductivity, the three key parameters primarily considered are the acid-injection displacement, the acid-injection time, and the acid-rock reaction temperature.

2.3.1. Injection Displacement and Time

The injection displacement (mL/min) at the laboratory experimental scale cannot be consistent with the injection displacement (m^3/min) at the field scale due to limitations in device injection capacity and rock sample size. The acid-injection displacement at the engineering scale is often downscaled to the laboratory experimental scale based on the Reynolds number similarity criterion. The conversion formula is as follows [17]:

$$q_1 = \frac{q_f h_1}{2h_f}\left(\frac{w_f}{w_1}\right)^{\frac{2n-2}{2-n}} \tag{1}$$

where q_l is the experimental acid-injection displacement, mL/min; q_f is the acid-injection displacement in the field, mL/min; w_l is the experimental fracture width, m; w_f is the field fracture width, m; h_l is the experimental fracture height, m; h_l is the field fracture height, m; n is the flow characteristics of the index.

The field acid-injection displacement is scaled down to the experimental acid-injection displacement based on the rheological parameter characteristics and acid-injection rate of self-generating acid and cross-linked acid in the Fuman Oilfield within the research area, as per Formula (1). The corresponding results are presented in Table 2.

Table 2. Conversion of field engineering scale into laboratory experimental scale displacement.

Self-Generating Acid (n = 0.125)		Cross-Linked Acid (n = 0.295)	
Field, m^3/min	**Laboratory, mL/min**	**Field, m^3/min**	**Laboratory, mL/min**
4.5	157	4.5	195
5.5	189	5.5	216
6.0	209	6.0	260

As shown in Table 2, the self-generating acid-injection displacement of Fuman Oilfield ranges from 4.5 to 6.0 m^3/min, which is equivalent to an experimental acid-injection rate of 157–260 mL/min. The experimental design of self-generating acid and cross-linked acid uses an average experimental acid-injection rate of 200 mL/min. The acid-injection rate of terminal gelled acid (uncross-linked cross-linked acid) is set at 100 mL/min to simulate the process of close acidizing fracture.

Acid-injection time is determined based on the standard acid-injection scale and acid-injection rate in the Fuman Oilfield. The duration for self-generating acid is 60 min, cross-linked acid requires 90 min, and gelled acid needs only 10 min.

2.3.2. Acid-Rock Reaction Temperature

Temperature exerts a significant effect on the acid-rock reaction of carbonate rocks, particularly in ultra-deep and high-temperature reservoirs.

Based on the typical injection volume (1000–2000 m^3) in single well acid fracturing at the Fuman Oilfield, the numerical simulation calculation method was employed to simulate the underground temperature field variations during different alternating stages with a certain total injection volume [26], which was used to determine the temperature of the acid-rock reaction. The distribution of each type of fluid in each alternating stage was evenly balanced based on the alternating series when there were changes in the alternate stages. The underground temperature change resulting from the alternating injection of slick water + fracturing fluid + self-generating acid + cross-linked acid + gelled acid (closed acidizing) at a total injection volume of 1750 m^3 is illustrated in Figure 3. In this process, slick water 450 m^3 and fracturing fluid 450 m^3 are used to cool the formation and create fractures. Self-generating acid with a volume of 300 m^3 is employed to etch the far end of the hydraulic fracture, while cross-linked acid at 450 m^3 is used to etch the middle section. Additionally, gelled acid 25 m^3 is applied for closed acidizing to enhance the conductivity near the wellbore. Finally, a displacement fluid of 75 m^3 displaces both surface and wellbore acids into the formation.

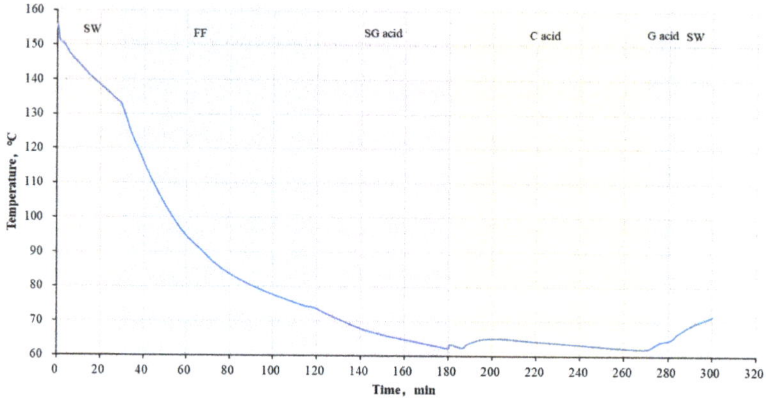

(a) One-stage alternating injection

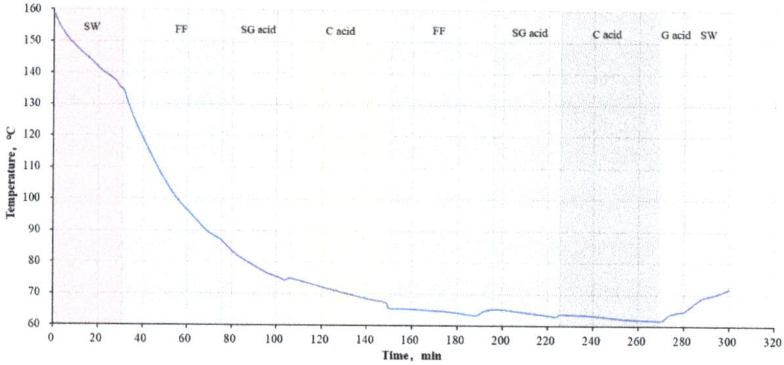

(b) Two-stage alternating injection

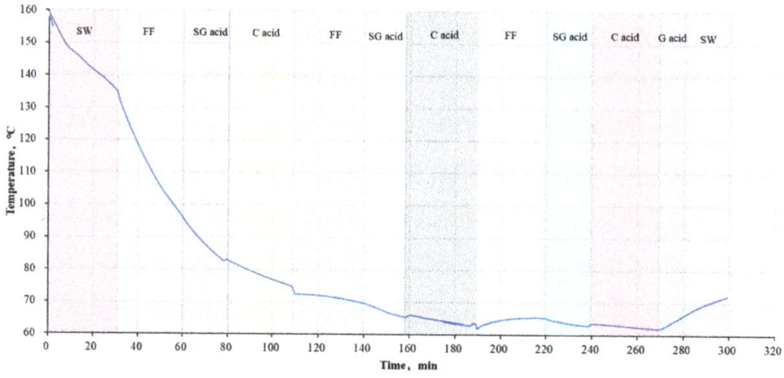

(c) Three-stage alternating injection

Figure 3. Variation of bottom temperature under different alternate stages. Note: SW is slick water, FF is fracturing fluid, SG acid is self-generating acid, C acid is cross-linked acid, and G acid is gelled acid.

The variation of underground temperature caused by different alternating injection modes is quite different, as depicted in Figure 3, under the condition of the same injection volume. In Figure 3a, it can be observed that when employing non-reactive pre-pad fracturing (i.e., one-stage alternating injection), the formation temperature during the slick water and fracturing fluid injection reaches 75 °C, which corresponds to the initial temperature for self-generating acid-rock reaction. Subsequently, upon complete injection of self-generating acid, the formation temperature drops to 60 °C, representing the initial temperature for the cross-linked acid-rock reaction. Similarly, from Figure 3b,c, we obtained fracture temperatures for two-stage and three-stage alternating acid-etching experiments, respectively, which are listed in Table 3.

Table 3. Acid-etching experimental scheme.

Rock No.	Acid Injection Mode	Alternative Stages	Temperature, °C			Acid Volume, L		
			SG Acid	C Acid	G Acid	SG Acid	C Acid	G Acid
1	FF + SG acid	/	75	/	/	31		
2	FF + C acid	/	/	75	/		31	
3	FF + SG acid +C acid (one stage)+ G acid	1	75	63	62	12	18	1
4	FF + SG acid +C acid (two stages)+ G acid	2	85 65	75 63	/ 62	6 6	9 9	 1
5	FF + SG acid +C acid (three stages)+ G acid	3	95 70 65	80 65 65	/ / 62	4 4 4	6 6 6	 1

2.3.3. Experimental Schedule

Based on the confirmed experimental acid-injection rate, time, and reaction temperature, the designed acid-etching fracture scheme is shown in Table 3. Experiments 1 and 2 simulate the effects of the single self-generating acid and cross-linked acid system on the acid-etching behavior. Experiments 3, 4, and 5 simulate the acid-etching behavior of fracturing fluid + self-generating acid + cross-linked acid + gelled acid (closed acidizing) under different alternate stages.

The Yijianfang Formation reservoir depth in the Fuman Oilfield ranges from 6500 m to 8200 m, with the closure pressure ranging from 40 MPa to 80 MPa. Sequential experiments were conducted on the acid-etching fracture followed by laser scanning to analyze the acid-etched fracture morphology. Conductivity tests were performed at intervals of increasing closure pressure by increments of 10 MPa until reaching a maximum pressure level of 80 MPa. The experimental procedure is illustrated in Figure 4.

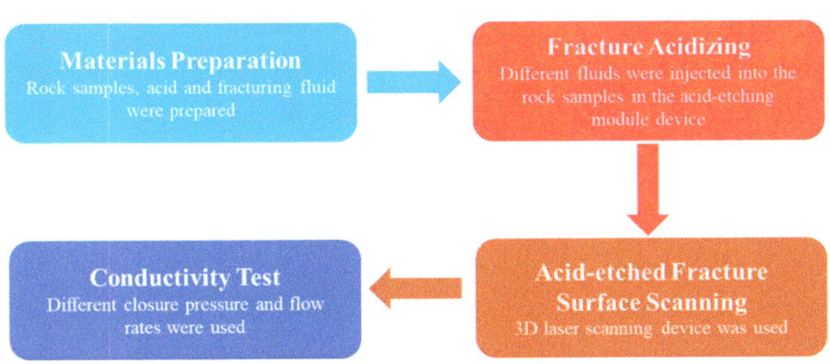

Figure 4. Variation of bottom temperature under different alternate stages.

3. Results and Discussion

3.1. Characteristics of Acid-Etched Fracture under Different Acid-Injection Modes

Based on the acid-etched fracture morphology and the quantity of rock plate dissolution, we investigated the acid-etched morphology, the distribution of fracture elevation, as well as the features of acid-rock reaction dissolution under conditions involving single acid injection and different alternating injection stages with various acid systems.

3.1.1. Morphological Changes of Hydraulic Fractures before and after Acid Etching

Table 4 shows the morphological changes of rock slabs before and after acid etching under different acid-injection model conditions. It can be observed from Table 4 that prior to acid etching, all rock slabs exhibit a uniformly smooth surface. The presence of bedding and texture variations in the rock samples was attributed to the influence of their respective bedding and joint characteristics.

Table 4. Changes of acid-etching fracture morphology under different acid-injection modes.

Acid-Injection Mode	Before Acid Injection	After Acid Injection	Height Colorimetric Bars
FF + SG acid	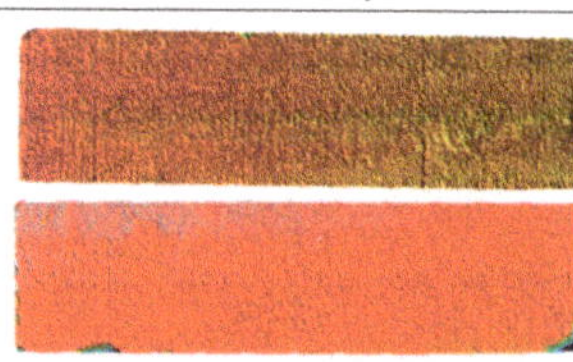	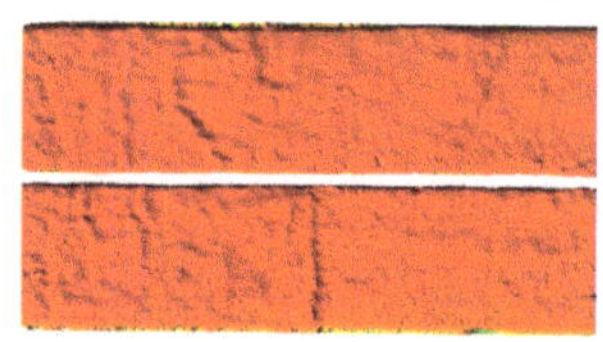	
FF + C acid		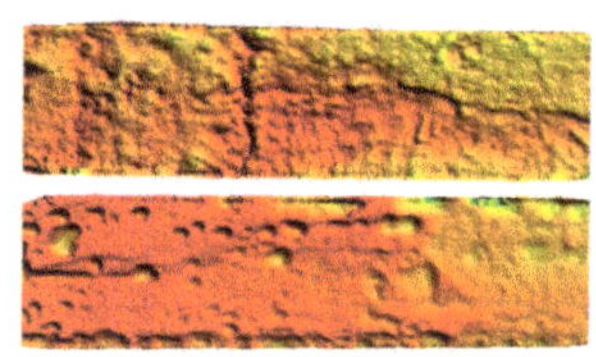	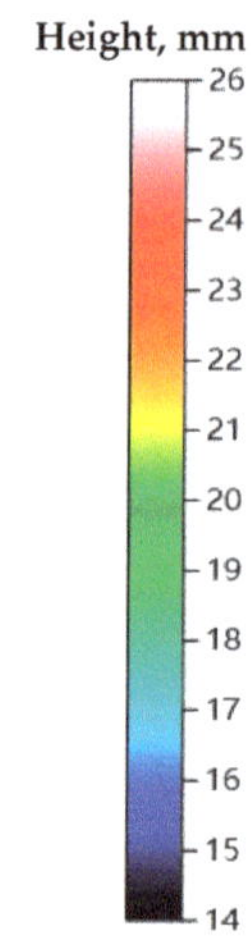
FF + SG acid + C acid (one stage) + G acid		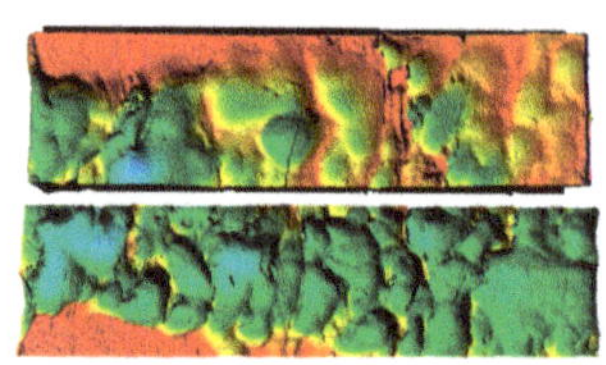	
FF + SG acid + C acid (two stages)+ G acid	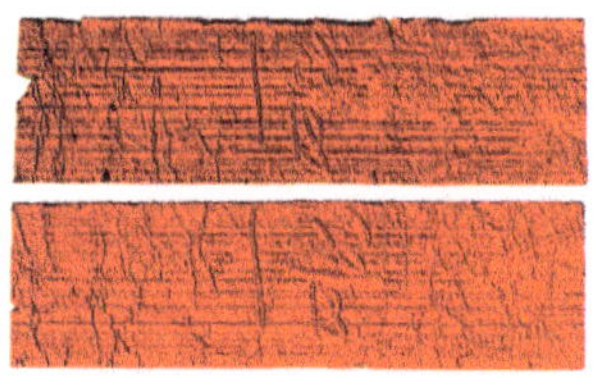	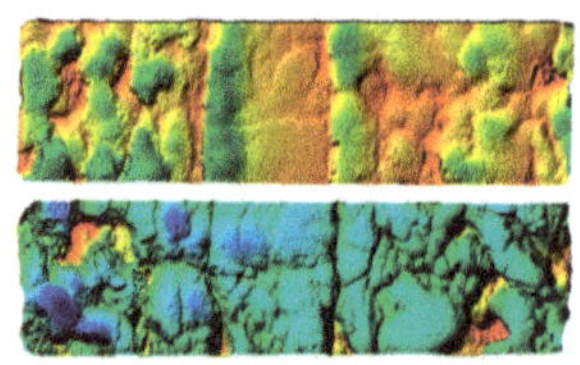	
FF + SG acid + C acid (three stages) + G acid		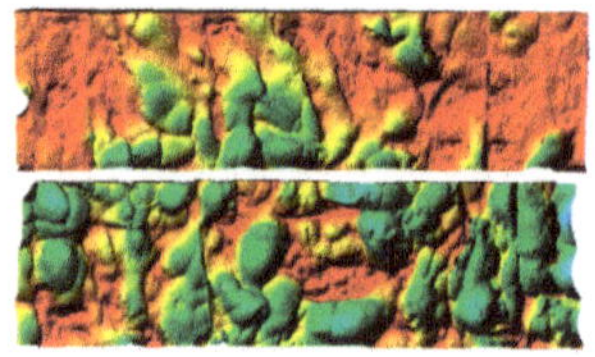	

When using the fracturing fluid in conjunction with self-generating acid-injection mode (FF + SG acid), a vertical groove emerged within the central region of the lower rock sample, perpendicular to the direction of acid flow. In the upper rock sample, erosion pits predominantly manifested near the inlet end within the middle section. This indicated that under this injection mode, the rock samples exhibited some degree of non-uniform etching, but the extent of non-uniform etching was relatively minimal. The underlying reason for this alteration was attributed to the limited reactive capacity of the self-generating acid, leading to restricted etching effects on the rock formation [17].

When utilizing the fracturing fluid in conjunction with cross-linked acid-injection mode (FF + C acid), noticeable non-uniform etching was observed on the rock samples. Point-like pits were present near the outlet end of the upper rock surface, while multiple grooves perpendicular to the direction of acid flow were evident on the lower rock surface. Following exposure to cross-linked acid alone, a predominant green coloration was observed on the lower rock plate, whereas the upper rock plate exhibited a predominantly red–green appearance, indicating more severe etching on the lower rock plate. Overall, compared to the injection mode of FF+SG acid, the injection mode of FF+C acid demonstrated a significantly stronger etching capability.

When employing the injection mode of fracturing fluid combined with self-generating acid and cross-linked acid, namely FF + SG acid + C acid (one-three stages) + G acid, the rock face appeared smooth prior to etching. However, after etching, varying degrees of etched grooves emerged on both upper and lower rock plates. In certain regions, natural fractures were formed alongside acid wormholes. As the alternating stage of the injection pattern increased from one to two, distinct fish-scale non-uniform morphology was observed on the rock sample after etching, which may be attributed to the obvious network filling fractured before etching. Furthermore, when the injection stage was increased to three, a more conspicuous fish-scale etching pit was formed.

Comparing the acid-etching morphological characteristics of hydraulic fractures under five different acid-injection processes, it was evident that the acid-etching extent was limited when employing the simple injection mode of fracturing fluid and acid. However, with the alternating injection mode of fracturing fluid and acid, the non-uniform dissolution of rock samples was significantly enhanced. Additionally, features such as dissolution grooves and pits appeared, and the acid fluid obviously communicated potential natural fractures and cavities.

3.1.2. Elevation Frequency Distribution Characteristics of Acid-Etched Rock Slabs

To analyze the impact of different acid-injection processes further quantitatively on acid-etched fracture behavior, the elevation frequency distribution of rock surfaces was selected for the quantitative characterization and analysis of the fracture surfaces before and after acid etching [27]. Elevation frequency distribution refers to the frequency of occurrence of heights at various points on the rock sample surface within a certain range. The histogram and its curve of the elevation frequency distribution visually depict the roughness of the rock sample surface.

Figure 5 illustrates the elevation frequency distribution characteristics after acid etching of rock plates under different acid-injection conditions. The elevation distribution range resulting from the two-stage alternating etching exhibited the widest span, indicating a stronger degree of non-uniform etching. In contrast, the elevation distribution range of the fracturing fluid + single acid-injection mode was narrower, with elevations mostly concentrated in the interval greater than or equal to 20 mm. This suggested that the etching extent of the fracturing fluid + single acid-injection mode was also lower compared to that of the fracturing fluid + multiple acid types alternating injection.

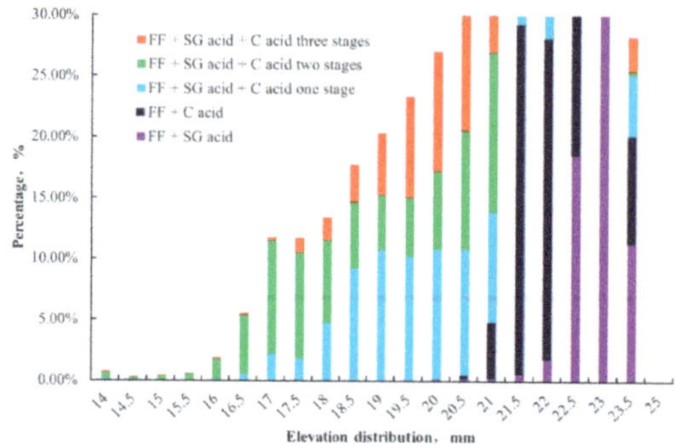

Figure 5. Elevation frequency distribution of acid-etched rock slabs under different acid-injection modes.

Figure 6 presents the cumulative frequency distribution of elevation for acid-etched rock plates under different acid-injection modes. Furthermore, it was observed that the frequency distribution of the rock plates etched with the fracturing fluid + single acid-injection mode was relatively narrow, and the frequency distribution curve exhibited a distinct steep characteristic, indicating a weaker degree of non-uniform etching in the acid-etched morphology. In contrast, the cumulative distribution curves of elevation frequency for acid-etched fracture surfaces formed by the fracturing fluid + different acid-injection modes were relatively flatter, with a broader distribution range. Among them, the elevation distribution curve formed by the two-stage alternating injection mode was the widest and flattest, indicating a stronger degree of non-uniform etching, consistent with the results in Table 4.

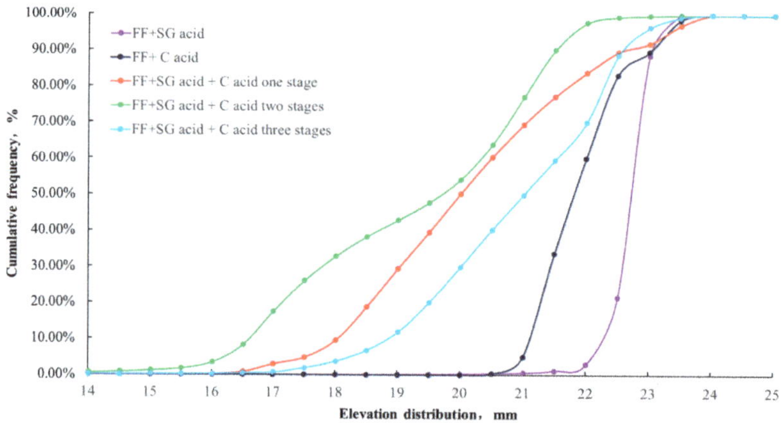

Figure 6. Cumulative frequency distribution of acid-etched slab elevation under different acid-injection modes.

3.1.3. Acid-Rock Reaction Characteristics

The overall acid dissolution quantity of the rock plates reflected the degree of acid-rock reaction during each set of acid-etching experiments. A higher dissolution quantity indicated a greater amount of rock consumed in the acid-etching process, which suggested a more intense acid-rock reaction. Figure 7 displays the variation in acid-dissolution quantity of rock plates under different acid-injection conditions. It is evident that the rock plates under two-stage alternating injection mode resulted in larger quantities of dissolved rock, indicating more intense and non-uniform etching. This non-uniform etching enhanced the roughness of the acid-etched fracture surface, thereby augmenting the area for the acid-rock reaction and consequently resulting in an escalation of corrosion. This substantial level of acid-induced corrosion and the extent of non-uniform etching were mutually reinforcing. This finding is consistent with the numerical characterization of acid etching and the elevation frequency distribution features in Table 4 and Figures 5 and 6.

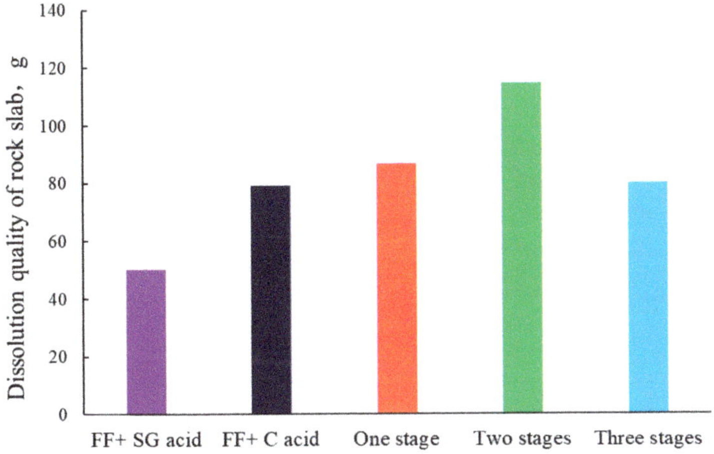

Figure 7. Amount of dissolution of slabs with different acid-injection modes.

3.2. Variation Characteristics of Fracture Conductivity under Different Acid-Injection Modes

In order to reveal the variation of the acid-etched fracture conductivity, the retention rate of conductivity at a certain closure stress is defined as the percentage of the conductivity in the initial conductivity by Equation (2) [17]:

$$R = \frac{(Kw)_i}{(Kw)_o} \times 100\% \tag{2}$$

where R is the conductivity retention rate, %; $(Kw)_i$ is the conductivity under closure stress i, D·cm; and $(Kw)_o$ is the initial conductivity, D·cm.

Figures 8 and 9 illustrate the variations in acid-etched fracture conductivity and conductivity retention rate under different acid systems and alternating injection modes.

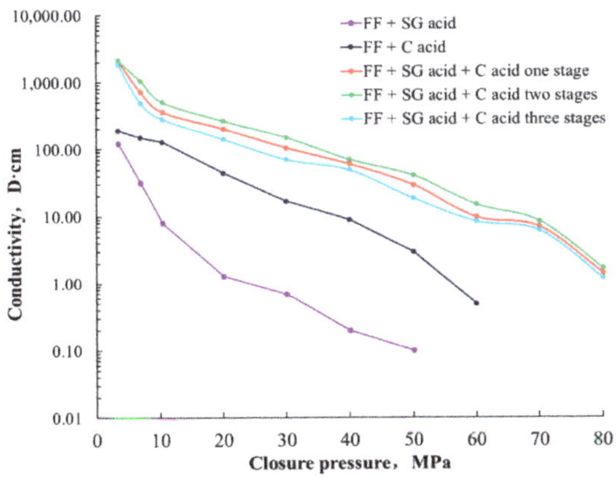

Figure 8. Conductivity under different acid-injection modes.

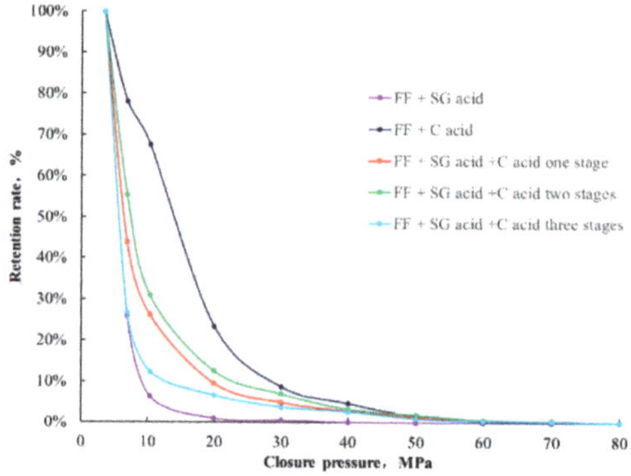

Figure 9. Comparison of changing rates of conductivity under different acid-injection modes.

Based on Figures 8 and 9, it is evident that significant differences exist in the conductivity and conductivity retention of acid-etched fractures under varying acid-injection conditions. In terms of conductivity, when the closure pressure was less than or equal to 60 MPa, the overall trend indicated that the conductivity under the alternating acid-injection mode was greater than that under the fracturing fluid + single acid-injection mode. Moreover, consistently displaying superior values, the two-stage alternating injection mode demonstrated optimal conductivity levels. When the closure pressure exceeded 60 MPa, the effectiveness of the fracturing fluid + single acid-injection mode in maintaining conductivity diminished significantly. Therefore, considering the conductivity factor, it was not advisable to employ the fracturing fluid + single acid-injection mode as pre-acid fracturing for deep carbonate reservoirs in the Fuman Oilfield.

Regarding conductivity retention, when the closure pressure was greater than 10 MPa, the acid-etched fracture conductivity was less than 50% of its initial value. This suggested that under the influence of normal closure pressure conditions, the acid-etched fracture gradually came into contact and consolidation, resulting in compression and reduction of the fracture flow channel. When the closure pressure exceeded 20 MPa, there was no significant change in conductivity retention rate. Overall, the mode of fracturing fluid + cross-linked acid exhibited a higher retention rate of conductivity, indicating a stronger ability of the acid-etched fracture to resist deformation under normal closure pressure. However, on the other hand, as indicated by Figure 8, when the closure pressure exceeds 60 MPa, the conductivity formed by the fracturing fluid + single acid-injection mode has essentially been lost. The reasons for the loss of conductivity under high closure stress are as follows: first, complete contact between the rock faces leaves no seepage space available; another contributing factor may be the significant deterioration in mechanical strength of rock samples during the acidizing process [28,29], leading to detachment and migration of mineral particles that gradually accumulate at fracture outlets, thereby impeding accurate measurement of conductivity. This phenomenon also occurs in real-life mining scenarios [30]. The comprehensive consideration of conductivity suggested that the two-stage alternating injection mode of fracturing fluid + acid was recommended for acid fracturing in deep carbonate reservoirs in the Fuman Oilfield.

4. Field Application

The 305H well is an appraisal well located in the Fuman Oilfield, with the target interval for stimulation being 7763.00–8410.00 m. During drilling, the well encountered a fracture-cavity body boundary directly (Figure 10). The spatial relationship between the wellbore and reservoir body is as follows: vertically, the initiation point of the fractures is 108 m away from the core of the cavity in the upward fracture height direction, 180 m away from the boundary of the cavity, and 10 m away from the boundary in the downward fracture height direction (Figure 10a). Horizontally, the cavity body extends in a northeast direction, similar to the direction of fracture propagation. Along the direction of fracture propagation, the initiation point is 134–202 m away from the cavity boundary and 104 m away from the core of the cavity (Figure 10b).

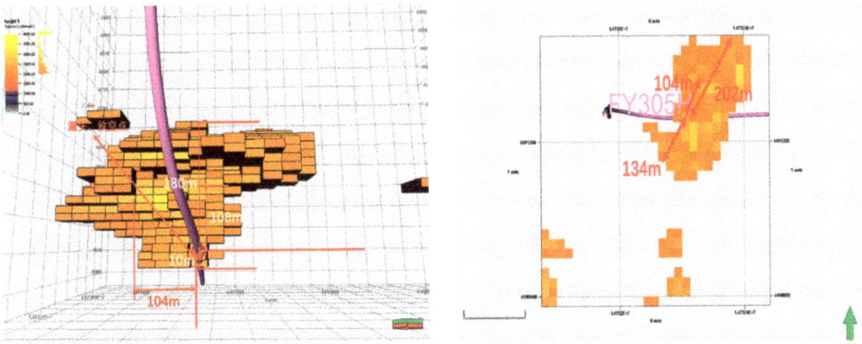

(**a**) Front view (**b**) Top view

Figure 10. The relationship between fracture-cavity and wellbore position in 305H well.

To ensure effective communication with the cavity body by employing acid fracture, there is a significant demand for the upper fracture height, which should be greater than 108 m, and the fracture length should be greater than 104 m. Figure 11 illustrates the variation in fracture geometry parameters calculated using fracturing simulation software 1.0 under different injection rates and injection volumes. To guarantee effective communication with the cavity body, it was recommended to use an injection displacement of $\geq$6.0 m^3/min and an injection volume of 1200–1400 m^3.

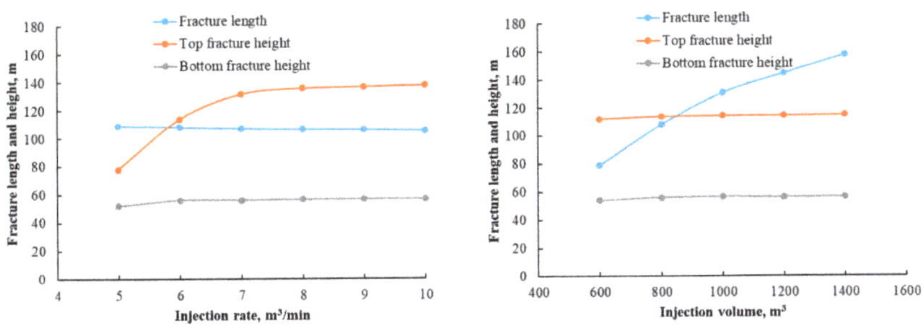

(**a**) Effect of injection displacement (**b**) Effect of injection volume

Figure 11. Influence of different injection parameters on geometric size of acid-fractured fracture.

Based on the experimental results of acid-etched fracture conductivity, the recommended acid fracturing injection mode was the two-stage alternating injection of FF + SG acid + C acid, followed by gelled acid for closed acidizing. According to the designed total injection volume of 1200–1400 m^3, the individual fluid volumes were as follows: fracturing fluid 390 m^3, gelled acid 200 m^3, self-generating acid 380 m^3, cross-linked acid 400 m^3, and displacement fluid 80 m^3.

The well successfully underwent a two-stage alternating acid-fracturing treatment in accordance with the design parameters (Figure 12). The shutdown tubing pressure curve depicted in Figure 12 appears to be nearly linear, which is a typical response of an acid-fracture communication pressure response for a large fracture cavity [31]. This indicates that the design objective for acid fracturing has been successfully achieved.

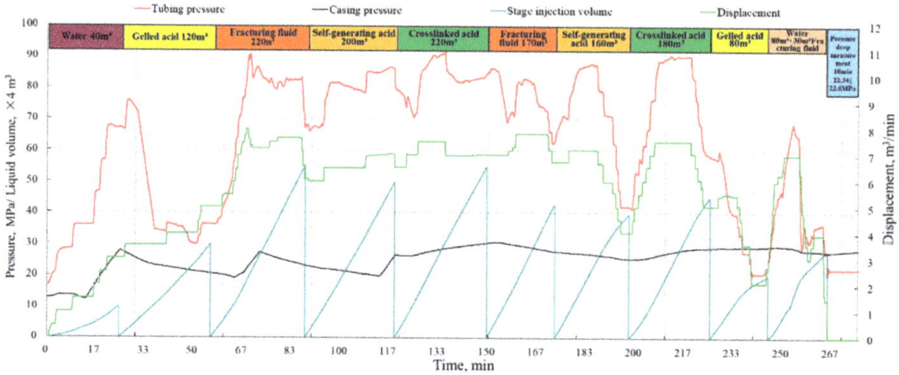

Figure 12. Construction curve of acid fracturing in FY305H well.

Following the acid-fracturing treatment, the daily oil production has reached 240 m^3/d, while the daily gas production stands at 85,686 m^3/d, resulting in an equivalent oil production rate of 265.6 t/d. This signifies a remarkable increase in production by 20.4% compared to neighboring wells employing three-stage alternating acid fracturing under similar conditions listed in Table 5. Furthermore, it is worth noting that the acid-injection process has been streamlined.

Table 5. Comparison of acid-injection parameters and production from test well and adjacent well.

Well	Fracturing Fluid m^3	Cross-Linked Acid m^3	Self-Generating Acid m^3	Gelled Acid m^3	Acid-Injection Mode	Equivalent Oil Production Rate t/d
305H	390	400	380	200	two stages	265.6
701	760	410	300	50	three stages	220.6

5. Proposal for Future Improvements

Although this paper conducted experimental research on the conductivity of different acid-injection modes, there are still several future research suggestions:

(1) The enhancement of rock pore and natural fracture permeability through acid solution should be prioritized, along with its subsequent impact on conductivity. The acid fluid infiltrates the rock pores and the natural fractures on the fracture wall during acid fracturing, lead to pore enlargement in the reservoir and increased width of natural fractures through acid dissolution. As a result, permeability is enhanced.

(2) The influence of rough hydraulic morphology and changes in rock mechanical strength on conductivity should be further emphasized, and a corresponding model for predicting conductivity should be established. This is also crucial for the design of acid pressure schemes and evaluating their effectiveness.

6. Conclusions

(1) The non-uniform etching degree of the hydraulic fractures is less pronounced in the fracturing fluid + single acid-injection mode compared to that under the fracturing fluid + multiple types of acid alternating injection mode. Furthermore, an increasing trend in the non-uniform etching degree is observed with an increase in injection stages. However, the two-stage alternating injection mode exhibits a higher degree of non-uniform etching.

(2) Conductivity generated under the fracturing fluid + multiple types of acid alternating injection mode is superior to that produced under the fracturing fluid + single acid-injection mode.

(3) From the perspective of conductivity, it is recommended to adopt the fracturing fluid + different types of acid alternating injection mode based on the closure pressure characteristics of the Fuman Oilfield. Field application results indicate significantly improved production enhancement and stimulation effects using this acid-injection mode.

Author Contributions: Writing—draft and investigation—experimental, J.L. (Jiangyu Liu); Investigation—discussion, D.R.; Conceptualization—S.F. and J.L. (Ju Liu); Investigation—field application and discussion, S.Q. and X.Q.; Reviewing and Conceptualization, B.G. All authors have read and agreed to the published version of the manuscript.

Funding: This research received no external funding.

Data Availability Statement: Data sharing is provided as required.

Conflicts of Interest: Authors Jiangyu Liu, Dengfeng Ren, Shaobo Feng, Ju Liu, Shiyong Qin and Xin Qiao were employed by the company PetroChina. The remaining author declares that the research was conducted in the absence of any commercial or financial relationships that could be construed as a potential conflict of interest.

References

1. He, D.F.; Jia, C.Z.; Zhao, W.Z.; Xu, F.Y.; Luo, X.R.; Liu, W.H.; Tang, Y.; Gao, S.L.; Zheng, X.J.; Li, D.; et al. Research progress and key issues of ultra-deep oil and gas exploration in China. *Pet. Explor. Dev.* **2023**, *50*, 1162–1172. [CrossRef]
2. Lei, Q.; Xu, Y.; Yang, Z.W.; Cai, B.; Wang, X.; Zhou, L.; Liu, H.F.; Xu, M.J.; Wang, L.W.; Li, S. Progress and development directions of stimulation techniques for ultra-deep oil and gas reservoirs. *Pet. Explor. Dev.* **2021**, *48*, 193–201. [CrossRef]
3. Aljawad, M.S.; Aljulaih, H.; Mahmoud, M.; Desouky, M. Integration of field, laboratory, and modeling aspects of acid fracturing: A comprehensive review. *J. Pet. Sci. Eng.* **2019**, *181*, 106158. [CrossRef]
4. Deng, J.; Mou, J.; Hill, A.D.; Zhu, D. A new correlation of acid-fracture conductivity subject to closure stress. *SPE Prod. Oper.* **2012**, *27*, 158–169. [CrossRef]
5. Zhang, L.; Zhou, F.; Mou, J.; Xu, G.; Zhang, S.; Li, Z. A new method to improve long-term fracture conductivity in acid fracturing under high closure stress. *J. Pet. Sci. Eng.* **2018**, *171*, 760–770. [CrossRef]
6. Bale, A.; Smith, M.B.; Klein, H.H. Stimulation of carbonates combining acid fracturing with proppant (CAPF): A Revolutionary Approach for Enhancement of Sustained Fracture Conductivity and Effective Fracture Half-length. In Proceedings of the SPE Annual Technical Conference and Exhibition, Florence, Italy, 20–22 September 2010.
7. Desouky, M.; Samarkin, Y.; Aljawad, M.S.; Amao, A.; AlTammar, M.J.; Alruwaili, K.M. Diammonium Phosphate Treatment for Sustained Hydraulic/Acid Fracture Conductivity in Chalk and Limestone Formations. *SPE J.* **2024**, 1–13. [CrossRef]
8. Desouky, M.; Aljawad, M.S.; Abduljamiu, A.; Solling, T.; Abdulraheem, A.; AlTammar, M.J.; Alruwaili, K.M. Enhancing Fracture Conductivity in Soft Chalk Formations with Diammonium Phosphate Treatment: A Study at High Temperature, Pressure, and Stresses. *SPE J.* **2023**, *28*, 3280–3290. [CrossRef]
9. Coulter, A.W.; Crowe, C.W.; Barrett, N.D.; Miller, B.D. Alternate stages of pad fluid and acid provide improved leakoff control for fracture acidizing. In Proceedings of the Society of Petroleum Engineers (SPE), New Orleans, LA, USA, 3 October 1976.
10. Li, X.; Chen, Y.; Yang, Z.; Chen, F. Large-scale visual experiment and numerical simulation of acid fingering during carbonate acid fracturing. In Proceedings of the SPE Asia Pacific Oil and Gas Conference and Exhibition, Jakarta, Indonesia, 17–19 October 2017.
11. Li, X.; He, Y.; Yang, Z.; Zhu, J.; Li, F.; Song, R. Fully coupled model for calculating the effective acid penetration distance during acid fracturing. *J. Nat. Gas Sci. Eng.* **2020**, *77*, 103267. [CrossRef]
12. Jiao, F.Z. Practice and knowledge of volumetric development of deep fractured-vuggy carbonate reservoirs in Tarim Basin, NW China. *Pet. Explor. Dev.* **2019**, *46*, 576–582. [CrossRef]
13. Petriz Munguia, J.M.; Gonzalez Valtierra, B.E.; Trujillo Hernandez, J.; Santos, S.; Campos Monroy, K. Acid-fracturing techniques as a good alternative to help improve field development assets. In Proceedings of the Abu Dhabi International Petroleum Exhibition and Conference, Abu Dhabi, United Arab Emirates, 11–14 November 2019.
14. Sarma, D.K.; Pal, T.; Kumar, D.; Lahiri, G.; Manchalwar, V.V. Application of Closed Fracture Acidizing for Stmulation of Tight Carbonate Reservoir in Mumbai Offshore. In Proceedings of the SPE Oil and Gas India Conference and Exhibition, Mumbai, India, 4–6 April 2017.
15. Mou, J.Y.; Zhang, S.C. Modeling acid leakoff during multistage alternate injection of pad and acid in acid fracturing. *J. Nat. Gas. Sci. Eng.* **2015**, *26*, 1161–1173. [CrossRef]
16. Zhang, K.; Chen, M.; Zhou, C.; Dai, Y.; Liu, F.; Li, J. Study of alternating acid fracturing treatment in carbonate formation based on true tri-axial experiment. *J. Pet. Sci. Eng.* **2020**, *192*, 107268. [CrossRef]
17. Gou, B.; Guan, C.C.; Li, X.; Ren, J.C.; Zeng, J.; Wu, L.; Guo, J.C. Acid-etching fracture morphology and conductivity for alternate stages of self-generating acid and gelled acid during acid-fracturing. *J. Pet. Sci. Eng.* **2021**, *200*, 108358. [CrossRef]
18. Zhang, Q.; Liu, P.; Xiong, Y.; Du, J. Self-Generated Organic Acid System for Acid Fracturing in an Ultrahigh-Temperature Carbonate Reservoir. *ACS Omega* **2023**, *8*, 12019–12027. [CrossRef]

19. Lungwitz, B.; Fredd, C.; Brady, M.; Miller, M.; Ali, S.; Hughes, K. Diversion and cleanup studies of viscoelastic surfactant-based self-diverting acid. *SPE Prod. Oper.* **2007**, *22*, 121–127. [CrossRef]
20. Navarrete, R.C.; Holms, B.A.; McConnell, S.B.; Linton, D.E. Laboratory, theoretical, and field studies of emulsified acid treatments in high-temperature carbonate formations. *SPE Prod. Facil.* **2000**, *15*, 96–106. [CrossRef]
21. Tariq, Z.; Hassan, A.; Al-Abdrabalnabi, R.; Aljawad, M.S.; Mahmoud, M. Comparative study of fracture conductivity in various carbonate rocks treated with GLDA chelating agent and HCl acid. *Energy Fuels* **2021**, *35*, 19641–19654. [CrossRef]
22. Sui, Y.; Cao, G.; Guo, T.; Li, Z.; Bai, Y.; Li, D.; Zhang, Z. Development of gelled acid system in high-temperature carbonate reservoirs. *J. Pet. Sci. Eng.* **2022**, *216*, 110836. [CrossRef]
23. Ren, Y.; Wang, D.; Feng, P.; Chen, J.; Shao, S.; Zhang, W.; Cui, B. An Integrated Experimental Method to Study Multistage Alternating Injection Acid Fracturing in Limestone Formation. In Proceedings of the SPE Middle East Oil and Gas Show and Conference, Manama, Bahrain, 9–21 February 2023.
24. Zhu, D.; Wang, Y.; Cui, M.; Zhou, F.; Wang, Y.; Liang, C.; Yao, F. Acid system and stimulation efficiency of multistage acid fracturing in porous carbonate reservoirs. *Processes* **2022**, *10*, 1883. [CrossRef]
25. Wang, Q. Origin of gas condensate reservoir in Fuman Oilfield, Tarim Basin, NW China. *Pet. Explor. Dev.* **2023**, *50*, 1128–1139. [CrossRef]
26. Guo, J.C.; Liu, Z.; Gou, B.; Zeng, M.Y. Study of wellbore heat transfer considering fluid rheological effects in deep well acidizing. *J. Pet. Sci. Eng.* **2020**, *191*, 107171. [CrossRef]
27. Gou, B.; Li, X.; Ma, H.Y.; Zhou, C.L. Effects of Morphology of Hydraulic Fractures on Acid Etching Behaviors and Fluid Diversion Capacity. *J. Southwest Pet. Univ.* **2019**, *41*, 80–90.
28. Jin, S.; Wang, X.; Wang, Z.; Mo, S.; Zhang, F.; Tang, J.; Xiong, H. Evaluation approach of rock brittleness index for fracturing acidizing based on energy evolution theory and damage constitutive relation. *Lithosphere* **2021**, *2021*, 2864940. [CrossRef]
29. Chen, W.; Yang, J.; Li, L.; Wang, H.; Huang, L.; Jia, Y.; Hu, Q.; Jiang, X.; Tang, J. Investigation of Mechanical Properties Evolution and Crack Initiation Mechanisms of Deep Carbonate Rocks Affected by Acid Erosion. *Sustainability* **2023**, *15*, 11807. [CrossRef]
30. Grifka, J.; Nehler, M.; Licha, T.; Heinze, T. Fines migration poses challenge for reservoir-wide chemical stimulation of geothermal carbonate reservoirs. *Renew. Energy* **2023**, *219*, 119435. [CrossRef]
31. Chen, Z.; Zhang, S. Evaluation of geological effects of acid-fracturing in deep carbonate reservoirs: Taking the fractured-vuggy carbonate reservoirs in Tahe oilfield as an example. *Oil Gas Geol.* **2004**, *25*, 686–691.

processes

MDPI

Article

Study of Acid Fracturing Strategy with Integrated Modeling in Naturally Fractured Carbonate Reservoirs

Xusheng Cao [1,2], Jichuan Ren [3,4,*], Shunyuan Xin [1,2], Chencheng Guan [3,4], Bing Zhao [1,2] and Peixuan Xu [3,4]

1 Petroleum Engineering Technology Research Institute, Northwest Oilfield Branch Company—Sinopec, Urumqi 830011, China; caoxsh0705.xbsj@sinopec.com (X.C.); xinshy1996.xbsj@sinopec.com (S.X.); zhaobincugb@126.com (B.Z.)

2 National Key Laboratory of Fractured Reservoirs Developing and Exploiting, Northwest Oilfield Branch Company—Sinopec, Urumqi 830011, China

3 School of Petroleum Engineering, Southwest Petroleum University, Chengdu 610500, China; yummxvvv@163.com (C.G.); 202221000832@stu.swpu.edu.cn (P.X.)

4 National Key Laboratory of Oil and Gas Reservoir Geology and Exploitation, Southwest Petroleum University, Chengdu 610500, China

* Correspondence: 202199010016@swpu.edu.cn

Abstract: Natural fractures and wormholes strongly influence the performance of acid fracturing in naturally fractured carbonate reservoirs. This work uses an integrated model to study the effects of treatment parameters in acid fracturing in different reservoir conditions. Hydraulic fracture propagation, wormhole propagation, complex fluid leak-off mediums, and heat transfer are considered in the modeling. The model is validated in several steps by analytical solutions. The simulation results indicated that natural fractures and wormholes critically impact acid fracturing and can change the predicted outcomes dramatically. The high permeability reservoirs with conductive natural fractures or low permeability reservoirs with natural fracture networks showed the highest stimulation potential in applying acid fracturing technology. The optimal acid injection rate depends on natural fracture geometry and reservoir permeability. This study also observed that obtaining a high production index is difficult because natural fractures and wormholes reduce the acid efficiency during acid fracturing. Building an acid-etched fracture system consisting of acid-etched natural fractures and hydraulic fractures may help us better stimulate the naturally fractured carbonate reservoirs. The paper illustrates a better understanding of the effects of the treatment design parameters on productivity. It paves a path for the optimal design of acid fracturing treatment for heterogeneous carbonate reservoirs.

Keywords: naturally fractured carbonate; acid fracturing; integrated modeling; stimulation strategy

1. Introduction

Carbonate reservoirs play a critical role in global hydrocarbon resources. Approximately 70% of the world's oil and gas resources are associated with carbonate reservoirs. These reservoirs are predominantly composed of carbonate rock. The main carbonate oil and gas production areas are distributed across regions, including the Middle East, North America, and Central Asia. These regions hold 96.8% of the world's recoverable carbonate rock oil and gas reserves (Li et al., 2018) [1]. Natural fractures are widely developed within these reservoirs, significantly influencing the flow capacity of the reservoirs [2–5]. However, they also lead to extensive filtration during the stimulation process, making hydraulic fracturing in such reservoirs prone to sand blockage, leading to construction failure. So, acid fracturing has become the common well stimulation technique mainly applied to these reservoirs.

After years of development, the current acid fracturing technique has evolved into a modified process system composed of multiple working fluids and technical measures.

Usually, a non-reactive viscous fluid (i.e., pad) is injected at high pressure to break the formation and create the fracture. Then, acid is injected after the pad fluid or as the first stage to etch the fracture walls in non-uniform patterns (Figure 1). Finally, slick water is injected as a final post-flush stage to reduce corrosion on downhole equipment and push the acid further into the fracture. After injection shutdown, the fracture faces do not entirely close because channels and asperities are created at the fracture faces. This provides a conductive path along which the reservoir fluid flows to the wellbore (Aljawad et al., 2019) [6].

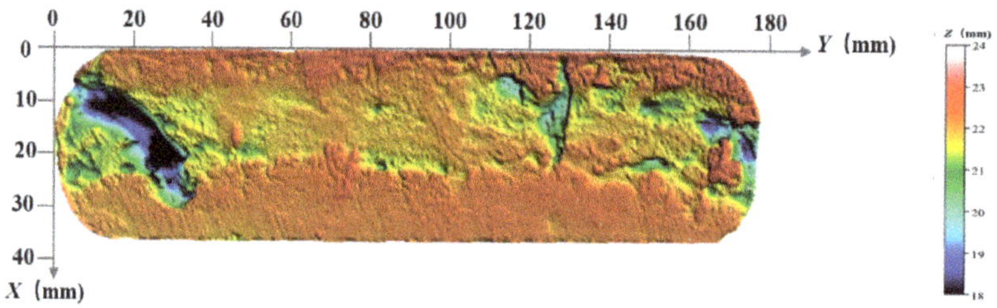

Figure 1. Non-uniform etching morphology on rock sample surface after acid etching.

During acid fracturing, the reactivity of the acid can lead to excessive fluid loss by creating wormholes and activating natural fractures, resulting in a significant decrease in the effective acid-etched fracture length and conductivity. However, at the same time, the acid filtrates into the natural fractures, which will also etch the walls of the natural fractures, increase the conductivity of the natural fractures, and form a relatively complex fracture network with the acid-etched artificial fracture. Under the influence of the above two processes, how to optimize acid fracturing parameters to obtain better simulation results is a very interesting question.

However, there is only a limited number of publications on acid fracturing simulation in naturally fractured carbonate reservoirs. Mou et al. (2012) [7] developed a mathematical model that assumed several natural fractures intersected with a fixed hydraulic fracture. The acid filtrated into the matrix through hydraulic fracture surfaces, and natural fractures were both considered. They revealed that the natural fracture width and density increased the acid leak-off. However, the effects of natural fracture leak-off on the acid etching were not studied. Recently, Ugursal et al. (2019) [8] presented a model that predicted acid fracturing performance in naturally fractured carbonate reservoirs. The fixed hydraulic fracture, which intersected by several transverse symmetric natural fractures, was also adopted in this model. Acid transport and acid/rock reaction in hydraulic and natural fractures were considered. This study revealed that the initial natural fracture width had a negligible effect on hydraulic fracture conductivity. Still, a higher initial natural fracture width led to stronger acid etching in natural fractures. Meanwhile, hydraulic and natural fracture conductivities increased as the acid injection rate increased. However, the effect of natural fractures on the hydraulic fracture geometry has yet to be fully understood, and the temperature is not considered in this model. To accurately assess the role of natural fractures during acid fracture, it is essential to investigate the effect of natural fracture properties on hydraulic fracture geometry.

When acid filtrates into the matrix, it dissolves the rock near fracture surfaces and creates wormholes. Early research (Hawkins, 1956) [9] showed that the length of the wormhole region is the most critical parameter that affects the acid leak-off. Until now, several models have been developed to predict wormhole propagation in carbonate reservoirs. However, only a few publications have considered wormholes during acid fracturing simulation. A recent review (Akanni et al., 2015) [10] listed seven different models, among which the two-scale continuous model is the most promising one to predict the propagation of wormholes [11–15], but it is also too computationally expensive to be used in the acid fracturing design. Dang et al. (2018) [16] presented a model that coupled a fixed hydraulic fracture with a two-scale continuous model. A simulation was carried out in a small reservoir domain with a width of less than 0.5 m. The results showed that the considering wormholes led to an uneven leak-off velocity profile caused by porosity heterogeneity and significantly reduced the acid penetration distance. However, parametric analysis was not carried out in this research. The semi-empirical model is a much better choice in field practice. This model predicts the propagation rate of the wormhole region instead of individual wormhole characteristics. The most famous ones include the volumetric model (Economides et al., 1993) [17], the Buijse and Glasbergen model (Buijse et al., 2005) [18], and the Furui model (Furui et al., 2012) [19]. Hill et al. (1995) [20] presented a leak-off model with a wormhole effect by combining Carter's leak-off model and the volumetric model. The results showed that the impact of wormholes on the overall fluid-loss coefficient is significant during acid fracturing in limestone reservoirs. These authors illustrate that it is essential to consider wormholes in acid fracturing simulation to obtain a more accurate conductivity profile.

Therefore, an integrated acid fracturing model for naturally fractured carbonate reservoirs was established in this study. An efficient and stable P3D fracture propagation model was adopted to calculate the hydraulic fracture geometry. A semi-empirical model was combined to calculate the leak-off change caused by wormholing. This model also accounts for acid flow, energy transfer, acid/rock reaction, and width change for hydraulic and natural fractures. With this model, we studied the fluid loss characteristics of natural fractures and acid-etched wormholes during the acid fracturing process, as well as the influence of hydraulic fracture geometric characteristics, etching morphology, and natural fracture etching morphology under different fluid loss conditions (permeability, natural fracture parameters). Based on the parameters of hydraulic fractures and natural fractures calculated by the model, an acid fracturing parameter optimization method that simultaneously considers the dissolution of hydraulic and natural fractures was established. This study revealed the mechanism of how acid fracturing worked in naturally fractured carbonate reservoirs and provided an important basis for the acid fracturing design in naturally fractured carbonate reservoirs.

2. Mathematical Model

The acid fracturing process of naturally fractured carbonate reservoirs is a combined result of fracture propagation, fluid flow, heat transfer, and acid transport and reaction, a complicated physic-chemical process. The multi-field coupling process among various physical fields is shown in Figure 2. In this model, the fluid is considered incompressible Newtonian fluid and stays in a laminar flow state for the acid injection rates typically for an acid fracturing treatment. A cell-based P3D fracture propagation model is implemented to calculate the fracture geometry during the treatment. A set of symmetric transverse natural fractures intersect the hydraulic fracture propagation plane vertically, and their height is considered equal to the pay zone (Figure 3). The hydraulic fracture and natural fracture interaction is regarded as "interact–fluid fill–cross" (Figure 4).

To precisely simulate the flow progress among hydraulic fractures, natural fractures, and matrix rock, individual models have been established for each of them. The matrix domain is a rectangular solid of length L, height H, and width W, which parallels the hydraulic fracture propagation plane in the length direction. The hydraulic geometry is approximated as an elliptical channel based on the P3D fracture propagation model, and the natural fracture geometry is approximated as a channel between two parallel plates. The reservoir is simplified as a symmetrical three-layer case, and the effect of gravity on fracture propagation and fluid flow is neglected.

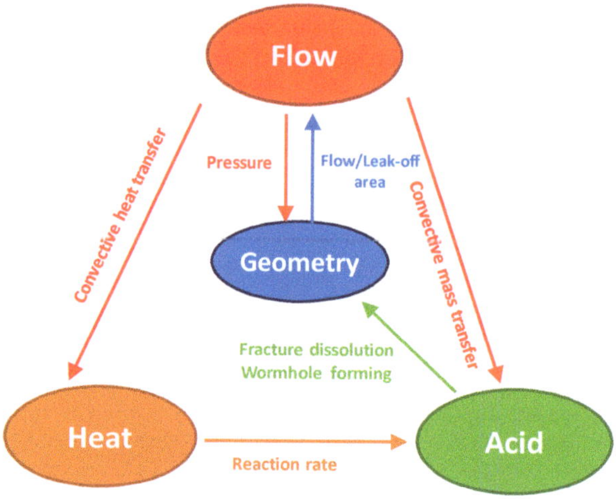

Figure 2. The schematic diagram of the multi-field coupling process.

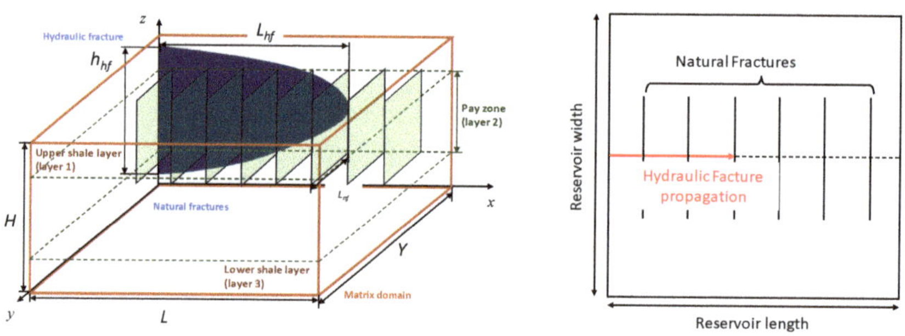

(**a**) (**b**)

Figure 3. The schematic diagram of the integrated acid fracturing model. (**a**) side view; (**b**) top view.

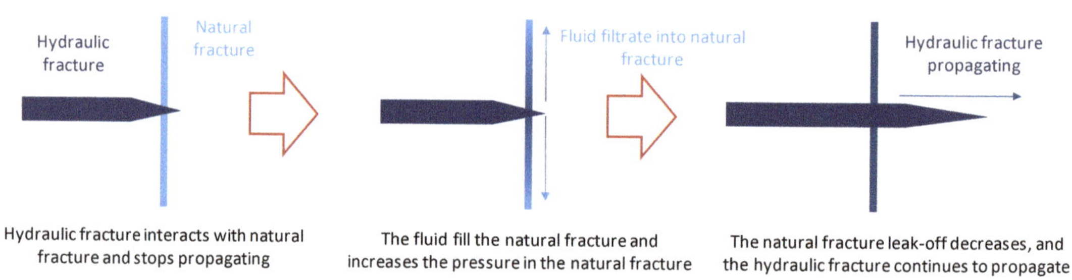

Figure 4. The schematic diagram of hydraulic fracture and natural fracture interaction.

2.1. Geometry Model

In the P3D fracture propagation model, a hydraulic fracture is considered to propagate within a barrier–reservoir–compartment formation and is divided into multiple units along the length; fluid flow is assumed to be essentially horizontal along the length of the fracture by ignoring the flow of fluid in the height direction. The pressure distribution within the hydraulic fracture during the fluid injection process is calculated based on the material balance equation (Equation (1)), which further calculates the fracture expansion position. Then, an analytical formula (Equation (2)) is adopted to calculate the crack width distribution (Economides et al., 1989) [21]. The mass balance in the hydraulic fracture is described by the equation:

$$\frac{\partial A_{hf}\overline{u_x}}{\partial x} + 2h_{hf}v_{L,hf} + \frac{\partial A_{hf}}{\partial_t} = 0 \quad (1)$$

where A_{hf} is the cross-sectional area of hydraulic fracture, m^2; $\overline{u_x}$ is the average flow velocity through the hydraulic fracture cross-section, m/s; h_{hf} is the hydraulic fracture height, m; $v_{L,hf}$ is the fluid leak-off velocity in the hydraulic fracture, m/s.

In a symmetrical case, the stress intensity factors at the top and bottom tips (K_{Iu} and K_{Il}) are equal. They can be simplified and written in terms of the pressure at the centerline of the hydraulic fracture p_{cp} and the closure stresses of pay zone σ_{pay} as follows:

$$K_{Iu} = K_{Il} = \sqrt{\frac{\pi h_{hf}}{2}}\left[p_{cp} - \sigma_{pay} - \frac{2\Delta\sigma}{\pi}\cos^{-1}\left(\frac{h_{pay}}{h_{hf}}\right)\right] \quad (2)$$

where h_{pay} is the pay zone height, m; $\Delta\sigma$ is the stress difference between the pay zone and shale layer; p_{cp} is the fluid pressure at the centerline of the hydraulic fracture, Pa; σ_{pay} is the closure stress of pay zone, Pa; K_{Iu} and K_{Il} are the stress intensity factors of the top and bottom tips of the hydraulic fracture, Pa·m$^{1/2}$. The width profile associated with p_{cp} is obtained as follows:

$$\begin{gathered} w_{hf}(z) = \tfrac{4}{G}\left(P_{cp} - \sigma_3\right)\sqrt{y\left(h_{hf} - z\right)} + \tfrac{4}{\pi G}\sum_{i=1}^{2}\left(\sigma_{i+1} - \sigma_i\right) \\ \left[\left(h_i - z\right)\cosh^{-1}\left(\tfrac{z}{|z-h_i|}\tfrac{h_{hf}-2h_i}{h_{hf}} + \tfrac{h_i}{|z-h_i|}\right) + \sqrt{z\left(h_{hf} - z\right)}\cos^{-1}\left(\tfrac{h_{hf}-2h_i}{h_{hf}}\right)\right] \end{gathered} \quad (3)$$

where G is the shear modulus, Pa; h_i is the hydraulic fracture height in the ith layer, m; σ_i is the closure stresses in the ith layer. When the hydraulic fracture interacts with a natural fracture, to obtain the criterion of fracture propagation, the critical stress intensity factor K_{Ic} needs to be calculated. Olson (2007) [22] provided the expressions of stress intensity factors as follows:

$$K_{Ic} = \frac{0.806E\sqrt{\pi}}{4(1 - v^2)\sqrt{\Delta x}}w_{tip} \quad (4)$$

where E is Young's modulus, Pa; w_{tip} is the width at the tip element of the hydraulic fracture, m; v is Poisson's ratio, dimensionless; K_{Ic} is the critical stress intensity factor, Pa·m$^{1/2}$.

2.2. Fluid Flow and Leak-Off Model

The P3D fracture propagation model gridding along the hydraulic fracture length direction uses the average flow velocity to calculate the mass balance (Figure 5). However, this velocity could not be used to calculate the heat transfer and reaction because the hydraulic fracture has a non-uniform geometry instead of a parallel plate. This will lead to changes in local heat flow and reaction characteristics. A 2-D flow field that can simulate fluid flow in a non-uniform fracture is needed to model the heat transfer and reaction correctly.

The fracture propagation model provides the geometry and pressure for the fluid flow solution at each time step. The hydraulic fracture domain can be re-gridding in the height direction with a grid that matches the matrix domain gridding, and the grid in the length direction remains the same as the fracture propagation model (Figure 5). According to the assumption of the P3D fracture propagation model, the fluid does not flow in the height direction. Therefore, considering the flow in the hydraulic fracture as multiple 1-D flows in the height direction, the mass balance is described using the equation (Dong et al., 2002) [23]:

$$\frac{\partial}{\partial x}\left(\frac{{w_{hf}}^3}{12\mu}\frac{\partial p_{hf}}{\partial x}\right) + v_{L,hf} = -\frac{\partial w_{hf}}{\partial t}. \quad (5)$$

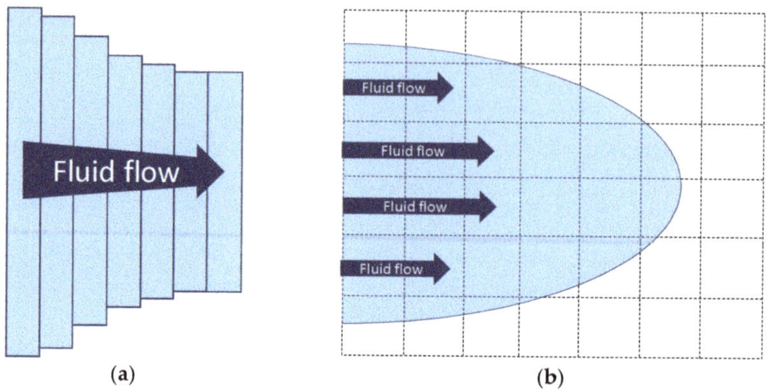

Figure 5. Hydraulic fracture domain and gridding methodology. (**a**) Gridding in P3D fracture propagation model; (**b**) Gridding in the fluid flow model.

The natural fractures are considered parallel plates, and the following equation describes the mass balance inside:

$$\frac{\partial}{\partial y}\left(\frac{{w_{nf}}^3}{12\mu}\frac{\partial p_{nf}}{\partial y}\right)+\frac{\partial}{\partial z}\left(\frac{{w_{nf}}^3}{12\mu}\frac{\partial p_{nf}}{\partial z}\right)+v_{L,nf}=-\frac{\partial w_{nf}}{\partial t} \tag{6}$$

where p_{hf} and p_{nf} represent the fluid pressures in hydraulic facture and natural fracture, Pa; w_{hf} and w_{nf} are the width of hydraulic fracture and natural fracture, m; μ is the fluid viscosity, mPa·s; $v_{L,nf}$ is the fluid leak-off velocity in the natural fracture, m/s.

Meanwhile, the fluid flow in the matrix domain is a three-dimensional seepage in porous media. Based on Darcy's law, the mass balance equation is written as follows:

$$\rho_{\mathrm{l}}\frac{\partial}{\partial x}\left(\frac{k_{\mathrm{mx}}}{\mu}\frac{\partial p_{\mathrm{m}}}{\partial x}\right)+\rho_{\mathrm{l}}\frac{\partial}{\partial y}\left(\frac{k_{\mathrm{my}}}{\mu}\frac{\partial p_{\mathrm{m}}}{\partial y}\right)+\rho_{\mathrm{l}}\frac{\partial}{\partial z}\left(\frac{k_{\mathrm{mz}}}{\mu}\frac{\partial p_{\mathrm{m}}}{\partial z}\right)=\rho_{\mathrm{l}}\phi C_{\mathrm{t}}\frac{\partial p_{\mathrm{m}}}{\partial t}+q_{\mathrm{f}} \tag{7}$$

where k_{mx}, k_{my}, and k_{mz} are the matrix permeability in x, y, and z direction, m^2; p_m is the fluid pressure in the matrix, Pa; ϕ is the rock porosity, dimensionless; C_t is the total compress coefficient, Pa^{-1}; ρ_l is the fluid density, kg/m^3; q_f is the fluid leak-off source term, kg·m^{-3}·s^{-1}.

During acid fracturing, acid leaks off into the matrix through natural fractures and hydraulic fracture walls (Figure 6). However, the dynamic hydraulic fracture geometry and wormhole propagation make the leak-off profile need to be calculated in every fracture propagation step. A small time step is required to maintain the calculation accuracy and stability of fracture propagation. On the other hand, calculating global fluid flow and heat transfer in the three-dimensional matrix domain in such a small time step is unaffordable. Therefore, two sets of time steps are adopted in this study: Δt_p is used to calculate hydraulic fracture propagation and leak-off from hydraulic fracture, set as 0.01–0.1 s. Δt calculates the global flow, heat transfer, and acid reaction, set as 5–60 s.

Therefore, the pressure-dependent velocity profile of leak-off through the hydraulic fracture wall is obtained in each fracture propagation time step:

$$v_{L,hf,m}(x,z,t_p)=\frac{k_{my}(x,z,t_p)}{\mu}\frac{\partial p(x,t_p)}{\partial y}. \tag{8}$$

The leak-off velocity from hydraulic fracture to natural fractures is calculated based on the mass balance, which includes natural fracture volume change and fluid leak-off through natural fracture to the matrix:

$$q_{L,hf,nf} = \int_0^{\Delta t}\int_0^{h_{pay}}\int_0^{L_{nf}} v_{L,nf}(y,z,t_p)dydzdt + \int_0^{\Delta t}\int_0^{h_{pay}}\int_0^{L_{nf}} \frac{\partial w_{nf}(y,z,t_p)}{\partial t}dydzdt. \quad (9)$$

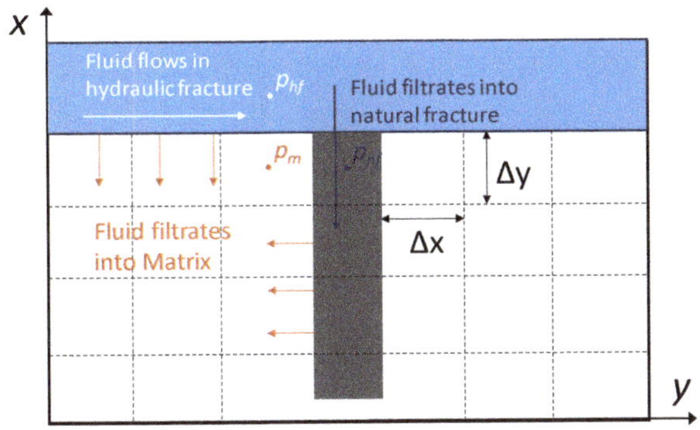

Figure 6. The schematic diagram of fluid leak-off.

The leak-off through natural fracture walls is calculated as follows:

$$v_{L,nf}(y,z,t_p) = \frac{k_{mx}}{\mu}\frac{\partial p_{nf}(y,z,t_p)}{\partial x}\bigg|_{x=0} + \frac{k_{mx}}{\mu}\frac{\partial p_{nf}(y,z,t_p)}{\partial x}\bigg|_{x=w_{nf}} \quad (10)$$

where $q_{L,nf}$ is natural fracture leak-off volume, m^3; L_{nf} is natural fracture length, m. Therefore, the accumulative leak-off volume in a hydraulic fracture grid in time step Δt can be calculated using the equation:

$$q_{L,hf}(x) = \int_0^{\Delta t}\int_0^{h_{hf}(x)} v_{L,hf,m}(x,z,t_p)dzdt + q_{L,hf,nf}. \quad (11)$$

2.3. Heat Transfer Model

The fracture etching profile can only be predicted if the temperature is known because the temperature strongly affects the in situ acid/rock reaction and mass transfer (Nolte and Economides, 2000) [24]. During the injection process, fluid will flow through the wellbore, hydraulic fractures, natural fractures, and finally, into the reservoir rock. During this process, the acid fracturing fluid will exchange energy with the reservoir rock, cooling the rock while heated. This process is the primary goal of acid fracturing temperature field simulation. To simulate the above procedure, a finite difference method developed by Eickmeier et al. (1970) [25] is implemented to calculate the wellbore heat transfer in this study. The temperature profile can be predicted once we obtain the velocity profiles in fractures and matrix. The heat transfer in the hydraulic and natural fractures is described by the following equations:

$$\rho_l c_l \frac{\partial\left(u_x w_{hf} T_{hf}\right)}{\partial x} + 2\rho_l c_l T_{hf} v_{L,hf} + 2\frac{\partial}{\partial y}\left[\lambda_l \frac{\partial\left(w_{hf} T_{hf}\right)}{\partial y}\right] = -\rho_l c_l \frac{\partial\left(w_{hf} T_{hf}\right)}{\partial t} \quad (12)$$

$$\rho_l c_l \frac{\partial\left(u_y w_{nf} T_{nf}\right)}{\partial y} + \rho_l c_l \frac{\partial\left(u_z w_{nf} T_{nf}\right)}{\partial z} + 2\rho_l c_l T_{nf} v_{L,nf} + 2\frac{\partial}{\partial x}\left[\lambda_l \frac{\partial\left(w_{nf} T_{nf}\right)}{\partial x}\right] = -\rho_l c_l \frac{\partial\left(w_{nf} T_{nf}\right)}{\partial t} \quad (13)$$

where T_{hf} and T_{nf} represent the average temperatures across the fracture width, K; c_l is the specific heat of fluid, J/(kg·K); λ_l is the thermal conductance of fluid, W/(m·K). u_x, u_y, and u_z are the flow velocities in the x, y, and z directions, m/s. It is considered that the temperature in the matrix domain obeys the assumption of local thermal equilibrium (Kaviany, 2012) [26]. Therefore, within the same matrix grid, $T_f = T_s = T_m$. Then, the heat transfer control equation in the matrix domain is as follows:

$$\rho_m c_m \frac{\partial T_m}{\partial t} + \rho_m c_m \left[u_x \frac{\partial T_m}{\partial x} + u_y \frac{\partial T_m}{\partial y} + u_z \frac{\partial T_m}{\partial z} \right] = \left[\lambda_m \frac{\partial^2 T_m}{\partial x^2} + \lambda_m \frac{\partial^2 T_m}{\partial y^2} + \lambda_m \frac{\partial^2 T_m}{\partial z^2} \right] + q_{h,m} \tag{14}$$

where T_m is the temperature of porous medium, K; ρ_m is the density of porous medium, kg/m^3; c_m is the specific heat of porous medium, J/(kg·K); λ_m is the thermal conductance of porous medium, W/(m·K); $q_{h,m}$ is the energy source term, J/(m^3·s).

2.4. Acid Model

After obtaining the temperature profiles in hydraulic and natural fractures, we can calculate the acid reaction parameters based on the Arrhenius equations. Then, combining the calculated velocities, acid mass balance in fractures is described by the convection–diffusion equation, and the mass transfer coefficient proposed by Robert (1975) [27] is combined with the acid rock reaction constant to construct the mass transfer reaction boundary conditions:

$$\frac{\partial (u_x w_{hf} C_{hf})}{\partial x} - 2\frac{k_c k_g}{(k_c + k_g)} C_{hf} - 2C_{hf} v_{L,hf} + \frac{\partial \left(w_{hf} C_{hf} \right)}{\partial t} = 0 \tag{15}$$

$$\frac{\partial (u_y w_{nf} C_{nf})}{\partial y} + \frac{\partial (u_z w_{nf} C_{nf})}{\partial z} - 2\frac{k_c k_g}{(k_c + k_g)} C_{nf} - 2C_{nf} v_{L,nf} + \frac{\partial \left(w_{nf} C_{nf} \right)}{\partial t} = 0 \tag{16}$$

where C_{hf} and C_{nf} are the average acid concentrations across the fracture width, mol/m^3; k_c is the acid reaction constant, m/s; k_g is the mass transfer coefficient, m/s.

During acid fracturing, the acid un-uniformly dissolves a fractured surface and forms wormholes, which lead to considerable acid leak-off. Due to the limitation of simulation scale and computing power, the global wormhole model is the best choice for simulating the wormhole growth during acid fracturing. This study adopts a semi-empirical wormhole model presented by Furui (2012) [19].

It is considered that the wormhole lengths (L_{wh}) in one grid are equal (Figure 7a), and the wormhole growth velocity can be predicted as follows:

$$v_{wh} = v_{i,tip} N_{Ac} \times \left(\frac{v_{i,tip} PV_{bt,opt} N_{AC}}{v_{i,opt}} \right)^{-\gamma} \times \left\{ 1 - \exp\left[-4 \left(\frac{v_{i,tip} PV_{bt,opt} N_{AC} L_{wh}}{v_{i,opt} L_{core}} \right) \right]^2 \right\}^2 \tag{17}$$

where N_{AC} is the acid capacity number; v_{wh} is the wormhole growth velocity, m/s; $v_{i,tip}$ is the average interstitial velocity in the wormhole front, m/s; $PV_{bt,opt}$ is the optimal breakthrough volume, dimensionless; $v_{i,opt}$ is the optimal injection velocity, m/s; L_{wh} is the wormhole length, m; L_{core} is the core length, m; γ is the coefficient that represents the effect of the fluid loss limiting the wormholing; in this study, it equals to 1/3.

Acid flow through the hydraulic fracture walls is considered as laminar flow, and $v_{i,tip}$ is calculated as follows:

$$v_{i,tip} = \frac{q}{\phi A_{wh}}. \tag{18}$$

where A_{wh} is the cross-sectional area perpendicular to the wormhole front, m^2; q is the acid flow rate, m^3/s.

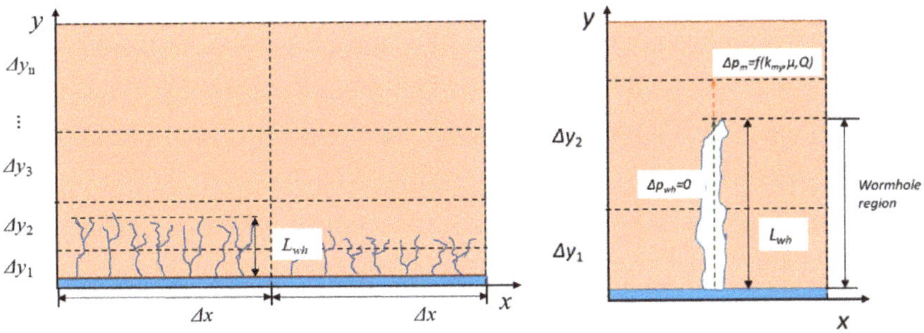

(a) (b)

Figure 7. The schematic diagram of the wormhole region. (**a**) Wormhole region; (**b**) Pressure drop in individual wormhole.

Because the flow capacity of wormholes is extremely strong compared to the matrix, the pressure drop in wormholes is considered to be zero (Figure 7b). Therefore, the permeability at the wormhole growth direction in the wormhole region should be set as infinite. However, this may cause calculation problems in numerical simulation; so the permeability of the elements in the wormhole region is set as a significantly larger value:

$$k_{wh} = 1\mathrm{e}^4 k_{my}. \tag{19}$$

For the grids at the tip of the wormhole region, their permeability is calculated as follows:

$$\begin{aligned} k_{wh} &= k_{my}\frac{\Delta y_{tip}}{\Delta y_{tip}-L_{wh,tip}}, k_{my}\frac{\Delta y_{tip}}{\Delta y_{tip}-L_{wh,tip}} \le 1\mathrm{e}^4 k_{my} \\ k_{wh} &= 1\mathrm{e}^4 k_{my}, k_{my}\frac{\Delta y_{tip}}{\Delta y_{tip}-L_{wh,tip}} > 1\mathrm{e}^4 k_{my} \end{aligned} \tag{20}$$

where k_{wh} is the permeability of the matrix grid with wormholes at the y direction, m^2; Δy_{tip} is the length of the grid where the wormhole tip is located, m; $L_{wh,tip}$ is the wormhole length at the element where the wormhole tip is located, m.

The etching distribution along the fractures can be calculated with the obtained acid concentration profile. This study considers that all acid filtrates through hydraulic fracture walls contribute to wormhole growth. Therefore, the change in the fracture width is calculated by the following equation:

$$\frac{\partial w_{hf}}{\partial t} = \frac{2\beta M \frac{k_c k_g}{(k_c+k_g)} C_{hf}}{\rho_r(1-\phi)}. \tag{21}$$

The wormholes on natural fracture walls are not considered. Therefore, the acid leak-off into the matrix is considered to contribute to the natural fracture width increase:

$$\frac{\partial w_{nf}}{\partial t} = \frac{\beta M\left[\eta v_{L,nf} C_{nf} + \frac{k_c k_g}{(k_c+k_g)} C_{nf}\right]}{\rho_r(1-\phi)} \tag{22}$$

where β is the dissolving power of acid on rock, kg rock/kg acid; M is the mole weight of the rock, kg/mol; ρ_r is the density of rock, kg/m^3; η is the fraction of the leaking acid that reacts at the fracture surfaces. In this study, the acid-etched conductivity is calculated by the correlations presented by Deng et al. (2012) [28], in which the geostatistical and rock mechanical parameters are considered.

3. Calculation Approach

The integrated acid fracturing model is developed in-house, with MatLab (2023a, academic version, MathWorks, Inc., Natick, MA, USA), preventing the need to transfer data to or from other commercial software. The procedures for calculating the fracture conductivity are as follows: After reading the input data, a calculated average leak-off velocity profile is adopted to calculate the hydraulic fracture geometry. Iterations between the fracture geometry and leak-off velocity are performed until convergence. After that, the temperature was first calculated based on an assumed temperature profile on the fracture wall, then updated according to the new wall temperature calculated from the matrix until convergence. The acid balance equations are solved using the calculated velocity and temperature profiles to obtain the acid concentration profile in hydraulic and natural fractures. After that, the etched-width and the wormhole region length are estimated, and then the fracture width and matrix permeability are updated. At last, the acid-etched conductivities of the overall fractures are obtained based on the etched-width when the acid injection is finished. The more detailed workflow is shown in Figure 8.

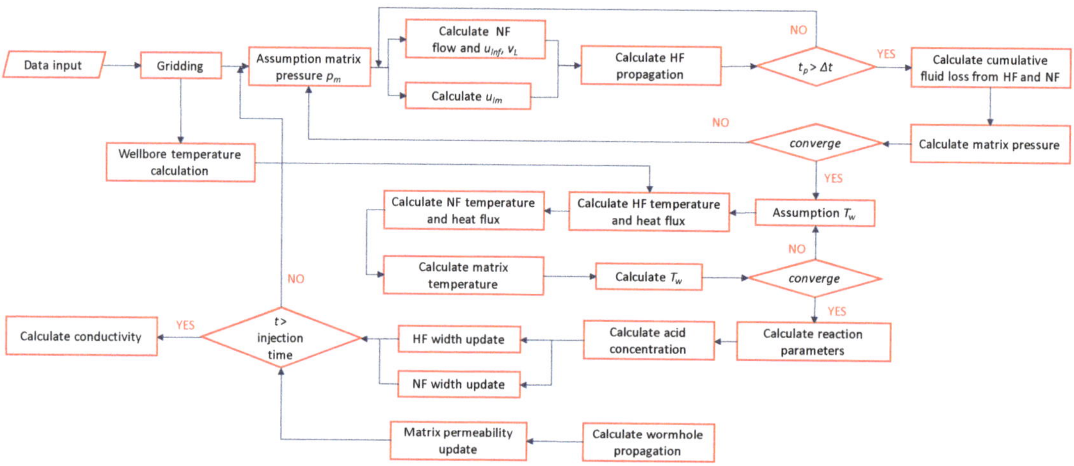

Figure 8. Workflow of acid fracturing simulation.

4. Model Validation

In this study, because the problem modeled is complicated, the presented model was simplified into several steps and validated with available analytical solutions. First, the fracture propagation model was validated with a PKN analytical solution derived by Nordgren (1972) [29] at a fixed height and with no leak-off conditions. Meanwhile, the analytical solution of the acid flow reaction (Schechter, 1992) [30] and the heat transfer (Terrill, 1964) [31] between two porous parallel plates were adopted to validate the acid concentration solution and the temperature solution of the model. As shown in Figure 9, the numerical simulation results matched well with the analytical solutions. Therefore, the validations were satisfactory.

During the fluid flow calculations, the fracture geometry and fluid pressure continue to be used while changing the grid. Therefore, fluid flow solutions need to be validated to ensure the conservation of mass during the calculation process. Figure 10 shows the calculation results in different grid systems. The match between the two models is within a 1% difference, and changing the grid only slightly influences calculation results.

The model was also validated against the acid fracturing model for the fractured carbonate reservoir established by Ugursal (2018) [32]. The validating case contains a constant hydraulic fracture and three natural fractures that intersect vertically and uniformly with hydraulic fractures. This model does not consider hydraulic fracture propagation but takes the acid flow front as the fluid loss range, and no temperature calculation is involved in this model. Furthermore, more natural fractures are planted in the established model to compare the dynamic leak-off between constant and propagating hydraulic fractures.

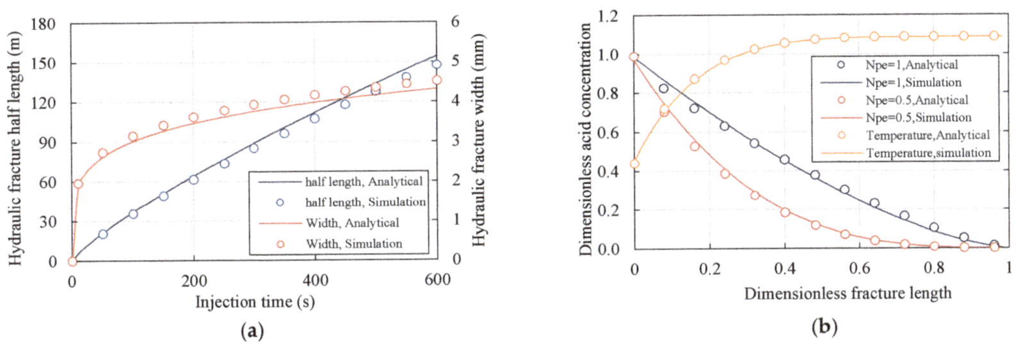

Figure 9. Comparison of solutions given by both the analytical solutions and numerical simulations. (**a**) Fracture geometry solutions; (**b**) Temperature and acid concentration solutions.

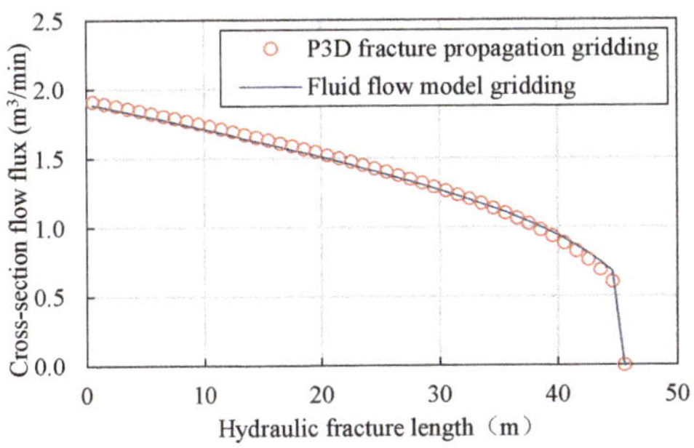

Figure 10. Flow flux solutions by different gridding.

The two models showed the same characteristic in the dynamic change of leak-off with injection time (Figure 11). When a hydraulic fracture interacts with a natural fracture, the acid will flow into and fill the natural fractures. It will cause a large amount of leak-off, which reduces the leak-off from the hydraulic fracture walls. As natural fracture is filled with fluid and its leak-off decreases, the leak-off through hydraulic fracture walls will gradually increase. This process continues circulating as hydraulic fracture propagates and interacts with new natural fractures. This results in leak-off in naturally fractured reservoirs exhibiting a jagged characteristic, which differs from the leak-off predicted by the classic Carter model.

Since Ugursal only planted three natural fractures in this simulation, the leak-off rate injection decreased with the increase in acid injection time after 6 min. However, in the established model, as the acid injection time increases, the leak-off rate presents a periodic change. This is because the wall area of hydraulic fractures continues to increase during the propagation, constantly interacting with new natural fractures, and the total fluid loss and injection volume are in a dynamic balance.

Due to the low flow capacity of hydraulic fractures at the initial propagation stage, it is difficult to build a high fluid pressure. Therefore, the leak-off rate of the first natural fracture is significantly lower than the results calculated by Ugursal's model. This leads to a faster interaction speed between hydraulic and natural fractures (Figure 12).

In both models, as the distance between the natural fracture and the entrance of the hydraulic fracture increases, the maximum leak-off rate of the natural fracture decreases, and the interaction interval increases. However, in the established model, the leak-off rate of the more distant natural fracture is more significant than that of Ugursal's model, although the speed of reaching the maximum leak-off slows down. This is because, compared to the constant hydraulic fracture, the width of the propagating hydraulic fractures continuously increases. This leads to a decrease in the flow pressure consumption, finally increases the fluid pressure at the entrance of the distant natural fractures, and a longer flow distance will slow down this process.

In summary, the results of the established model are consistent with those of existing models. By further considering hydraulic fracture propagation, the established model provided a more complete and acceptable description of the dynamic leak-off process in fractured reservoirs.

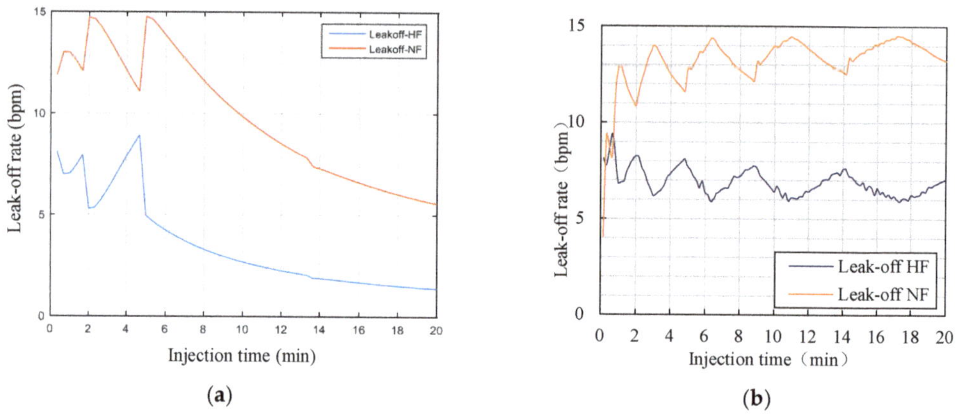

Figure 11. Comparison of total leak-off results between Ugursal's model and the established model. (**a**) Solutions from Ugursal's model (2018) [32]; (**b**) Solutions from established model.

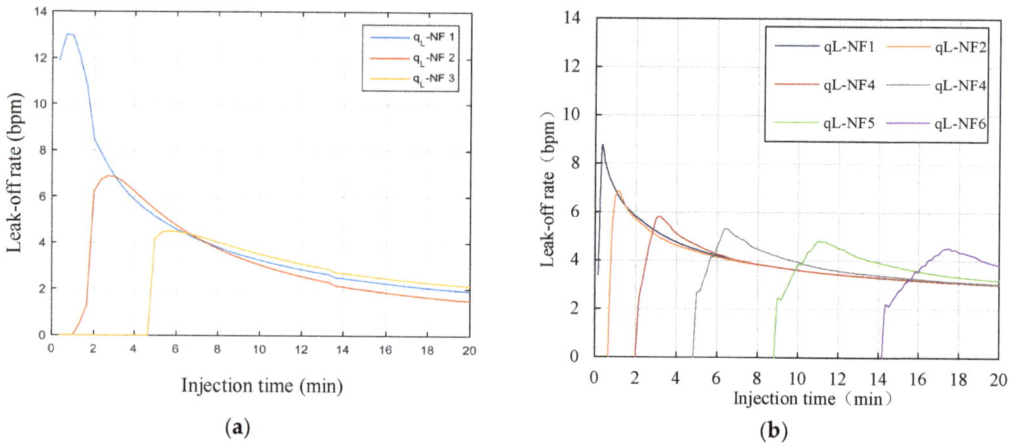

Figure 12. Comparison of leak-off in natural fractures between Ugursal's model and the established model. (**a**) Solutions from Ugursal's model (2018) [32]; (**b**) Solutions from established model.

5. Results

Based on the model proposed in the previous sections, we first studied the leak-off behavior of wormholes and natural fractures. Then, parametric studies were carried out to investigate how wormholes and natural fractures affect hydraulic fracture propagation and acid etching under various reservoir parameters. Then, the dimensionless production indexes under various treatment parameters were calculated based on four designed cases with different reservoir conditions. This led us to find the optimal treatment parameters and obtain the acid stimulation strategy under different reservoir situations. The input parameters for this study are given in Table 1.

Table 1. Input parameters for the simulation.

Input Data	SI Unit	Input Data	SI Unit
	Wellbore properties		
Inner tube diameter, r_0	0.062 m	Outer casing diameter, r_3	0.127 m
Outer tube diameter, r_1	0.073 m	Ground surface temperature, T_G	25 °C
Inner casing diameter, r_2	0.103 m	Steel thermal conductance, λ_s	45 W/(m·K)
Steel specific heat capacity, c_s	460 J/(kg·K)		
	Reservoir properties		
Reservoir pressure, P_e	4×10^7 Pa	Minimum horizontal stress in pay zone, σ_h	5×10^7 Pa
Total compressibility, C_t	2.2×10^{-9} Pa^{-1}	Minimum horizontal stress in shale layer, σ_{hs}	5.8×10^7 Pa
Reservoir fluid viscosity, μ_o	0.01 Pa·s	Reservoir fluid density, ρ_o	850 kg/m^3
Reservoir width, L_y	500 m	Reservoir length, L_x	500 m
Pay zone height, h_{pay}	30 m	Reservoir height, L_z	200 m
Limestone-specific heat capacity, c_r	850 J/(kg·K)	Reservoir temperature, T_e	439 K
Limestone thermal conductance, λ_r	1.57 W/(m·K)	Limestone density, ρ_r	2600 kg/m^3
Young's modulus, E	6×10^{10} Pa	Poisson's ratio, v	0.15
	Acid properties		
Acid density, ρ_l	1100 kg/m^3	Acid initial concentration, C_0	6020 mol/m^3
Acid viscosity, μ_l	0.03 Pa·s	Acid-specific heat capacity, c_l	4200 J/(kg·K)
Acid thermal conductance, λ_l	0.7 W/(m·K);	Frequency factor, k	0.003 m/s
Activation energy, Ea	16,000 J/mol	Reference diffusion coefficient, D_{e0}	5.6×10^{-9} m^2/s
Reference temperature, T_0	323 K		
	Mou-Deng conductivity correlation parameters		
Horizontal correlation length, λ_{Dx}	1	Vertical correlation length, λ_{Dx}	0.05
Normalized permeability standard deviation, σ_D	0.4	Closure pressure, σ_c	3×10^7 Pa

5.1. Natural Fracture Leak-Off Behavior

First, the leak-off behavior of natural fracture in different geometric parameters (50–250 μm width, 5–25 m length) and reservoir permeability (1, 10 mD) is simulated. During the simulation, the natural fracture spacing was considered equal to 1 m while maintaining a 6 m^3/min injection rate and 600 s injection time. Wormholes were not considered in this section, and the other input parameters are given in Table 1.

Figure 13 shows that the accumulative leak-off volume of a natural fracture does not always grow with a unilateral increase in width, length, or matrix permeability. It is observed that increasing the length of a natural fracture with a small width hardly improves its leak-off capacity. However, the magnitude of this effect grows as the width increases, and it shows a significant effect when the width exceeds 150 μm. On the contrary, an increase in the natural fracture width greatly improves the accumulative fluid loss volume under all natural fracture lengths. It is also observed that matrix permeability has a more obvious effect on fluid leak-off, with larger natural fracture widths and lengths. The effects of these three parameters on the fluid leak-off show a mutually constrained and non-linear characteristic.

The pressure profile in natural fractures was analyzed to further understand the mechanism of these parameters' influence on fluid leak-off. Figure 14 reveals that only a part of the natural fracture length plays a role in leak-off when the natural fracture width is small (Figure 14a). As the width increases, the pressure drop in the natural fracture decreases, and the effective leak-off length extends to the whole natural fracture when the width exceeds 150 μm. Furthermore, suppose we change the length of natural fractures with a large width. In that case, it can be seen that the pressure drop rate increases as the length gets longer (Figure 14b). It can be observed that the natural fracture width controls the inflow rate at the entrance and the pressure drop along the length. In contrast, the natural fracture length and matrix permeability determined the leak-off area and flow capacity from the natural fracture to the matrix. Therefore, only when the width, length, and matrix permeability increase, the accumulative fluid loss volume of natural fracture increases significantly.

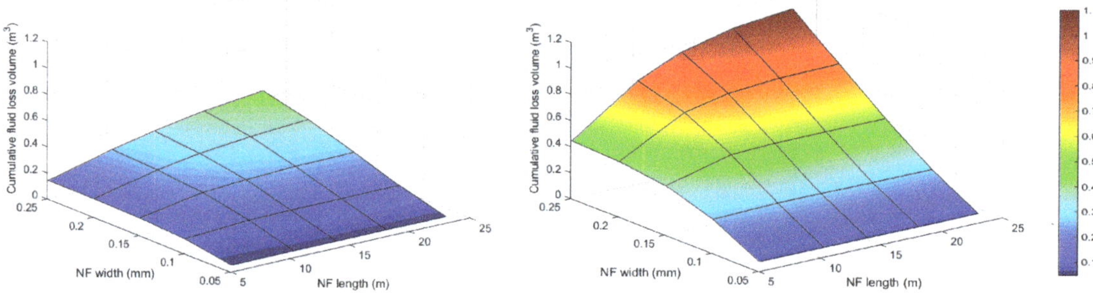

(a) (b)

Figure 13. The effects of natural fracture geometry and matrix permeability on leak-off (NF: natural fracture). (**a**) Matrix permeability is equal to 1 mD; (**b**) Matrix permeability is equal to 10 mD.

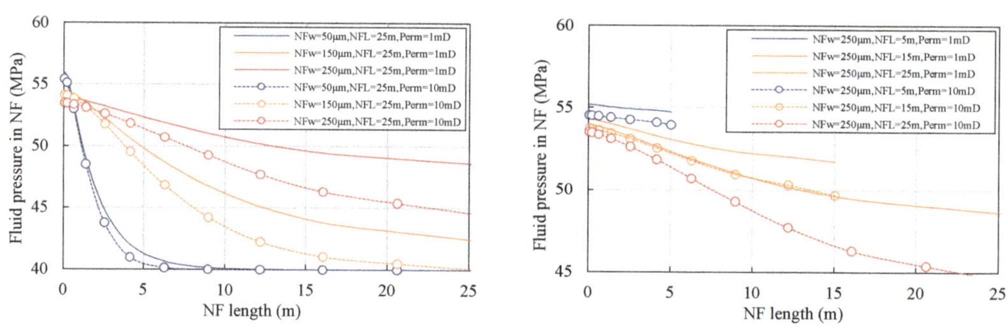

(a) (b)

Figure 14. Natural fracture geometry and matrix permeability effects on pressure profile in natural fracture. NFw: natural fracture width; NFL: natural fracture length; Perm: matrix permeability. (**a**) 50–250 μm width, 25 m length; (**b**) 250 μm width, 5–25 m length.

5.2. Wormhole Leak-Off Behavior

This section studied the effects of matrix permeability (1, 10 mD) on wormhole propagation and distribution. The former research showed that $PV_{bt,opt}$ is reduced with reservoir permeability, while there is no apparent relationship between $v_{i,opt}$ and reservoir permeability (Etten, 2009) [33]. Therefore, several cases were simulated with $PV_{bt,opt}$ varying from 0.048 to 0.026 to 0.005 with the matrix permeability ranging from 10 mD to mD 1 to 0.1 mD while $v_{i,opt}$ considered equal to 2.3×10^{-4} m/s. The injection rate and time are the same as in the last section, and other input parameters are given in Table 1.

Figure 15 reveals that the wormhole region length increases as the matrix permeability increases. Furthermore, Figure 15b also shows that the influence of wormhole propagation quickly becomes substantial at the first 200 s of the acid injection, where the cumulative leak-off volume difference (the cumulative leak-off volume considering wormholes minus cumulative leak-off volume without considering wormholes) increases sharply with a quick increasing grid permeability.

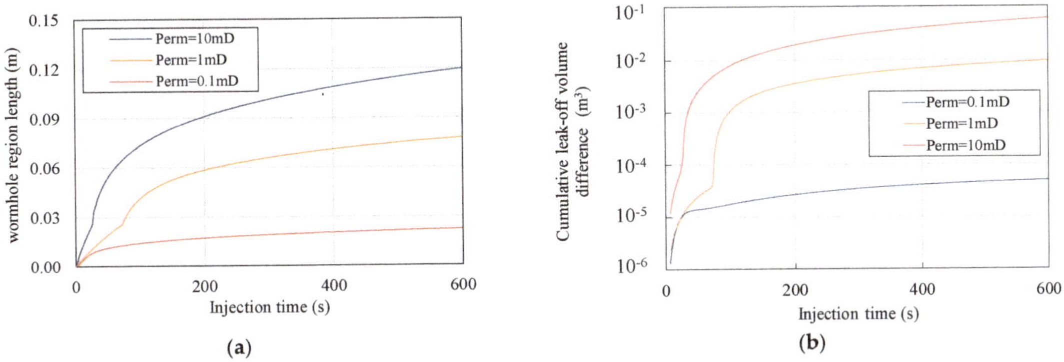

Figure 15. Matrix permeability affects the length and leak-off of the wormhole region. (**a**) Wormhole region length at the hydraulic fracture inlet. (**b**) Cumulative leak-off volume difference of an inlet grid with and without considering wormhole. The grid is located in a hydraulic fracture inlet, and the grid size is 1 m in length and 4 m in height.

It is also observed from Figure 16 that the wormhole length distribution is affected by both the matrix permeability and natural fractures. The case with lower matrix permeability has a shorter wormhole length and more extensive wormhole distribution than the cases with higher permeability. At the same time, the distribution length decreases with the increasing leak-off caused by the natural fractures.

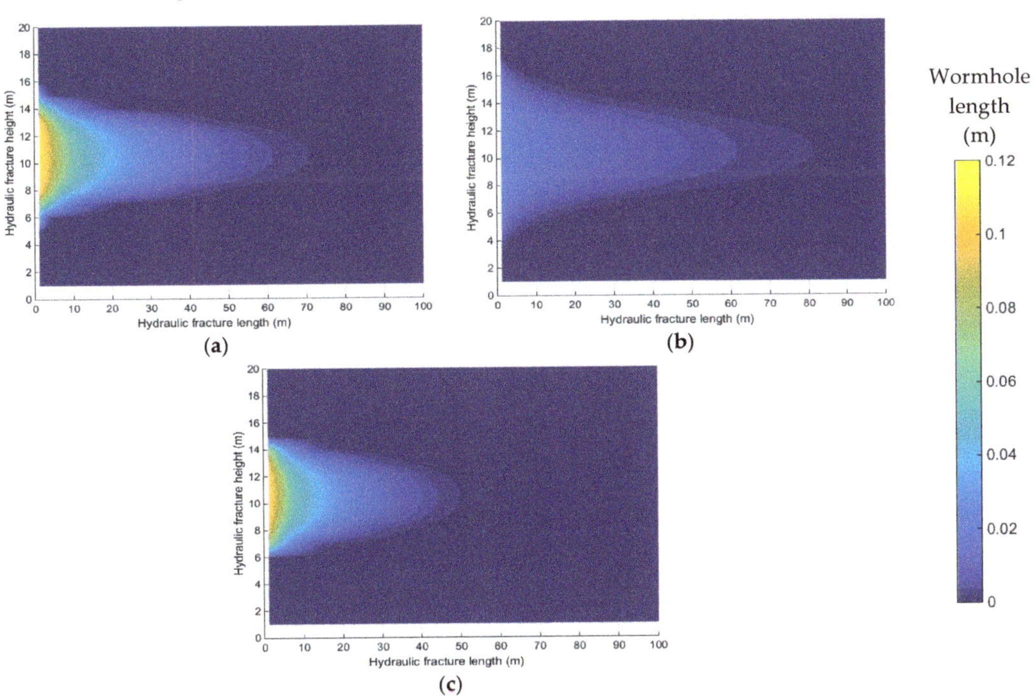

Figure 16. Matrix permeability and injection rate effect on wormhole distribution. (**a**) 10 mD permeability; (**b**) 0.1 mD permeability; (**c**) 10 mD permeability, with natural fractures.

5.3. Cumulative Leak-Off Volume and Hydraulic Fracture Propagation

We can see that the matrix permeability affects the leak-off behavior of both wormholes and natural fractures. So, the effects of wormhole and natural fractures on cumulative leak-off volume and hydraulic fracture propagation were investigated by simulating the cases with matrix permeability varying from 0.1 mD to 1 mD to 10 mD. The natural fractures were considered constant width and length of 100 μm and 10 m. Other treatment and reservoir conditions were the same as in the last section.

It is observed that the cumulative leak-off volume increases when considering wormholes and natural fractures but in different regularity. For the wormhole case, there is only a slight increase with 0.1 mD permeability (Figure 17a). As the permeability increases, the increase becomes substantial, and the influence becomes even more significant than the simulated natural fractures when the matrix permeability is 10 mD. The results also show that as the matrix permeability increases, the most considerable leak-off increment of natural fracture cases appears at 1 mD (Figure 17b). This is because, with the simulated reservoir condition, the leak-off through natural fractures is limited by the leak-off capacity in the 0.1 mD cases, but restricted by the number of natural fractures intersected in the 10 mD cases.

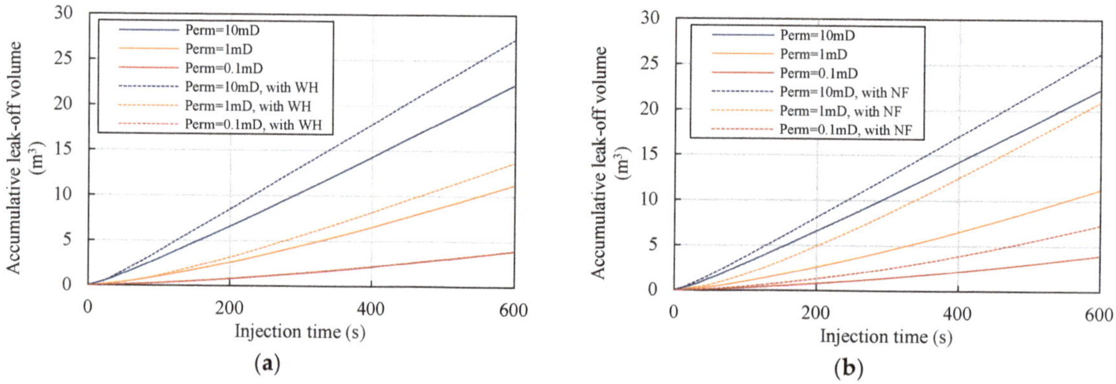

Figure 17. Total cumulative leak-off volume for various matrix permeabilities. (**a**) Wormhole cases; (**b**) Natural fracture cases.

It is observed that the increased acid leak-off caused by the natural fractures and wormholes limits the hydraulic fracture growth (Figure 18). When the permeability is 0.1 mD, the impact of natural fractures is relatively small on the hydraulic fracture geometry. When the permeability increases to 1 mD, the influence of natural fractures and wormholes on fracture geometry increases. At this time, the impact of natural fractures is more significant than that of wormholes, and the length of hydraulic fractures is dominated by natural fracture leak-off. When the permeability increases to 10 mD, the dynamic length of hydraulic fractures is shortened by 65% compared with the 1 mD case. When the reservoir permeability is relatively high, the leak-off seriously affects the propagation of hydraulic fractures. At this time, the effects of wormholes and natural fractures on hydraulic fracture geometry are similar. The hydraulic fracture propagation mechanism has changed from natural fracture leak-off control to competitive control by natural fracture and wormhole leak-off.

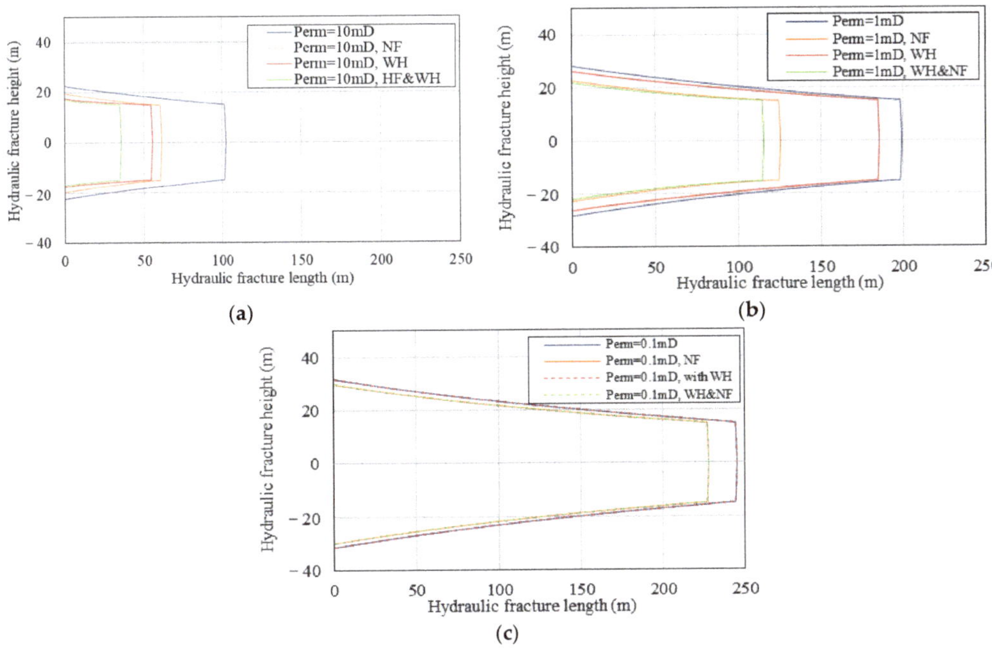

Figure 18. Wormholes and natural fractures affect hydraulic fracture propagation in different matrix permeabilities. (**a**) 10 mD permeability; (**b**) 1 mD permeability; (**c**) 0.1 mD permeability.

5.4. Acid Etching in Hydraulic Fracture

The effects of wormholes and natural fractures under different matrix permeability on the acid-etched width profile are studied by simulations with matrix permeability varying from 0.1 mD to 1 mD to 10 mD. Other treatment and reservoir conditions were the same as in the last section.

It can be seen from Figure 19 that the impact of natural fractures and wormholes on the etching pattern of hydraulic fractures are quite different. Since wormholes are mainly developed at the entrance of hydraulic fractures (Figure 19a), the leak-off caused by them is also concentrated in the same place. Therefore, it greatly changes the slope of the front section of the etching width curve, which leads to a rapid etching width decrease near the entrance of the hydraulic fracture in the 10 mD case. On the contrary, the natural fractures are evenly distributed throughout the reservoir, and the leak-off of natural fractures tends to change the slope of the profile curve across the entire fracture length (Figure 19b).

It is observed that when the matrix permeability is 0.1 mD, wormholes and natural fractures only slightly affect the acid penetration distance and the maximum dissolution width. With the increase in permeability, the influence of leak-off on etching gradually becomes stronger. It shows different characteristics (Figure 20). Since the chemical leak-off ability of wormholes increases by orders of magnitude with the increase in permeability (Figure 20b). When the permeability increases, the acid penetration distance decreases in the case of wormholes. While the permeability increases to 10 mD, the effective etching length is reduced to 19.3 m, which significantly limits the length of the etched hydraulic fractures (Figure 20a).

On the contrary, since the leak-off of natural fracture is controlled by the length, width, and permeability, increasing the permeability decreases the acid penetration distance. In the meantime, changes in leak-off will affect the propagation of hydraulic fracture, thereby affecting the number of natural fractures it encounters. Therefore, when the matrix permeability increases from 0.1 mD to 10 mD, the effective etching length shows a relatively stable downward trend. As the leak-off increases, net pressure in the hydraulic fracture decreases, leading to a smaller fracture width. A smaller hydraulic fracture width will increase the flow velocity in the fracture, leading to an increase in the mass transfer coefficient. Therefore, the maximum etching width of the fracture increases slightly with the increase in matrix permeability.

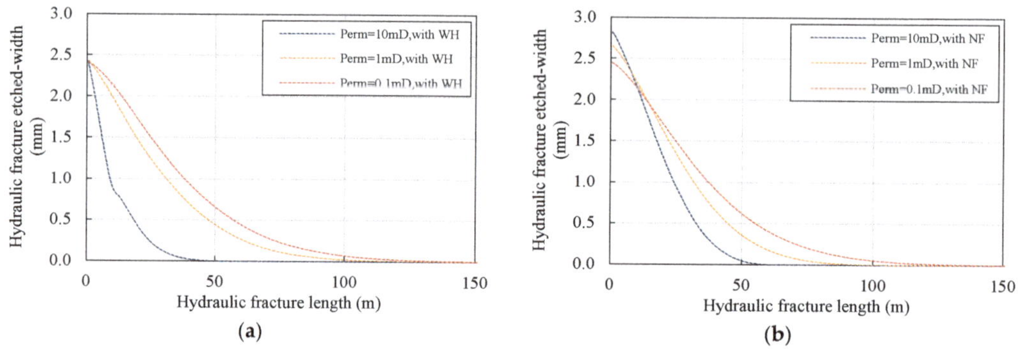

Figure 19. Acid-etched width profile for various matrix permeabilities. (**a**) Wormhole cases; (**b**) Natural fracture cases.

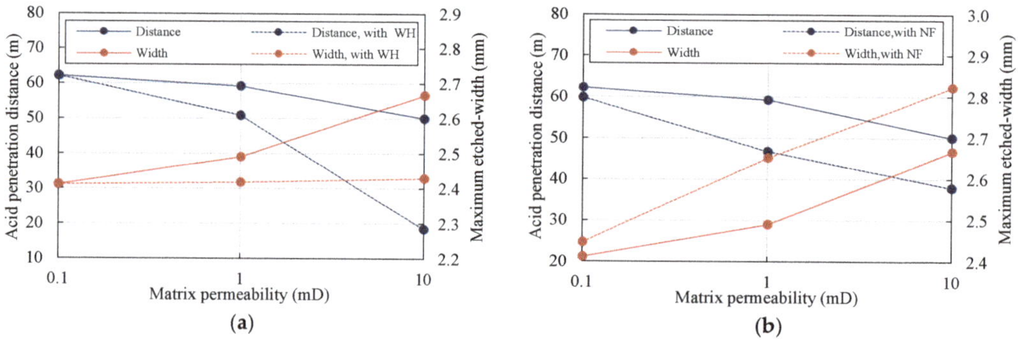

Figure 20. Comparison of acid-etching characteristics with different leak-off mediums. (**a**) Wormhole cases; (**b**) Natural fracture cases.

5.5. Acid Etching in Natural Fractures

The results presented in Figure 21 reveal that the natural fracture geometry and matrix permeability also substantially influence the acid etching in natural fractures. The etched-width in the natural fractures decreases sharply along the natural fracture length. It is observed that natural fractures with larger initial widths and longer lengths have both larger etched-width and longer effective etched-length compared to the natural fractures with smaller initial widths and shorter lengths because of the higher cumulative leak-off volume and flow velocity (Figure 21a).

Figure 21b reveals that the etched-width is larger for the natural fractures near the hydraulic entrance than the natural fractures farther from the hydraulic fracture inlet. The acid etching in natural fractures highly depends on the acid concentration profile in the hydraulic fracture. It is also observed that as the leak-off capacity of natural fractures decreases, the etching pattern of natural fractures developed from etching several natural fractures near the hydraulic fracture entrance to more homogeneous etching over a longer distance.

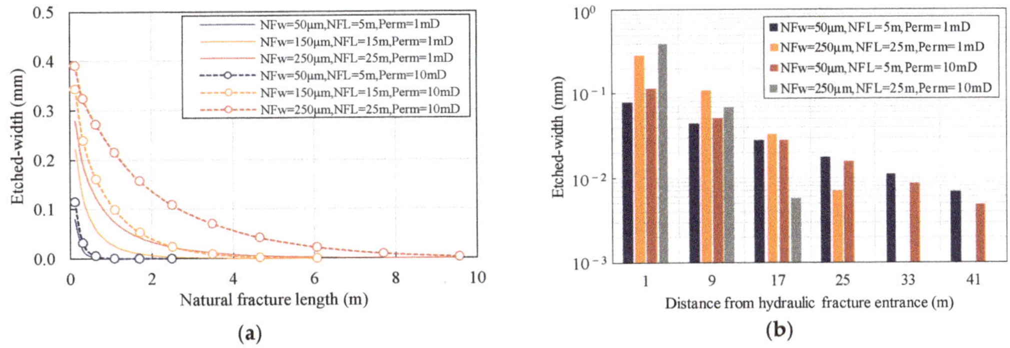

Figure 21. Natural fracture geometry, matrix permeability, and position effects on fluid natural fracture etched-width. (**a**) Acid etched-width in natural fracture; (**b**) Etched-width at the inlet of natural fractures.

5.6. Acid Fracturing Strategy Study

The geological parameters that affect the acid fracturing results mainly include reservoir permeability, natural fracture geometry, and distribution. To analyze the optimal treatment parameters in different reservoir situations, four cases were designed to represent different conditions (Table 2). Case 1 represented an extremely high leak-off condition caused by both high matrix permeability and conductive natural fractures. Case 2 replaced the conductive natural fractures in Case 1 with natural fracture networks with natural fractures of small width but high density. Case 3 represents the reservoir with less matrix permeability and less conductive natural fractures compared to case 1, which is used to study the acid fracturing performance in a poorer reservoir. Finally, Case 4, which had low matrix permeability and natural fracture networks with the highest natural fracture density, was proposed. Treatment parameters, including injection rate and volume, were set to 3–10 m^3/min and 100–400 m^3. The values of other parameters were adopted, as shown in Table 1.

Table 2. Designed parameters for different reservoir conditions.

	Permeability, mD	NF Width, μm	NF Length, m	Spacing, m
Case 1	10	250	25	1
Case 2	10	50	25	0.2
Case 3	1	150	15	1
Case 4	1	50	25	0.1

To properly evaluate the effect of treatment parameters on productivity, a production model for naturally fractured carbonate reservoirs, which was presented by Ugursal et al. (2018) [32], was adopted in this study to calculate the dimensionless production index. The calculated wormhole permeability and conductivities of hydraulic and natural fractures were adopted to solve this model. The results presented in this study used the constant flow rate boundary at the wellbore and no-flow at the outer boundaries, and the production in the index is calculated in a pseudo-steady state.

5.7. 10 mD Permeability: Case 1 and Case 2

Figure 22 shows the calculation results of the dimensionless productivity index of Case 1 and Case 2 under different acid injection rates and acid volumes. As the acid volume increases, the dimensionless productivity index increases. Moreover, each given acid volume corresponds to an optimal acid injection rate controlled by reservoir characteristics.

For example, in Case 1, the optimal acid injection rates under all acid volumes are 3 m^3/min, and increasing the acid injection rate will decrease productivity (Figure 22a). This is due to the extremely low acid efficiency caused by the strong leak-off. It is difficult to increase the effective etched-length of hydraulic fractures by increasing the acid injection rate.

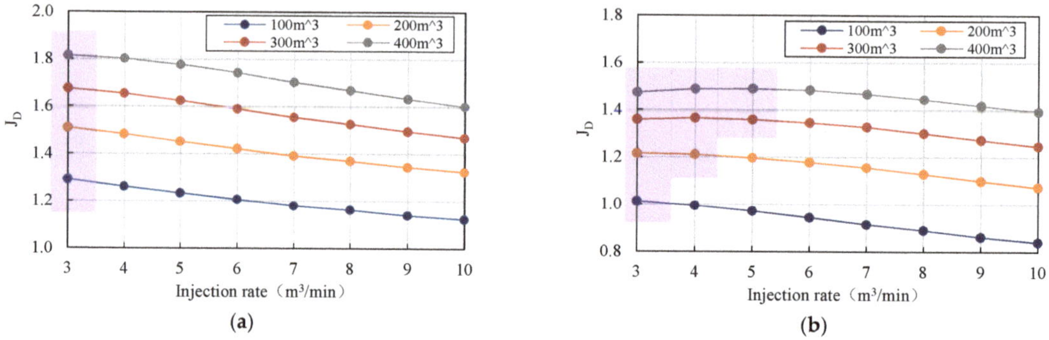

Figure 22. Dimensionless productivity index obtained from different injection rates and treatment volumes. NFw: natural fracture width; NFl: natural fracture length; NFs: natural fracture spacing. (**a**) 10 mD, NFw 250 µm, NFl 25 m, NFs 1 (Case 1); (**b**) 10 mD, NFw 50 µm, NFl 25 m, NFs 0.2 (Case 2).

However, under the conditions of Case 2, the width of natural fractures (50 µm) limits the leak-off of natural fractures, the acid efficiency increases, and the natural fractures are less conductive than in Case 1. Therefore, when the acid injection volume reaches 300 m^3 and 400 m^3, it is recommended to increase the acid injection rate to 4 m^3/min and 5 m^3/min (Figure 22b).

It can be seen from the comparison of the acid etched-width, obtained by the optimal acid injection rate with 400 m^3 injection volume in Case 1 and Case 2, that Case 1 needs higher fracture conductivity (Figure 23), which shows that under the condition of 10 mD permeability, compared with natural fracture network, natural fractures with higher fluidity have a higher contribution to productivity.

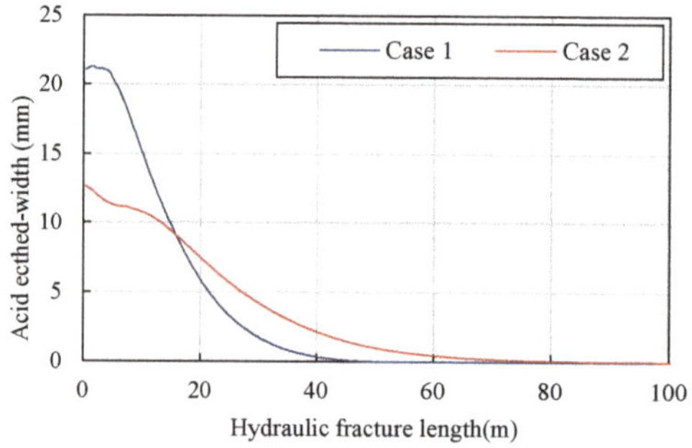

Figure 23. Acid etched-width profile at optimal injection rate and 400 m^3 acid volume in Case 1 and Case 2.

5.8. 1 mD Permeability: Case 3 and Case 4

Figure 24 shows the calculation results of the dimensionless productivity index of Case 3 and Case 4 under different acid injection rates and acid volumes. Under the 1 mD permeability condition, the optimal acid injection rate increases as the acid volume increases. However, the dimensionless productivity index shows a trend of first increasing and then decreasing. This is because the relatively low matrix permeability limits the leak-off of natural fractures and decreases the acid-etched width and length of the natural fractures, resulting in a decrease in their flow capacity after acid fracturing. A longer but less conductive hydraulic fracture is needed in such reservoirs compared to Case 1.

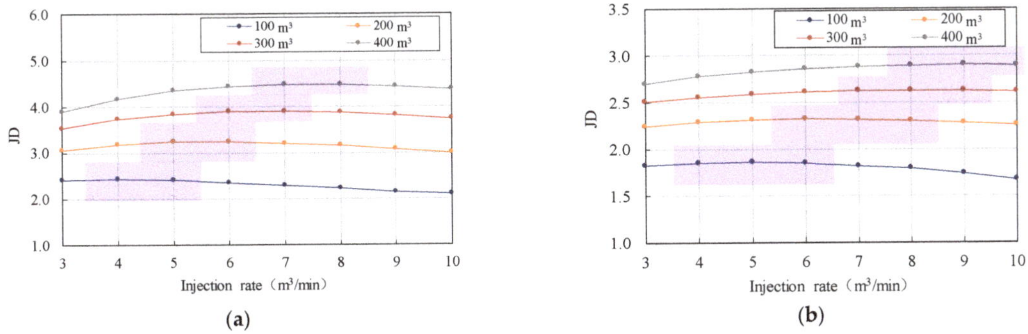

Figure 24. Dimensionless productivity index obtained from different injection rates and treatment volumes. (**a**) 1 mD, NFw 150 μm, NFl 15 m, NFs 1 (Case 3); (**b**) 1 mD, NFw 50 μm, NFl 25 m, NFs 0.1 (Case 4).

In Case 4, the acid volume and acid injection rate have the same effect on the dimensionless productivity index as in Case 3. However, the optimal acid injection rate continuously increases with the increased volume. It shows lower requirements for the flow ability of acid-etched hydraulic fracture. This is because the higher density (10 strips/m) and lower width (50 μm) of the natural fracture network limit the leak-off of natural fractures. Therefore, the acid could barely dissolve the natural fractures, which leads to a poor contribution during the production.

By comparing the acid-etched fracture width obtained by the optimal acid injection rate with the 400 m^3 acid volume in Case 3 and Case 4 (Figure 25), it can be seen that Case 3 requires higher fracture conductivity. This indicates that under the condition of low matrix permeability, natural fractures with high fluidity contribute more to productivity during production than natural fracture networks.

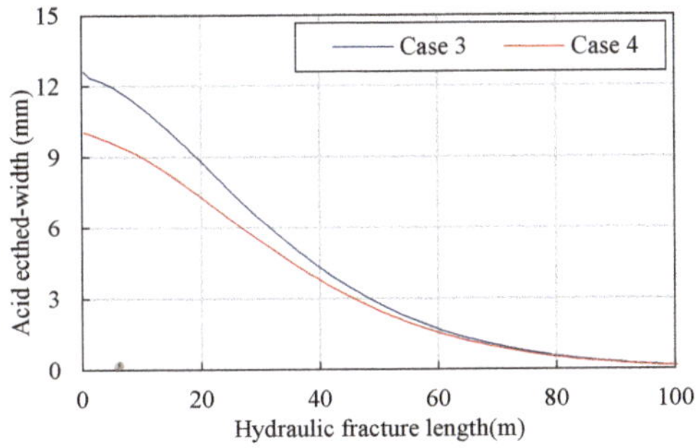

Figure 25. Acid etched-width profile at optimal injection rate and 400 m^3 acid volume in Cases 3 and 4.

5.9. Effects of Treatment Volume on Dimensionless Production Index

It can be seen that the acid-etched fractures still cannot achieve the conductivity that fully matches the reservoir flow capacity under the 400 m^3 acid volume. The low acid efficiency caused by leak-off leads to higher requirements of acid volume. In order to study the influence of the acid volume on productivity, the modeling parameters in Case 1 and Case 3 are adopted to simulate 100–3000 m^3 acid volume.

The results show that the dimensionless productivity Index increases with the increase in acid volume, but the relationship between them is non-linear. Moreover, the two cases show the same regularity: As the acid volume increases, the increase in dimensionless productivity index shows a rapid decline initially (acid injection volume < 1000 m^3) and then a trend of stability (Figure 26). However, due to the low reservoir permeability, Case 3's dimensionless productivity index is more significant than Case 1's in terms of value and growth rate, injecting more acid can get better earnings in such a reservoir. Therefore, the low permeability reservoir with developed natural fractures has more stimulation potential for acid fracturing.

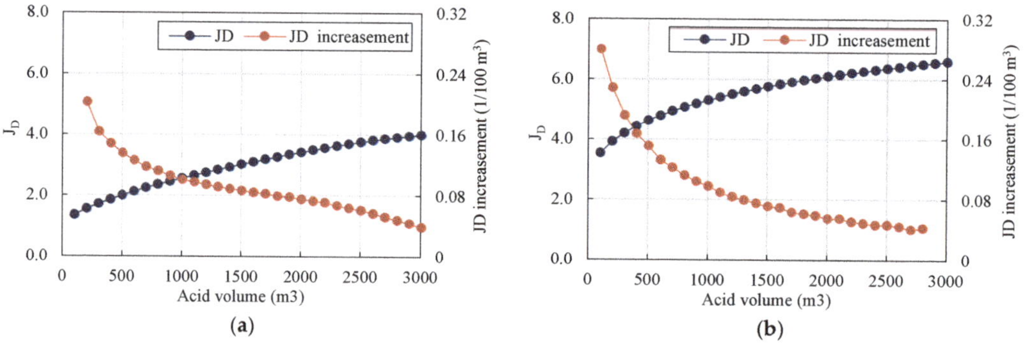

Figure 26. Dimensionless productivity index obtained from different treatment volumes. (**a**) Case 1, 6 m^3/min injection rate; (**b**) Case 3, 10 m^3/min injection rate.

Theoretically, the injection of more acid can improve the acid fracturing effect. According to the simulation results, more than 2000 m^3 of acid is required to obtain more satisfying productivity. However, engineering problems such as tubing corrosion and over-dissolution near the wellbore cannot be considered in the model, optimizing acid injection volume based on the results is challenging. Therefore, when optimizing the design on-site, comprehensive consideration should be given to factors such as economy and safety.

6. Discussions

It can be seen that the cases with more conductive natural fractures obtain higher optimal dimensionless productivity index in the same treatment conditions (Case 1, 1.81; Case 3, 4.43; versus Case 2, 1.48; Case 4, 2.82). This is because the acid can better dissolve the more conductive natural fractures and combine them with the acid-etched hydraulic fracture; they constitute a "tree" shape acid-etched fracture system and realize the volumetric control of the reservoir. The stronger the flow capacity of natural fractures, the longer the lateral control distance can be achieved (Figure 27).

Therefore, we can conclude that when acid fracturing naturally fractured carbonate reservoirs, preventing the acid leak-off from the natural fractures is not fully necessary. With the thoughts of obtaining an acid-etched fracture system and building volume control of the reservoir, we may first inject low-viscosity acid to deeply dissolve the natural fractures after the hydraulic fracture is created. Limited by the accumulative leak-off capacity of the natural, acid cannot keep entering the natural fracture; therefore, after stimulating the natural fractures, acid with high viscosity and low reaction rate can be injected to build a long and conductive acid-etched hydraulic fracture. This acid-etched fracture system may help us better stimulate the naturally fractured carbonate reservoirs.

Furthermore, due to the adoption of a simplified fracture model in the calculation process, the interconnected fractures are simplified to equivalent lengths, and the fracture network formed by the expansion of natural fractures is not considered. Consequently, the findings may not be directly applicable to reservoirs characterized by fracture networks resulting from acid stimulation under low in situ stress conditions.

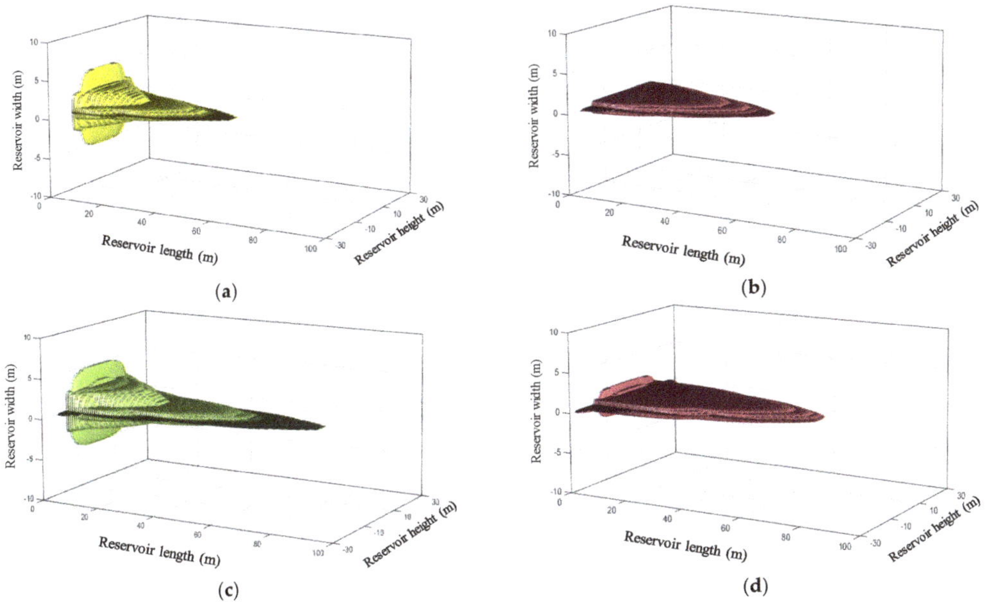

Figure 27. The schematic diagram of acid-etched fracture system in different cases. (**a**) 10 mD, NFw 250 μm, NFl 25 m, NFs 1; (**b**) 10 mD, NFw 50 μm, NFl 25 m, NFs 0.2; (**c**) 1 mD, NFw 150 μm, NFl 15 m, NFs 1 (Case 3); (**d**) 1 mD, NFw 50 μm, NFl 25 m, NFs 0.1 (Case 4).

7. Conclusions

This study presented an integrated acid fracturing simulation model for naturally fractured reservoirs, and the parametric analysis and productivity evaluation were performed under different reservoir parameters and treatment parameters. The following conclusions were made based on the results of this study:

1. Both geometric parameters and matrix permeability govern the acid leak-off of natural fractures, and it only increases significantly while all three parameters grow together. Therefore, natural fractures significantly affect the hydraulic geometry and acid-etching profile. It is necessary to take natural fractures into the simulation in naturally fractured reservoirs.
2. Wormholes significantly influence leak-off behavior and hydraulic fracture propagation. They also substantially change the etched-width profile near the hydraulic entrance where the wormholes are well developed. Therefore, wormholes must be considered in the acid fracturing simulation, especially in a high-permeability reservoir ($\geq$1 mD).
3. For all the simulated cases, there is an optimal acid injection rate that can obtain the highest productivity under various treatment volumes. The optimal parameters strongly depend on natural fracture geometry and reservoir permeability.
4. In high-permeability fractured reservoirs, creating a more conductive fracture is preferable. The optimal injection rate does not always increase with treatment volume but depends on natural fracture. The conductive natural fracture case shows the highest stimulation potential by acid fracturing, and a lower injection rate should be adopted to obtain better productivity.
5. In low-permeability fractured reservoirs, a higher injection rate is recommended to create a longer acid-etched fracture, and the optimal injection rate increases with treatment volume. The natural fracture network case shows the highest stimulation potential by acid fracturing, and the highest injection rate should be adopted to obtain better productivity.

Author Contributions: Conceptualization, B.Z.; Methodology, J.R.; Software, J.R.; Investigation, X.C., S.X. and B.Z.; Data curation, P.X.; Writing—original draft, J.R.; Writing—review & editing, C.G.; Project administration, X.C. All authors have read and agreed to the published version of the manuscript.

Funding: This research was funded by Youth Fund of the National Natural Science Foundation of China (52304047).

Data Availability Statement: The data presented in this study are available on request from the corresponding author. The data are not publicly available due to it involves the data from oil company.

Conflicts of Interest: Authors Xusheng Cao, Shunyuan Xin and Bing Zhao were employed by the company Northwest Oilfield Branch—Sinopec. The remaining authors declare that the research was conducted in the absence of any commercial or financial relationships that could be construed as a potential conflict of interest. The company Northwest Oilfield Branch—Sinopec had no role in the design of the study; in the collection, analyses, or interpretation of data; in the writing of the manuscript, or in the decision to publish the results.

Nomenclature

A_{hf}	cross-sectional area of hydraulic fracture, m^2;
C_t	total compress coefficient, Pa^{-1};
C_{hf}, C_{nf}	average acid concentration across the fracture width, mol/m^3;
c_l	specific heat of fluid, J/(kg·K);
c_m	specific heat of porous medium, J/(kg·K);
E	Young's modulus, Pa;
G	shear modulus, Pa;
h_{hf}	hydraulic fracture height, m;
h_{pay}	pay zone height, m;
h_i	hydraulic fracture height in the *i*th layer, m;
K_{Iu} and K_{Il}	stress intensity factors of the top and bottom tips of the hydraulic fracture, Pa·m$^{1/2}$;
K_{Ic}	stress intensity factor, Pa·m$^{1/2}$;
k_c	acid reaction constant, m/s; k_g is the mass transfer coefficient, m/s;
k_{mx}, k_{my}, and k_{mz}	matrix permeability in x, y, and z directions, m^2;
k_{wh}	permeability of the matrix grid with wormholes at the y direction, m^2;
L_{nf}	natural fracture length, m;
L_{wh}	wormhole length, m;
$L_{wh,\ tip}$	wormhole length at the element where the wormhole tip is located, m.
M	mole weight of the rock, kg/mol;
N_{AC}	acid capacity number;
p_{hf}	fluid pressure in hydraulic fracture, Pa;
p_{nf}	fluid pressure in natural fracture, Pa;
p_{cp}	fluid pressure at the centerline of the hydraulic fracture, Pa;
p_m	fluid pressure in matrix, Pa;
$PV_{bt,opt}$	optimal breakthrough volume, dimensionless;
q_f	fluid leak-off source term, kg·m^{-3}·s^{-1};
$q_{L,nf}$	natural fracture leak-off volume, m^3;
$q_{h,m}$	energy source term, J/(m^3·s).
T_{hf}, T_{nf}	average temperature across the fracture width, K;
T_m	temperature of the porous medium, K;
$\overline{u_x}$	average flow velocity through hydraulic fracture cross-section, m/s;
u_x, u_y, u_z	flow velocity in the x, y, z direction, m/s;
v	Poisson's ratio, dimensionless;
$v_{L,hf}$	fluid leak-off velocity in the hydraulic fracture, m/s;
v_{wh}	wormhole growth velocity, m/s;

$v_{i,\,tip}$	average interstitial velocity in the wormhole front, m/s;
$v_{i,opt}$	optimal injection velocity, m/s;
w_{hf}	hydraulic fracture width, m;
w_{tip}	width at the tip element of the hydraulic fracture, m;
μ	fluid viscosity, mPa·s;
β	dissolving power of acid on rock, kg rock/kg acid;
σ_{pay}	closure stress of pay zone, Pa;
$\Delta\sigma$	stress difference between pay zone and shale layer, Pa;
σ_i	closure stresses in the *i*th layer;
ϕ	rock porosity, dimensionless;
ρ_l	fluid density, kg/m^3;
ρ_m	density of porous medium, kg/m^3;
ρ_r	density of rock, kg/m^3;
λ_l	thermal conductance of fluid, W/(m·K);
λ_m	thermal conductance of porous medium, W/(m·K);
η	fraction of the leaking acid that reacts at the fracture surfaces;
Δy_{tip}	length of the grid where the wormhole tip is located, m.

References

1. Li, Y.; Kang, Z.; Xue, Z.; Zheng, S. Theories and practices of carbonate reservoirs development in China. *Pet. Explor. Dev.* **2018**, *45*, 669–678. [CrossRef]
2. Brenda, A.; Aguilera, R. Buildup Evaluation of a Tight Overpressured Naturally Fractured Carbonate Reservoir with the Use of a Semi-Empirical Model. *SPE Res. Eval. Eng.* **2023**, *26*, 1498–1513.
3. Chen, L.; Liu, F.; Yang, J. Horizontal well staged acid fracturing technology for deep and ultra-deep carbonate gas reservoirs in the Sichuan Basin. *Nat. Gas. Ind.* **2022**, *42*, 56–64.
4. Farouk, S.; Sen, S.; Abuseda, H.; El-Shamly, Y.; Salam, A.; Elhossainy, M.M. Petrophysical Characterization of the Turonian and Cenomanian Intervals in the Abu Gharadig Field, Western Desert, Egypt: Inferences on Reservoir Quality and Resource Development. *Nat. Resour. Res.* **2022**, *31*, 1793–1824. [CrossRef]
5. Yi, M.; Shen, X.; Jin, L.; Wang, J.; Huang, Z.; Shen, G. Forward Modeling of Natural Fractures within Carbonate Rock Formations with Continuum Damage Mechanics and Its Application in Fuman Oilfield. *Energies* **2022**, *15*, 6318. [CrossRef]
6. Aljawad, M.; Aljulaih, H.; Mahmoud, M.; Desouky, M. Integration of field, laboratory, and modeling aspects of acid fracturing: A comprehensive review. *J. Pet. Sci. Eng.* **2019**, *181*, 106158. [CrossRef]
7. Mou, J.; Zhang, S.; Zhang, Y. Acid leak-off mechanism in acid fracturing of naturally fractured carbonate oil reservoirs. *Transp. Porous Media* **2012**, *91*, 573–584. [CrossRef]
8. Ugursal, A.; Zhu, D.; Hill, A.D. Development of Acid Fracturing Model for Naturally Fractured Reservoirs. *SPE Prod. Oper.* **2019**, *34*, 735–748. [CrossRef]
9. Hawkins, M.F., Jr. A Note on the Skin Effect. *J. Pet. Technol.* **1956**, *8*, 65–66. [CrossRef]
10. Akanni, O.O.; Nasr-El-Din, H.A. Modeling of Wormhole Propagation During Matrix Acidizing of Carbonate Reservoirs by Organic Acids and Chelating Agents. In Proceedings of the SPE Annual Technical Conference and Exhibition, Dubai, United Arab Emirates, 26–28 September 2016.
11. De Oliveira, T.J.L.; De Melo, A.R.; Oliveira, J.A.A.; Pereira, A.Z. Numerical Simulation of the Acidizing Process and PVBT Extraction Methodology Including Porosity/Permeability and Mineralogy Heterogeneity. In Proceedings of the SPE International Symposium and Exhibition on Formation Damage Control, Lafayette, LA, USA, 15–17 February 2012.
12. Kalia, N.; Balakotaiah, V. Effect of medium heterogeneities on reactive dissolution of carbonates. *Chem. Eng. Sci.* **2009**, *64*, 376–390. [CrossRef]
13. Palharini, S.M.; Zhu, D.; Hill, A.D. Anisotropic-Wormhole-Network Generation in Carbonate Acidizing and Wormhole-Model Analysis through Averaged-Continuum Simulations. *SPE Prod. Oper.* **2019**, *34*, 90–108.
14. Palharini, S.M.; Hill, A.D.; Zhu, D. A New Up-Scaled Wormhole Model Grounded on Experimental Results and in 2-Scale Continuum Simulations. In Proceedings of the SPE International Conference on Oilfield Chemistry, Galveston, TX, USA, 8–9 April 2019.
15. Panga, M.K.R.; Ziauddin, M.; Balakotaiah, V. Two-Scale Continuum Model for Simulation of Wormholes in Carbonate Acidization. *AIChE J.* **2005**, *51*, 3231–3248. [CrossRef]
16. Dang, L.; Zhou, C.; Huang, M.; Jiang, D. Simulation of effective fracture length of prepad acid fracturing considering multiple leak-off effect. *Nat. Gas. Ind. B* **2019**, *6*, 64–70. [CrossRef]
17. Economides, M.J.; Zhu, D.; Hill, A.D.; Ehlig-Economides, C. *Petroleum Production Systems*, 2nd ed.; Prentice Hall: Upper Saddle River, NJ, USA, 2013.
18. Buijse, M.A.; Glasbergen, G. A semi-empirical model to calculate wormhole growth in carbonate acidizing. In Proceedings of the SPE Annual Technical Conference and Exhibition, Dallas, TX, USA, 9–12 October 2005.

19. Furui, K.; Burton, R.C.; Burkhead, D.W.; Abdelmalek, N.A.; Hill, A.D.; Zhu, D.; Nozaki, M. A Comprehensive Model of High-Rate Matrix-Acid Stimulation for Long Horizontal Wells in Carbonate Reservoirs: Part II—Wellbore/Reservoir Coupled-Flow Modeling and Field Application. *SPE J.* **2012**, *17*, 280–291. [CrossRef]
20. Hill, A.D.; Zhu, D.; Wang, Y. The Effect of Wormholing on the Fluid Loss Coefficient in Acid Fracturing. *SPE Prod. Oper.* **1995**, *10*, 257–264. [CrossRef]
21. Economides, M.J.; Nolte, K.G. *Reservoir Stimulation*, 3rd ed.; Prentice Hall: Englewood Cliffs, NJ, USA, 1989.
22. Olson, J.E. *Fracture Aperture, Length and Pattern Geometry Development under Biaxial Loading: A Numerical Study with Applications to Natural, Cross-Jointed Systems*; Special Publications; Geological Society of London: London, UK, 2007; Volume 289, pp. 123–142.
23. Dong, C.; Zhu, D.; Hill, A.D. Modeling of the Acidizing Process in Naturally Fractured Carbonates. *SPE J.* **2002**, *7*, 400–408. [CrossRef]
24. Nolte, K.G.; Economides, M.J. (Eds.) *Reservoir Stimulation*; John Wiley and Sons: Chichester, NY, USA, 2000.
25. Eickmeier, J.R.; Ersoy, D.; Ramey, H.J. Wellbore temperatures and heat losses during production or injection operations. *J. Can. Pet. Technol.* **1970**, *9*, 115–121. [CrossRef]
26. Kaviany, M. *Principles of Heat Transfer in Porous Media*; Springer Science and Business Media: Berlin/Heidelberg, Germany, 2012.
27. Roberts, L.D.; Guin, J.A. A New Method for Predicting Acid Penetration Distance. *SPE J.* **1975**, *15*, 277–286. [CrossRef]
28. Deng, J.; Mou, J.; Hill, A.D.; Zhu, D. A New Correlation of Acid-Fracture Conductivity Subject to Closure Stress. *SPE Prod. Oper.* **2012**, *27*, 158–169. [CrossRef]
29. Nordgren, R.P. Propagation of a vertical hydraulic fracture. *SPE J.* **1972**, *12*, 306–314. [CrossRef]
30. Schechter, R.S. *Oil Well Stimulation*; Prentice-Hall, Inc.: Englewood Cliffs, NJ, USA, 1992.
31. Terrill, R.M. Heat Transfer in Laminar Flow Between Parallel Porous Plates. *Int. J. Heat. Mass. Transfer.* **1965**, *8*, 1491–1497. [CrossRef]
32. Ugursal, A.; Schwalbert, M.P.; Zhu, D.; Hill, A.D. Acid Fracturing Productivity Model for Naturally Fractured Carbonate Reservoirs. In Proceedings of the SPE International Hydraulic Fracturing Technology Conference and Exhibition, Muscat, Oman, 16–18 October 2018.
33. Etten, J.; Zhu, D.; Hill, A.D. The Combined Effect of Permeability and Pore Structure on Carbonate Matrix Acidizing. In Proceedings of the EUROPEC 2015, Madrid, Spain, 1–4 June 2015.

processes

MDPI

Article

Study on the Evolution Law of Four-Dimensional Dynamic Stress Fields in Fracturing of Deep Shale Gas Platform Wells

Yongchao Wu [1], Zhaopeng Zhu [2], Yinghao Shen [1,*], Xuemeng Yu [2], Guangyu Liu [2] and Pengyu Liu [1]

[1] State Key Laboratory of Petroleum Resources an Engineering, China University of Petroleum (Beijing), Beijing 102249, China; wuyongchao7434@163.com (Y.W.)
[2] PetroChina Jilin Oilfield Company, Songyuan 138000, China
* Correspondence: shenyinghao@cup.edu.cn

Abstract: Compared with conventional gas reservoirs, deep shale gas reservoirs are characterized by developed faults and fractures, strong heterogeneity, high stress sensitivity, and complex in situ stress distribution. To address traditional 3D static models' inability to predict in situ stress changes in strongly heterogeneous reservoirs during fracturing, this study takes the deep shale gas in the Zigong block of the Sichuan Basin as an example. By comprehensively considering the heterogeneity and anisotropy of geomechanical parameters and natural fractures in shale gas reservoirs, a 4D in situ stress multi-physics coupling model for shale gas reservoirs based on geology–engineering integration is established. Through coupling geomechanical parameters with fracturing operation data, the dynamic evolution laws of multi-scale stress fields from single-stage to platform-scale during large-scale fracturing of horizontal wells in deep shale gas reservoirs are systematically studied. The research results show the following: (1) The fracturing process has a significant impact on the magnitude and direction of the stress field. With the injection of fracturing fluid, both the minimum and maximum horizontal principal stresses increase, with the minimum horizontal principal stress rising by 1.8–6.4 MPa and the maximum horizontal principal stress by 1.1–3.2 MPa; near the wellbore, there is an obvious deflection in the direction of in situ stress. (2) As the number of fracturing stages increases, the minimum horizontal principal stress shows an obvious cumulative growth trend, with a more significant increase in the later stages, and there is a phenomenon of stress accumulation along the wellbore, with the stress difference decreasing from 15 MPa to 11 MPa. (3) The on-site adoption of the fracturing operation method featuring overall flush advancement and inter-well staggered fracture placement has achieved good stress balance; comparative analysis shows that the stress communication degree of the 400 m well spacing is weaker than that of the 300 m well spacing. This study provides a more reasonable simulation method for large-scale fracturing development of deep shale gas, which can more accurately predict and evaluate the dynamic stress field changes during fracturing, thereby guiding fracturing operations in actual production.

Keywords: deep shale gas; hydraulic fracturing; 4D stress field

1. Introduction

China is rich in shale gas resources, with favorable exploration areas reaching 43×10^4 km^2. According to preliminary evaluation results from the Ministry of Natural Resources, China's total shale gas resources amount to 134.4×10^{12} m^3, with recoverable resources reaching 25×10^{12} m^3, ranking among the world's highest. Notably, deep shale oil and gas resources account for a considerable proportion [1,2]. Deep (3500–4500 m) to ultra-deep (>4500 m) shale gas has now become an important replacement field for China's shale gas development, with technically recoverable resources accounting for 56.63% of the total recoverable shale gas resources. Enhancing its exploration and development is of great strategic significance for ensuring national energy security. After more than a decade of theoretical innovation and practical development in shale gas exploitation, China has established technical capabilities for shale gas well extraction at depths shallower than 3500 m. Although significant progress has been made in supporting extraction technologies for depths of 3500–4000 m, commercial development remains distant, and the overall exploration level of deep shale gas is relatively low [3,4]. The Sichuan Basin and its surrounding areas contain 65.8% of China's total recoverable deep to ultra-deep shale gas resources, making it the primary exploration and development base for domestic deep shale gas [5–7]. However, unlike medium-deep formations, deep shale exhibits significantly increased formation temperature, pressure, and stress with greater burial depth, along with more complex geological structures and in situ stress conditions, leading to substantially greater challenges in fracturing stimulation [7–10].

In situ stress has a significant impact on the productivity of a single well. For horizontal wells in deep shale gas, if well logging is not performed, the calculation parameters for in situ stress cannot be obtained. Therefore, it is urgent to carry out research on the modeling of the original three-dimensional stress field to support the optimization of the fracturing scheme. During hydraulic fracturing, changes in formation pore pressure, combined with interactions with natural fractures and induced fracture networks, generate a perturbed stress field, leading to real-time variations in the near-wellbore stress distribution; Refs. [11,12]. This may even lead to stress reversal, which directly affects the propagation direction and morphology of the hydraulic fracture network and changes the flow efficiency of oil and gas from the rock matrix to the wellbore [13,14]. At the same time, it will also cause formation deformation and activate faults, generating a disturbed stress field that can lead to fault slip-induced casing deformation [15,16], posing safety risks. After fracturing, both the fracturing well and offset wells experience significant changes in stress orientation. If initial stress data are still used for subsequent fracturing simulations of adjacent stages or wells, the results may be unreliable [17]. Thus, it is essential to investigate the dynamic evolution of in situ stress during deep shale gas fracturing.

The generation of hydraulic fractures alters the magnitude and orientation of in situ stresses around the fractures, an effect termed the "stress shadow" effect. Sneddon [18] derived an analytical expression for the stress field distribution around fractures under plane strain conditions in an infinite elastic medium, demonstrating that the impact on the minimum horizontal principal stress near the fracture is more pronounced than on the maximum principal stress. Crouch [19] pioneered the use of the displacement discontinuity method to investigate displacement discontinuities and the surrounding stress distribution in fractured rock media. Roussel [20,21] integrated the displacement discontinuity method with the finite element method, proposing that dynamic stress field variations induce pore volume and rock matrix compression, thereby influencing fluid flow and fracture propagation. Zhang et al. [22] found that both stress contrast and fracture initiation angle affect the propagation path of fractures, causing the orientation of new fractures to gradually deviate and eventually reorient toward the maximum horizontal stress direction after overcoming stress barriers. Wang et al. [23] employed the extended finite element method to analyze stress interference phenomena in sequential and alternate fracturing, revealing its dependence on the fracturing sequence.

For a long time, modeling dynamic stress fields in hydraulic fracturing has been a research hotspot both domestically and internationally: Biot [24] was the first to propose a seepage-stress coupled 3D in situ stress model based on elastic porous media; later,

Geertsma [25] put forward the theory of pore-rock volume change to better understand the pressure–volume relationship of reservoir rocks, establishing connections between the porosity, rock volume compressibility of isotropic porous media, and the elastic and viscous deformation constants of rock matrix and materials; since simple isotropic models cannot describe real rock behavior, Lekhnitskii [26] proposed the theory of anisotropic elastic bodies; Pei et al. [27] established a 3D geomechanical–fluid flow coupling model for multi-layer reservoirs to analyze the impact of stress differences on infill well fracturing effects, though the model assumes steady-state fluid flow and static stress fields, ignoring the cumulative effect of stress shadows in multi-stage fracturing; Zhang et al. [28] proposed a workflow integrating microseismic monitoring and geomechanical modeling to study the correlation between induced seismicity and stress redirection in shale gas reservoirs; Glubok et al. [29] developed a physics-constrained machine learning algorithm that converts microseismic event distributions into near-real-time fracture propagation visualizations, reducing interpretation errors from 300% to 10%; Ibrahim et al. [30] used the Adaptive Neuro-Fuzzy Inference System (ANFIS) to predict horizontal in situ stress via logging data with a correlation coefficient (R) of 0.96; Khormall [31] applied artificial neural networks to analyze reservoir rock permeability changes caused by solid precipitation during water flooding; on this basis, Hatchell et al. [32–34] coupled reservoir simulators with seismic-driven geomechanical models to establish a 4D dynamic in situ stress model considering the time effect of production or injection processes; Zhu [35] constructed a 4D in situ stress evolution model for shale gas reservoirs, combining continuum mechanics with the Discrete Fracture Network (DFN) method to simulate stress field reconfiguration due to pore pressure decline during long-term exploitation; and Ma [36] proposed an integrated static and dynamic geomechanical modeling workflow for hydraulic fracturing optimization, conducting dynamic (4D) stress field simulations for tight oil reservoirs.

Existing studies have limitations in several aspects: analyses of stress fields mostly focus on static characteristics and fail to fully capture the dynamic reconstruction laws of stress fields during fracturing; existing 4D models mostly target isotropic reservoirs, ignoring the heterogeneity and anisotropy of reservoir geomechanical parameters and natural fractures, which leads to a significant deviation from the actual geological conditions of shale gas reservoirs; the application of machine learning methods in stress field prediction is still dominated by static or semi-static analyses, with insufficient dynamic adaptability and a lack of real-time linkage with fracturing operation parameters (such as injection rate and pressure), making it difficult to dynamically reflect stress evolution during the operation process; meanwhile, existing stress field studies mainly focus on stress changes caused by reservoir depletion in shallow shale gas and tight oil reservoirs, and there is a lack of research on real-time dynamic changes of stress fields during fracturing operations in deep shale gas.

Therefore, this paper first comprehensively considers the heterogeneity and anisotropy of geomechanical parameters and natural fractures in deep shale gas reservoirs and establishes a 4D in situ stress multi-physics coupling model for shale gas reservoirs based on geology-engineering integration. Then, by coupling geomechanical parameters with fracturing operation data, it systematically studies the dynamic evolution laws of multi-scale stress fields from single-stage to platform-scale during large-scale fracturing of horizontal wells in deep shale gas reservoirs, As shown in Figure 1. The research results can provide a certain scientific basis for optimizing fracturing design and improving shale gas recovery efficiency.

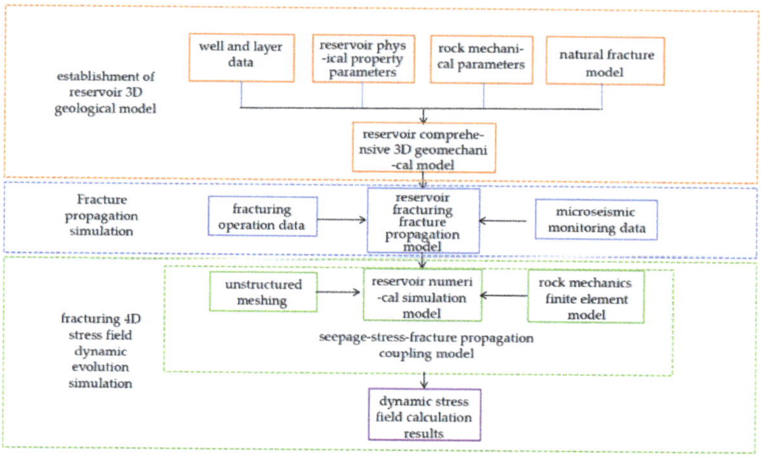

Figure 1. Simulation workflow for 4D stress field evolution in deep shale gas.

2. Regional Geological Overview and Geological Modeling

2.1. Geological Overview

The Panlongchang syncline is structurally located within the Ziliujing Depression of the Central Sichuan Uplift Belt, bounded by the Shengdengshan structure to the west and the Yundingchang structure to the east. With an area of 188.3 km^2, the syncline is delineated by secondary faults, featuring a major axis of 17.8 km and a minor axis ranging from 7.1 to 12.8 km. The structural low point at the base of the Longmaxi Formation reaches an elevation of −4100 m, with burial depths varying between 4000 and 4400 m. The workable area covers 169.2 km^2, while the deployable area spans 131 km^2. The study area is situated in the central part of the Panlongchang syncline. The syncline exhibits favorable fault stability, with only one weak-slip fault developed. The maximum horizontal principal stress exceeds the vertical principal stress, indicating a strike-slip stress regime. The location of well area is shown in Figure 2. Natural fractures in the Panlongchang syncline are predominantly oriented NE-SW and NW-SE. The NE-SW trending fractures include Fe1 through Fe13 (totaling 13 natural fractures), while the NW-SE trending fractures comprise Fw1 through Fw8 (totaling 8 natural fractures). The basic parameters of the study area are summarized in Table 1.

Table 1. The key petrophysical parameters of the study area.

Formation Pressure/MPa	Permeability/mD	Porosity/%	Young's Modulus/GPa	Poisson's Ratio	Minimum Horizontal Stress/MPa	Maximum Horizontal Stress/MPa
85.6~90.8	0.001~0.002	2.7~4.5	27~45	0.17~0.27	80~104	100~121

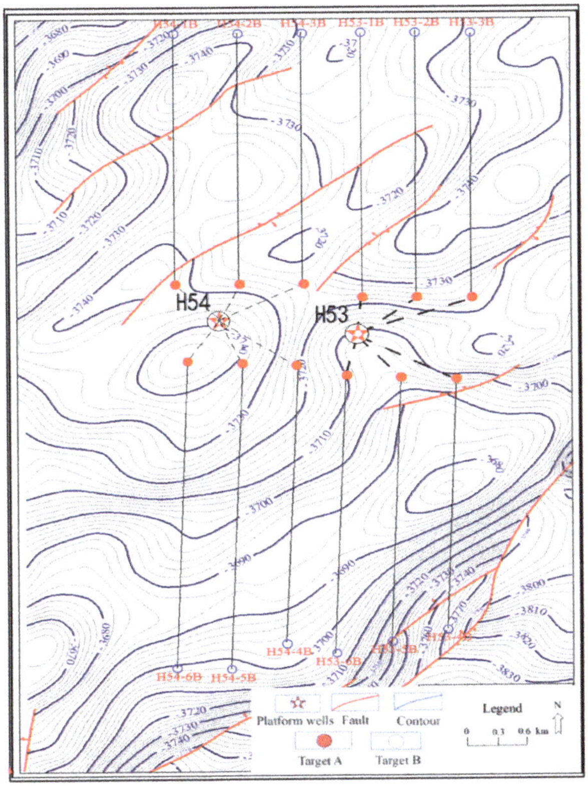

Figure 2. Reference location map of the study area.

2.2. 3D Geomechanical Modeling

Geological modeling serves as a crucial bridge connecting various geological understandings. The methodology integrates well data, test results, and seismic interpretation to establish structural models through refined stratigraphic correlation and fault/formation surface interpretation, while reservoir property models are built using conventional and specialized logging data (Figure 3). The study area contains two major NE-SW trending faults modeled with 3D Structural Framework technology to represent complex fault intersections, subsequently converted into a Pillar-Gridding corner-point grid fault model. Natural fractures were effectively characterized using ant-tracking seismic attributes in the Petrel integrated geological-engineering platform, demonstrating high consistency (Figure 4). Laboratory mechanical tests provided calibration for field logging data, enabling the development of accurate 1D geomechanical models with key parameters determined based on structural framework and formation properties. Employing Pillar Gridding technology within well-controlled favorable zones, the 3D structural model achieved 20 m $\times$ 20 m planar resolution with true north-oriented grids. The reservoir section features high-precision vertical gridding at 1 m intervals, ensuring robust simulation accuracy.

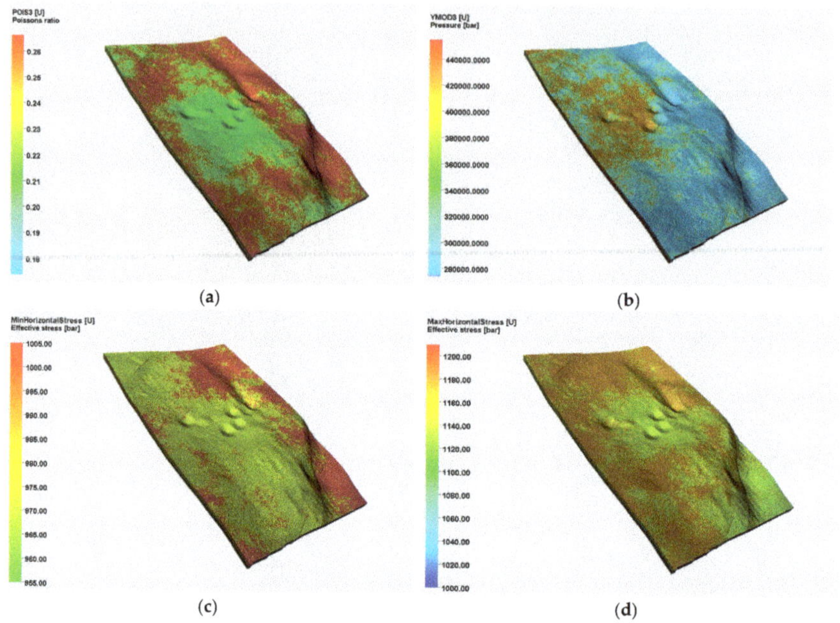

Figure 3. Attributes of three-dimensional geomechanical model: (**a**) static Young's modulus; (**b**) static Poisson's ratio; (**c**) minimum horizontal principal stress; (**d**) maximum horizontal principal stress.

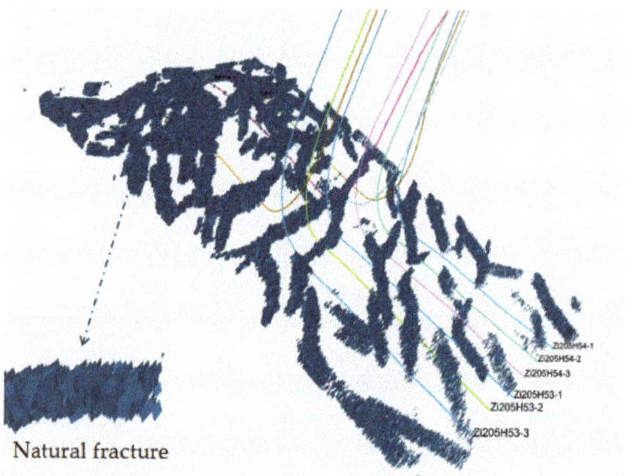

Figure 4. 3D natural fracture mode.

3. Methodology

3.1. Fluid–Solid Coupling Mathematical Model

In the process of oil and gas reservoir production, the most significant changes in the reservoir are caused by pore pressure increases resulting from stimulation measures such as fracturing and pore pressure depletion due to production. Based on the initial in situ stress field model and combined with regional formation pressure changes provided by reservoir simulation, the induced stress field considering production time is calculated to achieve four-dimensional in situ stress field modeling. In reality, the reservoir is a highly complex system. To solve the stress field in the reservoir, it is simplified here as a fluid–solid coupling problem in porous media.

For the rock matrix part (solid phase medium), the governing equations are as follows:

$$\sigma_{ij} = 2G\varepsilon_{ij} + \lambda\delta_{ij}\varepsilon_{kk} \quad (1)$$

where σ_{ij} denotes the total stress tensor, Lamé constants $\lambda = \frac{Ev}{(1+V)(1-2V)}$, shear modulus $G = \frac{E}{2(1+V)}$, and δ_{ij} is the Kronecker delta symbol.

For the fluid phase, the mass conservation equation for single-phase flow is expressed as follows:

$$\frac{\partial\left(\phi P_f s\right)}{\partial t} + \vec{\nabla}\cdot\left(p_f\vec{v}\right) = Q_m \quad (2)$$

In the equation, ϕ represents porosity; p_f denotes fluid density; S indicates saturation; t is time; $\vec{v}$ stands for flow velocity vector; $\vec{\nabla}$ is the gradient operator with respect to three-dimensional coordinates

The fluid velocity is described by Darcy's law:

$$\vec{V} = -\frac{k}{\mu}\vec{\nabla}\left(p - p_f gD\right) \quad (3)$$

where k represents permeability; μ denotes fluid viscosity; p is fluid pressure; g stands for gravitational acceleration constant; D indicates depth with surface as reference zero point. Under the condition of negligible fluid density, Equation (2) simplifies to the following:

$$\frac{\partial\vec{\zeta}}{\partial t} + \vec{\nabla}\cdot\left[\frac{k}{\mu}\vec{\nabla}\left(p - p_f gD\right)\right] = Q_V \quad (4)$$

where $Q_V = \frac{Q_m}{P_f}$ represents the volumetric fluid flow rate per unit time, characterizing the fluid volume flux through a unit cross-section; $\zeta = \phi s$ denotes the fluid content in porous media (dimensionless), defined as the ratio of fluid volume to total bulk volume.

According to the effective stress principle for porous media, the stress acting on the solid matrix is given by the following:

$$\sigma'_{ij} = \sigma_{ij} - \alpha\delta_{ij}p \quad (5)$$

where σ'_{ij} denotes the effective stress.

To address the limitations of the conventional Biot coefficient α being treated as a constant, this model introduces a time-dependent equation to characterize matrix–fluid interactions during fracturing:

$$\alpha(t) = 1 - \frac{K_{dry}}{K_{solid}}\exp(-\beta t) \quad (6)$$

where K_{dry} and K_{solid} represent the bulk modulus of dry rock and solid matrix, respectively, and β denotes the attenuation coefficient (range: 0.05–0.15 h^{-1}) [37] characterizing the dynamic stiffness variation of matrix caused by fracturing fluid imbibition. This enhancement enables the model to better capture the stress–seepage coupling characteristics of deep shale reservoirs.

Accordingly, based on poroelastic theory with total stress and fluid content as primary variables, the constitutive relations of porous media can be expressed as follows:

$$\sigma_{ij} = 2G\varepsilon_{ij} + \frac{2Gv}{1-2v}\delta_{ij}\varepsilon_{kk} + \alpha\delta_{ij}p \quad (7)$$

For porous media containing incompressible solid grains fully saturated with incompressible fluid (when α = 1.0, B = 1.0), Equation (4) becomes the following:

$$\frac{\partial \varepsilon_{kk}}{\partial t} + \vec{\nabla} \cdot \left[\frac{k}{\mu} \vec{\nabla}(p - \rho g D) \right] = -Q_V \tag{8}$$

Considering the equilibrium equations of the entire system:

$$\sigma_{ij,j} + f_i = 0 \tag{9}$$

where f represents external or body forces.

Based on the strain–displacement relationship, combining Equations (7) and (8) yields the following:

$$G\nabla^2 u_i + \frac{G}{1-2v}(u_{k,k})_{,i} + \alpha p_{,i} = f_i \tag{10}$$

$$\frac{\partial \varepsilon_{kk}}{\partial t} + \vec{\nabla} \cdot \left[\frac{k}{\mu} \vec{\nabla}(p - \rho g D) \right] = -Q_V \tag{11}$$

The system of equations composed of Equations (9)–(11), when supplemented with appropriate initial and boundary conditions, enables the complete solution for porous media behavior.

3.2. Fracture Propagation Criteria

During hydraulic fracturing, a main hydraulic fracture is generated along the wellbore. With the continuous injection of fracturing fluid, branch fractures are formed around the main fracture. The main fracture and branch fractures together constitute a complex fracture network, thereby improving the seepage capacity. The basic principle of fracture propagation is based on stress conditions and rock failure criteria. In the process of fracture propagation, two issues need to be clarified: whether the fracture propagates and extends, and the direction of propagation. The J-integral method is used to calculate the stress intensity factor at the fracture tip:

$$K_I^{eq} = \frac{1}{2}\cos\left(\frac{\theta}{2}\right)\{K_I(1+\cos(\theta)) - 3K_{II}\sin(\theta)\} \tag{12}$$

where K_I^{eq} is the equivalent stress intensity factor; K_I is the Mode I stress intensity factor; K_{II} is the Mode II stress intensity factor.

The propagation direction (angle θ) of the fracture is calculated according to the maximum circumferential stress criterion:

$$\theta = 2\arctan\frac{1}{4}\left(\frac{K_I}{K_{II}} - sign(K_{II})\sqrt{\left(\frac{K_I}{K_{II}}\right)^2 + 8}\right) \tag{13}$$

If there are natural fractures developed in the reservoir, the main hydraulic fractures will connect with the natural fractures during their propagation and extension. The Warpinski–Teufel criterion is a widely used failure criterion for tensile fractures at present:

$$p > \sigma_n \tag{14}$$

where p is the pore pressure in the natural fracture; σ_n is the normal stress on the wall of the natural fracture.

Whether shear slip occurs in the natural fracture is determined by the Mohr–Coulomb criterion, whose expression is as follows:

$$|\tau| > \tau_0 + C_f(\sigma_n - P) \tag{15}$$

where τ is the shear stress on the fracture wall; τ_0 is the cohesion of the rock; C_f is the wall friction factor.

Based on the theories of elasticity and fracture mechanics, the net pressures required for the generation of tensile fractures and shear fractures are as follows:

$$p_t > \frac{\sigma_H - \sigma_h}{2}(1 - \cos 2\theta) \tag{16}$$

$$p_s > \frac{1}{C_f}\left[\tau_0 + \frac{\sigma_H - \sigma_h}{2}\left(C_f - \sin 2\theta - C_f \cos 2\theta\right)\right] \tag{17}$$

where P_t is the net pressure required for the generation of tensile fractures; P_s is the net pressure required for the generation of shear fractures; σ_H and σ_h are the maximum and minimum horizontal principal stresses, respectively. The fluid loss rate of fracturing fractures is calculated by the Carter leak-off model:

$$v_L = \frac{2C_L}{\sqrt{t - t_0}} \tag{18}$$

where v_L is the leak-off velocity; t is time; t_0 is the initial time of leak-off; C_L is the Carter leak-off coefficient. Using the cubic law, the continuity equation for the flow of fracturing fluid in the fracture considering fluid leak-off is as follows:

$$\frac{\partial w}{\partial_t} + \nabla \cdot \left(-\frac{w^3}{12\mu}\nabla P_f\right) + \frac{2C_L}{\sqrt{t - t_0}} = 0 \tag{19}$$

where μ is the viscosity of the fracturing fluid; p_f is the pressure in the fracture. The criterion for determining the dynamic boundary position of pressure propagation caused by fracture propagation in volume-fractured horizontal wells under nonlinear seepage conditions is as follows:

$$\begin{cases} \left(\frac{\partial p}{\partial x} - G_a\right)\Big|_{r=Rtx} = 0 \\ \left(\frac{\partial p}{\partial y} - G_a\right)\Big|_{r=Rty} = 0 \end{cases} \tag{20}$$

$$G_a = \frac{1 - a}{b} \tag{21}$$

where Rtx and Rty are the dynamic boundary positions in the x and y directions, respectively; Ga is the critical start-up pressure gradient.

3.3. Research Methodology

During the long-term production of shale gas reservoirs, formation pressure continuously changes, and the reservoir stress state essentially represents the dynamic coupling between fracture–pore dual-medium seepage and matrix skeleton stress. Based on reservoir numerical simulation methods and geomechanical theory, a seepage–stress–fracture propagation coupling model is established, with research conducted using an integrated geological-engineering platform. The forms of coupled calculation can be divided into one-way coupling, iterative coupling, and full coupling [38]. Since one-way coupling has low calculation accuracy and full coupling has low calculation efficiency, the iterative coupling method is selected for calculation. The difference from conventional iterative coupling is that the seepage–stress–fracture propagation iterative coupling simulation solves the fracture morphology, pore pressure, stress, and strain at all time points in one iteration.

This methodology utilizes the Kinetix fracture propagation simulator, Intersect reservoir numerical simulator, and Visage finite element mechanical simulator to model fracture geometry, seepage field, and stress field, respectively. The process begins by generating an initial fracture network. During one iteration, pore pressure and saturation at each timestep are solved, followed by grid conversion. These results are then used as input conditions for the finite element model to calculate stress and strain data at each timestep. Subsequently, the computed fracture geometry, pressure, saturation, and stress data are updated to initiate the next iteration step. This cyclic computation is repeated to complete the coupled simulation of seepage–stress–fracture propagation across different development stages (Figure 5).

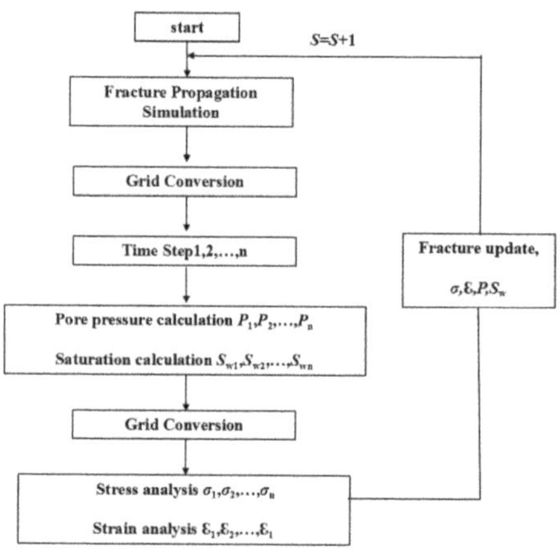

Figure 5. Coupling process of seepage–stress–fracture propagation.

4. Study on the Evolution Laws of 4D Dynamic Stress Fields

During hydraulic fracturing operations, the dynamic evolution of reservoir stress fields directly influences fracture propagation patterns, stimulation volume, and inter-well interference levels, representing a core scientific issue for geology-engineering integration optimization. Traditional static stress field models struggle to reveal the spatiotemporal evolution characteristics of stress states under fracturing disturbances, which constrains the precision of large-scale fracturing designs. This chapter systematically investigates based on multi-scale coupling modeling methods: (1) study on single-stage stress field evolution, analyzing the dynamic coupling mechanism between fracture propagation and near-field stress redistribution; (2) study on single-well stress field evolution, revealing the stress accumulation effects of multi-stage fracturing; (3) study on pad-scale stress field evolution, clarifying the stress interference boundaries and balanced control principles during multi-well simultaneous fracturing.

By establishing a four-dimensional dynamic response model of "fracture–stress–fluid" interactions, this research provides theoretical foundations for stress field regulation and engineering optimization pathways for efficient shale gas development.

4.1. Study on Single-Stage Stress Field Variation Patterns

4.1.1. Fracture Propagation

The simulated treatment pressure for Stage 7 of Well 54-2 shows good agreement with actual field data. The actual shut-in pressure was 66.31 MPa, while the simulated shut-in pressure was 63.2 MPa, yielding an error of 4.7% (Figure 6). As shown in Figure 7, the spheres represent the values of the microseismic monitoring signals. The results correlate well with microseismic monitoring data, demonstrating that the simulated fractures accurately represent actual hydraulic fracture parameters and geometry, thereby validating the model's accuracy.

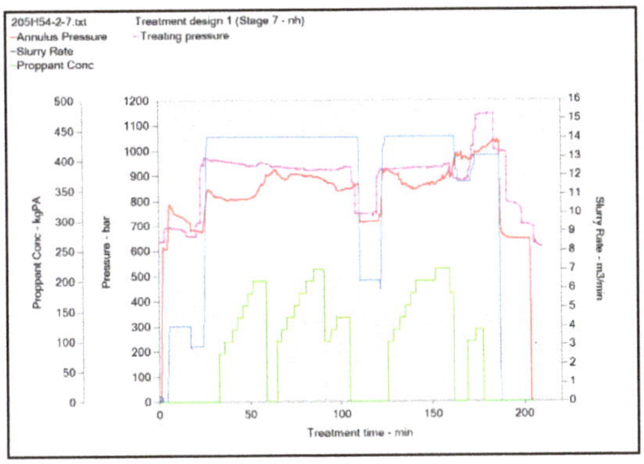

Figure 6. Fracturing treatment curve matching.

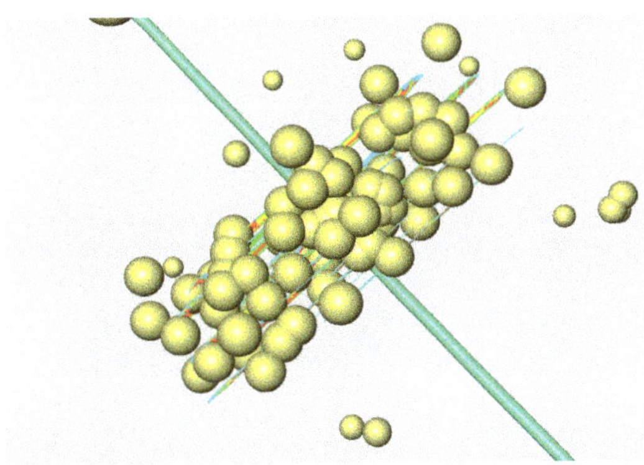

Figure 7. Microseismic monitoring comparison.

The hydraulic fractures propagate relatively symmetrically on both sides of the wellbore, with fracture complexity near the wellbore showing minimal variation compared to the fracture tips. The average hydraulic fracture length is 223 m, average height is 47 m, and average width is 8.54 mm.

4.1.2. Stress Field Variation

(1) Stress Magnitude Variation

Figure 8 shows the distribution of the minimum horizontal principal stress in the formation before and after fracturing in Stage 7 of Well 54-2. The figure clearly shows that after fracturing, the minimum horizontal principal stress in the formation around the fractures underwent significant changes, increasing by 3.4 MPa compared to the pre-fracturing state. Compare with the pre-fracturing conditions, the impact of fracturing on the surrounding formation can be divided into three zones. The first is the stress reversal zone, where the presence of hydraulic fractures causes stress reversal in a certain area around the fractures. In the formation close to the fracture faces, the minimum horizontal principal stress significantly decreases after fracturing. Additionally, there is an "elliptical" zone around the fractures where the minimum horizontal principal stress increases noticeably. Finally, on both sides of the fractures along the wellbore direction, there is a "triangular" zone where the minimum horizontal principal stress shows varying degrees of increase, with more pronounced stress elevation closer to the wellbore. After fracturing, the stress contrast decreases in regions far from the fractures, while it increases near the fracture faces. In comparison, the maximum horizontal principal stress shows a smaller increase of about 1.3 MPa, concentrated mainly at the fracture tips (Figure 9).

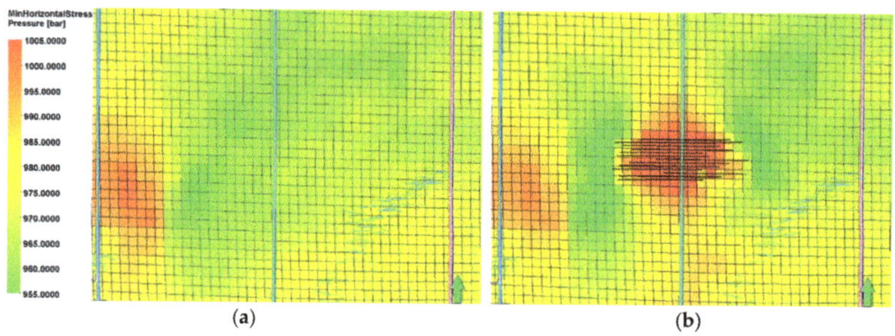

Figure 8. Minimum horizontal principal stresses of Stage 7 in well 54-2 (**a**) before fracturing and (**b**) after fracturing.

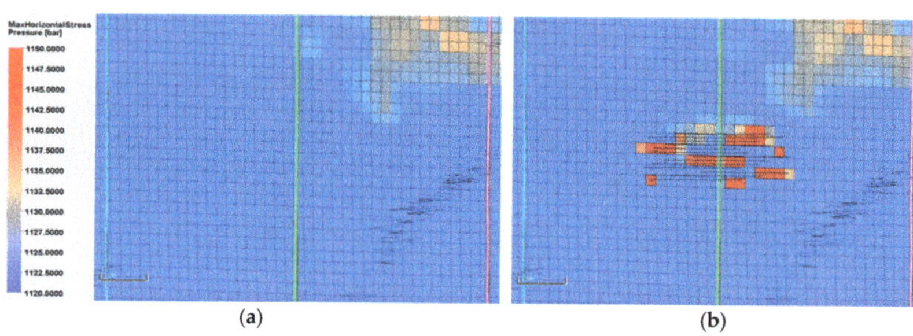

Figure 9. Maximum horizontal principal stresses of Stage 7 in Well 54-2 (**a**) before fracturing and (**b**) after fracturing.

(2) Stress Orientation Variation

After the fracturing operation, the opening of fractures causes changes in the displacement around the fractures, leading to the deflection of the directions of the maximum and minimum principal stresses. The degree of stress direction deflection is influenced by the net fracture pressure, which is related to parameters such as fracturing displacement and fluid volume. Figures 10 and 11 shows the directional changes of the formation principal stresses before and after the fracturing of the seventh section of Well 54-2. where the arrow directions indicate the azimuth angles of in-situ stress and the color bands represent the magnitude of in-situ stress. The maximum deflection of the maximum horizontal principal stress occurs in the middle of the fracture, in the area perpendicular to the fracture. In contrast, the maximum deflection of the minimum horizontal principal stress occurs at both ends of the fracture. Post-fracturing, the maximum horizontal principal stress, originally parallel to the fracture extension direction, starts to deflect towards a direction perpendicular to the fracture extension. Conversely, the minimum horizontal principal stress, originally perpendicular to the fracture extension direction, begins to deflect towards a direction parallel to the fracture extension.

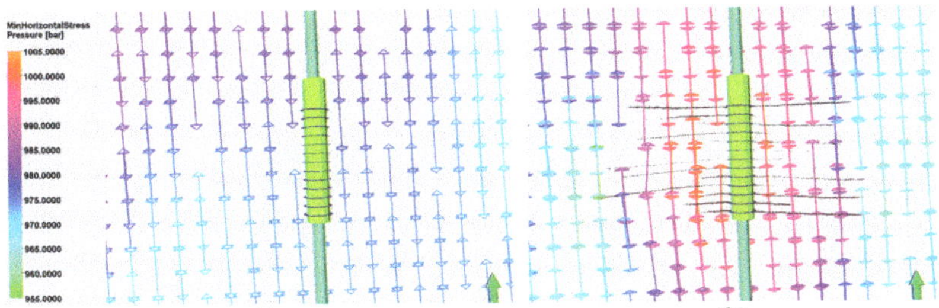

(a) (b)

Figure 10. Minimum horizontal principal stress orientation of Stage 7 in Well 54-2 (**a**) before fracturing and (**b**) after fracturing.

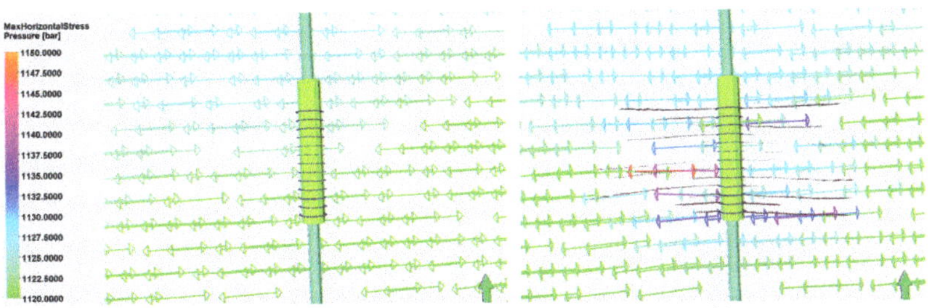

(a) (b)

Figure 11. Maximum horizontal principal stress orientation of Stage 7 in Well 54-2 (**a**) before fracturing and (**b**) after fracturing.

4.2. Single-Well Stress Field Evolution Study

Well 54-3 has abundant pre-fracturing seismic monitoring data and related information. Following the above methodology, fracture propagation and stress simulation were performed for Well 54-3. The main fracturing operation in Well 54-3 adopted large-stage multi-cluster, high-intensity, and multi-stage temporary plugging fracturing technology with the following perforation techniques: Stage 1 used coiled tubing perforation while the remaining stages employed cable-conveyed cluster perforation. The primary fluid system used was variable-viscosity slickwater, with low-viscosity slickwater accounting for approximately 90% of the total volume. During smooth operation phases, low-viscosity slickwater was predominantly used to minimize consumption of high-viscosity fluids. The proppant selection consisted of 70/140 mesh quartz sand mixed with 40/70 mesh ceramic particles at an 8:2 ratio. The perforation parameters for Well 54-3 were as follows: Stage 1 used coiled tubing perforation with a density of 8 holes per 0.5 m, each cluster having a 0.5 m perforation interval, totaling 40 holes; the remaining fracturing stages had 13 clusters with a density of 3 holes per 0.3 m, each cluster having a 0.3 m perforation interval, totaling 39 holes.

4.2.1. Fracture Propagation

The overall treatment pressure matching for Well 54-3 shows good agreement, with an actual average shut-in pressure of 65.75 MPa and a simulated average shut-in pressure of 68.06 MPa, resulting in an error of 3.5%. The results generally correlate well with microseismic monitoring data, demonstrating that the simulated fractures accurately represent the actual hydraulic fracture parameters and geometry (Figure 12). The average hydraulic fracture length is 267 m, average width is 5.3 mm, and average height is 47.8 m. Post-fracturing results indicate effective fracture geometry improvement, with sufficient fracture propagation achieved at all well locations. As fracturing fluid injection continued, energy gradually accumulated within the reservoir, ultimately inducing both tensile and shear failure in the formation.

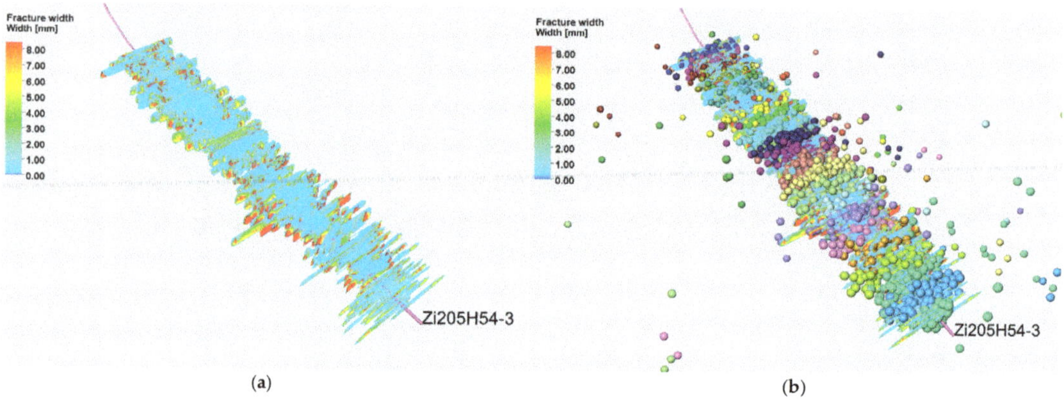

Figure 12. (**a**) Fracture propagation diagram of Well 54-3. (**b**) Comparison between fractures and microseismic monitoring in Well 54-3.

4.2.2. Stress Field Changes

(1) Stress Magnitude Changes

As the fractures in each fracturing stage propagate, the injection of formation energy causes corresponding adjustments in the in-situ stress in the adjacent areas of each fracturing stage. Figure 13 shows the changes in the minimum horizontal in-situ stress of Well 54-3 before and after fracturing. From the simulation results, it can be observed that in terms of the magnitude of stress change, due to the positive induced stress generated by the fracturing stages, the minimum horizontal in-situ stress increased significantly compared to the original stress during the fracturing process (ranging from 1.8 to 6.4 MPa), with an average increase of 4.2 MPa. As the number of fracturing stages increased, the minimum horizontal principal stress showed an obvious cumulative growth trend, and the increase amplitude became more significant in the later stages, with a phenomenon of stress accumulation along the wellbore. After 22 stages of fracturing, the cumulative increase in the minimum horizontal principal stress reached 6.4 MPa. Meanwhile, compared to the minimum horizontal principal stress, the maximum horizontal in-situ stress shows limited increase, ranging from 1.1 to 3.2 MPa (Figure 14). When considering both stress components, the horizontal stress contrast decreases after each fracturing stage. Figure 15 presents the variation in bi-directional stress contrast, which more visually demonstrates the differential impacts of fracturing on the maximum and minimum horizontal principal stresses. The figure shows that the pre-fracturing bi-directional stress contrast was about 15 MPa, which decreased to approximately 11 MPa after fracturing. Notably, significant stress reversal was observed in certain stages (e.g., Stage 11), where the minimum horizontal principal stress exhibited an anomalous decrease of about 1.1 MPa during fracturing. Through comprehensive analysis of microseismic event distribution characteristics and fracturing operation curves, this phenomenon is primarily attributed to the activation of natural fracture systems.

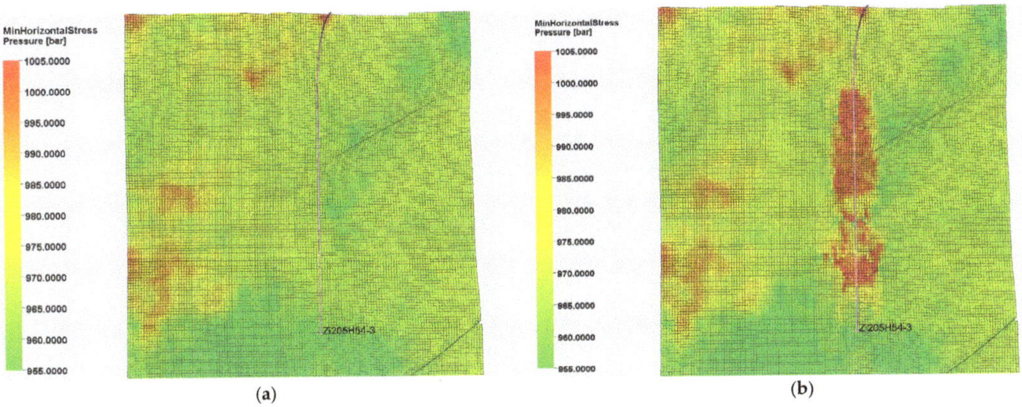

Figure 13. Minimum horizontal principal stress of Well 54-3 (**a**) before fracturing and (**b**) after fracturing.

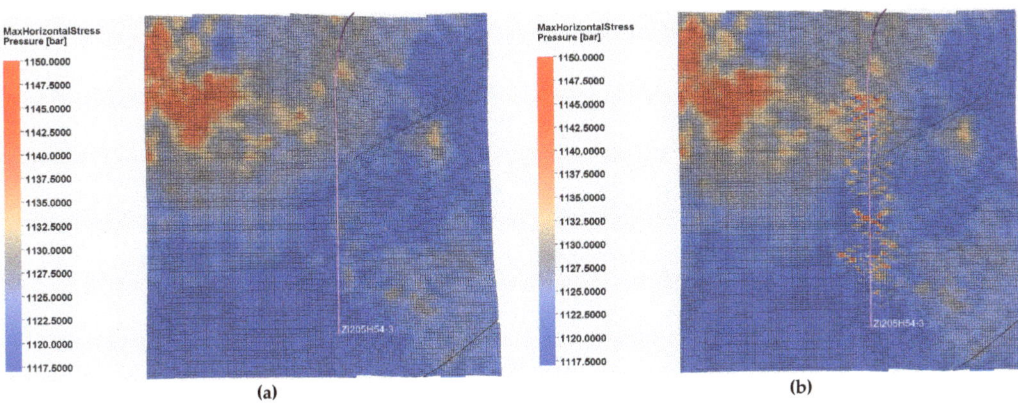

Figure 14. Maximum horizontal principal stress of Well 54-3 (**a**) before fracturing and (**b**) after fracturing.

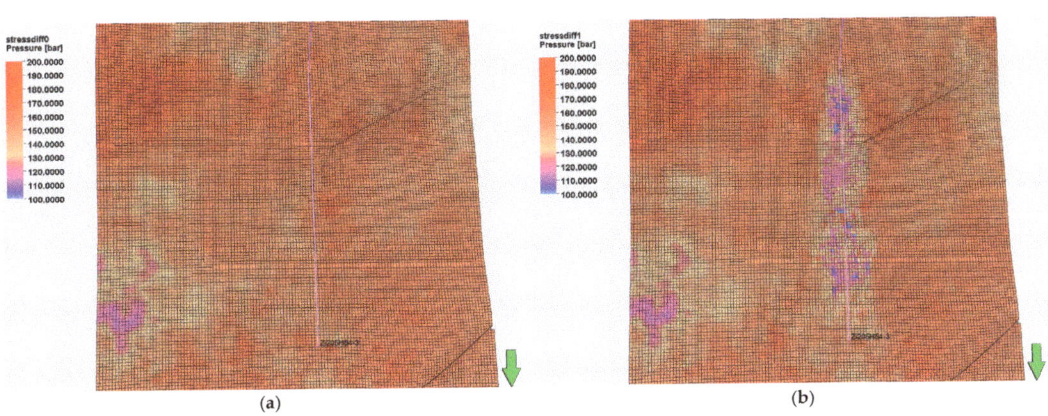

Figure 15. Horizontal stress difference of Well 54-3 (**a**) before fracturing and (**b**) after fracturing.

(2) Stress Direction Changes

Based on the simulation results, the minimum and maximum directions of horizontal in-situ stress before and after fracturing are plotted, as shown in Figures 16 and 17. Here, the arrow directions and color bands represent the direction and magnitude of horizontal in-situ stress, respectively. The analysis indicates that the morphology of the ultra-large-scale fracture network exhibits high complexity. During the hydraulic fracturing process, not only do the in-situ stress values change but the in-situ stress directions also show a certain degree of deflection. In general, the variation range of in-situ stress after fracturing is relatively small: the direction of the minimum horizontal principal stress changes by 0°~20°, and that of the maximum horizontal principal stress changes by 0°~10°. The influence is mainly within the range controlled by fractures, with the most obvious changes near the wellbore, and the directional variation decreases as the distance increases.

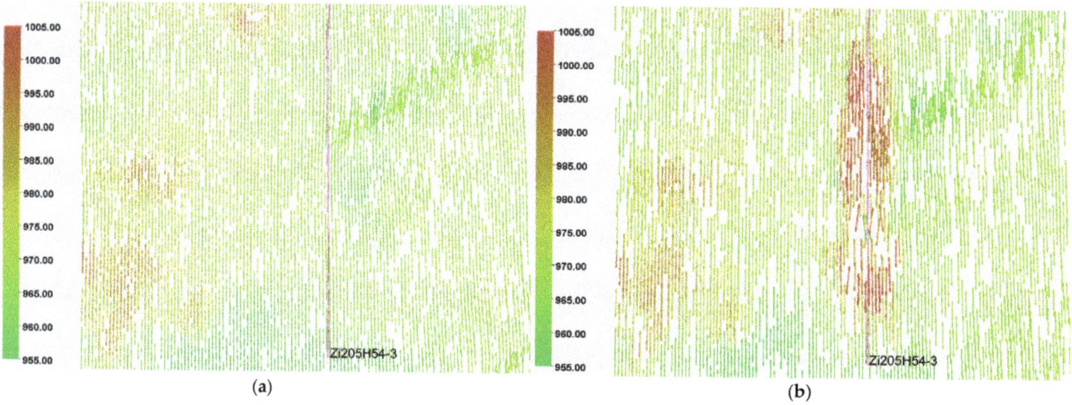

Figure 16. Orientation of minimum horizontal principal stress in Well 54-3 (**a**) before fracturing and (**b**) after fracturing.

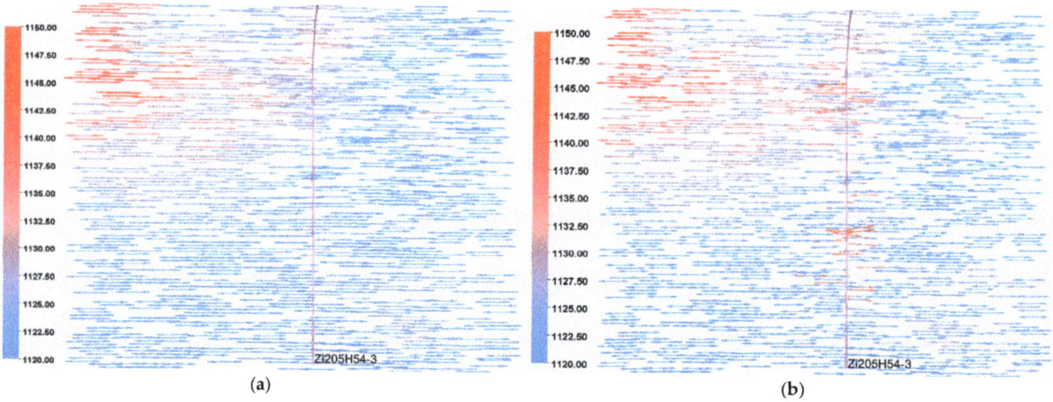

Figure 17. Orientation of maximum horizontal principal stress in Well 54-3 (**a**) before fracturing and (**b**) after fracturing.

4.3. Platform Stress Field Evolution Study

The single-well simulation results of Well 54-3 indicate the following: After large-scale fracturing, significant changes occurred in the stress state (magnitude and direction) and fracture system of adjacent stages and wells in the horizontal well. Based on these changes, if the original stress and fracture data continue to be used for subsequent fracturing simulations of neighboring stages and wells, significant errors will occur, greatly interfering with the selection of subsequent fracturing methods and optimization of fracturing parameters.

Therefore, based on the concept of dynamic fracturing stress fields and the integrated geomechanics approach, this study employs the actual field fracturing sequence to conduct more reasonable simulations of subsequent horizontal well fracturing processes. The three-dimensional model is updated in real-time according to the in-situ stress field and fracture conditions after each stage of fracturing. In field operations, a coordinated zipper fracturing approach with staggered inter-well fracture placement was adopted. A total of five wells—two from Platform 53 and three from Platform 54—were fractured simultaneously. Due to the influence of natural fracture development, casing deformation was observed in certain intervals. To mitigate risks, some stages were either skipped or combined into larger intervals for modified fracturing treatment. Ultimately, a total of 105 stages were successfully executed.

4.3.1. Fracture Propagation

The specific fracture morphology after platform fracturing simulation is shown in Figure 18. The simulation results show that the average opening exceeds 5 mm, and due to stress concentration effects, the maximum width of local main fractures can reach 8–10 mm, providing sufficient flow channels for proppant injection. The main fracture extension distance exceeds 200 m, and secondary branch fractures exhibit multi-level bifurcation, forming a complex fracture network system that effectively expands the stimulated reservoir volume (SRV). After fracturing, the vertical extension height of fractures in each horizontal well exceeds 40 m, with some intervals showing sudden height changes (up to 55 m) due to natural weak planes, indicating that fracturing energy has effectively penetrated thin interlayers.

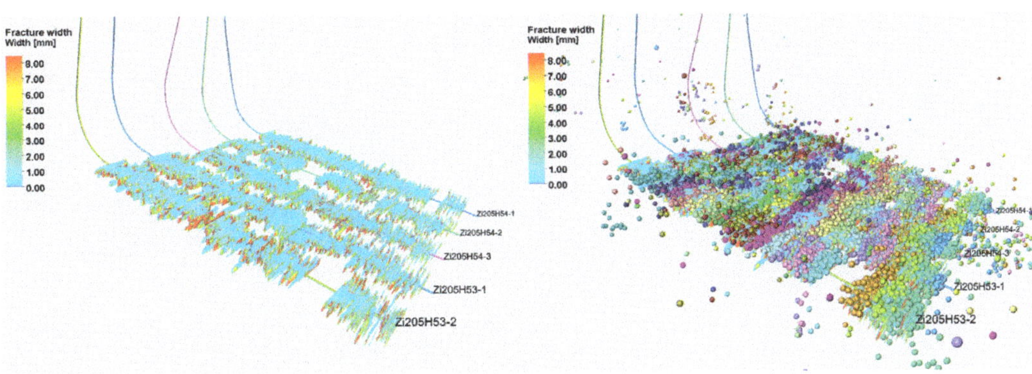

(a) (b)

Figure 18. (**a**) Platform fracture propagation simulation results. (**b**) Platform microseismic monitoring results.

4.3.2. Stress Field Changes

(1) Stress Magnitude Changes

As shown in Figures 19 and 20, during the fracturing process, the minimum horizontal principal stress in the platform area changes in real time, showing an overall increasing trend. The increase range of the minimum horizontal principal stress is approximately 1.1–7.6 MPa, with an average increase of 3.8 MPa, while the influence range is mainly distributed around the fractures. The maximum horizontal principal stress also changes simultaneously, showing an overall increasing trend, with a slightly smaller increase magnitude than the minimum horizontal principal stress, increasing by about 0.3–3.5 MPa, with an average increase of 1.7 MPa. The stress influence range after platform fracturing is relatively large, with an average stress change range of 290 m. Comparative analysis shows that a 400 m well spacing (Platform 54) has better stress balance and fewer stress communication phenomena compared to a 300 m well spacing (Platform 53), effectively balancing reservoir stimulation efficiency and inter-well interference control.

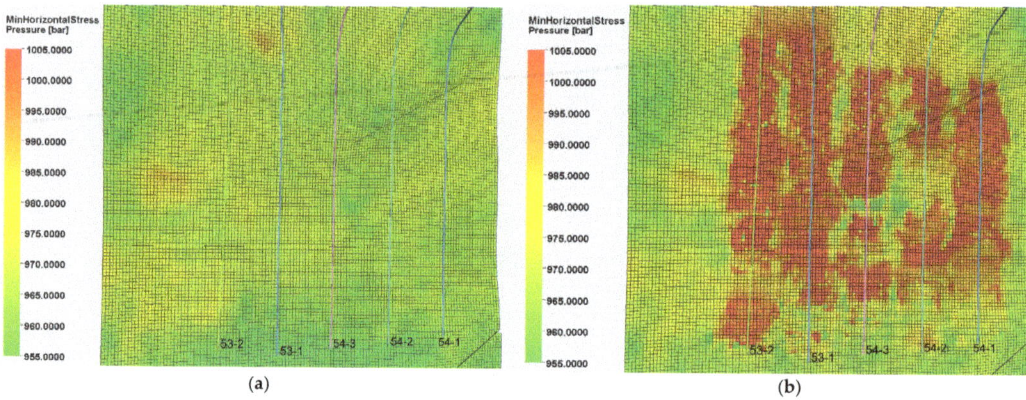

Figure 19. Minimum horizontal principal stress of the platform (**a**) before fracturing and (**b**) after fracturing.

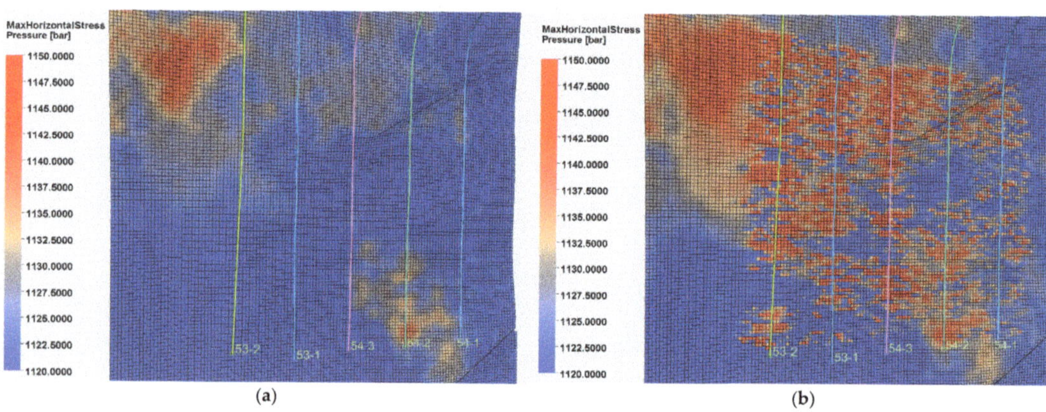

Figure 20. Maximum horizontal principal stress of the platform (**a**) before fracturing and (**b**) after fracturing.

(2) Stress Direction Changes

Meanwhile, significant changes have occurred in the stress directions of both the fractured well and adjacent wells. It is more reasonable to conduct subsequent fracturing simulations on adjacent stages and wells based on these changes. The minimum and maximum directions of horizontal in-situ stress before and after fracturing are plotted in Figures 21 and 22, where the arrow directions indicate the azimuth angles of in-situ stress and the color bands represent the magnitude of maximum in-situ stress. It can be seen that the ultra-large-scale fracture network is relatively complex. During the fracturing process, not only does the magnitude of in-situ stress change but the direction of in-situ stress also undergoes certain changes. Specifically, the greater the variation amplitude of pore pressure, the more obvious the deflection of stress direction. The direction of the minimum horizontal principal stress changes by 0°~45°, and that of the maximum horizontal principal stress changes by 0°~30°, forming a regional stress equilibrium zone.

Near the red high-stress areas, the local stress direction changes more significantly after fracturing, indicating obvious fracture deflection and high fracture complexity.

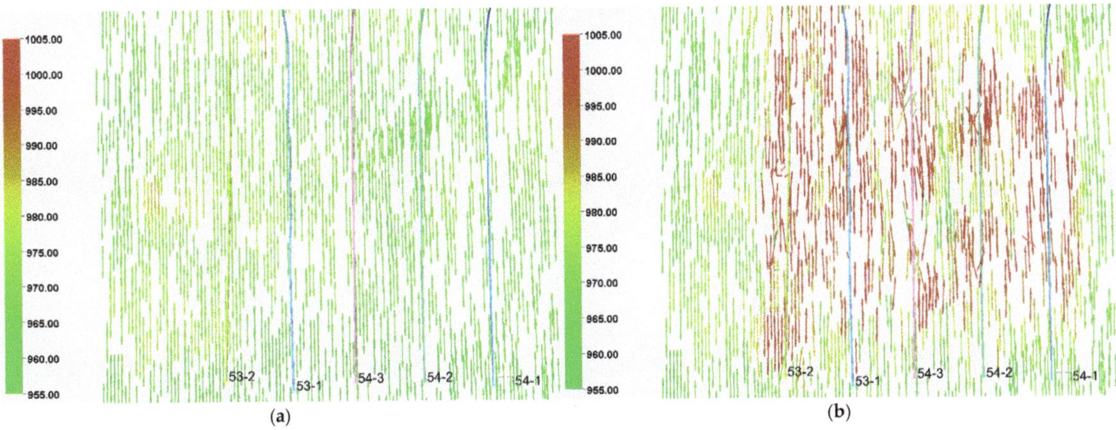

Figure 21. Orientation of minimum horizontal principal stress of the platform (**a**) before fracturing and (**b**) after fracturing.

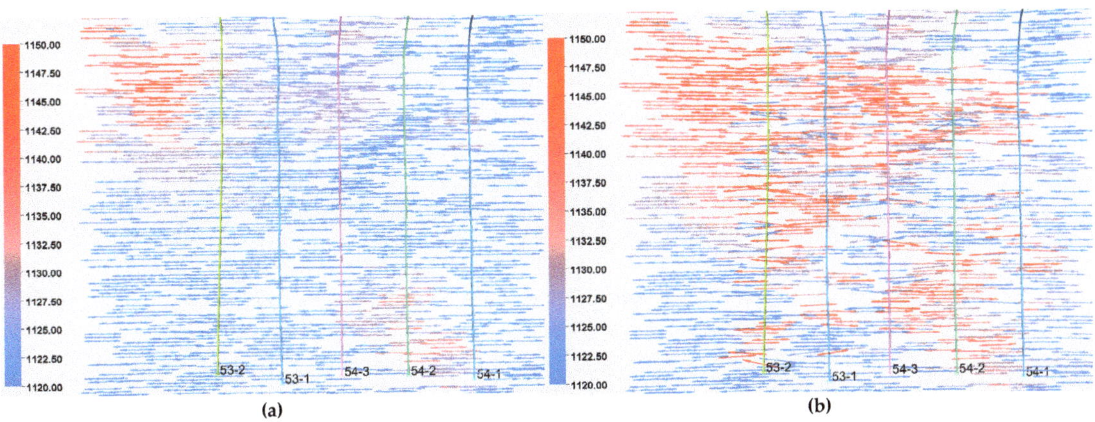

Figure 22. Orientation of maximum horizontal principal stress of the platform (**a**) before fracturing and (**b**) after fracturing.

Since the hydraulic fracturing stress field cannot be directly measured, the Instantaneous Shut-In Pressure (ISIP) is employed to characterize the stress field during fracturing, as illustrated in the Figure 23. After large-scale fracturing operations, the ISIP fluctuations observed in Platform 54 were significantly smaller than those in Platform 53, indicating a more balanced post-fracturing stress field in Platform 54.

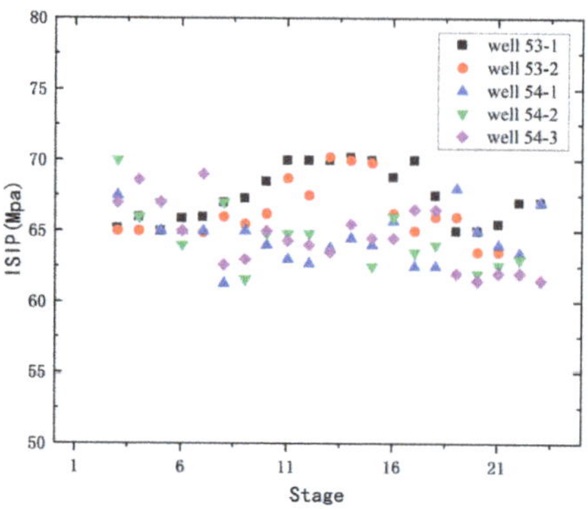

Figure 23. Instantaneous shut-in pressure.

5. Conclusions

Taking the deep shale gas in the Zigong block of the Sichuan Basin as the research object, this study comprehensively considers the heterogeneity and anisotropy of geomechanical parameters and natural fractures in shale gas reservoirs, and constructs a 4D in-situ stress multi-physics coupling model for shale gas reservoirs based on geology–engineering integration. By coupling geomechanical parameters with fracturing operation data, the dynamic evolution laws of multi-scale stress fields from single-stage to platform-scale during large-scale fracturing of horizontal wells in deep shale gas reservoirs are systematically studied. The research results are as follows:

(1) The fracturing process has a significant impact on the magnitude and direction of the stress field. With the injection of fracturing fluid, both the minimum and maximum horizontal principal stresses show an increasing trend, with the minimum horizontal principal stress increasing by 1.8–6.4 MPa and the maximum horizontal principal stress increasing by 1.1–3.2 MPa; moreover, near the wellbore, the direction of in-situ stress deflects obviously.

(2) As the number of fracturing stages increases, the minimum horizontal principal stress presents a significant cumulative growth trend, with a more prominent increase in the later stages. There is a phenomenon of stress accumulation along the wellbore, and the stress difference decreases from 15 MPa to 11 MPa.

(3) The on-site adoption of the fracturing operation method featuring overall flush advancement and inter-well staggered fracture placement has achieved good stress balance; comparative analysis shows that the stress communication degree of the 400 m well spacing is weaker than that of the 300 m well spacing.

(4) By studying the evolution laws of the stress field during fracturing and accurately identifying the distribution characteristics of the stress field, it is possible to guide the on-site real-time adjustment of the fracturing sequence, thereby improving the stress field balance and enhancing the fracturing stimulation effect. In future research, real-time in-situ stress measurement technology can be incorporated, and combined with the changes of the stress field during the production process, research on the stress field evolution throughout the whole life cycle of shale gas wells can be carried out.

Author Contributions: Conceptualization, Y.W. and Y.S.; methodology, Y.W.; software, Y.S.; validation, Z.Z., X.Y. and G.L.; investigation, X.Y.; resources, G.L.; data curation, P.L.; writing—original draft preparation, Y.W.; writing—review and editing, Y.S.; visualization, X.Y.; supervision, G.L.; project administration, P.L. All authors have read and agreed to the published version of the manuscript.

Funding: This research was funded by National Key R&D Program of China: (No. 2020YFA0710600).

Data Availability Statement: The original contributions presented in this study are included in the article. Further inquiries can be directed to the corresponding authors.

Acknowledgments: All authors express their gratitude for the valuable comments provided by the reviewers and editors.

Conflicts of Interest: Authors Zhaopeng Zhu, Xuemeng Yu and Guangyu Liu were employed by the PetroChina Jilin Oilfield Company. The remaining authors declare that the research was conducted in the absence of any commercial or financial relationships that could be construed as a potential conflict of interest.

Abbreviations

The following abbreviations are used in this manuscript:

P	the fluid pressure, MPa
σ_H	the maximum horizontal principal stress, MPa
σ_h	the minimum horizontal principal stress, MPa
σ_{ij}	the total stress tensor, MPa
G	the shear modulus, GPa
λ	the Lamé constants
ϕ	the porosity
p_f	the fluid density, kg/m^3
S	the saturation
t	the time, s
$\vec{v}$	the flow velocity vector
$\vec{\nabla}$	the gradient operator with respect to three-dimensional coordinates
k	the permeability, md
μ	the fluid viscosity, cp
g	the gravitational acceleration constant, m/s^2
D	the depth with surface as reference zero point, m
Q_V	the volumetric fluid flow rate per unit time, m^3/s
ζ	the fluid content in porous media
σ'_{ij}	the effective stress, MPa
σ_n	the normal stress on the wall of the natural fracture, MPa
K_{dry}	the bulk modulus of dry rock matrix, GPa
K_{solid}	the bulk modulus of dry solid matrix, GPa
β	the attenuation coefficient, m^{-1}
α	the ratio of the volume strain caused by pore pressure change to the pore pressure change, Pa^{-1}
f	the external or body forces, Pa
K_I^{eq}	the equivalent stress intensity factor, Pa·m$^{1/2}$
K_I	the Mode I stress intensity factor, Pa·m$^{1/2}$
K_{II}	the Mode II stress intensity factor, Pa·m$^{1/2}$
τ	the shear stress on the fracture wall, MPa
τ_0	the cohesion of the rock, MPa
C_f	the wall friction factor
P_t	the net pressure required for the generation of tensile fractures, MPa
P_s	the net pressure required for the generation of shear fractures, MPa
Rtx	the dynamic boundary positions in the x directions, m
Rty	the dynamic boundary positions in the y directions, m
Ga	the critical start-up pressure gradient, Pa/m

References

1. Guo, X.; Wang, R.; Shen, B.; Wang, G.; Wan, C.; Wang, Q. Geological characteristics, resource potential, and development direction of shale gas in China. *Pet. Explor. Dev. Online* **2025**, 52, 17–32. [CrossRef]

2. Zhao, S.; Zhao, Q.; Zhang, X.; Jiang, X.; Wu, H.; Yan, J. Key policy needs for the success of China's shale gas revolution. *Energy Rep.* **2024**, *11*, 4015–4020. [CrossRef]
3. Shen, C.; Wu, J.; Zeng, B.; Song, Y.; Yao, Z.; Dong, Y.; Du, Y. Measures and results of prevention and control on casing deformation and frac-hit in deep shale gas wells in southern Sichuan Basin. *Nat. Gas Ind. B* **2024**, *11*, 262–273. [CrossRef]
4. Tang, L.; Wang, P.; Zhao, Z.; Song, Y.; Chen, X.; Jiang, Z.; Jiang, S.; Li, Q.; Li, X. Overpressure origin and evolution during burial in the shale gas plays of the Wufeng-Longmaxi formations of southern Sichuan basin. *Geoenergy Sci. Eng.* **2024**, *236*, 212729. [CrossRef]
5. Zhang, J.; Li, Z.; Wang, D.; Xu, L.; Li, Z.; Niu, J.; Chen, L.; Sun, Y.; Li, Q.; Yang, Z.; et al. Shale gas accumulation patterns in China. *Nat. Gas Ind. B* **2023**, *10*, 14–31. [CrossRef]
6. Sun, C.; Nie, H.; Dang, W.; Chen, Q.; Zhang, G.; Li, W.; Lu, Z. Shale Gas Exploration and Development in China: Current Status, Geological Challenges, and Future Directions. *Energy Fuels* **2021**, *35*, 6359–6379. [CrossRef]
7. Zhao, J.; Ren, L.; Jiang, T.; Hu, D.; Wu, L.; Wu, J.; Yin, C.; Li, Y.; Hu, Y.; Lin, R.; et al. Ten years of gas shale fracturing in China: Review and prospect. *Nat. Gas Ind. B* **2022**, *9*, 158–175. [CrossRef]
8. Zhang, J.; Lin, L.; Tang, X.; Jiang, S.; Li, Y. The distribution patterns of organic-rich shale and exploration prospect of shale gas in China. *Adv. Resour. Res.* **2022**, *2*, 16–30.
9. Ma, X.; Wang, H.; Zhou, S.; Shi, Z.; Zhang, L. Deep shale gas in China: Geological characteristics and development strategies. *Energy Rep.* **2021**, *7*, 1903–1914. [CrossRef]
10. Li, S.-Z.; Zhou, Z.; Nie, H.-K.; Zhang, L.-F.; Song, T.; Liu, W.-B.; Li, H.-H.; Xu, Q.-C.; Wei, S.-Y.; Tao, S. Distribution characteristics, exploration and development, geological theories research progress and exploration directions of shale gas in China. *China Geol.* **2022**, *5*, 110–135. [CrossRef]
11. Zhang, P.; Meng, Z.; Jiang, S.; Chen, X. Characteristics of in-situ stress distribution in Zhengzhuang Region, Southern Qinshui Basin, China and its stress path during depletion. *Eng. Geol.* **2020**, *264*, 105413. [CrossRef]
12. Zhao, J.; Qiang, W.; Hu, Y.; Zhao, C.; Zhao, J. Prediction of Pore Pressure–Induced Stress Changes during Hydraulic Fracturing of Heterogeneous Reservoirs through Coupled Fluid Flow/Geomechanics. *J. Eng. Mech.* **2019**, *145*, 05019001. [CrossRef]
13. Altmann, J.B.; Müller, B.I.R.; Müller, T.M.; Heidbach, O.; Tingay, M.R.P.; Weißhardt, A. Pore pressure stress coupling in 3D and consequences for reservoir stress states and fault reactivation. *Geothermics* **2014**, *52*, 195–205. [CrossRef]
14. Haddad, M.; Eichhubl, P. Poroelastic models for fault reactivation in response to concurrent injection and production in stacked reservoirs. *Geomech. Energy Environ.* **2020**, *24*, 100181. [CrossRef]
15. Soltanzadeh, H.; Hawkes, C.D. Assessing fault reactivation tendency within and surrounding porous reservoirs during fluid production or injection. *Int. J. Rock. Mech. Min.* **2008**, *46*, 1–7. [CrossRef]
16. Bao, X.; Eaton, D.W. Fault activation by hydraulic fracturing in western Canada. *Science* **2016**, *354*, 1406–1409. [CrossRef]
17. Zhang, J.; Wang, X.; Xiao, H.; Ge, H.; He, J. Study on Fracture Propagation Rules of Shale Refracturing Based on CT Technology. *Processes* **2024**, *12*, 131. [CrossRef]
18. Sneddon, I.N. The distribution of stress in the neighbourhood of a crack in an elastic solid. *Proc. R. Soc. London. Ser. A Math. Phys. Sci.* **1946**, *187*, 229–260.
19. Crouch, S.L. Solution of plane elasticity problems by the displacement discontinuity method. I. Infinite body solution. *Int. J. Numer. Meth Eng.* **1976**, *10*, 301–343. [CrossRef]
20. Roussel, N.P.; Sharma, M.M. Quantifying Transient Effects in Altered-Stress Refracturing of Vertical Wells. *SPE J.* **2010**, *15*, 770–782. [CrossRef]
21. Roussel, N.P.; Florez, H.A.; Rodriguez, A.A. Hydraulic fracture propagation from infill horizontal wells. In Proceedings of the SPE Annual Technical Conference and Exhibition, New Orleans, LA, USA, 30 September–2 October 2013. [CrossRef]
22. Zhang, G.Q.; Chen, M. Dynamic fracture propagation in hydraulic re-fracturing. *J. Pet. Sci. Eng.* **2010**, *70*, 266–272. [CrossRef]
23. Wang, X.; Liu, C.; Wang, H.; Liu, H.; Wu, H. Comparison of consecutive and alternate hydraulic fracturing in horizontal wells using XFEM-based cohesive zone method. *J. Pet. Sci. Eng.* **2016**, *143*, 14–25. [CrossRef]
24. Biot, M.A. General theory of three-dimensional consolidation. *J. Appl. Phys.* **1941**, *12*, 155–164. [CrossRef]
25. Geertsma, J. The effect of fluid pressure decline on volumetric changes of porous rocks. *Trans. Aime* **1957**, *210*, 331–340. [CrossRef]
26. Lekhnitskii, S.G.; Fern, P.; Brandstatter, J.J.; Dill, E.H. Theory of elasticity of an anisotropic elastic body. *Phys. Today* **1964**, *17*, 84. [CrossRef]
27. Pei, Y.; Sepehrnoori, K. Investigation of parent-well production induced stress interference in multilayer unconventional reservoirs. *Rock. Mech. Rock. Eng.* **2022**, *55*, 2965–2986. [CrossRef]
28. Zhang, F.; Yin, Z.; Chen, Z.; Maxwell, S.; Zhang, L.; Wu, Y. Fault reactivation and induced seismicity during multistage hydraulic fracturing: Microseismic analysis and geomechanical modeling. *SPE J.* **2020**, *25*, 692–711. [CrossRef]
29. Glubokovskikh, S.; Sherman, C.S.; Morris, J.P.; Alumbaugh, D.L. Transforming microseismic clouds into near real-time visualization of the growing hydraulic fracture. *Geophys. J. Int.* **2023**, *234*, 2474–2487. [CrossRef]

30. Ibrahim, A.F.; Gowida, A.; Ali, A.; Elkatatny, S. Machine learning application to predict in-situ stresses from logging data. *Sci. Rep.* **2021**, *11*, 23445. [CrossRef]
31. Khormali, A.; Ahmadi, S.; Aleksandrov, A.N. Analysis of reservoir rock permeability changes due to solid precipitation during waterflooding using artificial neural network. *J. Pet. Explor. Prod. Technol.* **2025**, *15*, 17. [CrossRef]
32. Hatchell, P.; Bourne, S. Rocks under strain: Strain-induced time-lapse time shifts are observed for depleting reservoirs. *Lead. Edge* **2005**, *24*, 1222–1225. [CrossRef]
33. Herwanger, J. Seismic geomechanics: How to build and calibrate geomechanical models using 3D and 4D seismic data. In Proceedings of the Fourth EAGE CO2 Geological Storage Workshop, Stavanger, Norway, 23–25 April 2014; European Association of Geoscientists & Engineers: Utrecht, The Netherlands, 2014; p. cp-439-00001.
34. Onaisi, A.; Fiore, J.; Rodriguez-Herrera, A.; Koutsabeloulis, N.; Selva, F. Matching Stress-Induced 4D Seismic Time-Shifts with Coupled Geomechanical Models. In Proceedings of the 49th U.S. Rock Mechanics/Geomechanics Symposium, San Francisco, CA, USA, 29 June–1 July 2015; p. ARMA-2015-801.
35. Zhu, H.Y.; Tang, X.H.; Liu, Q.Y.; Li, K.D.; Xiao, J.L.; Jiang, S.; McLennan, J.D. 4D multi-physical stress modelling during shale gas production: A case study of Sichuan Basin shale gas reservoir, China. *J. Pet. Sci. Eng.* **2018**, *167*, 929–943. [CrossRef]
36. Ma, J.; Wang, J.; Fan, Q.; Jiao, S.; Chen, L.; Guo, C.; Hu, Z.; Liu, X.; Jiang, X.; Wang, Y. Integrated Workflow with Static and Dynamic Geomechanical Modelling for Hydraulic Fracturing Optimization. In Proceedings of the SPE/IADC Middle East Drilling Technology Conference and Exhibition, Abu Dhabi, United Arab Emirates, 23–25 May 2023. [CrossRef]
37. Li, C.; Singh, H.; Cai, J. Spontaneous imbibition in shale: A review of recent advances. *Capillarity* **2019**, *2*, 17–32. [CrossRef]
38. Wang, R.; Wang, R.; Zhang, J. High-performance FEM-FVM solid-fluid coupled dynamic analysis method for soil liquefaction. *Jpn. Geotech. Soc. Spec. Publ.* **2024**, *10*, 1322–1328. [CrossRef]

MDPI AG
Grosspeteranlage 5
4052 Basel
Switzerland
Tel.: +41 61 683 77 34

Processes Editorial Office
E-mail: processes@mdpi.com
www.mdpi.com/journal/processes

www.ingramcontent.com/pod-product-compliance
Lightning Source LLC
Chambersburg PA
CBHW061615220926
51850CB00005B/511
9783725853557